# The Soyuz Launch Vehicle
## The Two Lives of an Engineering Triumph

Christian Lardier and Stefan Barensky

Translator: Tim Bowler

# The Soyuz Launch Vehicle

## The Two Lives of an Engineering Triumph

 Springer

Published in association with
**Praxis Publishing**
Chichester, UK

Christian Lardier
Space Editor
*Air & Cosmos*
Paris
France

Stefan Barensky
Space Editor
*Science & Vie* and *European Voice*
Jaillans
France

Original French edition: *Les Deux vies de Soyouz* © Éditions Édite 2010

SPRINGER–PRAXIS BOOKS IN SPACE EXPLORATION

ISBN 978-1-4614-5458-8        ISBN 978-1-4614-5459-5 (eBook)
DOI 10.1007/978-1-4614-5459-5
Springer New York Heidelberg Dordrecht London

Library of Congress Control Number: 2012944347

Cover design: Jim Wilkie
Project copy editor: David M. Harland
Typesetting: BookEns, Royston, Herts., UK

Printed on acid-free paper

Springer is part of Springer Science+Business Media (www.springer.com)

# Contents

Foreword . . . . . . . . . . . . . . . . . . . . . . . . . . . . . . . . . . . . . . . . . . . . . . ix
Introduction . . . . . . . . . . . . . . . . . . . . . . . . . . . . . . . . . . . . . . . . . . . . xi
October 21, 2011 . . . . . . . . . . . . . . . . . . . . . . . . . . . . . . . . . . . . . . . . xiii

Part One – Soyuz in the East . . . . . . . . . . . . . . . . . . . . . . . . . . . . . . . . . 1
by Christian Lardier

Chapter 1 – The V-2's heir . . . . . . . . . . . . . . . . . . . . . . . . . . . . . . . . . . . 3
Taking over the V-2 missiles . . . . . . . . . . . . . . . . . . . . . . . . . . . . . . . . . 5
The gunners learn to use rockets . . . . . . . . . . . . . . . . . . . . . . . . . . . . . . 6
The Soviets in Germany . . . . . . . . . . . . . . . . . . . . . . . . . . . . . . . . . . . . 10
The Berlin and Nordhausen institutes . . . . . . . . . . . . . . . . . . . . . . . . . . 11
The Germans in the Soviet Union . . . . . . . . . . . . . . . . . . . . . . . . . . . . . 16
The decree of May 13, 1946. . . . . . . . . . . . . . . . . . . . . . . . . . . . . . . . . 18
Institute N°88 . . . . . . . . . . . . . . . . . . . . . . . . . . . . . . . . . . . . . . . . . . . 20
The first ballistic rockets . . . . . . . . . . . . . . . . . . . . . . . . . . . . . . . . . . . 22
Korolev's rockets. . . . . . . . . . . . . . . . . . . . . . . . . . . . . . . . . . . . . . . . . 25
The first strategic missile . . . . . . . . . . . . . . . . . . . . . . . . . . . . . . . . . . . 29
Constructing Baikonur . . . . . . . . . . . . . . . . . . . . . . . . . . . . . . . . . . . . . 37
The construction army. . . . . . . . . . . . . . . . . . . . . . . . . . . . . . . . . . . . . . 40
The decree of April 14, 1955 . . . . . . . . . . . . . . . . . . . . . . . . . . . . . . . . 41

Chapter 2 – Designing the Semyorka . . . . . . . . . . . . . . . . . . . . . . . . . . . 43
The Semyorka launch platform . . . . . . . . . . . . . . . . . . . . . . . . . . . . . . . 48
The launch . . . . . . . . . . . . . . . . . . . . . . . . . . . . . . . . . . . . . . . . . . . . . . 53
The tracking network . . . . . . . . . . . . . . . . . . . . . . . . . . . . . . . . . . . . . . 55
Development of the Semyorka. . . . . . . . . . . . . . . . . . . . . . . . . . . . . . . . . 56
The R-7A . . . . . . . . . . . . . . . . . . . . . . . . . . . . . . . . . . . . . . . . . . . . . . . 61
The first "Sputnik" . . . . . . . . . . . . . . . . . . . . . . . . . . . . . . . . . . . . . . . . 65

**Chapter 3 – The council of Chief Designers** . . . . . . . . . . . . . . . . . . . . . . . . . 71
The father of the Semyorka . . . . . . . . . . . . . . . . . . . . . . . . . . . . . . . . . . 71
Taming the fire . . . . . . . . . . . . . . . . . . . . . . . . . . . . . . . . . . . . . . . . . 82
The upper stage engines . . . . . . . . . . . . . . . . . . . . . . . . . . . . . . . . . . . 86
Father of the "Beep-Beep" . . . . . . . . . . . . . . . . . . . . . . . . . . . . . . . . . 88
Telemetry and trajectography . . . . . . . . . . . . . . . . . . . . . . . . . . . . . . . 92
Guidance systems . . . . . . . . . . . . . . . . . . . . . . . . . . . . . . . . . . . . . . . 94
Gyroscopes . . . . . . . . . . . . . . . . . . . . . . . . . . . . . . . . . . . . . . . . . . . 96
Launch platforms . . . . . . . . . . . . . . . . . . . . . . . . . . . . . . . . . . . . . . . 99

**Chapter 4 – Korolev's subsidiaries** . . . . . . . . . . . . . . . . . . . . . . . . . . . . . 105
Dmitri Kozlov . . . . . . . . . . . . . . . . . . . . . . . . . . . . . . . . . . . . . . . . . 106
"Progress" plant N°1 . . . . . . . . . . . . . . . . . . . . . . . . . . . . . . . . . . . . . 111
Plant N°24 "Motorostroitel" . . . . . . . . . . . . . . . . . . . . . . . . . . . . . . . . 118
TsSKB satellites . . . . . . . . . . . . . . . . . . . . . . . . . . . . . . . . . . . . . . . . 119

**Chapter 5 – The various versions** . . . . . . . . . . . . . . . . . . . . . . . . . . . . . . 137
The R-7/R-7A ICBM . . . . . . . . . . . . . . . . . . . . . . . . . . . . . . . . . . . . 137
The three-stage 8K72 . . . . . . . . . . . . . . . . . . . . . . . . . . . . . . . . . . . . 140
The four-stage 8K78 "Molniya" . . . . . . . . . . . . . . . . . . . . . . . . . . . . . . 146
8K78M . . . . . . . . . . . . . . . . . . . . . . . . . . . . . . . . . . . . . . . . . . . . . 153
8A91 . . . . . . . . . . . . . . . . . . . . . . . . . . . . . . . . . . . . . . . . . . . . . . . 157
8A92 . . . . . . . . . . . . . . . . . . . . . . . . . . . . . . . . . . . . . . . . . . . . . . . 157
8A92M . . . . . . . . . . . . . . . . . . . . . . . . . . . . . . . . . . . . . . . . . . . . . 157
11A55 and 11A56 . . . . . . . . . . . . . . . . . . . . . . . . . . . . . . . . . . . . . . 159
11A57 . . . . . . . . . . . . . . . . . . . . . . . . . . . . . . . . . . . . . . . . . . . . . . 160
11A58 . . . . . . . . . . . . . . . . . . . . . . . . . . . . . . . . . . . . . . . . . . . . . . 163
11A59 . . . . . . . . . . . . . . . . . . . . . . . . . . . . . . . . . . . . . . . . . . . . . . 163
11A510 . . . . . . . . . . . . . . . . . . . . . . . . . . . . . . . . . . . . . . . . . . . . . 165
11A511 . . . . . . . . . . . . . . . . . . . . . . . . . . . . . . . . . . . . . . . . . . . . . 165
Soyuz-2 "Rus" . . . . . . . . . . . . . . . . . . . . . . . . . . . . . . . . . . . . . . . . 175
Yamal . . . . . . . . . . . . . . . . . . . . . . . . . . . . . . . . . . . . . . . . . . . . . . 178
Aurora . . . . . . . . . . . . . . . . . . . . . . . . . . . . . . . . . . . . . . . . . . . . . . 179
Onega . . . . . . . . . . . . . . . . . . . . . . . . . . . . . . . . . . . . . . . . . . . . . . 181
Soyuz-2-3 . . . . . . . . . . . . . . . . . . . . . . . . . . . . . . . . . . . . . . . . . . . 182
Soyuz-1 . . . . . . . . . . . . . . . . . . . . . . . . . . . . . . . . . . . . . . . . . . . . . 184
Conclusion . . . . . . . . . . . . . . . . . . . . . . . . . . . . . . . . . . . . . . . . . . . 184

**Chapter 6 – The launch bases** . . . . . . . . . . . . . . . . . . . . . . . . . . . . . . . . 187
Baikonur . . . . . . . . . . . . . . . . . . . . . . . . . . . . . . . . . . . . . . . . . . . . 187
Plesetsk . . . . . . . . . . . . . . . . . . . . . . . . . . . . . . . . . . . . . . . . . . . . . 196

**Part Two – Soyuz in the West** . . . . . . . . . . . . . . . . . . . . . . . . . . . . . . . 205
**by Stefan Barensky**

**Chapter 7 – A fantasy launcher** . . . . . . . . . . . . . . . . . . . . . 207
Wave of panic . . . . . . . . . . . . . . . . . . . . . . . . . . . . . . . . . . . . 211
Phantom threat . . . . . . . . . . . . . . . . . . . . . . . . . . . . . . . . . . . 213
Misleading displays . . . . . . . . . . . . . . . . . . . . . . . . . . . . . . . . 216
Missiles on Red Square . . . . . . . . . . . . . . . . . . . . . . . . . . . . . . 219
CIA seeking intelligence . . . . . . . . . . . . . . . . . . . . . . . . . . . . . 221
Paranoia in Washington . . . . . . . . . . . . . . . . . . . . . . . . . . . . . 224
The great revelation . . . . . . . . . . . . . . . . . . . . . . . . . . . . . . . . 229
What's that rocket called? . . . . . . . . . . . . . . . . . . . . . . . . . . . . 232
From fantasy to icon . . . . . . . . . . . . . . . . . . . . . . . . . . . . . . . . 235

**Chapter 8 – East meets West** . . . . . . . . . . . . . . . . . . . . . . . . 237
Commercial efforts in the West . . . . . . . . . . . . . . . . . . . . . . . . 240
One world ends, another begins . . . . . . . . . . . . . . . . . . . . . . . . 243
Blossoming of the constellations market . . . . . . . . . . . . . . . . . . 246
Launchers to complement Ariane . . . . . . . . . . . . . . . . . . . . . . . 249
Powder fizzles out . . . . . . . . . . . . . . . . . . . . . . . . . . . . . . . . . 252

**Chapter 9 – The genesis of Starsem** . . . . . . . . . . . . . . . . . . . 257
New deal in Samara . . . . . . . . . . . . . . . . . . . . . . . . . . . . . . . . 260
A French proposal . . . . . . . . . . . . . . . . . . . . . . . . . . . . . . . . . 263
From Irene to Ikar . . . . . . . . . . . . . . . . . . . . . . . . . . . . . . . . . 265
Breaking out of the circle . . . . . . . . . . . . . . . . . . . . . . . . . . . . 267
New rules of the game . . . . . . . . . . . . . . . . . . . . . . . . . . . . . . 270
Objective Globalstar . . . . . . . . . . . . . . . . . . . . . . . . . . . . . . . 272
The politicians come on stage . . . . . . . . . . . . . . . . . . . . . . . . . 276
Arianespace faced with a dilemma . . . . . . . . . . . . . . . . . . . . . . 278
Starsem is born . . . . . . . . . . . . . . . . . . . . . . . . . . . . . . . . . . . 280

**Chapter 10 – Europeans on the steppe** . . . . . . . . . . . . . . . . . . 285
Accommodating satellites and teams . . . . . . . . . . . . . . . . . . . . 286
Seeking new customers . . . . . . . . . . . . . . . . . . . . . . . . . . . . . . 289
A resurrection and a death . . . . . . . . . . . . . . . . . . . . . . . . . . . 290
"One man's loss is another man's gain" . . . . . . . . . . . . . . . . . . . 293
Fregat makes its appearance . . . . . . . . . . . . . . . . . . . . . . . . . . 296
Market collapse . . . . . . . . . . . . . . . . . . . . . . . . . . . . . . . . . . . 303
Time for a change . . . . . . . . . . . . . . . . . . . . . . . . . . . . . . . . . 304

**Chapter 11 – Russians in the jungle** . . . . . . . . . . . . . . . . . . . 317
Glut of launch sites . . . . . . . . . . . . . . . . . . . . . . . . . . . . . . . . 318
Kourou in sight . . . . . . . . . . . . . . . . . . . . . . . . . . . . . . . . . . . 320
A tropical, cryogenic Soyuz . . . . . . . . . . . . . . . . . . . . . . . . . . . 321

Political games and stakes . . . . . . . . . . . . . . . . . . . . . . . . . . . . . . . . . . . . 323
Aussie poker . . . . . . . . . . . . . . . . . . . . . . . . . . . . . . . . . . . . . . . . . . . . . . . 326
From Elysée to Edinburgh. . . . . . . . . . . . . . . . . . . . . . . . . . . . . . . . . . . . . . 328
The tables are turned. . . . . . . . . . . . . . . . . . . . . . . . . . . . . . . . . . . . . . . . . 329
Meeting in Paris . . . . . . . . . . . . . . . . . . . . . . . . . . . . . . . . . . . . . . . . . . . . 330
Money makes the world go round . . . . . . . . . . . . . . . . . . . . . . . . . . . . . . . 332
A construction site in Guiana . . . . . . . . . . . . . . . . . . . . . . . . . . . . . . . . . . 335

**Chapter 12 – Soyuz, launcher of the future** . . . . . . . . . . . . . . . . . . . . . . . 349
From the Soyuz launcher to the Soyuz spacecraft in Guiana . . . . . . . . . . . . 352
Towards the 2,000th flight? . . . . . . . . . . . . . . . . . . . . . . . . . . . . . . . . . . . 365
3, 2-3, 1: the final countdown?. . . . . . . . . . . . . . . . . . . . . . . . . . . . . . . . . . 367
The mirage of a Euro-Russian spacecraft. . . . . . . . . . . . . . . . . . . . . . . . . . 368
Post-Soyuz in Russia. . . . . . . . . . . . . . . . . . . . . . . . . . . . . . . . . . . . . . . . . 370
Post-Soyuz in Europe . . . . . . . . . . . . . . . . . . . . . . . . . . . . . . . . . . . . . . . . 372

**Annexes** . . . . . . . . . . . . . . . . . . . . . . . . . . . . . . . . . . . . . . . . . . . . . . . . . 377
**Chapter notes**. . . . . . . . . . . . . . . . . . . . . . . . . . . . . . . . . . . . . . . . . . . . . 389
**Semyorka launches**. . . . . . . . . . . . . . . . . . . . . . . . . . . . . . . . . . . . . . . . . 405
**Bibliography**. . . . . . . . . . . . . . . . . . . . . . . . . . . . . . . . . . . . . . . . . . . . . . 465
**Index of names**. . . . . . . . . . . . . . . . . . . . . . . . . . . . . . . . . . . . . . . . . . . . 473
**Photo credits** . . . . . . . . . . . . . . . . . . . . . . . . . . . . . . . . . . . . . . . . . . . . . 485

# Foreword

Since October 4, 1957, with the launch of Sputnik, the planet's first artificial satellite, the Soyuz rocket has been an integral part of the history of space conquest and thus of the history of humanity.

More than fifty years and 1,750 launches later, Soyuz is still here and its recent introduction to French Guiana, South America, far from its birthplace on the Kazakh steppe has given the legendary rocket a second life.

The value of this remarkable book, co-authored by Christian Lardier and Stefan Barensky, lies in its complete and accurate description of the two lives of Soyuz, thus becoming a reference book chronicling the cooperative space endeavor of Europe and Russia.

The first chapters take us back to the early days of astronautics, at a time when technology was at the service of politics. From archives found in the Soviet Union and impressive documentation work the authors show the difficulty of designing a rocket in the immediate post-war period and the heavy price that Eastern engineers had to pay to get there.

That period was followed by the golden age of Soyuz, with the multiplication of launch pads at Baikonur and at Plesetsk and the numerous missions Soyuz carried out. On the one hand, scientific missions and manned flights were publicized world-wide and used as a powerful propaganda tool. On the other hand, the much more numerous military missions were kept highly confidential!

Then came the fall of the Soviet Union with all its consequences for Russian astronautics. That collapse raised the issue of relations with the West and the sometimes tortuous path that led to unexpected cooperation with Europe, now seen as exemplary, while the United States salvaged for their sole benefit almost all of the ex-Soviet space resources.

After that, Soyuz began its second life. The role of a few visionaries in Russia and in Europe, who decided to leave their respective isolation behind and bring Soyuz and Ariane together, is recalled in detail and cannot leave the reader indifferent. The authors analyze all phases of the implementation of this unprecedented cooperation, from the political agreements to the decision to launch Soyuz from Guiana, and of course the creation of Starsem.

But in the final analysis, what the book describes most profoundly is a formidable human adventure, with its political, technical and commercial ramifications. At a time when a new order was taking shape in the space sector, the players being the United States, Russia, Europe, and Asia, and when economic difficulties sometimes made it tempting to withdraw, this book reminds us that, in the global space sector, nothing is set in stone. That is the overall lesson to be learned from the two lives of the Soyuz rocket.

Jean-Yves Le Gall
Chief Executive Officer
Arianespace – Starsem

# Introduction

Throughout history mankind has built a number of machines which stand out from the rest. In the field of power plants, these include the steam engine, the first internal combustion engine, Frank Whittle's turbojet, and Robert Goddard's liquid-propellant rocket. In the field of transport vehicles we think of the first automobile built by engineer Nicolas-Joseph Cugnot, the locomotive by Richard Trevithick, the Wright brothers' first aircraft, Wernher von Braun's long range V-2 rocket, etc. Among automobiles, some have enjoyed a greater career than others, including the very popular VW Beetle (21,529,500 built) or the Citroen 2 CV (5,115,000 built). Among combat aircraft, we can point to the Ilyushin Il-2 "Sturmovik" (more than 36,000 built) or the MiG-21 (over 11,000).

In the realm of space launch vehicles, the most remarkable machine is without a doubt the rocket known today as the Soyuz (meaning "Union" in Russian). Developed after the Second World War, it became the first intercontinental ballistic missile in August 1957. It carried Earth's first artificial satellite aloft in October of that year. It enabled Yuri Gagarin to be the first man in space, in April 1961. It has become the "workhorse of space", launching all kinds of satellites and interplanetary probes, and all the cosmonauts. It is still in use, launching crews to the International Space Station.

In fact, this rocket has had several names: R-7, also known as Semyorka ("little seventh" in Russian), Sputnik, Luna, Vostok, Voskhod, Soyuz, Molniya, etc. In total, more than 1,750 units were produced and launched in the first 53 years of operations from the Baikonur and Plesetsk cosmodromes, while only 603 units of the American equivalent, the Atlas, were launched during the same period (including the new Atlas 5 version which has no technical relationship to the original Atlas missile). But the career of the extraordinary machine that is the Soyuz launcher – not to be confused with the piloted spacecraft of the same name – did not end there, since it began a second life at the Guiana Space Centre (CSG), Europe's space port, with the inaugural flight on October 21, 2011 carrying a pair of European navigation satellites into medium Earth orbit. This marked the starting point of an equatorial operation which should last for 20 years. At the same time, the manufacturer – the Samara Space Center (formerly TsSKB-Progress) – still envisages improving the launcher for new uses: the Soyuz-1 and Soyuz-2-3 versions were exhibited for the first time at the Paris Air Show in June 2007. At the current rate of about 15 launches per year, the number of launched units will one day pass the 2,000 mark!

This book is presented in two parts: Christian Lardier has chronicled the "first life" in Russia and Stefan Barensky has explored its "second life", covering Starsem, the Franco-Russian company and implementation by ESA at the French Guiana Space Centre.

Part one has been developed, with a maximum of historical rigor, from Russian sources. It provides a rather descriptive approach to purely technical issues. This is a first, as no single work, dedicated solely to this rocket, has been published before, including in Russia.

The second part tells a contemporary story gathered more from Western sources and interviews with key protagonists (Jean-Yves Le Gall, Victor Nikolaiev, Jérome Paolini, Michel Delaye, Charles Bigot, Jean-Marie Luton, Jean-Jacques Dordain). This part is more narrative and addresses the sensitive issue of the strategic choices that led to the establishment of the Soyuz in Guiana.

Finally, we wish to thank Mrs. Lucette Calaque, Jean-Marc Astorg, Philippe Clerc, Stéphane Corvaja, Patrick Eymar, Michel Guérin, François Maroquène, Pierre Marx, Boris Melioransky, Jacob Terweij, Jacques Tiziou, Timothy Varfolomeiev, Alexandre Chliadisky, and Jean-Charles Vincent.

# October 21, 2011

This is a unique sighting, in this remote launch pad on the edge of the Amazonian rain forest. A Russian-built launch vehicle, derived from a former Soviet nuclear weapon carrier, is standing on its launch table. Although no launcher has been seen more often on a launch pad, this one is different. Its flanks are covered with logos for Roskosmos, TsENKI, TsSKB-Progress and NPO Lavochkin, as well as those of the European Space Agency (ESA) and CNES, the French space agency, but these have been seen before on previous commercial flights. The white 4-m diameter fairing is reminiscent of the long-gone Ariane 4 launchers. It features large logos for Galileo, the European global navigation satellite system, and for Arianespace, the operator of the Ariane vehicles. There is also a commemorative marking for the Ariane Cities Community with the name of its nearest member – "City of Sinnamary". But what most amazes the observer is the thick cloud of condensation vapor that shrouds the vehicle, like cotton, while its tanks are filled with liquid oxygen. For the first time, on this morning of October 21, 2011, a Soyuz launch vehicle will lift-off in a 100% hygrometry environment, from a new launch pad within the premises of the Guiana Space Centre, Europe's spaceport in French Guiana.

It is a Soyuz-2.1b brought to the European ST-B standard by the addition of a Belgian-built "safety kit" to provide secured telemetry links capable of enabling the range safety officer to shut down the propulsion in case the vehicle veers off course and the onboard system fails to order flight termination. Only two months ago, on August 24, a more traditional Soyuz-U launched from Baikonur in Kazakhstan suffered an engine failure on its Block I stage and crashed into the Altai with its payload, a Progress-M freighter carrying supplies for the International Space Station. This was the first failure after 80 consecutive successes of the Soyuz-U and Soyuz-FG, so the international space community is holding its breath because the Soyuz launcher is the last remaining man-rated vehicle available to maintain the lifeline with the ISS and ensure crew rotation. It will eventually return to flight with success on October 30, and the ISS crew will be renewed before Christmas, but as the first Guianese Soyuz awaits launch on its equatorial pad, everybody is still anxious about the future of the largest manned orbital outpost in history. The Europeans authorized their own Soyuz launch to proceed only after a similar Soyuz 2-1.b vehicle was successfully launched from Plesetsk in northern Russia on October 2, carrying a Russian Glonass global positioning satellite. The Soyuz-2.1b features a different engine (RD-0124) on its modified Block I than the model that caused the Progress launch failure (RD-0110).

It is baptism of fire for the new Soyuz launch complex (ELS, for "Ensemble de Lancement Soyouz" in French). The facility was completed, tested and qualified by CNES in the first months of 2011. On March 31, after at the end of an acceptance review, it was handed over to ESA, which immediately turned it over to Arianespace to start its qualification process for launch operations. A dry rehearsal was conducted from April 29 through May 5, with the very first erection of a Soyuz vehicle on pad. The handover ceremony was replayed with Jean-Jacques Dordain, director general of ESA, Jean-Yves Le Gall, chairman and CEO of Arianespace, Yannick d'Escatha, president of CNES, and Vladimir Popovkin, head of Roskosmos on May 7.

In July, the Launcher Flight Readiness Review gave its green light to start the launch campaign for the first Soyuz mission from Guiana, to be designated VS01,[1] on August 16. The final assembly of the vehicle began on September 12, and the rollout from the MIK assembly building was performed on October 14. That same day, the vehicle was erected in upright position, enclosed in the new mobile gantry, and integrated vertically with its payload composite in preparation for a launch on October 20. Later the launch had to be postponed when a propellant line accidentally disconnected from the launcher during the very first fuelling operations. The joint European-Russian team was able to restore the vehicle to launch-ready configuration within 24 hours.

And here we are, on this foggy Guianese morning of October 21, 2011. Right on time, the very first Soyuz to be launched from South America lights the 32 chambers of its five main engines and their 12 vernier thrusters. The counterweights open the petals of the Tyulpan system that has supported the vehicle on pad, and the launcher rises in the cloudy sky. Some 3 hours 49 minutes later, the new Fregat-MT upgraded upper stage, with 900 kg of additional propellant, releases "Thijs" and "Natalia", the first pair of satellites for the Galileo constellation, into a circular orbit at an altitude of 23,222 km.

This textbook mission will be followed by a second one on December 16, this time into sun-synchronous orbit carrying the Pleiades 1 dual high-resolution remote sensing satellite for CNES, a quartet of French Elisa military electronic intelligence satellites, and the SSOT military observation satellite for Chile. The launch manifest is filling up rapidly, mostly with science and Earth observation missions, but it could also accommodate some small geostationary satellites that are unable to find a timely slot on the larger Ariane 5.

---

[1]   As long as Arianespace was operating only Ariane launchers, its missions were designated by V-(flight number). With the diversification of its fleet, the designation was changed. Since April 2011, Ariane launches have been designated VA, Soyuz launches are VS, and lightweight Vega launches are VV.

This marks the start of a new era for the very-long-lasting family of Semyorka vehicles, of which the Soyuz is the latest and most prominent member. With the announcement to the Duma, the Russian parliament, by Vladimir Popovkin, head of Roskosmos, that the development of the Soyuz proposed successor, the Rus-M, had been canceled because of budgetary issues, Roskosmos plans to rely on a mix of new vehicles (namely the Angara system under development by Khrunichev) and existing ones, like the Soyuz.

# Part One – Soyuz in the East

**by Christian Lardier**

# 1

## The V-2's heir

The name "Soyuz" represents both a launch vehicle and a piloted spacecraft. In fact, it was the capsule which gave its name to the launcher in 1966.

Initially, this launcher was an intercontinental missile: the R-7, designed by S. P. Korolev – better known as the Semyorka (i.e. "little seventh" in Russian). It fostered a long line of various versions: Sputnik (Fellow Traveler), Vostok (East) Voskhod (Sunrise), Molniya (Lightning) and Soyuz (Union). In each of these versions, it is the composite upper (third and fourth) stages which are different. In addition, industry and the military use a specific codification: 8K71, 8K72, 8K73, 8K74, 8K78, 8A91, 8A92, 11A57, 11A58, 11A59, 11A510, 11A511 and 14A14 for the launchers, and 8D74, 8D75, 8D76, 8D77, 8D78, 8D79, 8D711, 8D714, 8D715, 8D719, 8D727 and 8D728 for the engines. Thus, since the program started, 16 different versions have flown. If we add the projects which did not fly, then the number grows significantly.

The Semyorka was descended from the Second World War V-2, which first flew successfully on October 3, 1942. The V-2 was massively used by the German army between September 1944 and March 1945 (approximately 6,000 were manufactured and 3,200 were launched against London, Antwerp, Paris and other targets). It was a revolutionary 13-ton vehicle propelled by a liquid oxygen-ethyl alcohol engine with a thrust of 25 tons, unlike other missiles such as the Wasserfall (8-ton-thrust engine) that were propelled by storable nitric acid fuels.

Fragments of a German V-2 (including the engine) were recovered in Poland near Blizna – a training site where the Germans launched hundreds of V-2s – by a Soviet team from institute N°1 (NII-1) of the ministry of the aviation industry on August 5, 1944.

The expedition was led by Lieutenant-General I. A. Serov of the KGB and Major-General P. I. Fedorov, head of NII-1 from June 1944 to February 1945. The team included engineers Yuri. A. Pobedonostsev, M. K. Tikhonravov, N. G. Chernyshyov, R. E. Sorkine and A. G. Shekhman. Back in Moscow on September 24, 1944, the "Raketa" group was formed with the help of engineers who, starting in 1941, had worked on the first Soviet rocket-plane, the BI-1 (Bereznyaki-Isaiev) by designer V. F. Bolkhovitinov. In September-October, the engine designer V. P. Glushko came from Kazan, accompanied by N. L. Oumansky and G. N. List, to examine the

Tikhonravov and Chernyshyov in Germany in 1944.

engine. This group gave the authorities its assessment in November. A second expedition would attempt to return to Germany in February 1945, but the aircraft carrying them crashed near Kiev, leaving no survivors.

NII-1 was the new name for the jet propulsion institute, previously called RNII from 1933 to 1937, NII-3 from 1937 to 1942, and GIRT from 1942 to February 18, 1944. It had been reorganized to develop jet aviation in the Soviet Union. It was initially headed by the chief of the central institute of aviation engines (TsIAM) V. I. Polikovsky until the end of May 1944. The first deputy was G. N. Abramovich until 1948. Its leading subsidiary was the N°293 design bureau (OKB) of Bolkhovitinov at Khimki. This was taken over by M. R. Bisnovat in June 1946 and became the MKB Fakel, contractor for most of the Soviet surface-to-air and anti-missile missiles.

The propulsion work within the institute was divided into several areas: rocket engines with V. P. Glushko, A. M. Isaiev and L. S. Dushkin, turbojets with A. M. Liulka, ramjets with M. M. Bondariuk, and pulsejets with S. G. Rozental. Glushko, who worked on rockets in a specially equipped prison (a sharashka) in Kazan, was to develop an engine for N. N. Polikarpov's Maljutka rocket-plane. But Polikarpov died on July 30, 1944, and was replaced by V. N. Chelomei, who developed the Russian V-1 as the 10Kh missile. In May 1946, Glushko's group was transferred to Khimki's OKB-456 to develop the V-2 engine. As deputy to Bolkhovitinov, Isaiev built the engines for the BI-1 rocket-plane. In May 1948 his group was transferred to NII-88 (which had become TsNII Mach) as its sector N°9 (later KB KhimMach). Dushkin, who had worked with A. G. Kostikov on the 302 rocket-plane from 1942 to 1944, designed the RD-2MZV engine for Mikoyan's I-270 aircraft and the RD-2MZV-F for the Bisnovat aircraft N°5. Liulka left NII-1 to establish OKB-165 in Moscow in March 1946. Bondariuk, opened OKB-670 in October 1950. In addition, I. F. Frolov, head of the NII-1 aircraft division, built the 4302 aircraft with an Isaiev engine, and the 4303 with a Dushkin engine. However, all these rocket-planes were abandoned after 1947.

On November 30, 1946, academician M. V. Keldysh, only 35 years old at the time, took over the management of NII-1, where he stayed until June 12, 1948. He turned it into a research institute and all of the designers left, except Dushkin. It was a subsidiary of TsIAM from June 1948 to June 1952. Keldysh remained its scientific director until 1961 and became president of the Academy of Sciences of the USSR and also of its interdepartmental committee on space research (established by decree N°1388-618 on December 10, 1959).

## TAKING OVER THE V-2 MISSILES

At the end of the Second World War, the Soviet Union, in its progression through the German heartland, overran Peenemünde on the Baltic coast and the underground V-2 factory at Nordhausen in the Harz mountains, which remained within the territory of East Germany (GDR) for 46 years. But the Americans had already retrieved Wernher von Braun's team (118 members recruited under contract) and a hundred V-2s and spare parts, in the context of operations called Overcast and Paperclip. From 1946 to 1952, the US Army launched 70 rockets. But the Soviets did not remain empty-handed – they gained control of factories, rocket elements, and a number of German specialists.

In February 1945 a committee to oversee the recovery of German technology was formed by the state committee for defense (GKO). Technologies of interest to the Soviets concerned many ministries: weapons industry (cannons, rifles, optics, etc.), the aeronautics industry (aircraft, engines, equipment), the naval industry (ships,

Korolev and Tiouline in Germany in 1945.

Peenemünde in September 1945 from left to right: Sudakov, Sinelchikov, Fonarev, Korolev, Zakharov, Kharlamov, driver.

submarines, gyroscopes, etc.), the munitions industry (explosives, solid fuels), the machinery and instrumentation industry (launch facilities), communications (radar and electronics), the chemical industry, etc. A permanent commission headed by representatives of the industrial ministries operated on each army's front, with assistance from members of the army's main directorate (glavka) of war trophies. Peenemünde was on the Belarus front. The army arrived on May 5, with General A. I. Sokolov, head of armaments for the Katyushas groups. Stalin issued a decree ordering the minister of the aeronautics industry, A. I. Shakhurin, and the chief of branch N°2 of NII-1 in charge of solid rocket fuel, Yuri. A. Pobedonostsev, to take charge of the recovery of the Peenemünde facilities.

## THE GUNNERS LEARN TO USE ROCKETS

The armaments directorate of the Katyushas groups (GUV) relied on mortars of the guard (GMTch) of the main artillery directorate (GAU) of the ministry of defense. This entity evolved into branch N°4 of the GAU in June 1946, then the directorate of the deputy to the commander of artillery (UZKA) in April 1953, then the directorate of the chief of the armed forces rockets division (UNRV) in 1955, and finally the principal directorate of rocket munitions (GURVO) in 1960. The latter, which was part of the strategic rocket forces (RVSN), became the main client of the industry for intercontinental missiles and the military space program. Generals N. N. Kuznetsov, A. I. Sokolov, A. I. Semenov, A. G. Mrykine, G. A. Tiouline, Yuri A. Mozjorine, K. A. Kerimov, and A. G. Karas all played important roles in the development of the Semyorka.

Kuznetsov, head of GUV from 1941 to 1945, was in Germany from 1945 to 1946, and was then in charge of purchasing rockets in series until his retirement in 1963.

Sokolov, the party apparatchik, dealt with training before the war, while also taking evening classes at the institute of rail transport in Moscow. In 1941 he was sent to Chelyabinsk to oversee production of the Katyushas, and in 1943 he became deputy to Kuznetsov. In March 1945 he replaced Kuznetsov and went to Peenemünde. He soon returned to Moscow, but went back to Germany on October 14 to attend a V-2 launch organized by the allies as part of operation Backfire in Cuxhaven. At that time he was accompanied by G. A. Tiouline, Yuri A. Pobedonostsev, V. P. Glushko, and S. P. Korolev – all wearing Red Army uniforms. Then he headed the 4th directorate and UZKA. In late 1954 he was sent to the Dzerjinsky artillery academy to receive advanced training. A year later, he became head of N°4 (NII-4) Bolchevo institute, where he was in charge of strategic rockets and the first satellites. He was president of the state commission for the R-16 and UR-100 intercontinental missiles. He was given the Order of Lenin for the launch of the first Sputnik in 1957, the Lenin Prize in 1961 for implementation of the KIK tracking network, and the State Prize in 1967 for the qualification of the UR-100 missile. He submitted his doctoral dissertation in 1964, but this was not to the liking of General A. A. Vasiliev, head of GURVO, who judged he did not have the required level due to having not completed his studies at the institute of railway transport. His management responsibility was withdrawn and he was sent back to school. He was awarded his doctorate upon the second defense of his thesis. In 1969 he fell ill, and left the institute the following year. He died six years later.

Semenov completed studies at the Dzerjinsky artillery academy in 1937. He was,

Operation Backfire in Cuxhafen in October 1945, on the right: Glushko, Sokolov and Tiouline.

Operation Backfire. In the center: Tiouline, Glushko, Pobedonostsev; on the right: Korolev.

in turn, section head of the 4th directorate from 1946-1953, chief engineer of UZKA from 1953-1955, head of UNRV from 1955-1960, head of GURVO from 1960-1964, and then a member of the science and technology committee of the army chiefs of staff from 1964-1970.

Mrykine completed studies at the academy of chemical protection in 1934. He was head of the section from 1946-1953, head of development and research at UZKA from 1953-1955, deputy director of UNRV from 1955-1959, first deputy director of GURVO from 1959-1965 (during which he was president of the state commission for the R-14 rocket) and then worked at the ministry of general machines (MOM) from 1965-1972, where he presided over the state commission for a number of deep space missions (Luna, Venera, and Mars).

Tiouline, who graduated from the University of Moscow in 1941, was sent to the front before being assigned to the study of German war trophies. He was a member of the Soviet delegation at Cuxhaven. Then he headed the calculations bureau at the Nordhausen institute before becoming, in turn, section head of ballistics for the 4th directorate of GAU in 1946, sector head for the theory of flight and the laboratory of aerodynamics at NII-4 from March 1948 to March 1949, deputy chief for science at NII-4 in 1949, where he headed the creation of a network of tracking stations and a fleet of tracking ships. In 1959 he became head of NII-88. He was appointed deputy of the ministry of the defense industry (MOP) in June 1961, then first deputy in June 1963. He was first deputy of the ministry of general machines from 1965-1976. He was president of numerous state commissions for the Vostok and Voskhod manned space flights from 1962-1966, deep space probes from 1966-1972, and then Apollo-Soyuz from 1972-1975.

Mozjorine had just received his diploma from the Moscow aviation institute when war broke out. He was sent to the front in 1941, but then returned to the Jukowsky academy in 1942 to study. He stayed in Germany from June 1946 to February 1947, and then became an engineer in the theory of flight section of the 4th branch, being made section chief in 1951. He then became deputy director of NII-4 in 1955, where he participated in the creation of the KIK tracking network. He was chief of NII-88 from July 1961 to December 1990.

Kerimov (whose real name was Akhmedov) graduated from the industrial institute of Azerbaijan in 1942, and then from the Dzerjinsky artillery academy in 1944. He served as the Red Army representative in the Katyushas factory before being sent to Nordhausen to supervise telemetry devices. He worked at GAU and GURVO, where he was responsible for the purchase of rockets in series. In September 1960 he took over the 3rd directorate of GURVO, which managed the space program. In 1963, he became president of the state commission for the Zenit (optical reconnaissance) Molniya (telecommunications) and Meteor (meteorology) satellite programs. In 1966 he received the Lenin Prize for this work. In October 1964 he was appointed head of TsUKOS – the precursor of the space forces – where he headed the military space program. In March 1965, with the formation of MOM, he took charge of the 3rd glavka, devoted to space. He was president of the state commission for piloted space flights from 1966-1991 and for interplanetary flights from 1974-1991. For this work he received the State Prize in 1979 and the Hero of Socialist Labor Medal in 1987. In 1974 he left the ministry to become first deputy of TsNII Mach, where he remained until his retirement in 1991.

Karas graduated from the Odessa artillery school in 1938 and the Dzerjinsky

Special function brigade (BON) designers in 1946 seated, from left to right: Kaplun, Korolev, Riazansky, Chertok, Pilyugin, Pobedonostsev and others.

artillery academy in 1951. He became chief of staff at Kapustin Yar in 1953 and then at Baikonur in 1955. He worked as a consultant scientist at NII-4 in 1957, headed the KIK tracking network in 1960 and then TsUKOS from 1965 to 1979. He was also a member and president of several state commissions.

In addition, other generals took charge of ground infrastructures such as launch facilities, tracking networks, etc. In particular, V. I. Vozniuk was head of the rocket range at Kapustin Yar from June 1946 to April 1973, and A. I. Nesterenko was chief of NII-4 from 1946-1950, head of the rocket faculty of the Dzerjinsky artillery academy from 1950-1955, first director of the Baikonur cosmodrome in 1955-1958, and a member of the science and technology committee of the general staff from 1958-1966.

All these men were part of the artillery corps,[1] whereas in the United States the development of intercontinental missiles and military space activities was mainly in the hands of the US Air Force – in other words, pilots.


## THE SOVIETS IN GERMANY

Several groups of Russian specialists went to Germany after April 18, 1945. The first group was from the ministry of the aeronautics industry. This was led by General N. I. Petrov, head of the institute of aeronautical devices (NISO). It included G. N. Abramovich (deputy director of NII-1), K. N. Sourjine (deputy director of TsAGI), V. D. Vladimirov (deputy director of TsIAM), S. R. Ambartsumyan (deputy director of VIAM), and others. Each institute had a specialized subcommission.

The NISO group, which included B. E. Chertok of NII-1, was in charge of retrieving autopilots, radars, navigation and communications systems. Among other sites, the group visited the factories of Askania, Telefunken, and Lorenz. In May, the NII-1 engine group went to BMW at Basdorf.[2]

Specialists visited Peenemünde on June 1, and then went to Nordhausen, which the Americans left on July 14. Abramovich returned to Moscow on July 31 and presented his report to minister A. I. Shakhurin in August. He requested that work on rockets be suspended in order to devote all resources to jet aircraft. He was arrested along with the air force chief A. A. Novikov in April 1946 and was not released until after Stalin's death in 1953.

The minister of munitions, B. L. Vannikov, in charge of solid-fuel rockets and explosives, was made head of the inter-ministerial commission for the development of atomic weapons. Because of this new task, he left rocket work to the ministry of armaments, headed by D. F. Ustinov. On August 20, 1945, special committee N°1, headed by the minister of the interior, L. P. Beria, was created. This would become the first main directorate (glavka) of the council of ministers (PGU) and in 1953 the ministry of medium-sized machines. In order to develop the atomic bomb, German scientists were transferred to Sukhumi in the Soviet Union. These included Manfred von Ardenne, Gustav Hertz, Heinz Pose, Nikolaus Riehl, Max Steenbeck, Peter-Adolf Theissen, Robert Döpel, and Max Volmer. They participated in particular in the development of centrifuges for the production of uranium. As for the ministry of munitions, it became the ministry of agricultural machinery (MSKhM) in 1946. As a result, until 1953, solid propellant rockets were produced by the same ministry that made combine harvesters!

Finally, on May 13, 1946, rockets were placed under the responsibility of the minister of the armaments industry, D. F. Ustinov. He had been director of the Leningrad Bolshevik factory in 1938 before becoming minister in 1941. He held this post for 16 years, during which he led the Soviet national ballistic rocket program, including the Semyorka and the space program. He became president of the military industrial commission (VPK) in 1957, first deputy of the council of ministers in 1963, secretary of the central committee of the Communist Party of the Soviet Union (CPSU) in 1965, and then minister of defense from 1976 to 1984.

In 1946 his first deputy was V. M. Riabikov, who became president of the state commission for the launch of the R-7 and the first satellites in 1957. His deputies included I. A. Mirzakhanov and N. E. Nossovsky. Mirzakhanov was director of Podlipki cannon factory N°8 near Moscow from 1931 to 1938, which later became NII-88, then TsNII Mach, where the Semyorka was built. In 1940 he became chief director, and then deputy minister in 1946. But in February 1952, after difficulties in developing an anti-aircraft gun, he was arrested along with deputy defense minister N. D. Yakovlev (who was president of the state commission for the launch of the V-2 in 1947), and I. I. Volkotroubenko, the GAU chief. They were released after Stalin's death in 1953. Nossovsky was director of Podlipki cannon factory N°8 from 1938 to 1940. He was sent to Berlin to take charge of recovering German technology. He was assisted by N. N. Kuznetsov and L. M. Gaidukov.

## THE BERLIN AND NORDHAUSEN INSTITUTES

During the war, the Red Army used solid-fuel rockets called Katyushas, known as "Stalin's organ". They were developed by RNII and manufactured in series by many factories of the ministry of mortars under P. I. Parshin and also of the ministry of munitions under B. L. Vannikov. In 1941 they were entrusted to the "special" design office (SKB) of the Moscow Kompressor factory (in October 1941 part of the plant was relocated to Chelyabinsk in the Urals to put it beyond the reach of the advancing Germans). The SKB was headed by V. P. Barmine, who would later build launch facilities for Soviet rockets.[3]

On March 19, 1945, the ministry of munitions created the GTsKB-1 design office at factory N°568 in Moscow. Led by N. I. Kroupnov, its task was the development of rocket mortars of 20-30 km range. As such, it was a competitor to the ministry of the aviation industry's NII-1 subsidiary. On July 23, it was complemented by the GSKB-2 of the "Mastiajart" factory N°67 for missiles of 30-100 km range, while plant N°70 imeni Vladimir Ilyich GSKB-3 was to receive the German V-2 and other rockets. This was because minister Vannikov, whose son Rafael was at Nordhausen, saw the V-2 as the successor to the Katyushas. He planned to appoint Glushko as the main designer at GNII-70. The ministry of armaments entrusted the German missiles

Designers at Rabe de Bleicherode in February 1946, from left to right, seated: Voskressensky, ?, Bakulin, Korolev, Michin, Pobedonostsev; standing: Pilyugin, Mrykin, Brovko, Chijzhikov, Kharchev, Budnik.

The Council of Designers of 1946. From left to right: Chertok, Barmine, Riazansky, Korolev, Kuznetsov, Pilyugin, Glushko.

(V-2, Wasserfall, Schmetterling, Rheintochter, Taïfun, etc.) to Podlipki factory N°88 because this plant specialized in anti-aircraft systems, and these missiles, except for the V-2, were surface-air missiles. On November 30, P. I. Kostin was appointed main designer of the SKB. But in August 1946 he was replaced by K. I. Tritko, formerly chief engineer of the Volgograd Barricade factory.

Meanwhile in Germany, the flow of engineers investigating German technology intensified. On May 24, V. S. Budnik of NII-1 left for Berlin. On July 27 it was the turn of V. P. Glushko of OKB-16. On August 9, a group arrived in Germany with V. P. Mishin, N. A. Pilyugin, V. I. Kuznetsov, M. S. Riazansky, E. Y. Boguslavsky, L. A. Voskresensky, V. A. Rudnitsky, Florensky, Bakurine, Goriunov, and others. This group was split up for three locations: Berlin, Nordhausen and Prague. In August, Korolev, then in Kazan, returned to Moscow to the Tushino air show. Then he went to Germany on September 8. A group of ballistics specialists from the Dzerjinsky artillery academy led by Professor Y. M. Shapiro also traveled to Germany twice.

On July 8, 1945, GKO decree N°9475 created the commission for the study of German rockets. It was entrusted to General L. M. Gaidukov.[4] He was also an apparatchik who headed the managers' sector for general machines of the central committee in charge of the Katyushas production plants. Later, he was appointed head of the Nordhausen institute from February 1946 to January 1947. He then headed sector N°10 of "special" committee N°2 of the council of ministers in charge of rockets, and then held different jobs at the rocket armaments directorate (UZKA, UNRV, GURVO) until 1962. He worked until the end of his career at NII-4, where he headed the 7th "space" directorate before his retirement.

Gaidukov's deputy was Yuri A. Pobedonostsev, who had the task of bringing the top Soviet rocket specialists to Germany; though some of them were imprisoned in sharashkas (design-office-prisons), such as Glushko and Korolev. They were civilian engineers, but had to don military uniforms. Pobedonostsev was a former colleague and friend of Korolev's. He graduated from the Bauman technical school (MVTU) in 1930, worked at TsAGI, and became leader of the 3rd GIRD brigade in 1931 (wind tunnels and ramjets), and then joined RNII in 1933, where he defined the Kappa combustion stability test criteria for rocket mortars, for which he received the Stalin Prize in 1941. In February 1944 he headed section N°5 of NII-1, and then was head of NII-1 subsidiary N°2 from December 1944 to May 1946, and a member of the technical commission on German rockets from 1945-1946. From 1946-1947 he was chief engineer of NII-88. In 1946 he also worked as a professor at the academy of artillery sciences (AAN) and as a consultant to NII-4. He taught courses on rocket technology at MVTU in 1947, then taught at the "special" academy (pro-rector in 1950). In 1956 he returned to industry and worked on solid-fuel missiles at NII-125 until 1973. He died in the corridors of the international astronautical congress being held in Baku.

The Soviets created two institutes in Berlin and Nordhausen. One, headed by D. G. Diatlov, was located in the Berlin-Köpenick Gema plant. The chief engineer was V. P. Barmine. There were three design offices and six construction sectors made up of Russian and German engineers:

- KB-2, headed by E. V. Sinelchikov of V. G. Grabin's OKB, dealt with the Wasserfall surface-air missile.
- KB-3, headed by S. E. Rachkov of A. E. Nudelman's OKB, rebuilt the Schmetterling and Rheintochter missiles.
- KB-4, headed by N. A. Soudakov, was in charge of a 283-mm ramjet-powered shell.
- Sector N°5, headed by N. I. Kroupnov, head of GTsKB-1, dealt with the Taïfun R, the Kh-7 Rotkäppchen (also known as the Fritz X), and boosters for the Schmetterling and Rheintochter missiles.
- Sector N°6, headed by N. L. Oumansky of OKB-16 in Kazan, dealt with liquid-fuel engines (Wasserfall, Schmetterling, Rheintochter).

- Sector N°7, headed by V. A. Goviadinov, was in charge of equipment for radio guidance and radio detonators.
- Sector N°8, headed by P. K. Kliaritsky, dealt with calculation and stabilization (control surfaces, composites, etc.).
- Sector N°9, headed by V. A. Timofeiev of GSKB Kompressor, specialized in launch facilities.
- Sector N°10, headed by A. K. Polevik of GAU, was a chemical laboratory involved in the synthesis of liquid fuels (Tonka 250, Tonka 841, etc.).

In addition, the Berlin institute included a science and technology sector, a test sector, an experimentation plant and subsidiaries (including the Peenemünde center).

The underground plant for the production of the V-2, known as Mittelwerk, was located at Nordhausen and had used deportees from the Dora concentration camp

The state commission for the R-1 in October 1947 from left to right: ?, ?, ?, Ustinov, Yakovlev, Vetochkin, Korolev.

Participants in the R-1 launch campaign in October 1947 at Kapustin Yar. From left to right, third row: Lavrov, ?, Riazansky, Korolev, Voskressensky, Pilyugin, Chertok, Borissenko.

The Germans at Kapustin Yar in October 1947. From left to right: Karl Stahl, Johannes Hoch, Helmut Gröttrup, Fritz Viebach, Hans Vilter.

for labor. The institute established there in February 1946 by the Soviets was headed by L. M. Gaidukov and the chief engineer was S. P. Korolev.

It included:

- The Rabe institute (Raketenberg) of Bleicherode, headed by B. E. Chertok (deputy to Korolev).
- Factory N°1 (KB Olympia) at Sömmerda, 80 km east of Leipzig, headed by V. S. Budnik.
- Factory N°2 "Montania" of Nordhausen in charge of the V-2 engine, headed by V. P. Glushko.
- Factory N°3 at Kleinbodungen, headed by E. M. Kourilo.
- Factory N°4 of Zondershausen (steering system devices).
- Factory N°5, which was in fact the Peenemünde center (V. K. Shitov) and Lehesten testbed 150 km south of Nordhausen, headed by V. L. Chabransky.
- The calculation bureau (also known as the ballistics group) headed by G. A. Tiouline, which included S. S. Lavrov, R. F. Appazov, N. F. Guerrasiouta, and others.
- The Vystrel group, headed by L. A. Voskresensky, was responsible for test flights of the V-2 and had to be rebuilt at Nordhausen.

On July 15, 1946, in the village of Berk, 6 km from Zondershausen, the first brigade of rockets BON RVGK was placed under the direction of Major-General A. F. Tveretsky. Unfortunately, it was not able to launch the V-2 from German territory. After being transferred in August 1947 to Kapustin Yar it carried out eleven launches in October-November. The members of the brigade included B. A. Komissarov, N. N. Smirnitsky, A. I. Nosov and Y. I. Tregub. Komissarov became a representative of the ministry of defense at YoujMach, then chief of the 7th main directorate (glavka), deputy minister of the defense industry, and ultimately vice president of the VPK. Smirnitsky was head of GURVO from 1969-1976. Nosov went to Kapustin Yar, and then to Baikonur, where he became chief of operations for the Sputnik 1 launch in 1957. He was nominated deputy director of Baikonur for science, but died in the explosion of an R-16 missile on the launch pad on October 24, 1960. Tregub went to Kapustin Yar, where he headed flight testing of the S-25 surface-to-air missile before becoming deputy director of NII-2 in 1957, then head of the operational group of the directorate of manned spaceflight in Evpatoria in the Crimea from 1969 to 1973.

In Bleicherode, the Rabe had three sectors (launch, guidance and ballistics) where about 350 German specialists worked. It was headed by B. E. Chertok and the German Rozen-Plenter. N. A. Pilyugin was Chertok's deputy. In October 1946, 732 Russian specialists from ten ministries were working at Nordhausen. The plan was to assemble ten V-2s in Germany (five equipped with scientific apparatus) and five with radio-controlled lateral trajectory correction, and an additional ten in the Soviet Union using components brought from Germany. The first would form the N series and the others the T series. In the end, six were assembled at Kleinbodungen and five at NII-88. They were taken to Kapustin Yar in September 1947.

## THE GERMANS IN THE SOVIET UNION

German specialists left for the Soviet Union on October 22, 1946. Some of them went to institute N°88 (which became TsNII Mach) at Podlipki (today Korolev) and some to the island of Gorodomlia (today Ostachov) in Lake Seliger located near Kalinin (later Tver). In June 1948 they were all grouped in Gorodomlia.

In all, 152 experts (a total of 495 people with their families) worked under the direction of Helmut Gröttrup in collective 88, becoming group G (for Gröttrup), in August 1947. The specialists included Waldemar Wolff (ballistics), Josef Blass (engineer), Kurt Magnus (gyroscopes), Heino Zeise (thermodynamics), Franz Lange (radar), Erich Apel (testing), Wilhelm Schütz (measurements), etc. German engineers were paid like Soviet workers. For example, Gröttrup received 4,500 rubles, Magnus 6,000 rubles, Korolev 6,000 and Mishin 2,500.

In Gorodomlia, the establishment administered by F. G. Sukhomlinov became subsidiary N°1 of institute 88. Collective 88 included six groups: guidance systems (Hans Hoch and Kurt Blasig), guided anti-aircraft missiles (W. Quessel and Emil Mende), engines and thermodynamics (Karl Umpfenback), ballistics and aero-dynamics (Wernher Albring), production (Alois Jasper) and construction (Heinz Jaffke).

During their stay in Russia, the Germans participated in the reconstruction of the V-2, development of guided anti-aircraft missiles, a study of a 100-ton thrust engine, and the projects for the G-1 (R-10) with a range of 600 km in 1947-1948, G-2 (R-12) with a range of 2,500 km in 1948-1949, and the G-3 with a range of 8,000 km, G-1M (R-13), the G-4 (R-14) with a range of 3,000 km, and G-5 (R-15) with a range of 3,000 km in 1949-1950.

Twenty of them participated in the V-2 launch campaign at Kapustin Yar from October 18 to November 13, 1947. The state commission was chaired by Marshal

The V-2 rocket of 1947.

N. D. Yakovlev.[5] Technical direction was provided by S. P. Korolev, with as deputies V. P. Glushko (main designer of OKB-456), M. S. Riazansky (main designer of NII-885), V. I. Kuznetsov (main designer of NII-10) and V. P. Barmine (main designer of GSKB). In addition, the following organizations participated in trials: NII-20 (B. M. Konoplev and G. I. Degtiarenko groups), NII-6 became the institute of chemistry and mechanics of Moscow (payloads), plant N°15 became PO Polimer of Tchapaievsk (explosives), NII-137 became the institute of precision mechanics of Leningrad (detonators), NII-862 of Zagorsk (ignition systems), and plant N°686 became PO Projektor Moscow (electrical machinery on the ground).

Of the eleven launches, five were successful, but three rockets deviated wildly from the target and three others exploded in flight. Two rockets were equipped with cosmic ray detectors provided by the institute of physics of the Academy of Sciences (FIAN). From 1950 to November 1953 the Germans gradually returned to the GDR, except for twelve who remained in the Soviet Union. But institute N°88 was not the only one to benefit from German specialists' know-how. Indeed, there were 23 under the direction of Oswald Putze in Glushko's OKB-456, 54 at Riazansky and Pilyugin's NII-885, 20 at Kuznetsov's NII-10, 27 at NII-1, 13 at the institute of applied chemistry (GIPKh), seven at NII 20, and three at Barmine's GSKB. In addition, Germans were used in aviation: plant N°1 at Podberezie near Dubna which became OKB-256 Raduga (Baade and Rössing aircraft), factory N°16 at Kazan (engines), plant N°500 of Tushino (engines), and plant N°2 of Kuybyshev which became SNTK imeni N. D. Kuznetsov (engines). At S. L. Beria's SB-1, which was in charge of the Kometa guided missile, 270 Germans worked in sector N°36 and N°38. Later, SB-1 became NPO Almaz (Sokol district), leader in surface-to-air missiles and ABMs. In optics, nearly 300 specialists worked in factories in Moscow, Leningrad and Ukraine (including 30% at the Zenit factory in Krasnogorsk).

## THE DECREE OF MAY 13, 1946

An initial proposal to create a rocket industry in the Soviet Union was sent to Stalin on April 17, 1946 by L. P. Beria, G. M. Malenkov, N. A. Bulganin, B. L. Vannikov, D. F. Ustinov and N. D. Yakovlev.

The founding decree was finally signed on May 13. It created special committee N°2 of the council of ministers (the first such committee dealt with the atom and the third with radar). Special committee N°2 was headed by the deputy of the council of ministers, G. M. Malenkov from May 1946 to March 1947.[6] Decree N°1454-388 of May 10, 1947 entrusted the committee to the minister of defense, N. A. Bulganin. It was dissolved on May 15, 1949, and subsequent work was carried out by the minister of armaments, D. F. Ustinov. Then on July 1, 1953, it was handed over to the new SpetzMach directorate of the ministry of medium-sized machinery (MSM). The minister was V. A. Malyshev, and the director of SpetzMach was V. M. Riabikov (former deputy of Ustinov and head of the 3rd glavka, called TGU, in charge of the S-25 Berkut anti-aircraft system). On April 14, 1955, Riabikov was appointed head of the special committee for weapons of the army and the fleet consisting of three branches of the MSM. As such he was the president of the state commission for the R-7 and the first Sputnik. On December 6, 1957, the military industrial commission (VPK) was created and assigned to D. F. Ustinov. He then headed the nine industrial ministries which worked on armaments. The one devoted to rockets was the state committee for the defense industry. The VPK was successively headed by Ustinov in 1957, L. V. Smirnov in 1963, Yuri D. Maslyukov in 1985, and I. S. Belussov from 1988 to 1991.

In the Gosplan, G. N. Pachkov was responsible for managing a technical rocket section with 27 people. In the ministry of defense, rockets were assigned to the 4th directorate of GAU and the directorate of the navy rocket branch. In addition, on July 23, 1947, a rocket section was formed within the chiefs of staff (8 people). The 4th directorate (UZKA, UNRV, GURVO) was headed by A. I. Sokolov in 1946, A. I. Semenov in 1955, A. A. Vasiliev in 1965, and N. N. Smirnitsky from 1969 to 1976. This directorate should not be confused with the 4th directorate of the ministry of defense (4 GU MO) which was created after the suppression of committee N°2. The latter was headed by N. N. Kuznetsov in 1949, A. N. Sergeyev in 1950, then A. I. Semenov from 1951 to 1952. It was also dissolved, and another fourth directorate was created in August 1954 for the management of the S-25 Berkut anti-aircraft system. The latter was headed by P. N. Kuleshov in 1954, G. F. Baidukov in 1957, and E. S. Yurassov from 1972 to 1979.

The decision was made to create institute N°4 (NII-4) at Bolchevo and the N°4 firing range. A commission headed by General V. I. Vozniouk was responsible for choosing the site of the range: three sites were assessed in Uralsk-Osinki, Raïgorod, and Vladimirov. The third site, 130 km south-east of Volgograd, was adopted by decree N°2642-817 on July 26, 1947, and named Kapustin Yar.

The directorate of rocket armaments (URAV) of the Soviet navy was not created until 1948. It was headed by A. M. Brezinsky, B. V. Lipatov, I. G. Ivanov, and V. A. Sychev. The directorate created an institute, NII-4, in charge of missiles aboard ships and submarines. It was headed by N. A. Soulimovsky in 1948, A. T. Melnikov in 1960, and N. I. Boravenkov from 1962 to 1984. In 1965 it became institute N°28 and was in charge of the Boulava missile.

Thus, we see that the number 4 has a direct link with rockets at the ministry of defense: the 4th branch of GAU, institute N°4 of Bolchevo, and firing range N°4 at Kapustin Yar.

At the ministry of armaments, rockets were assigned to glavka N°7 (120 people) under the direction of S. I. Vetotchkine in 1946, A. S. Spiridonov in 1949 (for 6 months), I. G. Zubovitch in 1949, L. V. Smirnov in 1951, M. S. Riazansky in 1952, M. A. Soubbotchev in 1954, V. A. Kolytchev in 1955, L. A. Grishin in 1956 (died in the explosion of the R-16 at Baikonur on October 24, 1960), E. N. Rabinovich in 1958, B. A. Komissarov in 1961, then G. M. Tabakov from 1963 to 1965. On March 2, 1965, it became the ministry of general machines (MOM). Its ministers were S. A. Afanaseiev in 1965, O. D. Baklanov in 1983, V. K. Dogujiev in 1988, and O. N. Shishkin from 1989 to 1991.[7] It included 13 institutes, 19 design offices, and 25 experimental production series factories. The name "MOM" has been used in three contexts: from 1939 to 1941, from 1955 to 1957, and then from 1965 to 1991. It was renamed ministry of mortars during the war and under minister P. I. Parshin was responsible for producing the Katyushas. From 1955 to 1957, under minister P. N. Goremykin, it was in charge of solid-fuel missiles.

- The 1st glavka, which dealt with intercontinental ballistic missiles and launchers, was headed by P. A. Sysoiev (former director of KrasMach), S. F. Sigaiev, V. N. Konovalov, V. D. Kryuchkov, E. A. Verbine, V. N. Ivanov, and V. A. Andreyev. Deputies were E. N. Rabinovitch, A. V. Matveiev, L. E. Makarov, and others.
- The 2nd glavka, in charge of the submarine-launched ballistic missiles and engines, was headed by I. I. Abramov, V. N. Konovalov, N. B. Guerassimov, and S. F. Sigaiev. Later, the engines section became glavka N°12 headed by P. A. Gorchakov.
- The 3rd glavka, in charge of the space program, was headed by K. A. Kerimov in 1965, V. D. Vatchnadze in 1974, Yuri N. Koptev in 1977, and V. D. Ostroumov from 1989 to 1991.
- The 4th glavka, in charge of ground infrastructure (cosmodromes), was headed by P. P. Kotcherov, V. F. Matiachine, A. M. Mokine, and Yuri A. Fomin.
- The 5th glavka, which dealt with guidance systems, was headed by A. P. Zubov for 25 years.
- The 6th glavka, which handled gyroscopes, was headed by B. V. Balmont, V. A. Frolov, and others.
- The 7th glavka, in charge of the maintenance of ICBMs, was headed by A. S. Matrenine, A. V. Oussenkov, and V. A. Shuliakovsky.
- The 8th glavka, responsible for R&D and planning, was headed by K. P. Kolobenkov, B. V. Balmont, A. K. Vanitsky, I. P. Rumyantsev, A. I. Dunaiev, and V. F. Gribanov.

- The 9th glavka, responsible for the maintenance of submarine-launched ballistic missiles, was headed by S. S. Vanine, then V. I. Mikerine.
- The 10th glavka, in charge of radio systems and onboard instrumentation, was headed by O. F. Antoufiev.
- The 11th glavka, in charge of Energiya-Buran from 1976 to 1991, was headed by P. N. Potekhine (V. N. Khodakov was in charge of manned and international flights, while M. V. Sinelchikov took care of Energiya-Buran).
- The 12th glavka, in charge of conversion, was headed by S. A. Shumakov. It became glavka N°11 after Energiya-Buran was closed down, headed by S. I. Yunochev.
- Finally the 13th glavka, devoted to export and marketing, was formed in 1985 (Glavkosmos), headed by A. I. Dunaiev.

With the dissolution of the Soviet Union, MOM was replaced by the ministry of industry (minister A. Titkin) and the RosObcheMach organization. A year later, Yuri N. Koptev took charge of the Russian space agency, RKA, which became Rosaviakosmos in 1999. In 2004, General A. N. Perminov, former chief of the Baikonur and Plesetsk cosmodromes, took over the management of the Roskosmos federal agency. The first deputies were V. V. Alaverdov (1992-2002), N. F. Moisseiev (2002-2006), and V. A. Davydov (since 2006).


## INSTITUTE N°88

NII-88 was created to develop liquid-fuel rockets. The director, L. R. Gonor, was appointed on August 15, 1946. He was close to Ustinov, who had headed artillery factories during the war. In 1950 he was transferred to factory N°4 at Krasnoyarsk (later KrasMach). But he was arrested with N. D. Yakovlev in January 1953. He was released after Stalin's death and ran the subsidiary of TsIAM in Lytkarino until 1964.

The SKB, headed by K. I. Tritko, included:

- Sector N°3 for the V-2, headed by S. P. Korolev, appointed main designer on August 9.
- Sector N°4 for the Wasserfall (R-101), headed by E. V. Sinelchikov.
- Sector N°5 for the Schmetterling (R-102) and the Rheintochter, headed by S. E. Rachkov.
- Sector N°6 for the Taïfun (R-103), headed by P. I. Kostin.
- Sector N°7 for rocket fuselages (A. I. Lapchine).
- Sector N°8 for liquid-fuel rocket engines (N. L. Oumansky).
- Sector N°9 for liquid-fuel rocket engines (A. M. Isaiev). This group was transferred from NII-1 to NII-88 in May 1948. In late 1949, it absorbed the Oumansky operation which had not achieved satisfactory results.
- Sector N°10 for detonators (P. I. Melechine).
- Sector N°11 for launch facilities (I. G. Kisselev).
- Sector N°15 for artillery (G. D. Dorokhin).

- Sector M for materials (V. N. Iordansky).
- Sector T for fuels (N. V. Timochouk).
- Sector U for guidance systems (B. E. Chertok).
- Sector P for strength (V. M. Panferov).
- Sector A for aerodynamics (P. M. Golovinov).
- Sector I for flight tests (P. V. Tsybine). In August 1950, it merged with the Zagork test benches to form subsidiary N°2 of the institute (the future NII-229/NII KhimMach).

Organizations involved in developing the Wasserfall were NII-88, NII-885 (later the institute of space instrumentation: RNII Kosmitcheskogo Priborostroéniya), NII-20 (later NPO Anteï), NII-49 (later NII for control apparatus), NII-504 (which became the Moscow Impulse firm), NII-627 (later the institute of electromechanics), GSKB (which became the KBOM), and Moscow factories N°523 and N°528. Thirty flights took place during 1948-1949. But it was replaced by S. A. Lavochkin's 205 missile. The Schmetterling made 17 flights during October-December 1949. As for the Taïfun, it was developed in two versions: R-110 using liquid fuel (Chirok) and R-110, solid fuel (Strij). In 1950 the Strij was assigned to A. D. Nadiradze of KB-2. It was part of the RZS-115 system, abandoned in 1953. In March 1952 the Chirok was entrusted to D. D. Sevruk of OKB-3 NII-88, which would later build the Korchoun system and then the MMR-05 meteorological rocket.

At the ministry of agricultural machinery (formerly the ministry of munitions), the principal rocket directorate was N°6 (58 people), headed by A. V. Sakhanitsky, also the head of NII-1, created from GSKB-1 (later the institute of heat technology). It should not be confused with the NII-1 of the aviation industry, which became the Keldysh center. Subsidiary N°2 of the latter was transformed into KB-2 of factory N°67 in Moscow (which became NII-642/GNPP Vympel). The latter was headed by A. N. Voznessensky from 1946-1947, B. M. Saprykin from 1947-1948, V. M. Vinogradov from 1948-1951, N. I. Kroupnov from 1951-1955, and others. The directorate also worked on non-contact detonators, powder, etc. To do this, it used NII-504, plant N°70 imeni Vladimir Ilyich, N°512 (which would become Liuberetsky's NII-125/NPO Soyuz), factory N°121, factory N°737, factory N°73, and the Sofrino test range.

At the ministry of the aviation industry, the main directorate N°14 (50 people) was headed by A. I. Eremeiev. This ministry was responsible for rocket-planes, liquid-fuel engines, aerodynamic studies and flight testing. To do this, it made use of research institutes (TsAGI, TsIAM, VIAM, LII, etc.), NII-1 which would operate as a subsidiary of TsIAM from June 12, 1948 to March 10, 1952, OKB-293 (subsidiary N°1 of NII-1), M. R. Bisnovat, etc. On July 3, 1946, V. P. Glushko was appointed main designer of OKB-456 in Khimki. His team, which had been located in Kazan, moved in November-December. Previously, factory N°456 was a subsidiary of the Ilyushin aircraft manufacturer.

The ministry of the electro-technical industry, main directorate N°10 (71 people) was headed by A. A. Zakharov. This was responsible for radar, and for ground and onboard radio navigation. To carry this out it used NII-885 formed from the NII-20 laboratory of tele-mechanics, and factory N°1 of the ministry of defense. In June 1946 a liaison group for the ministry of the industry was created with directorate N°6 for rockets, while directorate N°10 became "special" directorate N°2 (24 people) in 1947 managing institutes N°627 (electro-mechanical institute with Machinoapparat of Moscow) headed by A. G. Yossifian, and N°686 (Projektor plant) headed by A. M. Goltsman, in charge of electrical equipment on board and on the ground.

The ministry of the naval industry, main directorate N°1 (59 people) was headed by V. N. Tretyakov. It dealt with gyroscopes, ship-borne radars, launch installations on ships, guidance payloads for underwater targets, etc. V. I. Kuznetsov, head of the ministry's SKB laboratory, was appointed chief constructor of rocket gyroscopes in sector N°2 of institute N°10 (later Altai NPO).

The ministry of the chemical industry, special directorate N°2 (25 people) was headed by V. V. Ofitserov. It was in charge of liquid fuels for rocket engines. A research area was created at the institute of applied chemistry (GIPKh) in Leningrad. Liquid oxygen was the responsibility of the oxygen directorate of the council of ministers (Glavkislorod), then headed by academician P. L. Kapitza (a future Nobel Prize laureate). In December 1945 factory N°28 in Moscow became VNIIKiMach (Kislorod Machinostroenié) for the production of liquid oxygen. The subsidiary of Balashikha would become VNIIKriogenMach in 1963. On August 17, 1946, Kapitza, who opposed L. P. Beria in the atomic bomb project, was relieved of all functions and replaced by M. K. Sukov.

The ministry of machinery and instrumentation, SpetzMach (21 people), was managed by K. K. Glukharev. It handled launch facilities, compressors, pumps, etc. V. P. Barmine was appointed chief constructor of the Moscow Kompressor plant's GSKB.

Finally, the decree of May 1946 also dealt with training. Five hundred specialists were selected within the institutes of the ministry of higher education and other ministries to work on rocket technology. Moreover, by the end of 1946 the plan was drawn up to form the first class with at least 200 people from technical schools and at least 100 people from universities.

## THE FIRST BALLISTIC ROCKETS

Korolev returned to Moscow on January 20, 1947, with orders to work at Podlipki. There, he undertook the completion of a Soviet version of the V-2: the R-1 under the decree of April 14, 1948. The Glushko engine was an RD-100 delivering 26 tons of ground thrust. The chamber pressure was 16 bars with a specific impulse of 203 s. The Messina telemetry system was replaced by the Brazilionit system. The guidance system included a horizon gyro (GG-1), a vertical gyro (GV-1), and an integrator gyro (IG-1). The engine fired for 65 s and gave the rocket a range of 270 km (the same as the V-2).

At the same time, the construction of the Kapustin Yar range began under the direction of Major-General V. S. Kossenko. The first officers arrived on August 20, 1947. Work focused on the construction of an integration building (wooden hangar),

The geophysical R-1D of 1951.

a static test bench, a launch area, and tracking stations. Personnel lived in railway carriages. A second test bench was created in Khimki on May 24, 1948. A third was built at subsidiary N°2 of NII-88 at Zagorsk and the first static firing took place on December 18, 1949.

The president of the state commission for the R-1 was S. I. Vetotchkine. The R-1 was launched nine times in September-October 1948. Only one flight succeeded; that on October 10. The failures were due to several causes, but the pyrotechnic ignition of the RD-100 was the real problem. This was modified for the second test campaign in September-October 1949. Out of 20 launches, 17 succeeded. Meanwhile, in May, six units were used in the R-1A geophysical version by the institute of physics of the Academy of Sciences (FIAN). A scientific study program was initiated on December 30, 1949. It resulted in versions R-1B, R-1V, R-1D and R-1E which flew until 1956. The capsule was modified to carry dogs on vertical trajectories. The first passengers were Dezik and Tsygan on July 22, 1951.

On November 25, 1950 the R-1 was finally declared operational. Three sites were envisaged for series production: Zlatoust, Kiev and Dnepropetrovsk. Plant N°385 in Zlatoust in the Urals became a backup for Korolev's OKB on December 14, 1947. The SKB factory was assigned to develop new versions of the R-1.[8] It studied a lightened version of the R-1 with tanks made of laminate or wood (versions 50R and

Left to right: the R-1 rocket of 1948; the R-2 rocket of 1949; the R-3 project of 1949; the R-5 rocket of 1953; and the R-11 rocket of 1953.

50RA) to achieve a longer range, but this project was abandoned. In 1952 it focused on the 8B51 tactical missile in competition with D. D. Sevrouk's Korchoun (derived from the German Taïfun). At the same time, the plant was producing Isaiev engines and launch facilities for the BM-14 (Katyushas). And in June 1955 it took charge of production of the R-11, the ancestor of the well-known "Scud" and ballistic missiles launched from submarines. This rocket was developed by Korolev in 1950-1953. It used a storable propellant device which (as with the Lavochkin 205 ground-to-air missile) inherited features from the Wasserfall. It weighed 5.3 tons and had an 8.3-ton-thrust Isaiev engine. It was therefore 2.5 times smaller than the V-2, but with the same range. On July 13, 1955, this was declared operational and series production began in Zlatoust. The main designer was V. P. Makeiev of Korolev's OKB. He was only 31 years old at the time, but as an activist in the communist party he had been a Communist Youth instructor from 1950-1952. He was the craftsman of the R-11FM, R-13, R-21, R-27 and R-29 submarine missiles, as well as the R-17 tactical missile that superseded the R-11.

The second site studied was in Kiev, but minister Ustinov refused to locate such a strategic plant in the capital of the Ukraine. Finally, plant N°586 of Dnepropetrovsk (later YoujMach) was chosen. It was an automobile manufacturing plant that would become the world's largest rocket production plant. N. S. Khrushchev would say that it produced rockets "like sausages". But going back to 1951, decree N°1528-768

The R-11A rocket of 1958.

of May 9 assigned the production of the R-1 to this plant. The goal was to make 70 units in 1951, 230 in 1952, 700 in 1953, then 2,500 starting in 1954. A team from Korolev's OKB headed by V. S. Budnik went to the site. Glushko's OKB sent N. S. Shniakin and his team. They established an SKB. Budnik was assisted by Shniakin and A. P. Elisseiev (former deputy to Sinilchikov). In addition, the guidance system was produced by plant N°897 of Kharkov (later to become PO Kommunar). There, Pilyugin's deputy, A. M. Ginzburg, opened a design office. The team also produced systems for the R-2, R-5, R-7 and R-11 rockets, and others. The first R-1 made by Dnepropetrovsk flew in November 1952.

## KOROLEV'S ROCKETS

Decree N°1401-370 of May 7, 1947 defined the plan for future rocket development, and concerned the following companies: NII-88 KB-2, plant N°51, NII-1 (MIT), and GSKB-47. In addition to the R-1 with a range of 270 km, it proposed developing rockets with 600 and 3,000 km range. These were Korolev's R-2 and R-3 projects. Decree N°1175-440 of April 14, 1948 focused on the development of numerous types of vehicles, including long range ballistic missiles at NII-88 (R-1, R-2, R-101), aeronautical and naval missiles of KB-2 (Shuka, Krab, etc.), OKB-51 (16KhA, 10KhM, 15KhM) and OKB-293 (Snars 250).

Korolev strove to improve the range of ballistic rockets. The R-2 was 350 kg heavier than the R-1, but the range was double. It had integrated tanks and an RD-101 engine with a ground thrust of 37 tons. The pressure in the chamber was 21 bars with a specific impulse of 210 s. For telemetry it had the Don system. The guidance system had been modified: the gyro-integrator was replaced by an automated range recorder, and a BRK-1 lateral guidance system was introduced for better accuracy. It was produced by plant N°285 of Kharkov (later to become Monolit PO). There, O. D. Baklanov moved up from simple worker to management. He was 25 years old when the plant produced the Semyorka system. He became chief engineer in 1963, director in 1972, deputy minister in the ministry of general machines in 1976, first deputy in 1981, minister in 1983, and thereafter secretary of the central committee of the CPSU from 1988 to 1991. His downfall came when he participated in the failed coup against Gorbachev in August 1991.

Five R-2 launches took place at Kapustin Yar in September-October 1949, two of which failed. A second series of twelve launches took place in September-December 1950. Unfortunately, they all failed: five R-2s were lost during the active flight phase and the payloads of the other seven were destroyed in the final phase as a result of overheating. Of a series of 13 launches in 1951, only one failed and that was due to a manufacturing defect. The R-2 was therefore declared operational on November 30, 1951 and three days later series production was approved at factory N°586. The first series batch were launched in August-September 1952 and 12 out of 14 reached their

The R-2A rocket of 1957.

The R-5A rocket of 1958.

targets. There was also a geophysical version. Thirteen rockets were launched from May 1957 to September 1960 and carried dogs to an altitude of 200 km.

The R-3 was to carry a payload of 3 tons for a distance of 3,000 km. The 2.8 m diameter rocket (compared to 1.65 m for the previous rockets), weighed 70 tons at take-off. It was equipped with a single-chamber 120-ton-thrust (60-bar) engine. Two constructors were in competition: Glushko with the RD-110 and A. I. Polyarny of NII-1 (NIITP) with the D-2. The use of kerosene instead of ethyl alcohol as fuel was studied because it delivers more energy. The higher combustion temperature made it necessary to improve the chamber's cooling system, since the cooling properties of kerosene are 1.5 times less than those of alcohol. The payload, which was to re-enter the atmosphere at a speed of 4,500 m/s, was covered with thermal protection capable of withstanding a temperature of 1500°C.

Glushko had been studying the RD-110 since 1947. He initially made a welded spherical chamber derived from that of the V-2 engine, using a closed-circuit cooling system, but its performance was unsatisfactory. As for A. I. Polyarny of NII-1, he failed to perfect his overly innovative engine. Glushko had to give up the spherical chamber for a cylindrical shape. In 1948 he made the KS-50 "Lilliput" demonstrator, a cylinder 60 mm in diameter. For the first time, an inner wall of copper was brazed onto a steel chamber. In 1949 he developed the experimental ED-

The KS-50 Lilliput experimental chamber.

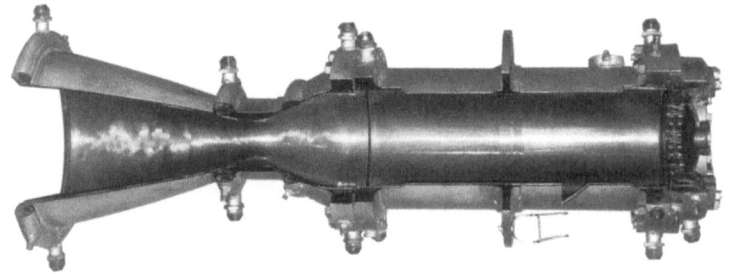

The ED-140 experimental chamber.

140, 7-ton-thrust chamber. It was a 240-mm cylinder using regenerative cooling (double wall) and, for the first time, tangential seepage of coolant onto the inner wall of the chamber (film cooling). It was originally scheduled to test the R-3 engine injection head, but in fact it was used to develop the 600-mm cylindrical chamber of the 60-ton-thrust engine. From 1952 onwards, several units were produced with different variants of elements designed to eliminate high-frequency instabilities that appeared as the diameter of the chamber was increased.

Furthermore, the development of the Topaz guidance system by B. M. Konoplev from NII-20 also posed problems. In 1951 Korolev decided to make a more modest version: an R-3 with a 1.5-ton payload bay. It weighed 23.4 tons at take-off and was equipped with a 40-ton-thrust engine. But it only had a range of 935 km, which was inadequate because a late version of the V-2 called the R-5 could fly 1,200 km. The latter was designed in October-November 1951. It weighed 28.5 tons at take-off and was equipped with an RD-103 engine with a ground thrust of 43 tons. The chamber pressure was 24 bars with a specific impulse of 220 s. Development took two years, with the first flight test campaign occurring in March-May 1953. The president of the state commission was General P. A. Degtiarev. Out of eight launches, six reached their targets. A second campaign was held in October-December 1953. Out of seven launches, six were successful. Finally, the third campaign ran from August 1954 to February 1955 and four out of 19 shots had range setting problems. In January 1955 the rocket was declared operational and series production began in Dnepropetrovsk. It was produced in several geophysical versions (R-5A, R-5B, R-5V) that flew from

1958 to 1975, as well as the R-5R and M-5RD versions used to test elements of the Semyorka.


## THE FIRST STRATEGIC MISSILE

On April 10, 1954, a decree called for equipping the R-5 with a nuclear warhead. The R-5M (8K51) rocket thus became the first strategic missile in the Soviet arsenal. The first A-bomb was tested successfully on August 29, 1949, at Semipalatinsk in Kazakhstan. The piston-engine Tupolev Tu-4 (a copy of the American B-29 Superfortress) carried the RDS-3 bomb as early as October 1951, while the Tu-16 jet became operational in October 1953.

The R-5M was equipped with an RD-103M (8D71) engine delivering a 44-ton thrust. The payload bay was developed by sector N°8 of OKB-1. The nuclear charge was provided by KB-11 of Arzamas[9] and its N°1 Moscow subsidiary (which would become institute N°25).[10] The test campaign was again managed by General P. A. Degtiarev. The R-5M made a total of 24 flights from January-November 1955 with a conventional payload, of which 21 flights were successful, the first of these being on

The R-5M rocket of 1956.

January 21. Then five flights were made for operation "Baikal" in January-February 1956. The first four rockets carried a model of a nuclear warhead, and the fifth, on February 2, delivered a 0.3-kiloton RDS-4 nuclear warhead to the target in the Aralsk area. The head of the state commission was General P. M. Zernov, deputy minister of medium-sized machines.[11]

The R-5M was deployed in several regiments based at Pervomaysk in 1959-1961, Gvardeysk in 1959-1966, Kolomyia in 1960-1968, and Voroshilov in 1961-1967. It was fired from Kapustin Yar between January 1957 and October 1961 for nuclear tests: four with explosions of 10 kilotons and one of 40 kilotons in the atmosphere, and two with explosions of 1.2 kilotons in space.

Starting on February 7, 1951 Korolev worked simultaneously in three directions: the development of a single-stage missile of 3,000 km range (theme N-1), a storable propellant missile (theme N-2), and an intercontinental missile (ICBM) of 5,000 to 10,000 km range with a 3-ton nuclear warhead (theme N-3). The latter was divided into two parts: T-1 for a two-stage ballistic missile, and T-2 for a winged rocket (a cruise missile powered by a ramjet at Mach 3). These devices had to be capable of carrying the RDS-6s hydrogen bomb, which was first tested on August 12, 1953 at Semipalatinsk. Three delivery systems were pursued: the strategic bomber, the cruise missile and the intercontinental missile. Two bombers were developed: the Tu-95

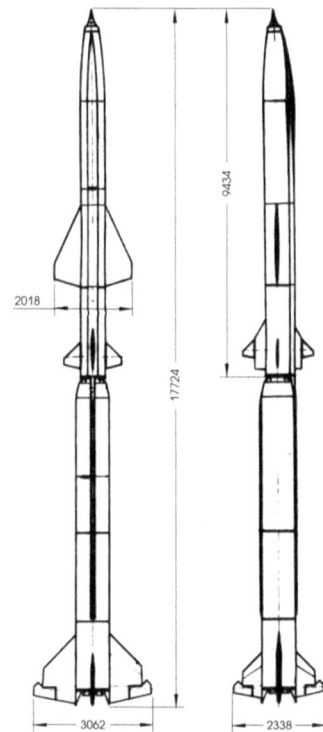

The EKR winged rocket project of 1953.

turboprop and the Myasishchev M-4 turbojet. The Tu-95 encountered development difficulties and did not enter service until September 1957, but the M-4 was flying by January 1953 and was declared operational in 1956.

Two decrees were issued on February 13, 1953. Decree N°442-212 dealt with the ICBM (T-1), the EKR experimental cruise missile (T-2), the development of the R-5, and the reassignment of the 2,000-km-range missile to OKB-586 in Dnepropetrovsk, which was formed on April 10, 1954 under the leadership of M. K. Yangel, who was director of NII-88 in 1952-1953. Decree N°442-213 defined the research work in the field of long range rockets for the period 1953 to 1955.

Korolev's EKR project made use of an R-11 as a booster for a winged missile powered by M. M. Bondariuk's RD-040 ramjet, which had been submitted to M. V. Keldysh, S. A. Khristianovich and M. M. Bondariuk on January 31. It was scheduled to make its first flight in 1954. But decree N°956-409 of May 20, 1954 assigned development of intercontinental winged rockets to aircraft designers Lavochkin (La-350) and Myasishchev (M-40). Lavochkin worked on the Buria missile, which made its first flight in June 1957 and its 18th and final flight in December 1960, while Myasishchev worked from April 1953 to November 1957 on the Buran missile that was never completed.

The ICBM was to carry 3.0 tons for a distance of 8,000 km. The initial design thrust of the first stage was set at 180-200 tons, and that of the second stage at 45-50 tons. The final project was to be handed over in the second quarter of 1953 with flight testing expected in 1955. However, Stalin died on March 5, 1953 and on June 26 Beria was arrested. In October 1953 the minister of medium-sized machines, V. A. Malyshev, ordered that the mass of the thermonuclear warhead be increased to 5.5 tons. Korolev increased the mass of the vehicle from 190 to 250 tons and the thrust from the then-envisaged 273 to 370 tons. Flight tests were delayed because of changes in performance, as well as by the unavailability of a wind tunnel at NII-88 and test beds at NII-229.

Korolev was considering two options, one involving stacked stages (pencil type) and the other involving stages assembled in parallel (bundle type). The idea of the bundle was proposed by M. K. Tikhonravov on July 14, 1948, in his presentation entitled "The path to the achievement of long range rockets" delivered at the annual meeting of the academy of artillery sciences (AAN) in the presence of its president, General A. A. Blagonravov, and Korolev. In this presentation, Tikhonravov spoke of the possibility – using the technology at that time – of reaching the 1st cosmic speed with a rocket stage and creating an artificial satellite of the Earth. A second report (N°207) was published in December 1948 on rocket stages. The study was published in Rocket Technology N°3 of 1949 and in AAN Newsletter N°6 in 1949. He then presented a "bundle" of three Korolev R-3 rockets capable of a range of 3,000 km. And then he presented "Rockets in a bundle and prospects for their use" in the first science and technology conference of NII-4 on March 15, 1950. His work showed that such a rocket could place a satellite into orbit. He also referred to the possibility of sending a man into space. After that, Tikhonravov was relieved of his duties, becoming a consultant to Korolev, and was replaced by his deputy Krasnov.

In 1949 Korolev proposed a bundle of three R-2s, and later a bundle of three R-3s.

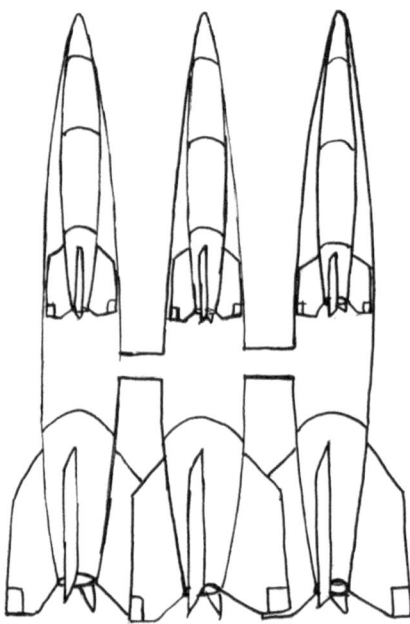

Tikhonravov's design for a staged rocket.

Lavochkin's Buria winged missile.

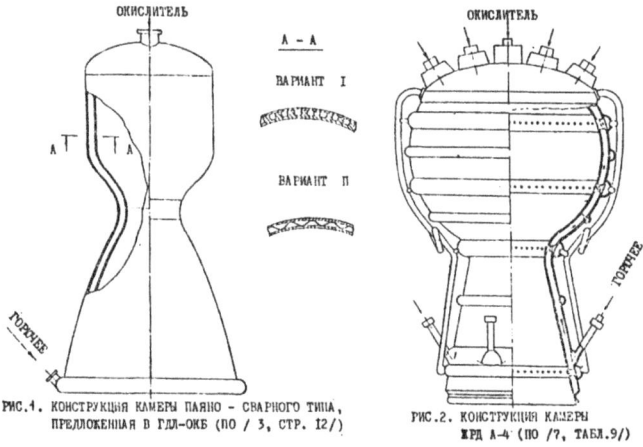

ОКИСЛИТЕЛЬ

A – A

ВАРИАНТ I

ВАРИАНТ II

ГОРЮЧЕЕ

РИС.1. КОНСТРУКЦИЯ КАМЕРЫ ПАЯНО – СВАРНОГО ТИПА, ПРЕДЛОЖЕННАЯ В ГДЛ–ОКБ (ПО / 3, СТР. 12/)

ОКИСЛИТЕЛЬ

ГОРЮЧЕЕ

РИС.2. КОНСТРУКЦИЯ КАМЕРЫ ЖРД A-4 (ПО /7, ТАБЛ.9/)

Comparison of V-2 and RD-107 chambers.

Test engines.

In 1951, M. V. Keldysh, S. S. Kamynine, and D. E. Okhotsimsky at the institute of mathematics of the Academy of Sciences (MIAN) calculated the ballistic capabilities of staged rockets, including Tikhonravov's bundle concept. They also resolved the issue of the movement of a rocket around its center of mass taking into account the sloshing of propellant in its tanks. And they determined the optimal pitch control of stage rockets.

In 1952 Glushko designed the single-chamber RD-105 and RD-106 engines for the first and second stages of the R-7. The chamber had an internal diameter of 600 mm. The pressure was 60 bars. The specific impulse was 302 s for the RD-105 and 310 s for the RD-106. They both delivered a thrust of 65 tons. The RD-105 was also intended to equip the four boosters of V. M. Myasishchev's Buran winged missile. Tests revealed combustion instability in the chamber. It was decided to use a smaller

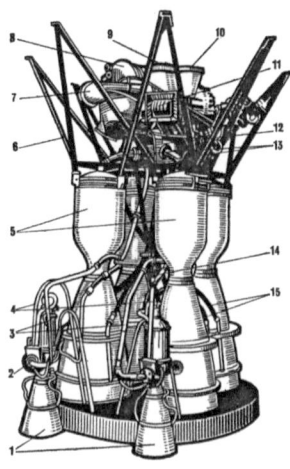

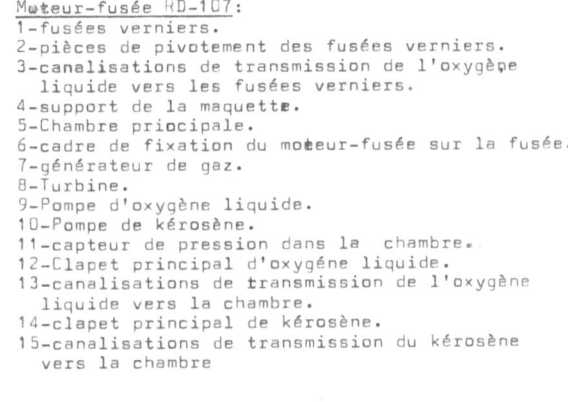

```
Moteur-fusée RD-107:
1-fusées verniers.
2-pièces de pivotement des fusées verniers.
3-canalisations de transmission de l'oxygène
   liquide vers les fusées verniers.
4-support de la maquette.
5-Chambre principale.
6-cadre de fixation du moteur-fusée sur la fusée.
7-générateur de gaz.
8-Turbine.
9-Pompe d'oxygène liquide.
10-Pompe de kérosène.
11-capteur de pression dans la chambre.
12-Clapet principal d'oxygène liquide.
13-canalisations de transmission de l'oxygène
   liquide vers la chambre.
14-clapet principal de kérosène.
15-canalisations de transmission du kérosène
   vers la chambre
```

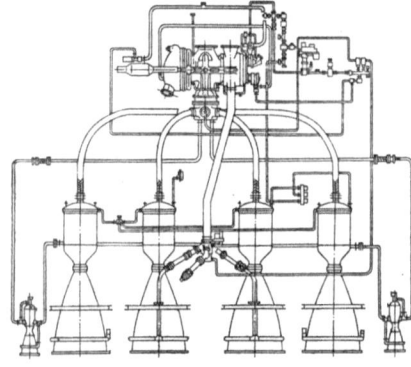

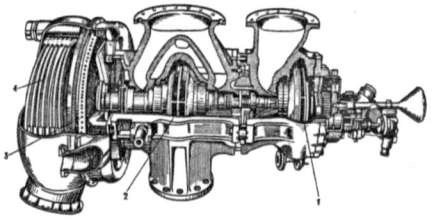

```
Groupe de turbopompes du RD-107:
1-pompe de kérosène.
2-pompe d'oxygène liquide.
3-turbine.
4-échangeur thermique.
```

```
Schéma du moteur-fusée RD-107.
```

Diagram of the RD-107.

chamber to overcome this problem. In fact, owing to an increase in the mass of the R-7, it was evident that the engine would require four chambers of 430 mm diameter. This would not be the first engine to have four chambers. Isaiev had this idea for the 8-ton-thrust engine equipping the Lavochkin surface-to-air missile in 1950. It was an assembly of four 2-ton chambers. For the Buria missile he developed a 17-ton-thrust engine comprising four chambers.

In 1953, given the performance of Glushko's engines, the bundle was defined to have four lateral boosters each delivering a thrust of 60 tons and a second stage with a thrust of 60 tons that ignited at take-off along with the first stage to produce a total thrust of 300 tons to propel a vehicle weighing 260 tons.

Glushko then designed the 8D72 and 8D73 experimental engines. These had four chambers of 25-ton-thrust and a pressure of 60 bars. The chamber underwent 250 tests from December 1954 to March 1955. Then it was equipped with the turbopump

The RD-107 at the 1967 Paris Air Show.

RD-107 vernier engines.

from the 8D71 (RD-103M) and underwent 39 test firings in April-June 1955. From July-December 1955 trials involved a group of two chambers. Finally, from January-December 1956 the tested engines had four chambers. This led to development of the 8D74 (RD-107) and 8D75 (RD-108). The ground thrust of the RD-107 was 83 tons with a specific impulse of 256 s. The figures for the RD-108 were 76 tons and 248 s. This time the tests were conclusive.

The simultaneous ignition of all these engines required their proper functioning to be verified prior to launch. It was therefore necessary to go through a preliminary procedure. The ignition was carried out in three phases: 40% of the power, then 70% and finally 100%. Furthermore, steering the vehicle could no longer be achieved by adjusting deflectors in the exhaust flow, it would require vernier engines. Glushko refused to develop these because he wanted to concentrate on high-thrust engines. Therefore M. V. Melnikov's engine sector was created within Korolev's facilities. Melnikov developed the S1-35800 engine with a thrust of 3 tons. There were many early problems. In particular, the throat and inner wall burned during combustion. By the spring of 1957, over 500 test firings had been carried out on 285 chambers. The second phase involved combined tests with the RD-107 and RD-108 engines on the test bench at OKB-456. High-frequency oscillations appeared both during the preliminary phase and during transition to the main phase. OKB-456 then proposed

RD-110 and RD-105 engines.

a new fuel supply design which yielded positive results. While the last tests were being carried out in 1957, OKB-456 made improvements to the vernier engine which became the D166-000.

At take-off, the R-7 fired twenty main chambers and twelve verniers (a total of 32 nozzles) which was the equivalent of twenty V-2s taking off at the same time.

## CONSTRUCTING BAIKONUR

The R-7 could not be launched from Kapustin Yar since the four lateral boosters could conceivably fall on inhabited areas in the region of Saratov. Moreover, the rocket's guidance system relied on two guidance stations that had to be located at 250-300 km from the launch pad, with a third at 300-500 km along the trajectory – which would have placed the latter either in the Caspian Sea or Iran. Consequently, in late 1953 a request was made to the council of ministers for a new launch site for the R-7, Buria, and Buran missiles. On March 17, 1954 the decision was finalized. A state commission led by General V. I. Vozniouk, director of Kapustin Yar, selected four candidates for the new site: Ioshkar-Orla in the Republic of Maris; the Dagestan region west of the Caspian Sea; east of the city of Kharabali in the

Astrakhan region (near Kapustin Yar); and the Kazakh desert where there were two options at the rail stations of Baikhoja and Tyuratam. Evaluations were carried out between March and December 1954.

On January 1, 1955 the committee selected Tyuratam ("sacred place" in Kazakh). On February 4 the military industrial commission forwarded the recommendation to the government.[12] Construction costs related to technical facilities, launch platforms and transportation (roads and railways) were estimated at 100 million rubles. On February 9 Marshal M. I. Nedeline became deputy minister for special weapons and rocket technology. On February 12 Marshal N. A. Bulganin, chairman of the council of ministers, signed secret decree N°292-181 ordering construction of polygon N°5 (Taiga object) of the ministry of defense in 1955-1958. Flight testing of the Buria and Buran missiles at limited range took place from launch facility N°4 at Kapustin Yar, but these projects were canceled before the second step at intercontinental range could be pursued.

The rail station at Tyuratam was built in 1901 to service a water tower supplying steam locomotives. It comprised a two-story brick house and several homes made of clay bricks for the staff of the water facilities and railway. But the authorities decided to name the launch facility "Baikonur", despite the real town of this name being 300 km to the north-east. This was intended to mislead Western intelligence services. The first mailing address was Tashkent 90, then Kzyl-Orda 50. The village was named Zarya, then Leninsk from January 28, 1958. During the visit of General de Gaulle in June 1966, the city was provisionally named Zvezdograd, or the "City of the Stars". The facilities were a main area of 2,474,973 hectares located approximately 100 km from the towns of Djoussaly and Novo-Kazalinsk in the regions of Kzyl-Orda

The Tyuratam station in 1955.

Tent city in 1955.

Arrival of the first soldiers in 1955.

and Karaganda, with the first stage segments of the R-7 impacting in the area of Akmolinski near Lake Tengiz in Kazakhstan, and the warhead impacting in Kliuchi, near Cape Ozerny in the region of Kamchatka some 8,000 km away.

## THE CONSTRUCTION ARMY

Baikonur was built by the army engineering corps of the ministry of defense and not, as some claim, using prisoners from the gulags. Within this corps, construction was assigned to the main directorate of special constructions (GUSS).[13] Its achievements include launch facilities at Kapustin Yar, Baikonur, Plesetsk, Semipalatinsk, Sary-Shagan and Novaya Zemlya, more than 300 rocket complexes, over 20 airfields, over 10 complexes for space monitoring and early warning systems (SKKP and SPRN), 16 science centers, and more. The GUSS regiment in charge of Baikonur was the 130th UIR (unit N°12253) which was already in the Tashkent region (Uzbekistan). It was led by General G. M. Shubnikov in 1951-1965 and by General I. M. Gurovich in 1965-1975.

The first soldiers of the military engineering corps arrived on January 12, 1955. Tyuratam station was located on the Moscow-Tashkent railway line, near the River Syr-Darya. At that time, the only indication on maps was by-pass station N°103. The nearest airfield was Djoussaly, which was used until construction of the facilities was completed.

Several construction sites started up simultaneously: the development of a rail and road network to serve different construction zones, the launch platform (Zone 1), the launcher integration building and living quarters (Zone 2), the KAZ fuel plant (Zone 3), the radio center's transmitter and receiver (Zones 4 and 5), the weather station (Zone 6), the anti-aircraft defense station to protect the site from spy planes (Zones 7 and 8), the industrial zone near the Tyuratam station (Zone 9), the living quarters (Zone 10), the thermal power plant (Zone 11), the repair station (Zone 12), the cemetery (Zone 13), Lastochka airfield (Zone 15), water reservoirs (Zone 17) and the IP-1 tracking station (Zone 18). The construction teams first set up living quarters in the industrial area near the station. They initially lived in railway sleeper carriages and tents. Then the first prefabricated wooden buildings (SR-2s) arrived by train. The residential area included the first barracks, staff headquarters, officers' mess, canteen, store, post office, sauna (Russian bath), and other amenities. The industrial zone included the concrete plant, sawmill, carpentry shop, and a garage for vehicle repairs.

On May 5, ground was first broken to construct the village (Zone 10) located between the station and the river. There, a real town would develop with the building of the staff headquarters, the Hotel Tsentralnaya, the Univermag store, housing, schools, a theatre, the "Orion" swimming pool, and other facilities. The KAZ plant, built in 1956-1957, produced up to 300 tons of oxygen and liquid nitrogen per day (kerosene arrived by train). Liquid nitrogen was gasified to pressurize the tanks of the rocket. A dirt airfield was built in 1957-1958, while the thermal power plant was completed in October 1958. The anti-aircraft defenses were dismantled in 1960 after the Americans abandoned U2 spy plane overflights of the USSR.

## THE DECREE OF APRIL 14, 1955

On July 1, 1953 rockets were placed under the responsibility of the GlavSpetzMach directorate at the ministry of medium-sized machines. V. M. Riabikov was in charge in 1953 and S. M. Vladimirsky from 1954 to 1955.

On April 14, 1955, decree N°720-435 created the special committee for weapons of the army and navy. It drew on three directorates of the ministry of medium-sized machines: GlavSpetzMach under S. M. Vladimirsky, GlavSpezMontaj under V. M. Riabikov, and the directorate of transport machines under G. N Pachkov. The committee, headed by V. M. Riabikov,[14] was placed under the direct responsibility of the deputy chairman of the council of ministers, M. V. Khrunichev (who would later lend his name to the center that produced the Proton rocket). This was responsible for developing Korolev's R-7, Lavochkin's Buria and Myasishchev's Buran missiles, the K-20 missile of the Tupolev-Mikoyan team, and the S-25 (Berkut), S-50 and S-75 anti-aircraft systems. Decree N°378 of February 27, 1957 expanded the committee's remit to include work on radars.

The committee coordinated the activities of several ministries. The ministry of the defense industry dealt with ballistic rockets, surface-air missiles, and engines. The ministry of the aviation industry was responsible for winged rockets, air-air missiles, guidance systems, aerodynamics, and liquid-fuel engines. The ministry of the naval industry was responsible for naval missiles and gyroscopes. The ministry of general machines produced solid-fuel rockets. The ministry of the radio industry managed radars and guidance systems. For ground resources, four ministries were concerned: the ministry of machinery and instrumentation, the ministry of heavy machinery, the ministry of transport machinery, and the ministry of the power industry. KB-1 (later NPO Almaz), OKB-2 (which became MKB Fakel), as well as plants N°41 Moscow (later Avangard), N°82 Tushino (TMZ), and N°464 of Dolgoprudny (DNPP) were transferred to the ministry of the defense industry. The same decree established the post of first deputy for rockets in four ministries: S. I. Vetochkin in the defense industry, S. M. Lechenko in the aviation industry, D. G. Diatlov in general machines, and S. M. Vladimirsky in the radio industry.

Vetotchkine was chief of the 7th glavka in 1946, deputy minister in 1949, first deputy of the 3rd glavka in 1951, then deputy director of GlavSpetzMach from 1953 to 1955. He was first deputy for rockets until 1958, and then first deputy of the VPK from 1958 to 1966. Lechenko was director of factories N°23 (ZIKh) and N°82 (TMZ) from 1946 to 1954, then head of the 6th glavka of the ministry of the aviation industry. He was deputy for rockets until 1957, first deputy of the ministry from 1957 to 1963, and then head of the institute of aeronautical technology (NIAT) from 1964 to 1974. Diatlov was head of NII-1 from 1947 to 1950, head of the 6th glavka of the ministry of agricultural machinery in 1952, head of the technical directorate of the ministry of the defense industry in 1953, and head of NII-24 from 1954 to 1955. He was first deputy of NII-642 until 1956, and then its director from 1956 to 1957. Vladimirsky was director of factory N°703 (radar), director of NII-108 in 1946, head of the 6th glavka of the communications industry in 1947, deputy minister in 1949, deputy to Beria on the special committee in 1952, chief engineer and then head of

KB-1 in 1953, head of GlavSpetzMach from 1954 to 1955, first deputy for the radio industry until 1957, and then deputy until 1968.

Decree N°1239-630 of August 31, 1956 established the composition of the state commission to conduct flight tests of the R-7: V. M. Riabikov (president), Marshal M. I. Nedeline, deputy minister of defense for armaments (president's deputy), S. P. Korolev (technical director), General A. I. Nesterenko (head of the cosmodrome), G. N. Pachkov, a deputy of Riabikov, General A. G. Mrykine, deputy at GURVO, S. N. Shishkin, deputy director of the directorate of nuclear warheads at the ministry of medium-sized machines, and General I. T. Boulitchev, deputy director of the army communications corps, along with constructors V. P. Glushko, M. S. Riazansky, N. A. Pilyugin, V. I. Kuznetsov, V. P. Barmine. The secretary of the commission, Colonel A. A. Maximov, would later direct the space forces from 1979-1989. The commission met on April 10, 1957 to prepare the first flight of the R-7 which took place on 15 May.

# 2

# Designing the Semyorka

The Semyorka, the first Soviet multi-stage rocket, featured the peculiarity of having parallel, rather than superimposed stages. It therefore had to be launched from a specially designed platform which, to date, remains unique in the world of rockets. Decree N°956-408 of May 20, 1954 officially initiated the development of the intercontinental R-7 (8K71) rocket. On June 28, decree N°1281-573, "Concerning the R&D plan for special purposes", defined the work required for the development of the rocket. On July 24 the R-7 project was completed. It was then submitted to the evaluation commission,[15] and adopted on November 20.

The 260-ton rocket could carry a 5.4-ton payload a range of 8,240 km. It stood 33.6 m tall and was configured as a bundle consisting of four lateral blocks of 20.9 m attached to a central core of 26.5 m.

At that time, it had to be integrated and set on four launch pads of the same type as the R-5 (vertical integration). But this version of the platform did not satisfy Korolev. Indeed, calculations showed that in strong winds, the rocket, which had a diameter of 10.3 m at the base, suffered deformations (buckling of the blocks) and risked toppling over. The ensuing debate went unresolved for a long time. Korolev investigated 37 different variants before arriving at the final version. One called for installing walls around the rocket, and these were referred to as the "Great Wall of China". But finally, in late 1954, it was decided to hold it in a vertical position using supports located at the level of the attachment of the boosters to the central body. Thus the rocket was suspended from four "petals", rather than being set on a launch pad. Consequently, the vertical integration was replaced by horizontal integration in the MIK, some 2.5 km from the pad. Two days prior to launch, the rocket would be transported on an 8U213 rail car supplied by TsKBTM and the Podiemnik factory. During the process of erection, the rocket was positioned very precisely using 8Ch15 optical-mechanical viewing systems from the Kiev arsenal plant, which had been developed by the main designer S. P. Parniakov, who headed the design bureau from 1956 to 1987. This system would evolve into the 8Ch19 and 19M for the 8K72 and 8K73, 8Ch23 and 23M for the 8K78 and 8K78-E6, and then the 11Ch115 universal system for the 8K78M, 8A92M and 11A511U.

The cone-shaped boosters were lengthened from 19.6 m to 20.9 m. The lower part

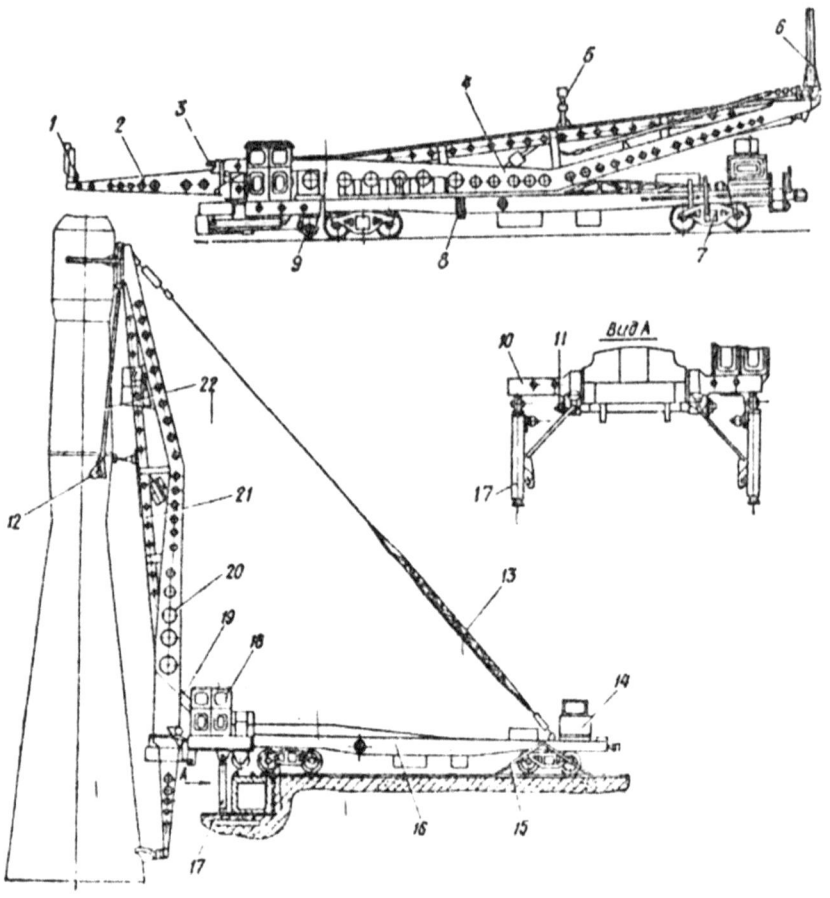

The 8U213 transport vehicle.

was attached to the central core using struts, while the top was inserted into a slot that received the thrust. When the boosters reached a given speed the "Separation 1" command was issued. The vernier engines went into the neutral position, and 112 s into the flight the main engines reduced their thrust from 100% to 84%. At +0.3 s, the strut break-away command was issued. At +0.2 s, the engines were shut down and the nozzle was opened from the kerosene tank. Its angle of 45° on the longitudinal axis made it possible for the booster to move away from the central core. Then, at the top, a contact triggered the pyrotechnic cartridge which unlocked the valve of the nozzle from the oxygen tank. Gas (nitrogen and oxygen) created thrust upon exiting the nozzle. The boosters moved away while the central core continued its trajectory.

Initially, the vehicle was steered by jet deflectors on the RD-107s of the boosters and four of M.V. Melnikov's vernier engines on the RD-108 of the central core, but the jet deflectors were eventually replaced by vernier engines. The Melnikov engines were replaced with Gluchko engines from December 1958 (in time for the phase III trials of the R-7). In addition, prior to take-off, each booster was equipped with a lateral rudder.

To synchronize fuel consumption from the different tanks, B. N. Petrov of the institute of automatic and telemechanics (IAT) of the Academy of Sciences and A. S. Abramov, the main designer of OKB-12, developed the SOB system (which became SOBIS in the phase III trials of the R-7). Petrov would later become an academician, secretary of the mechanics and guidance process section, president of Interkosmos from 1966 to 1980, and vice president of the Academy of Sciences. In 1964 OKB-12 merged with NII-25 to become the institute of instrumentation (NIIP).

The R-7 guidance system used the Pilyugin inertial method and the Riazansky radio method. For the former, the rocket had a horizontal gyro (GG) N°I-11-1 A-3, a vertical gyro N°I-55-1 and speed regulation sensors (DRS) N°GV I-12-6-3 and I-12-7-3. For the latter, the goniometer with an antenna 7 m in diameter was developed by G. Y. Guskov at the central radio institute (TsNII-108/TsNIRTI), and radio stations (RUP-A and RUP-B) were built in Tartugai and Togyz. For telemetry, two systems were developed concurrently: RTS-7 from SKB-567 and the Tral from OKB MEI, with the latter being selected.

The rocket was modified in mid-1956, with an increase in mass to 273.5 tons (+ 10 tons of fuel) and the thrust was augmented to 403 tons (+ 38 tons). For a warhead of 5.37 tons the dry mass was 26.5 tons. The warhead was a sharp cone 7.2 m in height

Booster upper straps.

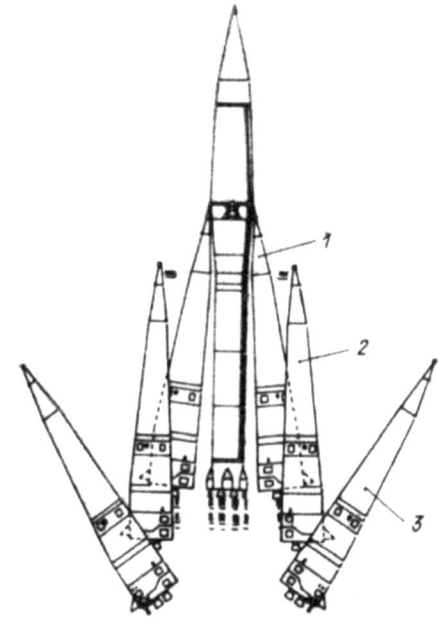

Diagram of stage separation.

Booster lower straps.

covered with silica-based material. The Arzamas KB-11 nuclear charge had a yield of 3 megatons. To test the different elements, Korolev decided to use a modified R-5. The various versions of the M-5R (three launches from May 31 to June 15, 1956) and the M-5RD (five launches from February 16 to March 23 and five more from July 20 to August 18, 1956) were used to test the radio guidance system, the engine synchronization system, the system of speed control, the stabilization system, the Tral telemetry system, the Fakel control system, and the payload bay (silicon and asbotextolit).

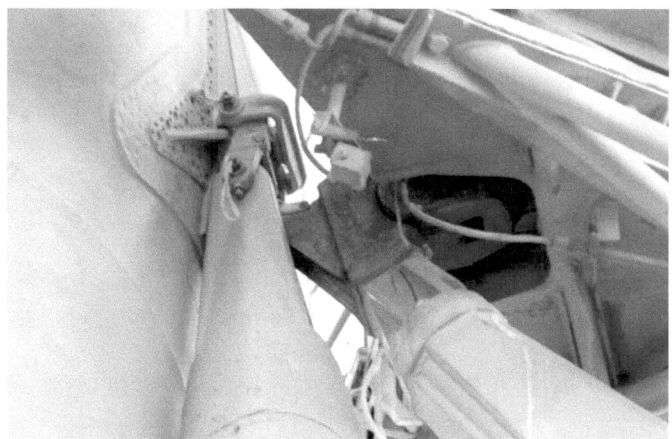

Booster upper straps.

The RD-108 engine.

## THE SEMYORKA LAUNCH PLATFORM

At KBOM, the platform was developed by B. I. Khlebnikov with sections N°2 of V. A. Rudnitsky, N°10 of A. N. Vasiliev, N°3 of Y. L. Troitsky, N°4 of M. M.Sidorov, N°7 of A. G. Chekhtman, N°8 of V. D. Grafov, and N°9 of A. V. Pridantsev.
　Subcontractors were:

- The central design bureau of heavy machinery (TSKB TM) for the means of transport and erection of the rocket.
- The design bureau of the national institute of oxygen machines (OKB VNII KiMach) for liquid oxygen systems, liquid nitrogen supply, and nitrogen gasification.
- The OKB of the Ural rail car factory for rail cars to supply oxygen and liquid nitrogen.
- The OKB of the Zhdanov heavy machinery plant for hydrogen peroxide supply cars.
- The Sumsk pump factory for kerosene supply cars.
- The design bureau of PKB-12 for the fueling control systems.
- The design bureau of chemistry and transport machines (SPKB, later KBTKhM) for fire protection equipment.
- The VNISI for lighting devices.
- The KB of the Kiev arsenal for optical-mechanical viewing systems.
- The KB of the Frunze factory in Sumsk for the stationary compression and gas supply station.
- The OKB of the Kalinin rail car factory for rail cars to transport rocket components and air-conditioning cars for the payload.
- The design bureau of transport machines (GSKB DorMach, later KB TM), for load handling equipment, lifts, and transport dollies.
- The VNII StroïDorMach for maintenance kits.
- The TsPI-31 military institute for technical buildings (MIK) and the launch platform.

In late 1956, static and dynamic tests of the launch platform were conducted at the Leningrad metal plant (LMZ). To do this, a shaft 19 m in diameter was built at the LMZ to test the complete system with a full-size rocket model. This "technological" model 8K71SN was available in December 1956. The work was conducted in six months instead of the one year originally planned. In early 1957, when the launch devices were completed, they were dismantled and transported to Baikonur for final assembly. Technical direction was provided by Y. L. Troitsky and his assistant M. I. Bakhtioukov. There were three test cycles: the synchronization of the petals, the complete program with the installed rocket system (dry run) and testing of simulation of launch with filling of tanks (wet run).
　The platform (Zone 1) comprised the flue (also known as the "stadion") for the ejection of hot gas, the concrete platform set on four concrete pylons and, above that, the metal platform with four petals, the umbilical mast and service towers. The flue was a pit measuring 250 × 100 × 50 m; the equivalent of one million cubic

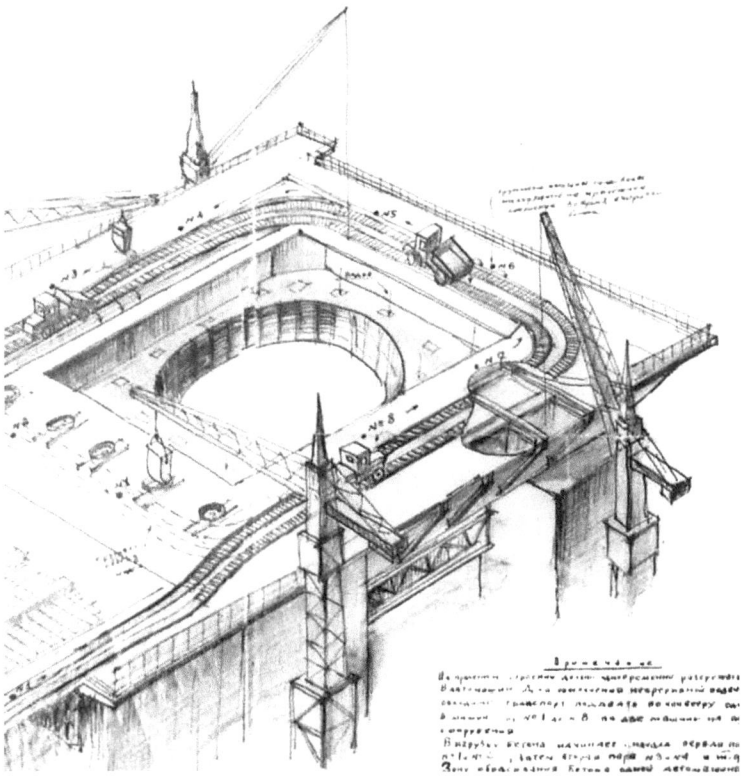

The concrete platform.

meters of earth. Excavation work began on July 20, 1955. Explosives were used to loosen a soil composed of sand and clay. The sand was reused in the concrete for the launch platform. Geological analyses showed that there was water underground, so it was necessary to build a drainage system with a pumping station. The flue was completed on April 4, 1956. The concrete work began on April 19. After the foundations were poured, the four corner pylons for the launch table were poured simultaneously. From April to July 1956, some 30,000 cu. m of concrete were poured. Starting on August 20, the table was built on the pylons. After this was completed on September 1, the metal launch platform was installed. At the center was a bay 16 m in diameter to contain the bottom of the rocket. It was set on four bearing points, set 7 m below, and the jaws were to hold the vehicle stable against wind gusts.

The launch system consisted of a base ring which was turned according to the azimuth targeted by the rocket. On the ring, the four petals were installed with huge counterweights. They supported the rocket at the level of the upper attachment points of the lateral blocks, forming a constraining belt at the top of the "pyramid".

This 8U215 unit was built by KBOM.

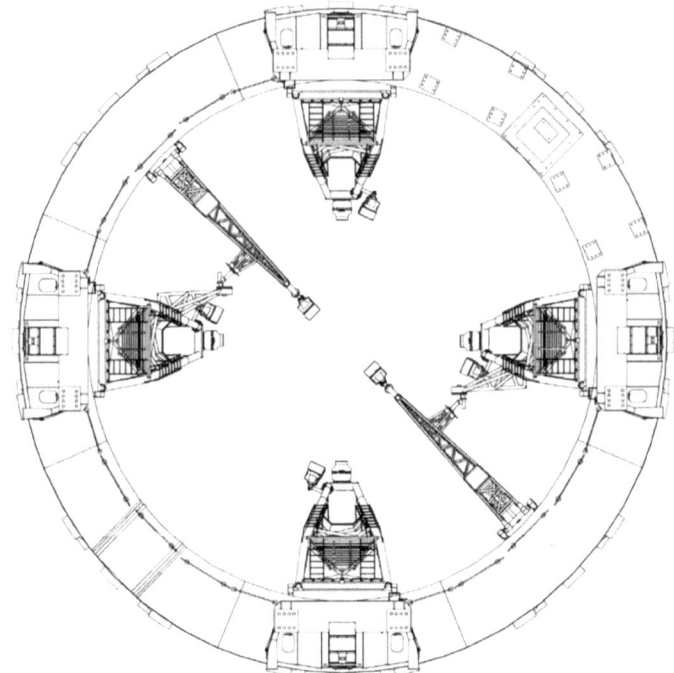

Diagram of the lower part.}

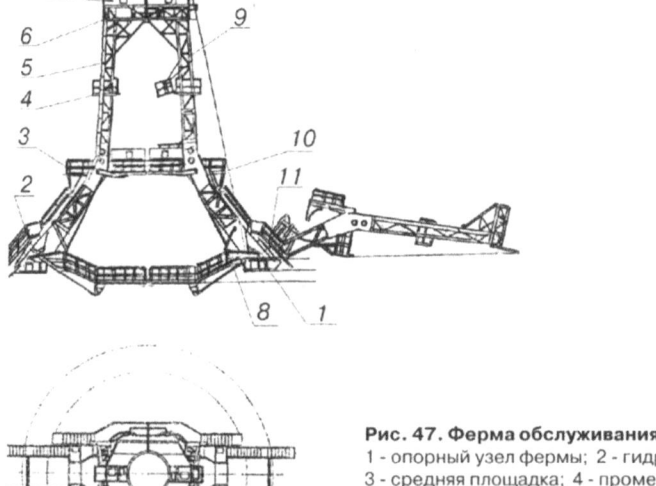

**Рис. 47. Ферма обслуживания 8Т119:**
1 - опорный узел фермы; 2 - гидроцилиндр подъема;
3 - средняя площадка; 4 - промежуточная площадка;
5 - несущая конструкция; 6 - верхняя площадка;
7,10 - щит ветрозащиты; 8 - тяга; 9 - площадка для
прохода на опорный пояс стартовой системы;
11 - рычажно-тяговый механизм

Diagram of the services towers.

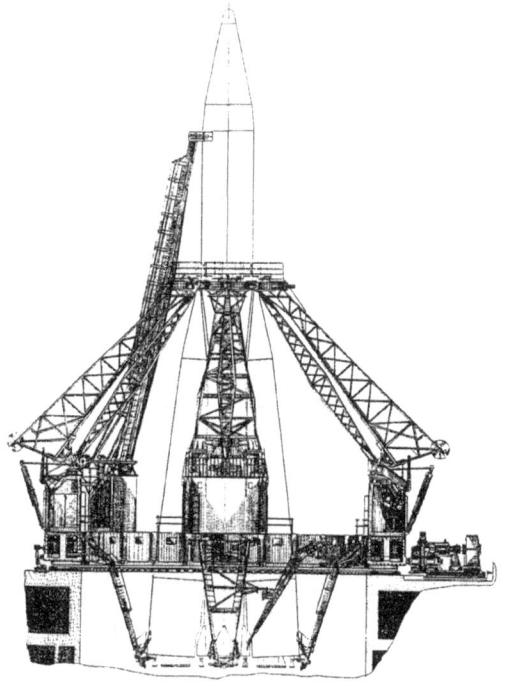

The 8U215 platform.

*A-A (2, 1150)*

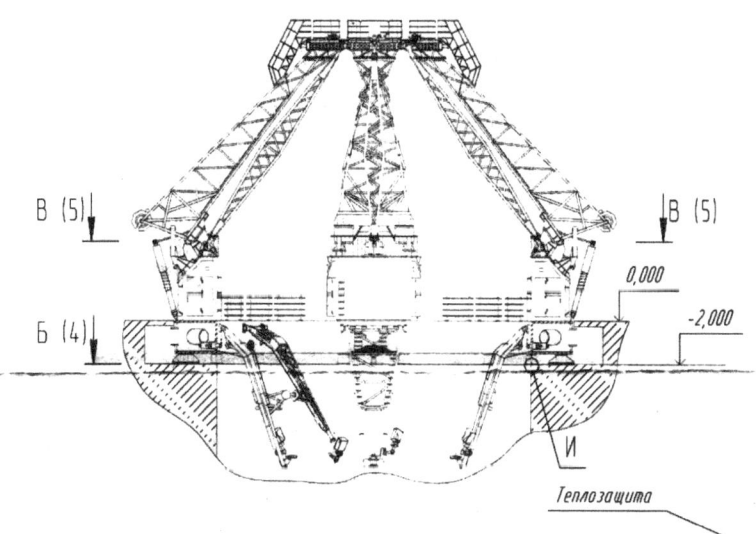

Diagram of the 8U215 platform.

Service towers varied radically, depending on the version of the Semyorka. For the R-7, it was an 8T119 built by the imeni Kirov lift and transport device factory and SKB. For the R-7A, the various elements became 8U0213, 8U0215, 8U0216 and 8T0119. For the Vostok (8K72) and Voskhod (11A57), the 8U213 rail car was modified for manned space flight: it became the 8U213FO equipped with an elevator and a cabin that wrapped around the piloted spacecraft. In December 1963, with the introduction of the Soyuz (11A57) and the Soyuz spacecraft (11F615), the car was replaced by an 11U219 and the service tower by the 11T11. The two arms swung up around the rocket (32.19 m high) with twelve platforms at levels to facilitate work on the rocket. The top platform was used by cosmonauts to enter their spacecraft. The umbilical mast was used for the supply of electricity, telemetry and fuel.

Beneath the concrete slab, several floors of underground facilities housed control and measuring systems and equipment for the supply of electricity, fuel and gases. The rocket was constantly cooled in order to limit evaporation of liquid oxygen. This refrigeration equipment was installed in a separate room. Beneath the rocket itself, an 8U216 mobile service cabin (on rails), supplied by the SKB machine factory of Novo-Kramatorsk (NKMZ), facilitated work and the connection of the fuel filling system. The fuel was brought by rail in special cars from the nearby fuel factory.

The KBOM's projects included underground installations for the R-7 in silos or buried in a mountain, but this went no further than drawings.

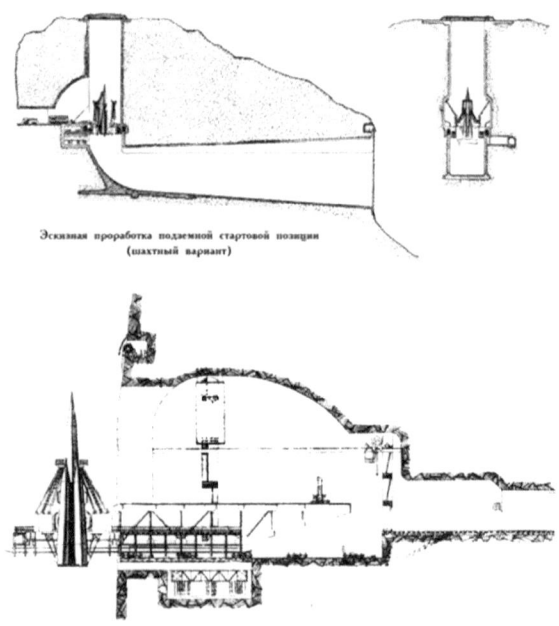

Эскизная проработка подземной стартовой позиции
(шахтный вариант)

Launch project from a mountain.

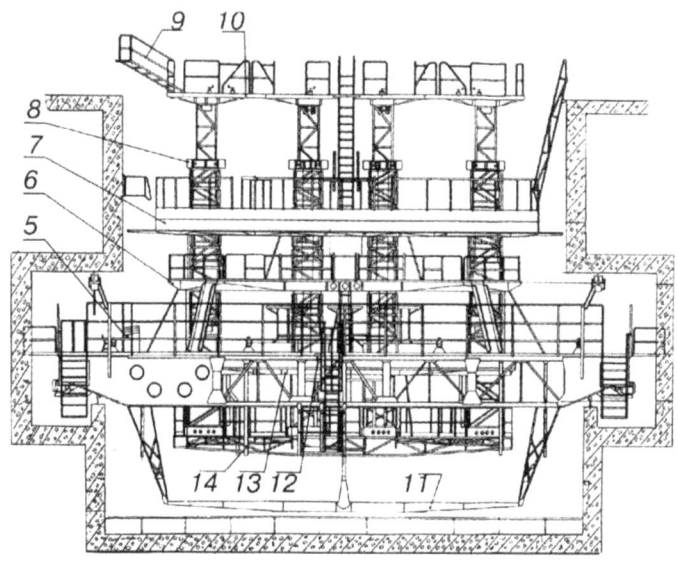

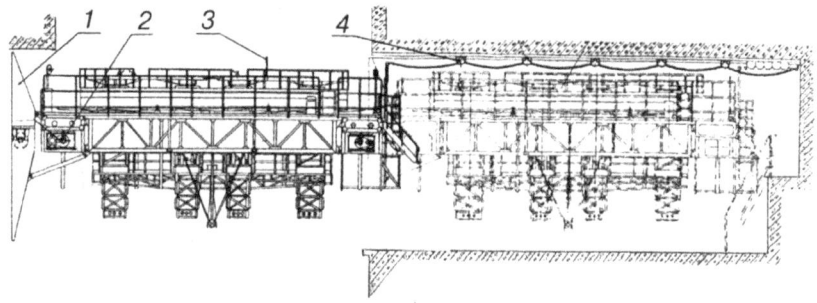

**Рис. 48. Кабина обслуживания 8У216 (пунктиром показана кабина, убранная в нишу стартового сооружения):**
1 - привод механизма передвижения; 2 - штора теплозащиты; 3, 4 - кабель питания; 5 - пульт управления; 6 - площадки центральные; 7 - средний ярус площадок; 8 - двухсекционная колонная; 9 - трапы входа на кабину; 10 - площадка кольцевая; 11, 12 - механизм поворота круга; 13 - круг поворотный; 14 - металлоконструкции

Diagram of service cabin.

## THE LAUNCH

The countdown began at T-7 hours. Fueling began an hour later. All non-essential personnel were evacuated from the site. Those who remained wore a red armband. At T-3 hours the flight program was entered into the memory of the guidance system. The thermal air and fluid control systems were unplugged at T-1 hour; the thermal regime would be maintained by inertia. At T-45 minutes the two service towers were lowered to the horizontal position at each side of the rocket. Only the umbilical mast with cables and tubing remained. At T-30 minutes the remaining staff

Bunker.

View of the lower part.

evacuated the platform to either the launch bunker or the observation point. The bunker was 200 m west of the complex, 7-8 m beneath the ground. There were five rooms inside. Two periscopes extended through the ceiling (one was used by the military launch director and the other by the main rocket designer's assistant for tests). It was connected to all operational units of the launch facility and the tracking stations.

At T-5 minutes, the "Key to launch" order was given, and an operator inserted the enabling key into the lock on the launch console. The final operations involved the simultaneous ignition of the 32 nozzles of the R-7. The synchronized sequence began with the "Control 1" order. Telemetry information, recorded on magnetic tape, was used to determine the parameters of onboard systems. Then came "Blow" (nitrogen was used to purge the pipes and combustion chambers), "Vent" (to eliminate the pressure produced by fuel vapor in the tanks), "Key on drainage" (close the drainage valves), "Launch" (start pressurizing the nitrogen tanks), and "Control 2" (verify that all systems are operating). At that time, the gyroscopes of the guidance system were released. On the order "Ground-aboard", the umbilical mast was separated and the rocket became autonomous (using its onboard supplies). Fuel and oxidizer valves were opened and fuel arrived in the combustion chambers. Upon the "Ignition" order, pyrotechnic igniters were activated. The turbopumps came into operation progressively for preliminary, intermediate, and main thrust. When main thrust was reached, the four lower attachment jaws, which ensured that the action of the wind or irregular thrust would not affect the initial orientation, released the rocket. Under the force of the thrust, the four petals opened and the rocket lifted off. The active flight phase lasted nine minutes.

## THE TRACKING NETWORK

The decree of May 20, 1954 and the directive of the chief of staff on July 27, 1955 designated the positions of the tracking stations and the locations of the impact areas of the stages. Nine tracking stations were to be located along the rocket's trajectory. Construction began in February-March 1956, and was completed on September 27, 1956.

- IP-1 at 1.5 km from the launch platform (Zone 18). At the moment of launch, Tral and RTS-5 type mobile antennas were positioned 12 km from the platform. They also contained the IP-1D station created in August 1957 for Sputnik.
- IP-2 and 3 were 25-30 km from the platform. They were built between March and August 1956. Equipped with KTh-41 cine-theodolites, these buildings were decommissioned in 1962.
- IP-4 and 5 were in desert zones 105-120 km from the platform.
- IP-6 and 7 were 500 km away, on the steppes in the Karaganda and Kostanay regions. IP-6, on the left of the flight path, was 12 km from Amangeld and 350 km from the Amankaragai railway station. For its construction, all the materials, devices, propellants, and products were sent by road from the station. In winter, loads were transported by sled. Firewood for families was delivered by an Li-2 plane using a dirt airstrip 200 m away. Potable water was located 18 km away. The buildings were constructed in May-October 1956.
- IP-8 and 9 were 800 km away in the Karaganda and Tselinograd regions. Even-numbered stations were on the left and odd numbers on the right of the flight path. The 43rd ONIS unit was stationed in the payload's impact area ("Kama" region, which later became "Kura") of Klioutchi, near Ust-

Kamchatsk (Kamchatka). It included Zone 20 and six stations. Three were within a rectangle of 29 × 42 km (IP-15, 16 and 17), while the other three (IP-12, 13, 14) were 80-150 km away.

In addition, there were two radio guidance stations (RUP), designated A and B, located 250 km from the platform on either side of the flight path.

The Ural 1 computer arrived in Baikonur in October 1957. It was installed in zone 10 in January-February 1958, and was operational by November 1958 to process trajectography and ballistics data.

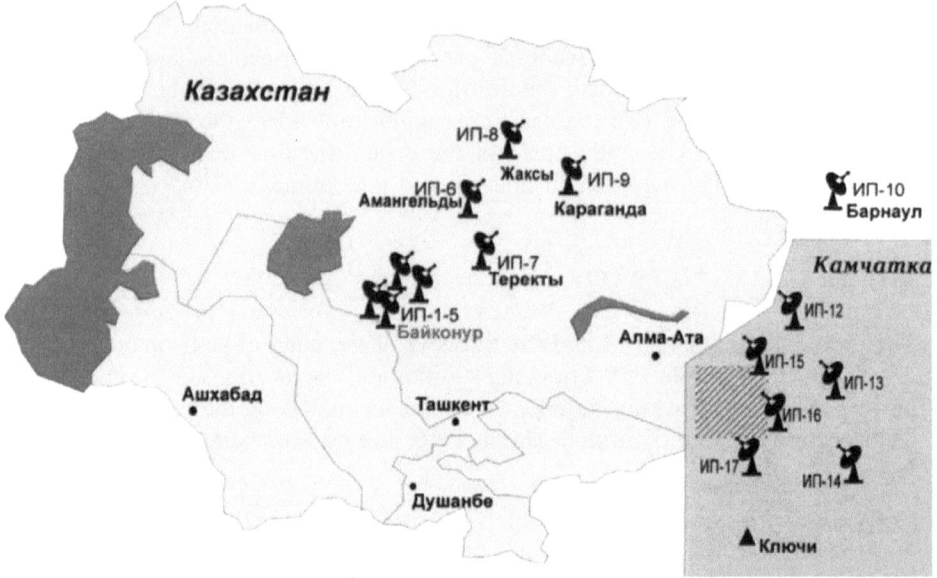

The tracking station network in 1957.

## DEVELOPMENT OF THE SEMYORKA

Engine tests were carried out at NII-229, Zagorsk. From July 1956 to March 1957 there were five tests using lateral blocks (August 15, September 1 and 24, October 11 and December 3), three trials of the central block (December 27, January 10 and 26) and two trials of the full bundle (February 20 and March 30).

From April-July 1956 three models of the R-7 (M1-1S, M1-2SN and M1-3) were built. In December 1956 the first flight model of M1-4SL was ready for factory testing. The M1-5 rocket arrived at Baikonur on March 3, 1957. It weighed 274.2 tons, measured 34.2 m in height, and carried 2.8 tons of measurement apparatus. The 8K71SN model was used in the LMZ plant in 1956, then in Baikonur from December 1956 to February 1957.

On May 15, 1957 the first R-7 (8K71 N°1-5) took off from Baikonur. The rocket was destroyed after 103.6 s of flight due to a fire in the central stage caused by a fuel leak prior to the pump of the RD-108. On June 11 the second rocket (8K71 N°M1-6) was removed from the launch platform after three launch attempts. The launch was aborted because it took too long to achieve the necessary pressure in the combustion chamber of the RD-108. A mistake had been made during installation of the nitrogen supply valve for the pipework to feed oxidizer to the central stage. On July 12 the third rocket (8K71 N°M1-7) was destroyed in flight due to a short circuit in the automatic stabilization unit 32.9 s into the flight, resulting in a loss of stability of the rocket. On the two following launches, the warhead was covered with protective asbotextolit. On August 21 the fourth rocket (8K71 N°M1-8) achieved success. But the warhead was destroyed at an altitude of 10 km while traveling at 5 km/s, prior to falling on Kamchatka. This meant that the USSR had its first ICBM and the door was open for the launch of the first artificial satellite of the Earth. On September 7 the fifth rocket (8K71 N°M1-9) was successful, but the warhead was again destroyed in flight. It was decided to shorten the warhead to 5.5 m and make its nose rounded.

After two successful flights, the green light was given for the launch of the first two satellites. On October 4, Sputnik 1 (83.6 kg) was successfully put into orbit by the 6th 8K71PS N°M1-1, and on November 3, the second satellite, Sputnik 2 (508.3 kg) was launched by the 7th 8K71PS N°M1-2.

The second phase of flight tests included six launches from December 1957 to July 1958. On December 31 the eighth rocket (8K71 N°M1-11) was removed from the launch platform after its second launch was aborted. After repairs were made, this rocket took off on January 30, 1958. But the separation of the lateral blocks damaged propellant tubing on the central block, preventing it from giving nominal thrust. The destruction system did not work. The payload did not separate from the central block, and impacted 80 km from the target point. This was the first time that a warhead had reached the ground. Following this flight, the warhead separation system was tripled. On March 12 the repaired second rocket (8K71 N°M1-6) was again placed on the launch platform. But it was again removed due to the failure of the oxygen valve to an RD-107 engine. On March 29, the warhead of the ninth rocket (8K71 N°M1-10) fell to Earth in Kamchatka. This was a success like that of August 21, but this time the warhead survived re-entry. On April 4, the tenth R-7 (8K71 N°M1-12) impacted 80 km from the target (18.2 km to the right), but one of the radio guidance antennas failed 142 s into the flight.

On April 27 and May 15, 1958, the space version (8A91) was launched on two occasions with Sputnik 3. The first launch was a failure, but the second succeeded. On May 24, rocket N°B1-3 equipped with an experimental warhead failed due to the destruction of the turbopump of the RD-108 engine. For the first time, the signal of the Tral G1 transmitter placed on the payload was received after passing into plasma. That day, launch preparation lasted 21 hours. On July 10, the B1-4 rocket exploded on the Baikonur launch platform, which was repaired with components from a launch pad under construction at Plesetsk north of Moscow. This was the final launch of the second phase of R-7 flight tests.

The third phase of flight testing involved eight rockets manufactured in Podlipki

The R-7 of 1957.

Again, the R-7 of 1957.

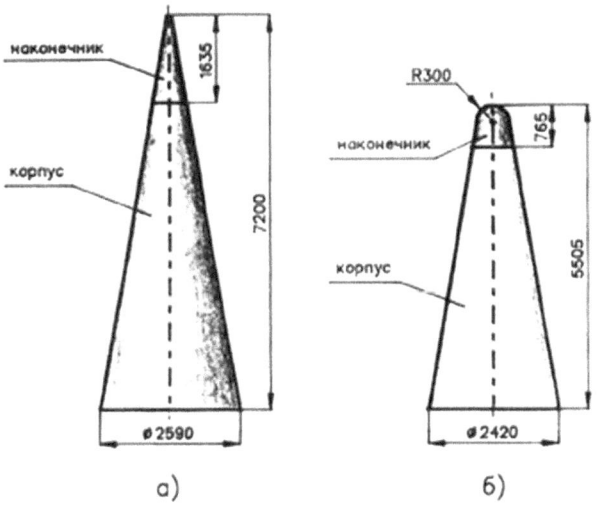

Comparison of two R-7 warheads.

Liftoff of the R-7 in 1957.

by OKB-1 and eight in Kuybyshev by Progress factory N°1. These 278-ton vehicles incorporated modifications: the inertial unit was transferred to the equipment bay, the SOB became the SOBIS, nitrogen pressurization was decreased by 15%, and the new OKB-456 vernier engines were installed. The flights ran from December 24, 1958 to November 27, 1959, with 10 out of 16 rockets reaching the target.

The first launch on December 24 ended in failure. Owing to a problem with the hydrogen peroxide reducer in a lateral block, the fuel was consumed and the block separated 117 s into the flight. The destruction command was triggered 130 s into the flight and the rocket impacted 672 km from the launch site. On February 21, 1959 rocket N°81 (produced in Kuybyshev) was withdrawn from the launch platform because of the failure of an RKD-4 relay. This was the third aborted launch. But on March 17 it took off with warhead N°13. Eight days later, the N°I3-18 rocket was launched with warhead N°15 and landed 17 km from the target (5 km to the right). On March 31, rocket N°I3-20 failed 280 s after launch due to a problem with the hydrogen peroxide pump. The central block landed in the Oymyakon region of Siberia. On May 9, rocket N°I3-21 was launched carrying warhead N°17. It was a complete success, landing 6.6 km from the target (1.1 km to the right). On May 31, rocket N°I3-22 landed 1,890 km further than planned due to the non-extinction of the engines. It fell into the ocean near the Aleutian Islands. On June 9, following a relay failure, the engine of a lateral block did not shut down and rocket N°I3-23 exceeded the target by 2,175 km; this was considered to be a partial success. On July 18, rocket N°I3-24 flew successfully, landing 1.51 km from the target (0.91 km to the right). Twelve days later, rocket N°041082 (Kuybyshev series) flew successfully during a test launch by military unit N°13973 (42nd BSS), which later deployed to platform N°41 at Plesetsk. After three postponed launches, rocket N°I3-25 flew successfully on August 14. Then on September 18 rocket N°I1-1T flew successfully.

The "equator" program was then initiated, with the announcement of the landing coordinates for the warheads in the Pacific Ocean. On October 22, R-7 N°267432 reached the maximum range of 8,635.5 km. Following a postponement, the second launch was carried out on October 25 using R-7 N°267434. On November 2, R-7 N°267431 was successfully launched towards the Kura polygon. On November 21, R-7 N°I2-1-T was launched successfully, but an oxygen pump cavitation occurred 298.72 s into the flight, causing the payload to fall 28.3 km short of the target. This test launch was by unit N°14003 (48th BSS), which subsequently deployed to platform N°16 at Plesetsk. On November 27, R-7 N°I3-33 concluded phase three of the flight testing by landing 1.75 km from the target (0.77 km to the right).

On January 20, 1960 the R-7 was declared operational. On June 4 of that year, R-7 N°L1-9 flew successfully. This test launch was by unit N°14056 (70th BSS), which later deployed to platform N°43 at Plesetsk. On February 27, 1961, R-7 N°L2-1 was launched from Zone 31.

## THE R-7A

On July 2, 1958, a decree was issued for the development of the R-7A (8K74). This version weighed 276 tons, was 31.0 m tall, carried a 3-ton nuclear warhead, and had a range of 12,000 km. The new inertial system performed the functions (other than trajectory direction) previously carried out by the eliminated radio correction system. In addition, the SOBIS and the engines had been improved.

Flight tests ran from December 23, 1959 to July 7, 1960. The warhead of R-7A N°I1-1 reached the Kura polygon on December 23. The second, R-7A N°I1-2, flew successfully 12,000 km into the Pacific Ocean on January 20. On January 24, R-7A N°I1-3 failed after 31 s, when a vernier engine exploded in a lateral block and started a fire. But on January 26 the R-7A flew successfully (Pacific Ocean), and this was followed by launches on January 31 (N°I1-4), March 18 (N°L1 5), March 24, July 5 (Pacific Ocean) and July 7 (Pacific Ocean). With seven out of eight launches being successful, the rocket was declared operational on September 12, 1960.

The first launch of an R-7A from Zone 31 at Baikonur was on January 14, 1961, and was a success. This zone included a MIK and the second launch platform, built between December 1958 and August 1960. It was similar to Zone 1, but the MIK was closer to the launch pad. Zone 31 was associated with residential Zone 32 and nuclear warhead storage Zone 37 (1960-1967). On February 13, R-7A N°L1-3T lifted off from Zone 31. The rocket came from the Kuybyshev Progress factory. On April 14 an R-7A launched from Zone 31 came down 2,380 km away. On June 15,

Lavochkin La R-7A of 1959.

Warhead of the R-7A.

R-7 N°E15001-06 was launched from Zone 31 successfully. Then on July 4 and 5 a simultaneous launch of R-7A N°L2-4 and R-7A N°L2-2 was carried out from Baikonur by unit N°13973 of Plesetsk. It was a success, with aerial explosion of the warheads. On September 21 R-7 N°E15003-03 reached the Pacific Ocean.

Liftoff of the R-7A in 1959.

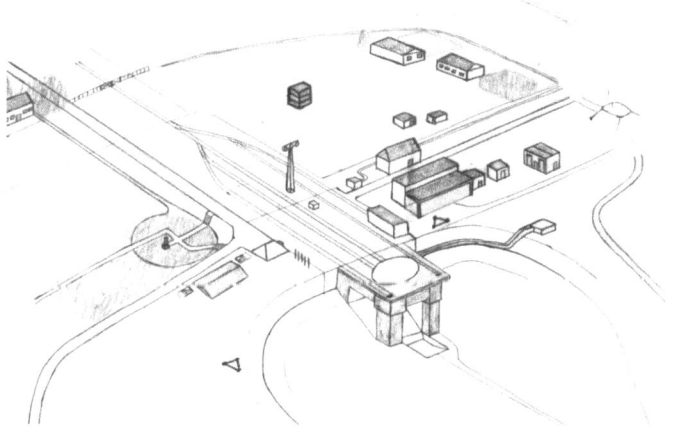

Drawing of Zone Nº1, the "Gagarin Zone".

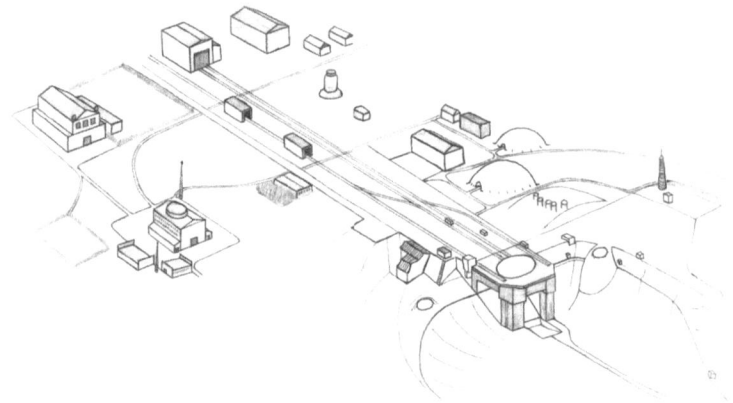

Drawing of Zone N°31.

1e version          2e version          3e version (R7A)          version Sputnik

The R-7 versions.

On July 27, 1964 and December 14, 1965, unit N°13973 conducted two R-7A test launches, the first from Baikonur and the second from Plesetsk. As for unit N°14056, it launched two R-7As from Plesetsk on December 21, 1965 and July 25, 1967. That was the last flight of the R-7A. It was on maximum alert during the Cuban missile crisis in 1962 (operation "Anadyr" of September 11 to November 21) and also during the "Prague Spring" in 1968. The R-7A was withdrawn from the arsenal in 1968. In September 1969 two test launches were planned for the R-7A from Plesetsk for the Berkut program, but they were canceled.

## THE FIRST "SPUTNIK"

On May 26, 1954 Korolev wrote to the council of ministers to inform them that the missile was capable of putting a satellite into orbit. Indeed, the theme of the artificial satellite (ISZ) had been studied by M. K. Tikhonravov's group at NII-4 from 1950 to 1953. And from 1954 to 1955 he continued this work at the request of Korolev. The first step was the definition of the constituents of a simple automatic satellite (PAS). Its purpose would be to study the near cosmos and transmit data via radio. The use of solar energy (solar panels) was studied for the power supply. One variant included an orientation system based on a gyroscope with stellar correction capabilities. The second step was to clarify the characteristics and parameters of onboard systems. There were two variants – of identical contour and size – for the study of the upper atmosphere and cosmic space. The first was the version using data transmission by radio. The second was a guided system, with a special cassette to return results to Earth. As early as 1955, B. V. Rauschenbach of institute N°1 of the aviation industry (NII-1) studied the type of guidance system that would later be used on Luna 3. In addition, the guided variant could accommodate either a large number of scientific instruments or a cabin for an animal with a life-support system derived from the one that had been developed in 1951 for geophysical rockets.

In 1953 M. V. Keldysh became interested in the idea of a satellite. On March 16, 1954 he gathered together M. K. Tikhonravov, I. M. Yatsounsky and G. Y. Maximov from NII-4 and applied mathematicians D. E. Okhotsimsky, T. M. Eneiev and V. A. Egorov from the institute of mathematics. On August 10 the council of ministers submitted a proposal to address theoretical issues concerning space flight to V. A. Malyshev (minister of medium-sized machines), B. L. Vannikov (first deputy), M. V. Khrunichev (first deputy) and K. N. Rudnev (deputy minister of armaments). On August 19, at the request of the Academy of Sciences, E. F. Ryazanov's group from sector 3 of OKB-1 (L. M. Miloslavsky, V. I. Froumson, V. A. Nikolaiev) began to investigate the launch of a satellite. Three solutions were proposed: a simple satellite launched by the R-7 (a 1.5-ton satellite equipped with a Block D central stage of 10 tons), a small satellite launched by a modified M-5, or construction of a heavy, 700-ton R-7 to launch a large satellite. Ryazanov argued for the first solution, Keldysh for the second, while the third was purely conjectural. The first solution gave rise to the D1 projects (with scientific equipment), D2 (with a dog in a capsule) and D3 (with other devices).

In September 1954 the Academy of Sciences created a "Tsiolkovsky" gold medal to be awarded to a first recipient in 1958. In December, academician L. I. Sedov was appointed by the Academy of Sciences to head its "permanent interdepartmental commission for coordination and inspection of organization work and realization of interplanetary travel". Sedov was a specialist in aero-hydro-mechanics. He held the chair of hydromechanics at the University of Moscow. He had worked with Keldysh at the Steklov institute of mathematics at TsAGI in 1931, at NII-1 in 1947 (deputy director) and at TsIAM from 1949 to 1955 (deputy director). The scientific secretary of the commission was A. G. Karpenko.

On June 25, 1955, in the annual report of the Academy of Sciences, of which Korolev had been a corresponding member since 1953, he raised the possibility of sending scientists to an altitude of 100 km in 1956. He talked of satellites, citing the research of M. K. Tikhonravov, who continued to work on the satellite question with engineer I. V. Lavrov of sector N°3. They sent Korolev a report on this subject on July 16, recommending the transfer of Tikhonravov's NII-4 group to OKB-1, which was done on November 1, 1956 by the creation of sector N°9. It then began to work on satellites, lunar probes, a spacecraft, and a satellite station.

On July 29, the United States announced that it would launch a satellite for the International Geophysical Year, which would run from July 1957 to December 1958. On August 3, Sedov told the 6th international astronautical federation congress in Copenhagen that the Soviet Union would do the same. On August 8, the government asked M. V. Khrunichev, deputy chairman of the council of ministers, and V. M. Riabikov, chairman of the special committee for armaments of the army and navy, and therefore responsible for rocket programs, to take the necessary steps to initiate a satellite program.

On August 30 a meeting was held in the Academy of Sciences by M. V. Keldysh, A. V. Toptcheiev, M. A. Lavrentyev, S. P. Korolev, V. P. Glushko, and others. They developed a scientific program focusing on the study of the ionosphere, cosmic rays, the magnetic field, the luminescence of the atmosphere, the sun and its effects on our planet. A special commission was established for the creation of Object D (the D1 project) with M. V. Keldysh as president and G. A. Skuridin as scientific secretary. The same day, at a meeting with V. M. Riabikov, Korolev presented a project to use a three-stage version of the R-7 to send a probe to the Moon. The upper stage would have to use either an oxygen/kerosene or a fluorine/ethylamine monoxide propellant mixture, and the payload weight would be 400 or 800 kg.

At a lecture given at the Bauman technical school (MVTU) on September 25, 1955, Korolev presented the project to launch a capsule carrying a man to an altitude of 100 km. Five versions were studied: a V1B rocket payload, a V1E rocket payload, an air-brake configuration, a jet-rotor configuration, and a winged configuration.

On January 7, 1956 in his annual report, Korolev indicated that research on the satellite had begun. On January 30, decree N°149-88 called for the creation of Object D with a mass of 1.4 tons, equipped with 200-300 kg of scientific instruments. The launch was planned for 1957-1958. On February 25 the technical specifications were defined and the definitive project was completed on September 25.

On August 31, 1956 V. M. Riabikov became chairman of the state commission for the R-7. His deputy was Marshal M. I. Nedeline, the deputy minister of defense for armaments. Korolev was technical director. The first flight of the R-7 was scheduled for May 15, 1957. But in 1956 it appeared that Object D would be late for a launch that was expected in the second quarter of 1957. Consequently, on January 5, 1957,

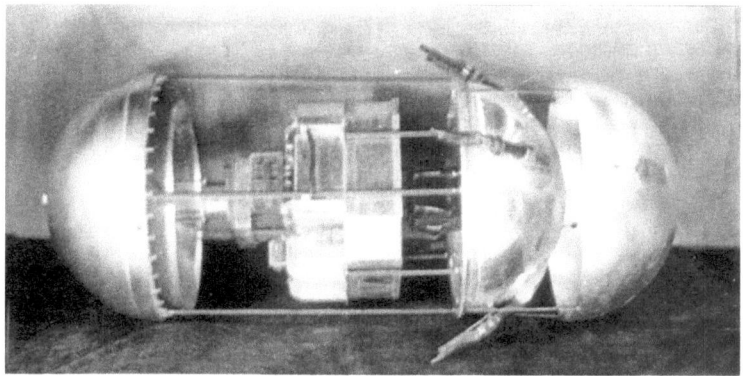

Composition of Sputnik 1.

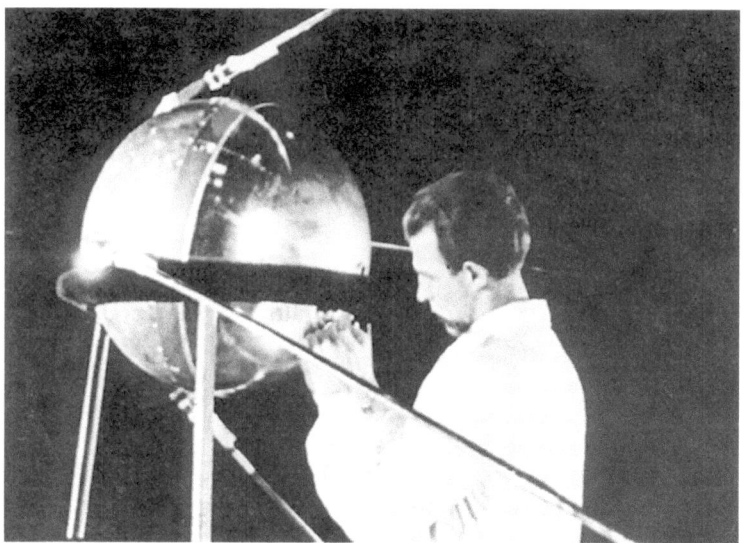

Preparing Sputnik 1.

Korolev instead proposed to launch a small spherical satellite (450 mm in diameter) weighing 40-50 kg into a 225-500 km orbit. It would be equipped with a transmitter which would have 7-10 days of autonomy. On February 7 the decree was published. A "crash program" was approved. In his annual report on January 24, 1957, Korolev had indicated that the satellite project was underway and that work had begun on the preliminary satellite (PS) in late 1956. To limit the risk of failure, two other satellite launch variants were studied, one using an R-5 for the first stage and an R-11 for the second stage, and the other (which was assigned to OKB-586) using an R-12 with an upper stage; but it was not necessary to develop either of these.

On February 15 the transmitter specifications were provided to M. S. Riazansky at NII-885: it would weigh 3.5 kg, and transmit intermittent signals of 0.4 s duration

Launch of Sputnik 1.

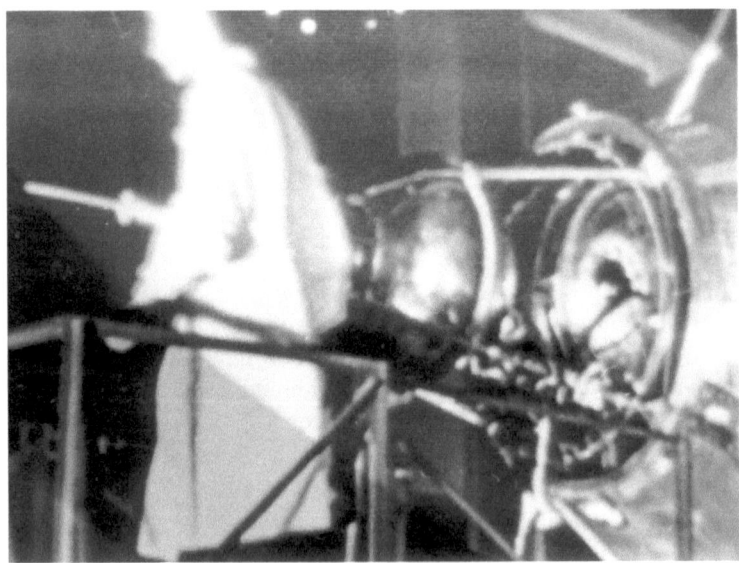

Sputnik 2.

Preparing Sputnik 2.

Sputnik 3.

at 20 and 40 MHz with a power of 1 watt. It would be powered by batteries developed by N. S. Lidorenko of the institute of power sources (later NPP Kvant). Tests were performed from a helicopter on May 5. The development of the PS was completed on June 24, and it was taken to Baikonur. On August 31 it underwent integration tests on the R-7. On September 20 the state commission set the launch date of October 7. On September 29, Korolev went to Baikonur and promptly advanced the launch by three days in order to reduce the risk of being overtaken by the Americans.

The launch took place October 4 at 22:28 (Moscow time). The rocket (272.8 tons) placed PS-1, named "Sputnik" into orbit. It was a 580-mm sphere with four antennas and weighed 83.6 kg. It emitted its famous "beep-beep" and re-entered the atmosphere on January 4, 1958.

On November 3, Sputnik 2 carried Laika, a female dog. Scientific information was correctly received during three orbits. The dog died after five hours due to excessive temperature (41°C) in the hermetic cabin. This was considered a partial success, because the payload failed to separate from the rocket stage and fell back into the atmosphere on April 14, 1958.

On December 18, 1957, the government awarded a large number of decorations for the launch of the first Spuntik.[16]

Object D, which had been intended to be the first satellite, was successfully placed into orbit on May 15, 1958, as Sputnik 3.

# 3

# The council of chief designers

## THE FATHER OF THE SEMYORKA

Sergei Pavlovich Korolev was born on December 30, 1906 (Julian calendar) or January 12, 1907 (Gregorian calendar) to a family of teachers – Pavel Yakovlevich Korolev and Maria Nikolaievna Moskalenko – in Zhitomir, Ukraine. The family moved to Kiev on June 28, 1908. When his parents separated in 1910, Sergei lived with his mother's family in Nejin near Chernygov, Ukraine.

From August 1914 to 1917 young Sergei lived in Kiev, then moved to Odessa where he entered middle school. It was during this time that the October revolution occurred. Sergei's parents provided his education between 1918 and 1922, since his step-father G. M. Balandin was an engineer and his mother a teacher of Russian and French. After that, he studied at the vocational school of construction in Odessa from 1922 to 1924. There he met Ksenia M. Vintsentini, whom he married on August 6, 1931. She graduated from the institute of medicine of Kharkov in May 1930 and became a doctor. On April 10, 1935 they had a daughter, Natasha.

Korolev's first love was gliders. He joined the Ukraine and Crimea society of aviation and air navigation in 1923 and built his first K-5 glider at the age of 17. His dream was to enter the Jukovsky military academy of aeronautics engineers (VVIA), but it only prepared senior cadres and he was too young. So on August 26, 1924 he joined the aviation section of the faculty of mechanics of the Kiev polytechnic institute. On September 7 of that year he participated in the second glider meeting of Koktebel in Crimea. The polytechnic institute then granted him accommodations at N°25 Bogoutovsky Street. But the Kiev aeronautics section closed down, and in November 1926 he transferred into the third year of the aeromechanics section at the mechanical engineering faculty of the Moscow Bauman technical school (MVTU).

From 1926 to 1936 Sergei lived with his parents at N°38, rue Oktiabrskaya, in the Rochi Marinoi district. In 1927 the paramilitary organization Osoaviakhim (OAKh) was created for the development of aviation and chemistry. In March he attended the Moscow glider school (where he earned his glider pilot certificate on November 2, 1929). In May 1927 he began working in P. D. Grigorovich's OPO-3 group, which was developing seaplanes at factory N°22 in Fili. In September he participated in the fourth glider meeting at Koktebel. That year, the family built a dacha in Barvikh, 25 km from Moscow. In September 1928 he returned to Koktebel for the fifth glider meeting. The following month, while still studying in the fifth year of MVTU, he was appointed brigade commander in French engineer Paul-Aimé Richard's OPO-4 group, which was developing the TOM-1 seaplane at factory N°28. In 1929 he took flying lessons in an Avro 504 at the central airfield. In conjunction with S. N. Lyushin, he designed the Koktebel (or SK-5) glider, for which the components were built in the workshops of the air fleet academy and assembled at the Dux plant. The glider was flown at the sixth glider meeting in October. The pilot, K. K. Artseoulov, set a distance record. The aircraft's remarkable quality made it the subject of several articles in the press. In 1929, Korolev avidly read K. E. Tsiolkovsky's book on space exploration using jet engines. In the fall he allegedly went to Kaluga to meet Tsiolkovsky, but this visit has never been confirmed and some historians doubt that it took place.

On December 28 Korolev earned his degree in aeromechanical engineering. His thesis topic, directed by aircraft builder A. N. Tupolev, was the SK-4 two-seater light aircraft built at factory N°28 with funding from the Osoaviakhim. In July 1930, following graduation, he joined TsKB factory N°39 "Menjinsky" as brigade leader of engine equipment. This operated a design-office-prison (or "sharashka") TsKB-39-OGPU from 1929 to 1931. He then undertook to build his second glider, the SK-3 "Krasnaya Zvezda" (Red Star). The technical committee, headed by S. V. Ilyushin, accepted the project and it was built for the seventh glider meeting of Koktebel in October 1930, where Korolev performed the world's first glider looping.

In November, Korolev came down with typhus, was hospitalized, and underwent a trepanation with three hours of anesthesia. His convalescence lasted two months, and then in March 1931 he went to Kislovodsk. In August the OKB and factory N°39 merged with TsAGI. Every Sunday during the autumn and the winter of 1931-1932, Korolev and his friend P. V. Flerov, trained beginner glider pilots at the Planernaya station. In addition, from 1931-1934 he designed the SK-6 and SK-7 gliders and the SK-8 airplane.

Korolev's second great love was rockets. In early 1931 he met F. A. Tsander, with whom he created the GIRD jet propulsion study group at OAKh in October. The first project was an RP-1 jet-propelled gliders – specifically a Cheranovsky BiCh 11 flying wing powered by a Tsander OR-2 engine. In May 1932 Korolev replaced Tsander as head of the MosGIRD, which included four brigades at the time: Tsander's brigade producing 02 and 10 engines for the RP-1 and the GIRD-10 rocket; Tikhonravov's brigade making 03, 07 and 09 engines for the RP-2, GIRD-07 and GIRD-09 rockets, and for the 05 rocket (with a Glushko ORM-50 engine); Pobedonostsev's brigade, dealing with ramjets; and Korolev's brigade working on RP-1, RP-2 rocket gliders and the GIRD-06 winged rocket (09 engine). Korolev proceeded with the launch of the first Soviet GIRD-09 rocket on August 17, 1933.

In October the gas dynamics laboratory (GDL) in Leningrad and GIRD merged to form the institute of jet propulsion (RNII),[17] directed by I. T. Kleimenov and deputy Korolev. But on January 17, 1934, following a disagreement with the head of the institute, Korolev left his position to become simply chief designer of winged

Launch of the GIRD-10 in 1933.

Korolev at GIRD in 1931.

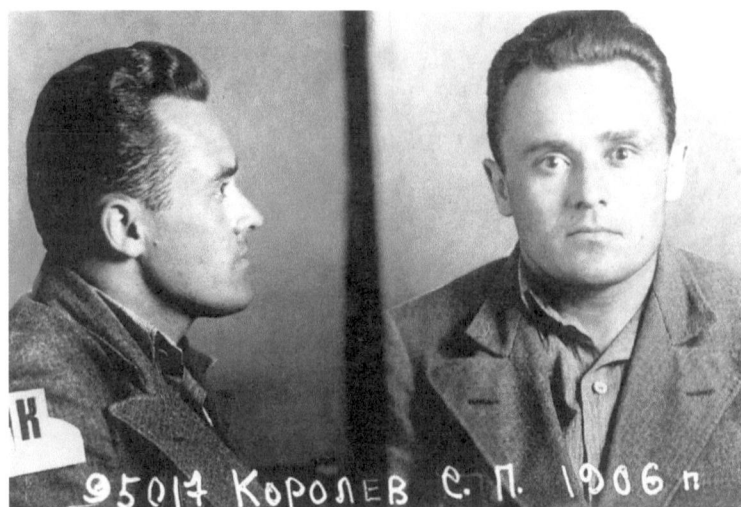

Korolev imprisoned in June 1938.

Tikhonravov, Kostikov and Dushkin.

rockets in the 8th section of the 2nd sector headed by his friend E. S. Chetinkov; he was replaced at the institute by G. E. Langemak. In March 1934, Korolev participated in the national conference on the study of the stratosphere, and published his book on stratospheric rocket flight. In early 1935 he replaced Chetinkov. In March 1936 the section became the 5th sector. There, Korolev made rockets, including the 06/2 (or 216), 06/3, 06/4, 201, and 212 using liquid propellants, and the 48, 217-1 and 217-2 solid-fuel rockets, as well as the 218 (RP-318) jet-powered glider. For the latter he used the SK-9 glider he had designed in 1935. However, in his four years of work at RNII none of Korolev's rockets were satisfactory – they were not integrated into the arsenal because they were a decade ahead of their time. On May 13, 1938 the 212 rocket engine exploded. On May 26 and 28 the ignition was successful. But on May 29 another failure resulted in Korolev receiving a head injury that required him to be hospitalized for trauma. After three weeks of convalescence, he returned home. A week later, on June 27, he was arrested. The repeated failures were used as a pretext to imprison him.

The Stalinist purges had begun in February 1937 with the fall of Marshal M. N. Tukhachevsky, who was executed in June 1937. Because he was responsible for the creation of RNII, his close collaborators suffered the same fate. In particular, N. Y. Ilin, head of the gas dynamics laboratory from 1931 to 1932, was executed in 1937. Suspicion and denunciation had become commonplace. At RNII, hostilities started at a meeting of the party cell, whose secretary was N. G. Belov, in March 1937. Critics focused on the management of Kleimenov, who had suspended work on the liquid oxygen rocket in November 1936 to give priority to nitric acid. L. S. Dushkin had therefore switched to nitric acid. The team of L. K. Korneyev and A. I. Polyarny, both former deputies of Tsander, continued the work on GAU's KB-7 from August 1935 to 1939 but their results were not satisfactory and Korneyev was arrested for having wasted money. A. G. Kostikov's work had also been halted by Kleimenov. They had every reason to be resentful, and denounced Kleimenov as an "enemy of the people". Worse still, he had worked in Berlin! On August 30, 1937, Kleimenov was transferred to TsAGI where he became deputy director of the propeller engines sector, and L. E. Schwartz replaced him as acting director.

In September, during a meeting of the communists of RNII, the mismanagement of Kleimenov and Langemak, Glushko's anti-social behavior, and Korolev's poor results were evoked. On October 14, B. M. Slonimer, a veteran of the Spanish Civil War who knew nothing about rocket technology, became director of the institute, which was renamed NII-3. Kostikov was then appointed chief engineer. Kleimenov and Langemak were arrested on November 2 and executed in January 1938. Glushko was arrested on March 23 and replaced by Dushkin. On June 27 it was Korolev's turn, and he was replaced by A. N. Dedov.

Judges made accusations with no basis in fact, and confessions were forced out of those already imprisoned, such as Kleimenov, Langemak and Glushko. For example, during the interrogation of June 5, Glushko had only spoken of Korolev in terms of some erroneous advice he had given concerning a torpedo motor and an unprepared ignition cartridge he had supplied which subsequently exploded.

On July 20, NKVD received a letter criticizing Korolev's work that was signed by A. G. Kostikov, L. S. Dushkin, A. N. Dedov, and M. P. Kalianova (secretary of the Komsomol, who replaced N. G. Chernyshyov as head of the section of chemists). At the trial they formed the committee of experts. He could receive no help from K. E. Tsiolkovsky (died 1935), Marshal M. N. Tukhachevsky (executed in 1937), R. P. Eideman (head of the Osoaviakhim, executed in 1937) or Y. I. Alksnis (chief of the air force, executed in 1938).

On September 25, Korolev's name appeared on a list of 74 prisoners sentenced to death, signed by Stalin and V. M. Molotov, L. M. Kaganovich and K. Voroshilov. On September 27 the sentence was commuted to 10 years in prison. Nevertheless, 59 of the 74 were executed. Korolev had truly come close to death. In the end, he was sent to Lyubianka from June 27 to October 10, 1938, to the Novocherkass camp from October 10, 1938 to June 1, 1939, and then to the Kolyma camp from June 1939 to December 1939. During the journey to the camp he was prisoner $N°1442$. He was transported by rail to Vladivostok, and then by boat to the Bay of Nagaiev and the Madiak gold mine where he did hard labor with pick and shovel. As evidence of how he was treated during that time, his head was scarred for life, while scurvy cost him half of his teeth. During this time, his mother and wife struggled with the authorities to have him released. They received support from test pilot M. M. Gromov and pilot V. S. Grizodubova, both deputies, who wrote to the minister of the interior L. P. Beria. During the revision of the trial on May 25, 1940, the charges were based on nothing, while the defense attempted to show the absurdity of the situation. On July 10, 1940 his sentence was transformed into eight years of imprisonment, which meant to June 1946.

Clearly his not being a CPSU member played a determining role in the case. For example, the two initiators of the purges, Kostikov and Korneyev had been members of the party since 1922 and 1917 respectively. Korneyev had been one of the leading communists of the GIRD (with N. I. Efremov, G. I. Ivanov, P. S. Alexandrov, etc.). Korolev and Glushko would actually not join the party until 1953 and 1956. K. I. Trunov, head of the sector at the NII VVS, was arrested and jailed from 1937-1941 and then worked with Korolev in Kazan from 1942-1946, becoming his deputy from 1946-1947, whereupon he was fired because he refused to join the party.

After the arrests of Kleimenov, Langemak, Glushko and Korolev, the institute was directed by B. M. Slonimer, A. G. Kostikov, F. N. Poida (successor to N. G. Belov as party secretary in 1938), L. E. Schwartz and Y. A. Pobedonostsev. But the arrests continued. The purge would later focus on Schwartz, Pobedonostsev and A. V. Pallo (Korolev's right-hand man, who tested the Glushko engine). But they avoided prison.

Kostikov's role in the purges was undeniable. He took credit (along with General V. V. Aborenkov of GAU, and I. I. Gvai, head of the NII-3 sector) for creating the legendary Katyushas, filing a patent for their invention on February 19, 1940. But Kostikov, a graduate of the Jukowsky academy in 1936, was a specialist in liquid engines and had nothing to do with solid propellants. Due to this usurpation, he was given the rank of major general, appointed a corresponding member of the Academy of Sciences in 1943, and received the Hero of Socialist Labor Medal in 1941 (medal $N°13$) and the Stalin Prize in 1942. In June 1991, in marking the 50th anniversary of the declaration of the Second World War, the Russian government posthumously awarded, the Hero of Socialist Labor Medal to the true creators of the Katyushas: Tikhomirov (creator of the gas dynamics laboratory), Petropavlovsky (designer of rocket mortars), Kleimenov and Langemak (the first directors of RNII), Slonimer (director of NII-3 from 1937-1941) and V. N. Luzhin (designer of rocket mortars). Luzhin was sentenced to eight years in prison in 1940 for having destroyed a portrait of Stalin at the end of a banquet in a restaurant in Moscow; he was replaced by D. A. Shitov.

In 1942 Kostikov proposed the project for rocket-plane 302P developed with M. K. Tikhonravov (the airframe), L. S. Dushkin (liquid-fuel engines) and V. S. Zuyev (ramjet). This aircraft was a competitor of V. F. Bolkhovitinov's BI-1. But while the BI-1 first flew on May 15, 1942, the 302P did not make its glide tests until 1943. On February 18, 1944 RNII became NII-1. Kostikov was arrested on March 15, 1944 but released on February 28, 1945. On August 1 he joined institute N°24 (later NIMI) where he was responsible for jet-propelled mortars and became head of office N°5 in 1946. In April 1947 he was sent to East Germany as a representative of the Academy of Sciences. In December 1948 he became president of the Academy commission on gas turbines and jet engines at the Krjijanovsky institute of energy. Kostikov died on December 5, 1950, and his obituary was published in Vestnik issue N°1 of 1951.[18]

During this time, work on the RP-318 jet-powered glider was continued by A. Y. Cherbakov (factory N°1), A. V. Pallo, L. S. Dushkin, and test pilot V. P. Fedorov (died in 1943 during the test flight of the Ilyushin DB-3). The first flight occurred on February 28, 1940, at Podlipki airfield (where RKK Energiya is located today).

Korolev left the gulag in December 1939. He returned to Moscow on February 28, 1940. He was first detained in the Butyrki prison, then transferred on September 18 to the Tupolev OKB in Bolchevo, which had been turned into a sharashka, prior to being transferred to Radio Street in Moscow (TsKB-29 of the NKVD). On July 13, 1941 he was transferred to factory N°166 in Omsk, Siberia, where he asked to join Glushko's group in Kazan. In November 1942 he was transferred to OKB-16, where he became Glushko's deputy for flight testing of the Petlyakov Pe-2RU equipped with an RD-1 rocket engine. In December, he proposed a project for an RP

Korolev and Gaidukov in Germany in 1946.

Tsiolkovsky 90th anniversary in 1947 from left to right: Kosmodemiansky, Vorobiev, Tikhonravov, Korolev.

rocket-plane powered by an RD-1 engine. On July 27, 1944 he was released after six years of imprisonment. He took an apartment in Liadova Street, where he stayed for one year. In late 1944 he designed the D-1 and D-2 solid-fuel rockets and the La-5VI rocket-plane. In April 1945 he began to give courses at the Kazan aviation institute. On May 12, during the flight of a Pe-2RU, the engine exploded, damaging the tail of the aircraft. The pilot urged bailing out by parachute, but Korolev refused because he wanted to identify the cause of the damage. When the pilot finally landed the aircraft, Korolev's face was burned around the eyelids and eyebrows – he could have lost his sight that day.

In June 1945 Korolev proposed opening a design bureau in NII-1 to develop its rockets but he was sent to Germany, where he remained until January 1947. While there, he was designated main designer for the V-2 at NII-88. In 1947 he was elected a corresponding member of the academy of artillery sciences (AAN). He taught at MVTU in the context of further training for engineers (VIK), and attended the 90th anniversary ceremonies for Tsiolkovsky on September 17, 1947. In 1949 he married Nina Ivanovna Kotenkova-Ermolaieva, widow of aircraft builder V. G. Ermolaiev. In 1952, his first wife, Ksenia, married E. S. Chetinkov, formerly with GIRD and RNII.

On April 26, 1950 the SKB was split into Korolev's OKB-1 and Isaiev's OKB-2. There was also an OKB-3 led by D. D. Sevruk from March 1952 to December 1958. When the director of NII-88, K. N. Rudnev, left in 1952, Korolev was not suggested as his replacement. Instead M. K. Yangel, head of sector 5 of OKB-1 and member of the party, was selected. Korolev then decided to attend the party school. He was a candidate in March 1952. Initially, Medkov, the party secretary at NII-88, objected because Korolev had not yet been rehabilitated (that did not occur until April 18, 1957). Nevertheless, he was admitted to the CPSU on July 15, 1953. His sponsors were D. I. Kozlov, party secretary at OKB-1 and chief designer of the R-7, and Y. A.

Pobedonostsev, his former collaborator at GIRD and RNII. In October 1953 Korolev was elected a corresponding member of the Academy of Sciences. In October 1961 Korolev and Glushko were elected as delegates to the 22nd congress of the CPSU.

Korolev was appointed deputy director of NII-88 for R&D on July 9, 1954. Decree N°4912 of August 13, 1956 separated OKB-1 from NII-88. At that time, it included 1,939 people, while the experimentation plant had 10,190 people. On April 20, 1956 Korolev received the Hero of Socialist Labor Medal for the R-5M rocket. The first flight of the R-7 took place on May 15, 1957. On September 15-17 of that year he attended the centennial ceremonies for Tsiolkovsky in Kaluga, with M. K. Tikhonravov, M. S. Riazansky and K. I. Trunov. On September 16 he published an article in Pravda under his real name, because at that time he was not yet regarded as a secretive man. In fact, he only became so after the launch of the first Sputnik, and an article that he published on December 10 was under the pseudonym "Professor K. Sergeyev".[19] Similarly, V. P. Glushko published an article on Tsiolkovsky under his real name in the 9th issue of Vestnik of the Academy of Sciences, but would later use the pseudonym "Professor G. Petrovich". Korolev attended a scientific conference on September 18-19 in Moscow. The report of this anniversary event was published by A. A. Blagonravov in Vestnik N°12. It mentions the presentations of S. P. Korolev and V. P. Glushko, both corresponding members of the Academy of Sciences. Later, Korolev helped with the construction of the Tsiolkovsky Museum in Kaluga where congresses took place after September 1966.

On December 18, 1957 Korolev received the Lenin Prize to mark the achievement of launching Sputnik. He was proposed for the Nobel Prize, but Khrushchev, not wanting to reveal the name of the "chief designer", refused. In 1958 Korolev was

Tsiolkovsky's 100th anniversary in 1957: From left to right, front row: Glushko, Korolev, ?, Blagonravov, ?, Riazansky, ?.

State commission of May 1957.

State commission for Sputnik 1 in October 1957.

elected academician and member of the presidium of the Academy of Sciences. He received the first gold Tsiolkovsky medal. From 1958 onward, relations between Korolev and Glushko began to deteriorate because Korolev wanted to continue the development of liquid oxygen rockets while Glushko wanted to focus on storable propellant missiles. Korolev therefore turned to the engineers M. K. Yangel, V. P. Makeiev and V. N. Chelomei.

In 1958 Korolev began working on solid-fuel rockets with B. P. Zhukov at NII-125. In July 1959 he absorbed V. G. Grabine's TsNII-58 which was conveniently in the same neighborhood. This artillery institute (TsAKB) was created in November 1944 (TsNIIAV in 1945) and in 1954 was assigned to the ministry of medium-sized machines for the production of fast-neutron nuclear reactors. A. P. Alexandrov was appointed director, with Grabine as head of sector. In 1955 it was transferred back to the defense industry as TsNII-58, with Grabine resuming management until 1959.

Mishin with the three K's in 1958. From left to right: Mishin, Keldysh, Kurchakov and Korolev.

Awarding the Socialist Labor Hero Medal for Gagarin's flight in June 1961.

Korolev then developed the RT-1 (8K95), RT-15 (8K96), RT-25 (8K97) and RT-2 (8K98) missile series jointly with the KB-7 plant of the Leningrad arsenal and Mach KB of factory N°172 imeni Lenin of Motovilikhinsk near Perm, which then merged with the Machinostroïtel (PZKhO) subsidiary to become NPO Iskra. In 1959 Korolev built the R-9 missile which became the only liquid oxygen intercontinental ballistic missile to stay in service for 15 years. A variant of the orbital bomb (GR-1/8K713) was developed, as well as an anti-satellite version (8K513), but neither of these were completed.

With respect to satellites, Korolev developed the Vostok manned spacecraft which led to the Zenit spy satellites (whose series production was transferred to Kuybyshev in 1964), the Luna, Venera and Mars interplanetary probes (transferred to

Korolev and Tiouline in 1965.

Lavochkin in 1965), the Molniya communications satellites (transferred to Rechetnev in 1965), and then the Voskhod and Soyuz manned spacecraft. In 1960 he designed the N-1 (nossitel, meaning "carrier") rocket whose first stage had 30 LOX-kerosene engines designed by the aeronautics manufacturer N. D. Kuznetsov. But work on this project did not really start until 1964. In December 1964 the drawings for the Block G and D stages, as well as the LOK spacecraft and LK lunar module were completed. They were adopted by an evaluation commission headed by M. V. Keldysh. On February 10, 1965 the N1-L3 program was elaborated: it planned a first flight of the system in 1966 and the lunar landing of a cosmonaut in 1967-1968 to mark the anniversary of the revolution. In addition, the lunar flyby program was adopted on August 26, 1965. This called for a lighter Soyuz capsule (7K-L1) with two cosmonauts to be launched in 1966-1967 by a Proton rocket developed by V. N. Chelomei. However, due to his premature death on January 14, 1966 Korolev would not see any of these grandiose projects come to fruition. He was scheduled for a 5-minute operation to remove a polyp in his colon, but the operation, conducted in person by the Minister of Health, Boris Petrovsky, developed complications and Korolev died after 5 hours on the operating table.[20]

## TAMING THE FIRE

Valentin Petrovich Glushko was born on September 2, 1908, in Odessa, Ukraine, where he grew up. He read books by Jules Verne at the age of 13, then began to take an interest in astronomy and joined an amateur club. On finding Tsiolkovsky's books at the public library, he wrote to him on September 26, 1923, receiving a response a short time later. In 1924 he finished secondary school and left for Leningrad to enter the faculty of physics and mathematics at the university. There, he met the editor Y. I. Perelman, author of a well-known book on interplanetary travel. The topic of his engineering diploma was an electro-thermal rocket engine. In 1929 he joined the gas dynamics laboratory to develop that project.

But Glushko quickly redirected his focus to liquid-fuel engines and designed the ORM-1. Owing to the relative complexity of this engine, it was not completed until 1931. In the meantime, he built a simple engine named the ORM. It was the first to be tested in the Soviet Union, ahead of Tsander's OR-2 which was not tested until March 1933. Glushko then took charge of the gas dynamics laboratory's liquid-fuel rockets sector and designed the RLA-1, RLA-2 and RLA-3. He directed the ORM-1 to 52 from 1931-1933. The ORM-50 delivered 150 kg of thrust and was intended for the GIRD 05 rocket, and the ORM-52 with 250-300 kg of thrust was for the RLA. In September 1933 he joined RNII and directed the ORM-53 to 70 engine development (nitric acid/kerosene), the ORM-101 and 102 (tetranitromethane/kerosene) and the GG-1 to 3 gas generators from 1933-1937. The ORM-57 had a chamber producing 500 kg of thrust, while the ORM-58 achieved a record thrust of 600 kg using two chambers. The thrust of the ORM-65 was variable in the range 50 to 175 kg, and was used on the 212 winged rocket and on Korolev's 218 rocket glider.

In December 1935 Glushko and G. E. Langemak (of RNII) published the book "Rockets, their mechanisms and uses". But on March 23, 1938 Glushko was arrested and replaced by L. S. Dushkin. Glushko was accused of having been close to I. T. Kleimenov and Langemak, of having written a book with Langemak (of which the RNII stock was destroyed at the request of A. G. Kostikov), and committing acts of sabotage. On August 15, 1939 he was sentenced to eight years in a prison camp, but benefited from the clemency of the court, which sent him to sharashka factory N°82 in Tushino in the autumn of 1939,[21] where he worked to improve the performance of

Glushko in Germany in 1945.

Gushko.

the "100", otherwise known as the Pe-2 Petlyakov, a plane produced in series at the Kazan aircraft plant. In 1940, the group was transferred to factory N°16 in Kazan where it became special sector N°28.[22] In November 1942 Korolev arrived in OKB-16. In 1943 S. M. Alekseyev (S. A. Lavochkin's deputy) came from Gorky in order to meet Korolev, D. D. Sevruk, N. L. Umansky and V. A. Vitka in Kazan. Finally, the RD-1KhZ engine was installed on a Pe-2 aircraft (which made 110 flights from August 1943 to June 1945), Sukhoi Su-7, Yakovlev Yak-3, and Lavochkin La-7R and La-120R. The La-7R flew at a speed of 742 km/h and the La-120R at 805 km/h.

During the war, Glushko's parents were detained in the blockade of Leningrad. His father was arrested and died in prison in 1948. His sister Galina disappeared. His brother Arkadi fought on the front lines.

On 27 July 1944 Glushko was released at the same time as Korolev. On December 7 he was appointed main designer of OKB-SD and on May 1, 1945 he became head of the chair of rocket engines at the Kazan aviation institute. Then he was sent to Germany from July to December 1945 and again from May to December 1946. In October 1945 he and Korolev took part in operation Backfire at Cuxhaven. On July 3, 1946 the ministry decided to transfer Glushko's team from factory N°16 (Kazan) to factory N°456 (Khimki). Glushko was the main designer, his deputies were D. D. Sevruk and G. S. Giritsky, and the head of the SKB was N. S. Chniakine. The move was made in November-December. Glushko developed engines derived from that of the V-2. From 1947 to 1954 he taught courses at MVTU. In 1953 he was elected a corresponding member of the Academy of Sciences. He joined the party in 1956. On April 20, he received his first Hero of Socialist Labor Medal for the R-5M. In 1957 he received the Lenin Prize for the achievement of Sputnik. In 1958 he became a full member of the Academy of Sciences and received the Tsiolkovsky gold medal. He began to focus on electrical and nuclear propulsion. He also set aside development of LOX-kerosene engines to concentrate on storable propellant engines. For this reason, he worked more often with M. K. Yangel and V. N. Chelomei than with Korolev. On June 17, 1961 he received his second Hero of Socialist Labor Medal for Vostok 1. But he did not participate in the N1-L3 lunar program, and for quite a long time he refused to develop cryogenic engines. In 1967 he received the State Prize for the Proton launcher developed by Chelomei. Then on May 22, 1974, he became director and general designer of NPO Energiya which combined OKB-1, OKB-456, their experimentation plants and their subsidiaries. His first deputies were Y. N. Trufanov, V. M. Kliutcharev, V. P. Radovsky and S. P. Bogdanovsky. At the time, there were six chief designers: K. D. Bushuyev on Apollo-Soyuz, Y. P. Semenov for Salyut orbital stations, Y. P. Kolyako for the heavy launcher, I. S. Prudnikov for the Zvezda lunar program, I. N. Sadovsky for the space shuttle, and V. P. Radovsky for engines. In 1976 Trufanov was replaced by I. N. Sadovsky, and in 1977 Glushko turned over his functions as director to V. D. Vachnadze.

Glushko had important relationships with four women during his lifetime and had four children. His first companion, when in Leningrad, was Suzanne Georgievskaya who wrote children's books. But when he moved to Moscow in 1934 he developed a relationship with Tamara Sarkissova, sister-in-law of B. S. Petropavlovsky, director of the gas dynamics laboratory in 1930-1931. When he married her in January 1938 she was seven months pregnant with Evguenia. They

Glushko at the Franco-Soviet PVH flight of 1982.

accompanied him to Kazan during the war. While in Germany, he had a girlfriend. They had a child that did not survive. Then he met Magda Maxovna Esmin, with whom he lived at N°43 Gorky Street in 1947. They had two children: Elena in 1948 and Yuri in 1952. In 1953 they moved to N°1 Vosstaniya Street. In 1972, when 64 years old, he had a relationship with Lydia Dmitrievna Perychkova, a young employee of OKB-456, who bore him a son, Alexandre. In 1975 he moved with Magda into a large eleven-room apartment at N°2 Serafimovich Street. From 1966-1989 he was a member of the supreme soviet. And from 1976 to 1989 he was a member of the central committee of the CPSU. In 1984, along with a group of researchers, he received the State Prize for his work on the thermodynamic properties of substances and products of combustion from 1965-1982. Then he developed the Energiya-Buran space transport system. Energiya flew without Buran in 1987 and with Buran in 1988. On April 8, 1988 he suffered a heart attack and remained in the hospital from April to June. He went into convalescence in July, but his health deteriorated. He celebrated his 80th birthday in hospital and died on January 10, 1989.[23]

### THE UPPER STAGE ENGINES

Semyon Arievich Kosberg was born on October 14, 1903 to a family of artisan iron-workers at Slutsk in the Minsk region. At 16 he worked with his father and did his secondary studies from 1922-1924. He then served in the army from 1925-1926. He studied at the Leningrad polytechnic institute from 1927-1929, then at the Moscow aviation institute from 1929-1931. He joined the CPSU in 1929. Like L. R. Gonor and B. E. Chertok, Kosberg was Jewish. But in the 1950s he was not targeted by the Stalinist repression. He then joined TsIAM, where he was responsible for direct fuel supply systems (NV) for aircraft engines. In 1940 he became deputy main designer and KB leader of factory N°33. But in October 1941 the plant was evacuated and split into two: Kosberg's part in Berdsk in the Novosibirsk region became the OKB of factory N°296 imeni Dzerjinsky. He took charge and developed NV systems for the Shvetsov, Mikulin, Klimov and Dobrynin engines. In April 1946 the OKB was transferred to factory N°265 in Voronezh, thereby becoming the OKB of factory N°154 (which was renamed KB KhimAvtomatiki – CADB in English – Voronezh mechanical plant). By this time aviation had become synonymous with jet propulsion and turbojets. The plant made injectors, pumps, control assemblies and powder or liquid starter units until 1954. Then the focus shifted to rocket engines. The first D-154 engine, with 4 tons of thrust, was a "JATO" monopropellant booster intended for the Mikoyan E-50A aircraft. In 1956 a derived version was developed. This D-7 had a thrust of 1.2 tons and was intended for I. I. Toropov's K-7 air-air missile. Starting in 1956 Kosberg built oxygen-alcohol engines: the SK-1 (RD-0101) for the Mikoyan E-50A, SK-1K (RD-0102) for the Yakovlev Yak-27V, and SK-2 (RD-0103) for the Sukhoi T-3. In parallel, from 1957-1962, Kosberg developed nitric acid-kerosene engines for surface-to-air missiles: the RD-0200 with 0.6-6.0 tons of thrust for the 5V11/400/DAL Lavochkin missile, and

Kosberg.

the RD-0201 with 3.0-6.0 tons of thrust for the Grushin V-1100. In 1958 activity refocused on space. The type RD-0100 engines burned oxygen-kerosene while the RD-0200s used nitric acid-kerosene.

On February 20, 1958, Korolev asked Kosberg to develop the engine for the third stage of the R-7, Block E. The order took the form of decree N°343-166 on March 20. This was produced in nine months using RD-0102 assemblies and a combustion chamber developed by M. V. Melnikov of sector 12 of OKB-1. With 44.9 bars of pressure in the chamber it delivered a thrust of 5 tons and a specific impulse of 316 s. It had several names: RO-05, 8D714, RD-0105 and RD-448. Fifty-eight static tests were conducted with 27 engines in NII KhimMach of Zagorsk from May-September 1958. However, on September 10 an engine exploded after 312 s. The first flight on September 23 was a failure. In addition, 33 experiments were carried out on tanks with nitrogen and gaseous helium pressurants. Finally, helium was selected. In 1959-1960, the engine was modified to improve reliability in manned flight. This entered service in December 1960 as the 8D719/RD-0109 with 5.56 tons of thrust.

In April 1958 Kosberg designed the 8D715/RD-0106 engine with 30 tons of thrust for the second stage of Korolev's R-9 rocket. Development began in February 1959 and two models were ready at the end of the year. The decree on the R-9 was issued on May 13, 1959, and the first flight took place on April 9, 1961 (8K75 N°E10308). The engine was improved for the R-7's Block I. It became the RO-07, 8D715K, RD-0107 or RD-461. Starting in October, 1960, it was used on the four-stage version of the R-7 called Molniya (8K78). The improved 8D715P/RD-0108 replaced it on the Voskhod and Molniya launchers from November 11, 1963 to October 4, 1965, then the final 11D55/RD-0110 version was used on the Molniya-M, Voskhod and Soyuz launchers.

Kosberg received his doctorate of technical sciences in 1959. He was awarded the Lenin Prize in 1960 and the Hero of Socialist Labor Medal in 1961. At that time, he began to develop integrated flow engines (staged combustion or closed cycle, gas-liquid configuration) as opposed to the derived flow (gas generator, or open cycle liquid-liquid). The first Soviet engine of this type was developed by NII-1 in 1958. Then the concept was applied to Melnikov's S1-5400, Kuznetsov's NK-9, Glushko's RD-253, Kosberg's RD-0202, and Isaiev's 3D40. Starting in 1961 Kosberg supplied engines for Chelomei's rockets: the RD-0202 to 0207 for the UR-200, the RD-0208 to 0214 for the UR-500 (Proton), and the RD-0216 to 217 for the UR-100.

On December 28, 1964 Kosberg was involved in an automobile accident and died of his injuries on January 3, 1965. A. D. Konopatov replaced Kosberg and directed operations from 1965-1993. He continued production of many engines using LOX-kerosene (including the RD-0124 of the Soyuz-2 rocket), LOX-hydrogen (including the RD-0120 of the Energiya heavy lifter, and RD-0146 of the Angara and Rus M), nitrogen-peroxide-UDMH (including the RD-0228 to 0230 of the R-36M, RD-0255 to 257 of the R-36M2, RD-0243 to 245 of the R-29RM, etc.), the RD-0410 nuclear engine (started in 1965), the RD-0600 gas dynamic laser (started in 1970), the tri-propellant RD-0750 engine (started in 1993), and the 58L hypersonic ramjet (started in 1994). Konopatov received the Hero of Socialist Labor Medal in 1966, the State Prize in 1970, and the Lenin Prize in 1976. He has been a member of the Academy of Sciences since 1991.

Since 1993 the head of KB KhimAvtomatiki has been V. S. Rachuk, a graduate of the polytechnic institute of Voronezh. He began working in the company in 1964 and became main designer in 1981. He is a doctor of technical science and a professor, and was awarded the State Prize in 1997.[24]

Series production was ensured by the mechanical engineering plant of Voronezh (VMZ), KrasMach of Krasnoyarsk (KMZ), the "Stalin" plant in Perm (later Perm Motor), the Ust-Katav plant, the "Krasnay Oktiabr" (Red October) Leningrad plant, the "Kalinin" factory (aggregates) in Perm, and the Ufa engine plant (supplying vernier engines). Created in 1928, VMZ first produced M-11 piston-engine aircraft. It became factory N°154 in March 1941, and was evacuated to Andijan (Uzbekistan) in October 1941, where it produced more than 30,000 engines. It was moved back to Voronezh in 1946 and began rocket engine production in 1958.[25] The plant gained its own design bureau (OKBM) in 1960, became a subsidiary of the OKB of Ivtchenko in 1963 (modifications of the AI-14), then gained independence in 1966 and was directed by I. M. Vedeneiev from 1960-1973 and thereafter by A. G. Bakanov.

## FATHER OF THE "BEEP–BEEP"

Mikhail Sergueievich Riazansky was born on April 5, 1909, in Leningrad, but was taken to live in Baku in Azerbaijan, an oil capital at the time. At age 14, he went to Moscow where he became a ham radio operator and learned Morse code. He joined the Komsomols in 1925 and became a member of the ham radio society. There, he

Riazansky in Germany in 1945.

was employed as a technician in the workshops. In August 1928 he was sent to the Gorky radio laboratory (NRL). During his stay in that city, he met his wife Cécile, with whom he had two sons: Vladimir in 1935 and Nikolai in 1945.

In 1931 Riazansky studied at the Leningrad electrotechnical institute. At the same time, he worked on a project for the OTB design bureau (now TsNII Granit). But he contracted tuberculosis and had to go to Bashkortostan to recuperate. In September 1933 he returned to Moscow to study at the energy institute (MEI), completing his studies in 1935. In December 1934 he began working as an engineer in the OTB in Moscow which became institute N°20 of the department of the power industry, then NII-244 in 1954, the radio engineering institute of Yausky in 1964, VNIIRT in 1972, and then NPO Skala (which should not be confused with NII-20 of the department of the defense industry, which became NPO Antei). In 1939 Riazansky participated in the design of the Pegmatit (RUS-2s) radar, for which he received the Stalin Prize in 1943. He was sent, together with other experts of NII-20 (E. Y. Boguslavsky, N. I. Belov, etc.) to study the telemetry systems of the V-1 and V-2 in Germany, Austria and Czechoslovakia. He worked at Nordhausen with Korolev, B. E. Chertok, N. A. Pilyugin and others from July 1945 to December 1946.

Special technical NII (NII ST) was created on May 13, 1946, primarily with

Riazansky.

personnel from NII-20, PKB-886, and factory N°528's SB-10. Riazansky set up his experimentation plant on the site of factory N°1 of the ministry of defense (NKO). This factory, called ZATEM, had been created in 1934 to manufacture sights for anti-aircraft guns. In 1936 it became factory N°205. It was evacuated to Saratov in 1941. The "Krasnaya Zarya" (Red Dawn) plant was transferred to the same site from Leningrad in 1942. Then it became factory N°1 of Aviamotornya Street in Moscow. From May to November 1946 the director was M. E. Salmanov, and from 1946-1949 it was N. D. Maximov. In October 1946 the institute received 54 German experts. And in December, Riazansky was appointed chief engineer and main designer. His deputies were N. A. Pilyugin for the guidance system and E. Y. Boguslavsky for the Messina telemetry system. The Hawai-Viktoria lateral-correction radio system was developed by M. I. Borissenko of PKB-886, which merged with NII-885 in 1947. But the unreliable Messina system was replaced by the Brazilionit, an improved 8-channel version developed by G. I. Degtiarenko of NII-20. However, competition from NII-885 and the OKB MEI was such that NII-20 was overwhelmed. At that time the institute totaled five sectors: Riazansky's radio-guidance sector, Pilyugin's autonomous guidance sector, Borissenko's lateral radio correction sector, V. A. Goviadinov's sector for guidance of anti-aircraft missiles, and Boguslavsky's telemetry sector. The latter produced the Don system (STK-1) for the R-2, for which he received the Stalin Prize in 1950.[26]

On February 19, 1949, a fire destroyed a large part of the institute's production workshops. The director was replaced by P. V. Kozlov, who served from 1949-1954, with G. S. Saveliev taking over from 1954-1955. There was also a reorganization of the institute in April 1950. B. M. Konoplev arrived from NII-20 to take over radio guidance, which was entrusted to the new sector N°15. Konoplev, a graduate of the University of Moscow, first worked at the institute of theoretical geophysics of the Academy of Sciences before joining NII-20 in 1943, where he was the main designer of the Topaz guidance system for Korolev's R-3 rocket (3,000 km range). Then he worked at NII-885 from 1950-1955 but was in conflict with Pilyugin. Consequently, he went to NII-695 (now MNIIRS) to design communication systems for long range bombers. In 1959 he opened OKB-692 in Kharkov to develop systems for inertial navigation. He built the system for M. K. Yangel's R-16 rocket, but unfortunately on October 24, 1960 he died in the explosion of the first model on the launch platform in Baikonur.

Similarly, in 1950 work on ground-air missiles was halted after unsatisfactory tests of the Wasserfall and the Schmetterling. Instead, the S-25 Berkut system was entrusted to KB-1 of the ministry of the defense industry (now NPO Almaz). And on January 30, 1951, Riazansky became chief engineer of NII-88. Then, on June 18, 1952, he was appointed director of the 7th glavka of the department of the defense industry, succeeding L. V. Smirnov (who became director of the YoujMach plant at Dnepropetrovsk, then president of the military industrial commission from 1963 to 1985). In 1954 Riazansky returned to NII-885 as deputy director for science and main designer. A year later, he became director, and held that position for 10 years. He completed the guidance systems for the R-5 and R-7 missiles that Konoplev had developed between 1950 and 1955. He received the Hero of Socialist Labor Medal for the R-5M (nuclear version of the R-5) in April 1956, and in December 1957 was awarded the Lenin Prize for developing the transmitter for the first Sputnik. In 1958 the father of the beep-beep received his doctorate in technical sciences and became a corresponding member of the Academy of Sciences. However, he did not receive a second Hero of Socialist Labor Medal for the Gagarin flight in July 1961. This was awarded to all members of Korolev's council of chief designers except Riazansky and V. P. Barmine. Why? It would appear that L. P. Beria and L. I. Brezhnev did not like them!

In 1952 the institute spun off several organizations. Sector N°4 became institute N°648 (now NII for precision instrumentation), sector N°12 became SKB-567, later to become the Evpatoria center for communicating with deep space probes (TsDKS), and sector N°16, which undertook modeling, became sector N°2 of SKB-245 (now NII Argon) and developed computers, including the Strela, Ural and Argon. In 1953 PKB-886 became an autonomous organization (now NII Kulon) in charge of radio-detonators. Radio guidance systems had been developed for the R-3 rockets (Topaz), R-5 (BRK-2 "Lena"), R-7 (BRK-71) and R-9 (Vega phased system). RUP stations had been installed at Baikonur and Plesetsk, but these were abandoned in the early sixties to switch to inertial guidance systems. This led to a reorganization in 1963. Complex N°2 was expanded by absorbing A. V. Belussov's SKB-567 and factory N°192 Radiopribor, while complex N°1 and the NII-885 experimentation plant, V. I. Kuznetsov's NII-944, and E. I. Eller's NII-346 together formed the new institute N°944 which managed both the activities of Pilyugin (inertial navigation systems) and those of V. I. Kuznetsov (gyroscopes). After two years spent under Pilyugin, Kuznetsov regained his independence in 1965 as NII-885 was split into the institute for instrumentation (NIIP) under Riazansky and the institute of automated systems and instrumentation (NIIAP) under Pilyugin.

By then the institute was no longer involved with radio guidance and autonomous guidance, only radio communications. There were sectors for manned space flight (main designer E. N. Galine), interplanetary communications (E. Y. Boguslavsky), defense, navigation and geodesy (N. E. Ivanov), telemetry (V. B. Kharine), lasers (V. D. Chargorodsky and V. P. Vasiliev), remote sensing, meteorology and television from space (A. S. Selivanov) and satellite communications. For manned space flight the institute built the Saturn MS and Kvant tracking complexes. For interplanetary communications it dealt with TsDKS and created the P-2500 antenna for the RT-70 radio telescopes at the Evpatoria and Ussuriysk stations. In navigation and geodesy, it worked on the Sfera, Tsikada, Cospas, Nadezhda, Glonass, and other programs. In telecommunications it built the Arion repeaters for the Luch series of geostationary relay satellites.

In late 1965 Riazansky became deputy director of the institute, which from then on was under the auspices of the ministry of general machines. He was replaced as director by L. I. Gussev (who remained in post until 2001).[27] In February 1977 the institute became NPO Radiopribor,[28] with Riazansky as deputy director general and also main designer. The same year, Kuznetsov's company became NPO Rotor. In 1984 the institute opened a subsidiary of the main satellite control center of the ministry of defense at Golytsino, and this became GNPP Orion in 1992. Then in 1985 the institute for cosmic instrumentation (NII KP) was created and, a year later, the institute of precision instrumentation (NII PP).

Riazansky had lost his wife in 1981 and then his son Vladimir in 1982. At the end of 1986 he was ill and had to leave the institute. In April 1987 he joined the ministry of general machines' science and technology committee, but he died on August 5. He was replaced by V. A. Grichmanovsky.

In May 2003, Riazansky's nephew Sergey Nikolayevich, a doctor at the institute of medical-biological studies of Moscow (IMBP), was selected to join the group of cosmonauts at the Star City. The Riazansky dynasty therefore continues to explore the cosmos.

## TELEMETRY AND TRAJECTOGRAPHY

Alexei Fedorovich Bogomolov was born May 20, 1913 in Sitskoe in the Smolensk region. At the age of 16 he got a job as an electrician in Moscow. He studied at the institute of energy (MEI) from 1932 to 1937 and then worked as an engineer for the TeploElektroProjekt company in Moscow. In 1938 he was an aspirant, then assistant at MEI. During the Second World War he served in the air defense on the front. In 1946 he returned to MEI on the radio-technical equipment faculty. He defended his thesis in technical science in 1949. The following year he was appointed dean, and in May 1955 became head of the chair.

On April 24, 1947, decree N°1327-340 created the special works sector in MEI under the ministry of education. Its director and main designer, Vladimir Kotelnikov, was president of the Interkosmos council of the Academy of Sciences from 1980 to 1991. He developed the Indikator telemetry and trajectory measurement system for Korolev's R-2 in 1948, which became the radio control system for precision firing (RKT) that was produced in series. The onboard transmitter was produced by factory N°567, and the ground receiver by factory N°304. In 1954 Kotelnikov was appointed to direct the institute of radio and electronics (IRE) of the Academy of Sciences, and he was succeeded by A. F. Bogomolov, his deputy for two years. Bogomolov then developed the Tral radio-telemetric system, Binokl trajectory measurement system, Irtysh goniometer, and Fakel onboard transmitter for the Semyorka. They were tested on the M5RD in 1956 before being produced in series in factory N°304 in Kuntsevo, factories N°528 and N°567 in Moscow, and factory N°797 in Lvov. Then he built the telemetry system for Sputnik 2 and Sputnik 3, for which he was awarded the Hero of Socialist Labor Medal in 1957 and became doctor of technical sciences. On March 18, 1958, decree N°838 transformed the sector into an OKB. He opened a testing station at Bear Lakes, 26 km from Moscow on the road to Chelkovo. A number of antennas were built, including TNA-9 (9 sq. m), TNA-150 (150 sq. m), TNA-200 (200 sq. m, 25 m in diameter), TNA-400 (400 sq. m, 32 m in diameter), TNA-57 (12 m in diameter) for the Orbita telecommunications system, and TNA-1500 (a 64 m diameter radio-telescope). Most of the antennas were produced in series at the Gorky machinery plant.

At that time, Bogomolov was a full member of the main designer's council, and

Bogomolov.

At the launch of Sputnik 1, from left to right: Bogomolov, Riazansky, Pilyugin, Korolev, Glushko, Barmine, Kuznetsov.

provided the Tral P systems for the Block E of the Luna and Vostok missions. He received the Lenin Prize in 1960 and the Order of Lenin in 1961. Based on the Tral, he created the Kosmos system, of which over 2,000 units were launched. He also developed the Almaz, Rubin, and Kama systems, as well as Topaz 10 (Vostok) and Topaz 25 (Voskhod) space television systems. By the end of 1965 OKB MEI had over 1,000 employees. By 1967 the workforce had grown to 3,000 people. The OKB had built, among other things, the Kontakt orbital rendezvous system for the N1-L3 lunar program from 1965-1970, the Polious V radar for the Venera 15 and Venera 16 probes from 1981-1983, the telemetry for the Tselina O and D intelligence satellites, and later for the Tselina 2 series and a control center for India's satellites at Bear Lakes.

Bogomolov received the State Prize in 1979 and again 1986. He was elected a corresponding member of the Academy of Sciences in 1966, then academician in 1984. He retired in 1988 and received the A. S. Popov gold medal the following year. He died on June 2, 2008.[29]

## GUIDANCE SYSTEMS

Nikolay Alekseyevich Pilyugin was born on May 18, 1908 in Krasnoe Selo near Leningrad. He studied while working as a laborer at TsAGI from the age of 18. He followed courses at MVTU from 1930 to 1935 and joined TsAGI as an engineer, becoming laboratory director in 1937. He entered the institute of flight testing (LII) of Stakhanovo in 1941, but was evacuated to the Kazan aircraft plant. He defended his thesis in technical science on a Sperry autopilot in 1943. On returning to Moscow in July 1944 he entered NII-1 as laboratory director in B. E. Chertok's sector. He went to Germany in April 1945 and became the first deputy and chief engineer of the Rabe institute at Bleicherode in August. In 1947 he returned to Moscow to work as an engineer at NII-885, where he was responsible for V-2 angular stabilization. He was Riazansky's deputy for autonomous guidance systems and developed the first inertial system for the R-5M strategic rocket, for which he was awarded the Hero of Socialist Labor Medal in 1956. Then he developed the guidance system for the R-7 and gained the Lenin Prize in 1957. In 1958 he was elected a corresponding member of the Academy of Sciences and received the Tsiolkovsky gold medal. Then he built the guidance system of the Vostok, Voskhod and Soyuz manned spacecraft. For this he received his second Hero of Socialist Labor Medal in 1961. In March 1963 his department left NII-885 to create a new organization on the site of institute N°346 in Ziouzino. This was designated NII-944 from 1963-1965, then became the institute of automation and instrumentation (NIIAP) of the ministry of general machines in

Pilyugin in Germany in 1945.

Pilyugin.

1965. But NII-944 regained its autonomy in 1965 and NII-346 was recreated as the institute of navigation in 1966 (later Delfin TsNII). It built most of the guidance systems for Soviet satellites and rockets.

In 1966 Pilyugin was elected to the Academy of Sciences and in 1967 became a member of its presidium. That year, he received the State Prize for the UR-100 missile developed by V. N. Chelomei. He was a professor from 1970, and held a chair at the MIREA institute in Moscow. He died on August 2, 1982. His successors were V. L. Lapygin (1982-1998), Y. V. Trunov (1998-2001), then E. L. Mejeritsky. The first deputies were G. P. Glazkov (1947-1961) and V. L. Lapygin (1961-1982). NIIAP had a number of subsidiaries. In 1951 a second production line was started at factory N°205 (which became the Korpus firm) in Saratov. In 1958 a subsidiary opened in Ostachov on the premises of branch 1 of NII-88 where the German team of Helmut Gröttrup worked (now the Zvezda firm). Finally the Sosensk instrumentation plant near Kaluga was established in 1975 for guidance of military rockets and space launch systems.

## GYROSCOPES

Viktor Ivanovich Kuznetsov was born on April 27, 1913 in Moscow. In 1933 he entered the Leningrad industrial institute (later polytechnic institute), where he took courses from L. G. Loïtsiansky (author of a book on theoretical mechanics with A. I. Lourié in 1933 and on mechanics of liquids and gases in 1950, and a corresponding member of the international academy of astronautics), B. I. Kudrevich (inventor of the Soviet gyrocompass), and E. L. Nikolai (a specialist in the dynamics of the gyroscope gimbal). He did his study-internship in factory N°212 (which was NII-303 in 1949, TsNII Elektropribor in 1966, NPO Azimut in 1974, and then TsNII Elektropribor once again in 1991). While there he worked with N. N. Ostriakov, a gyrocompass specialist who was killed in a car accident in Berlin in 1946. At that time, the only gyrocompasses in the world were produced by the American firm Sperry and the German firm Anschütz-Kämpfe. In 1938 Kuznetsov received his diploma and began his career by creating a directional stabilizer for guns on the Kirov and Maxim Gorky cruisers, for which he received the Stalin Prize in 1943. In 1940 he moved to NII-10 (later NPO Altair) in Moscow, where he directed the gyroscope sector. On October 20 he set off for Berlin as an expert to purchase equipment for Soviet ships. The trip, originally planned for two months, lasted eight. He was in Germany when war with the Soviet Union was declared on June 22, 1941. To return to his country, he was obliged to pass through Austria, Yugoslavia, Bulgaria and Turkey, finally arriving in Moscow on August 14. In October the institute was evacuated to Sverdlovsk, where Kuznetsov built various systems, including aircraft radio guidance systems, sighting systems for fighter aircraft, and stabilizers for tanks. In 1946 he received the Stalin Prize for the second time.

On April 2, 1942 the SKB of the ministry of the marine industry was created in Moscow to develop acoustic torpedoes (it became MNII-1 in 1946, NII Agat in

Kuznetsov in Germany in 1945.

Kuznetsov.

1971, and then NPO Agat in 1977). The associated plant which occupied the former site of factory N°251 became N°706 (now MZEMA). On June 13, 1945 construction sector factory N°212 became a Leningrad subsidiary of SKB. When Kuznetsov returned to Moscow in 1943, he joined this sector, which was led by V. N. Tretyakov, to direct the gyroscope laboratory.

In Germany the first inertial platforms were installed on the A-3 missile in 1937, the A-5 in 1939, and the A-4 in 1942. In May 1945, Major Kuznetsov was sent to Berlin with a group of rocket specialists. He was accompanied by Z. M. Tsetsiour. After being promoted to colonel, he returned to Germany from August 1945 to early 1946 and worked with S. P. Korolev, V. P. Mishin, B. E. Chertok, M. S. Riazansky, N. A. Pilyugin, and others. He visited the Kreiselgeräte firm, the Siemens plant (the V-2 gyro) and the Askania plant (copy of a Sperry gyrocompass). Then he returned to Moscow. In April, N. N. Khlybov had joined the laboratory. He tested submarine gyrocompasses at factory N°706 in July and August 1946.

Meanwhile, on May 13 the decree creating the rocket industry had been issued and Kuznetsov became the main designer of gyroscopes for long range rockets. His laboratory became sector N°2 of NII-10 (led by V. D. Kalmykov from 1941-1949, who was minister of the radio industry from 1954 to 1974). In 1953 the sector was turned into SKB-10, headed by A. K. Bauline with Kuznetsov as main designer. For the V-2 (first flight in 1947), the R-1 (first flight in 1948) and the R-2 (first flight in 1950), Kuznetsov built the horizon gyro (GG-1), the vertical gyro (GV-1), and the integrator gyro (IG-1) based on German technology. From 1951 to 1953 Kuznetsov developed gyroscopes for the R-11 and R-5 rockets (GG-5, GV-5, etc.). In 1953, the

SKB began to develop floating gyroscopes. For Korolev's R-7 and R-9 rockets the guidance system combined radio-guidance and an autonomous inertial system. But radio-guidance was abandoned in 1963. In September 1955 the organization became the institute of gyroscopic stabilization (NII GS) of the ministry of the naval industry (NII-944). A. Y. Ichlinsky, who had worked with Kuznetsov from 1946-1948, then returned as scientific director. Elected academician in 1960, Ichlinsky left NII-944 to take direction of the institute of mechanical problems of the Academy of Sciences in 1964.

NII-944 built the gyroscopes for most Soviet satellites and rockets. Kuznetsov was awarded the Hero of Socialist Labor Medal in 1956 for the R-5M and the Lenin Prize in 1957 for the first Sputnik. In 1958 he became doctor of technical sciences and a corresponding member of the Academy of Sciences. On October 24, 1960, while he was near the R-16 missile, he moved away for a cigarette in the company of a few people (including M. K. Yangel and A. G. Mrykine). It was then that the rocket exploded, killing 76 people instantly (including Marshal M. I. Nedeline and deputy minister L. A. Grishin), and 16 others who later died in hospital. In 1961 Kuznetsov received the Hero of Socialist Labor Medal for Vostok 1.

In 1966 the company became the institute of applied mechanics (NII PM) of the ministry of general machines and series production was handled by factory N°706. In 1951 part of the production activities were transferred to factory N°897 in Kharkov. Production also started in Saratov factory N°205, PO Zvezda in Ostashkov (1958), the electromechanical plant of Berdsk (1959), the electromechanical NII of Miass (1959), the Omsk electromechanical plant (1960), the Tomsk plant (air suspended gyroscopes), the Metallist plant of Serpukhov, Ramenskoe apparatus plant, factory N°308 of Kiev, etc.

In 1967 Kuznetsov received the State Prize for the UR-100 missile. The following year he became a full member of the Academy of Sciences. Then he was a professor at MVTU starting in 1973. The institute was transformed into NPO Rotor in 1978, and was successively led by K. I. Mikhailov, V. V. Lapchine, R. R. Kiriouchine, and Y. S. Sarymov. It included NII PM, factory N°706, and the Zvezda subsidiary which was the leader in floating gyroscopes. V. I. Kuznetsov died on March 22, 1991[30] and with the collapse of the Soviet Union the NPO was dissolved. Afterwards the NII PM became part of TsENKI.

## LAUNCH PLATFORMS

Vladimir Pavlovich Barmine was born on March 17, 1909 in Moscow to Pavel Ivanovich and Maria Issaievna. After secondary school he passed three competitive entrance exams for the Bauman technical school (MVTU), Lomonosov mechanical engineering institute and the Mendeleiev chemical-technological institute in 1926. In the first semester, he took courses in two institutes. But in the second he attended only Bauman, where he graduated in 1930 with an engineering diploma relating to cooling systems. He then joined the Kotloapparat plant, which produced refrigerators and in 1931 became the Kompressor factory. At the same time, he gave courses in thermodynamics at MVTU. He married Lydia Ivanovna, with whom he had two sons: Vladimir born in 1934 and Igor in 1943. He subsequently remarried Svetlana Alexandrovna.

In 1933 he began directing the study group on compressors at the design bureau. In 1935 he developed the UG-160 compressor used for the refrigeration system for Lenin's mummy in the mausoleum of Red Square. At the end of the year he was sent to the United States as a specialist in a group from the department of machinery. On its return in May 1936, this delegation was a natural target for the Stalinist purges. Barmine was the only member of the delegation not to be imprisoned as an "enemy of the people". In late 1940 he was appointed main designer. When the war broke out on June 22, 1941, it was decided that the Kompressor plant would produce launch platforms for the Katyushas solid-fuel rockets. Barmine was then appointed head of the SKB and deputy to A. G. Kostikov, who was the main designer of Katyushas in NII-3. However, the two men were unable to work together. Kostikov complained to the secretariat of the central committee, calling for Barmine's ouster. In 1937-1938 Kostikov had already been instrumental in the execution of I. T. Kleimenov and G. E. Langemak (his seniors at NII-3) and the imprisonment of S. P. Korolev and V. P. Glushko (fellow engineers). But minister P. I. Parshin formed a commission which concluded that Kostikov should concentrate on NII-3 with Barmine as main designer. During the war approximately 3,000 launch platforms were sent to the front. For this, Barmine received the Stalin Prize in 1943.

In mid-1945, Barmine was sent to Germany to study the German secret weapons (V-2, V-1 Wasserfall, Schmetterling). He was appointed chief engineer of the Berlin

Barmine in Germany in 1945.

Barmine.

institute. On May 13, 1946, the decree creating the rocket industry in the USSR transformed Barmine's SKB into GSKB SpetzMach, in charge of development and production of launch facilities for ballistic rockets. It first built launch complexes for the V-2 and its Russian counterpart, the R-1, developed by Korolev. Then it built facilities for Korolev's R-2, R-5 and R-5M. The latter, carrying a nuclear warhead, was the first strategic rocket in operational service. Barmine received the Hero of Socialist Labor Medal in 1956 for this achievement.

In 1951 Barmine was put in charge of the launch platform for Lavochkin's V-300 missile. Derived from the German Wasserfall, this rocket was part of Moscow's anti-aircraft defense system (S-25 Berkut). In 1953 he received the order for the launch installation of the intermediate R-12, developed by M. K. Yangel's design bureau in Dnepropetrovsk, Ukraine. In 1954 Barmine received the order for Korolev's R-7 intercontinental missile launch facility, which was built at Baikonur between 1955 and 1957. The first successful flight occurred in August 1957, followed by the launch of the first Sputnik in October. For this, Barmine received the Lenin Prize in 1957. In 1960-1961 a second platform was assembled in Baikonur, followed by four others at Plesetsk. These were maintained on alert for two months during the Cuban missile crisis in September-October 1962.

Barmine was elected a corresponding member of the Academy of Sciences in 1958 and became a full member in 1966. He became a doctor of technical sciences in 1959, and in 1960 was granted the title of professor at MVTU. In 1961, unlike his comrades on the council of chief designers – S. P. Korolev, V. P. Glushko, N. A. Pilyugin and V. I. Kuznetsov – but like M. S. Riazansky, Barmine did not receive the second Hero of Socialist Labor Medal for Gagarin's flight, only the Order of Lenin.

In 1958 Barmine was ordered to build the launch installation of Yangel's R-14 intermediate range missile. Following a decree of May 1960, he developed the first silos for the R-12U and R-14U missiles (respectively first launched in late 1961 and early 1962). The R-12U silo was transformed for the launch of the Kosmos satellite (Mayak 2 platform and two Dvina silos in Zone 86 at Kapustin Yar). In November 1962 he got the order for the launch installation of Korolev's R-9A intercontinental missile. This was the only launch vehicle to be deployed in silos, even though it used a mixture of non-storable kerosene-liquid oxygen, unlike the Yangel and Chelomei missiles which used storable UDMH-$N_2O_4$. An accident in a Baikonur R-9A silo in October 1963 caused the death of eight soldiers. Then Barmine began working with V. N. Chelomei on the UR-100 and UR-500 intercontinental missiles. In April 1963 he received the order for the launch platform and silos for the UR-100 of which a thousand units would be deployed in the Soviet Union, earning him the State Prize in 1967.

In April 1962 Barmine was ordered to build the launch facility for the UR-500. This was conceived as an intercontinental missile to carry a warhead with a yield of 50 megatons, but when this type of weapon was halted the project was abandoned. M. V. Keldysh, president of the Academy of Sciences, saved the development of the rocket by assigning it a scientific payload: the Proton satellites to study of cosmic rays. Then in 1964 the UR-500 was selected as the launcher for the Almaz military orbital space station. This was the basis for the Salyut civil and military stations, as well as the modular Mir station and the Zvezda module of the International Space Station. In 1965 the UR-500 was selected to launch the L-1 (Zond) spacecraft which was to carry pairs of cosmonauts on circumlunar trajectories. For this, a four-stage version was developed and tested in 1967. It was this same version that launched the second generation interplanetary probes (Luna, Venera and Mars) and geostationary telecommunications satellites. Today a dozen units are launched each year. For this work Barmine received the State Prize in 1985.

In 1964 Barmine received the order for launch facilities for the N-1/L-3 program that was to land a cosmonaut on the Moon. After four failed launches between 1969 and 1972 the program was abandoned in 1974. For the November 1967 Galaktika planetary exploration program Barmine designed facilities for Chelomei's UR-700 super-rocket that was intended for manned missions to the Moon (LK-700) and to Mars (MK-700), but the UR-700 was abandoned. In March 1968, in the context of Galaktika, Barmine received the contract for the Kolumb lunar base project. A report was filed in December 1969 and work continued for a few years, then abandoned. In addition, in 1969 KBOM received an order for drilling equipment to take samples from Venus and the Moon. For this, it opened a subsidiary in Tashkent. The LB-09 device was used by the Luna 23 and Luna 24 probes which returned their samples to Earth, and the VB-09 device was used on the Venera 13 and Venera 14 probes which examined their samples in situ. For this work Barmine was awarded the State Prize in 1977. In December 1971 he got an order to build silos for endo-atmospheric and exo-atmospheric missiles of the A-135 anti-missile system to defend Moscow. This ABM work was in line with that done on the S-25 Berkut system in the 1950s. Then in 1974 he studied launch facilities for RLA-120, RLA-130, RLA-140 and RLA-150 super-rocket projects. In 1976 he received the order for the Energiya-Buran launch facilities. The first flight of the Energiya rocket took place on May 15, 1987, while Buran flew on November 15, 1988. Upon the collapse of the Soviet Union in 1991 the entire program was abandoned.

In 1975 KBOM expanded with a department to develop instruments for space technology experiments (Splav, Zona, Konstanza, Kachtan, etc.). The following year Barmine began to work on ground facilities for cryogenic aircraft. The first of these, the Tupolev Tu-155, was equipped with NK-88 liquid hydrogen engines. The first flight took place in April 1988. Then in 1989 the aircraft was converted to liquefied natural gas (LNG). In 1994 the decision was made to equip the Tu-156 with NK-89 LNG engines, but the program was canceled.

Vladimir Barmine died on July 17, 1993. He was replaced by his son Igor, who was born on January 12, 1943 in Moscow. Like his father, he graduated from MVTU (1966). He worked in V. P. Glushko's research bureau (KB EnergoMach) until 1974 then joined his father's company, where he was involved in the planetary exploration program, space technology experiments and launch facilities prior to taking over. He became a doctor of technical sciences in 1989, professor in 1994, and received the State Prize in 1983. He was replaced by A. V. Pravoslavnov in February 2009.

# 4

# Korolev's subsidiaries

At the end of the 1950s the government began to increase the number of companies dedicated to the production of rockets. Thus, in April 1955 it decided to build plant N°139 in Miass near Chelyabinsk to produce R-5M and R-7 missiles in series. But in March 1958 this project was abandoned and the plant became a duplicate of NII-88 and of its subsidiary in Zagork (later NII-229) and absorbed by V. P. Makeiev in October 1959. In February 1958, plant N°1001 KrasMach, in Krasnoyarsk, was affected by this move. The minister Dimitri Ustinov wanted an underground factory created to manufacture Korolev's rockets. In fact, an underground nuclear plant had already been built in the secret city of Krasnoyarsk-26. In the Soviet Union there were many so-called secret cities, or ZATOs, with numbers, such as Arzamas-16, Chelyabinsk-70, Sverdlovsk-44, Tomsk-7, Penza-19, Krasnoyarsk-26, and Zlatoust-36. A decree issued on April 1, 1959 called for the creation of subsidiary N°2 of Korolev's OKB adjacent to the KrasMach plant. The director was M. F. Rechetnev, chief constructor of the R-11M. Like Makeiev, he was 35 at the time. He succeeded him and made new versions of the R-11. Rechetnev's version was produced in series by plant N°47 in Orenburg. But in October 1959 the underground plant project was abandoned. The subsidiary would therefore not build the R-7 and R-9. Instead, in 1960 the subsidiary was assigned to produce M. K. Yangel's R-14 (which served as the basis for the Kosmos 3M space launcher). In December 1961 subsidiary N°2 was transformed into OKB-10. It specialized in telecommunications, navigation, and geodesic satellites, of which it has supplied more than 1,000! Similarly, on January 2, 1958 the government decided to transfer production of the R-7 and R-9 to Progress plant N°1 of Kuybyshev. On July 14, 1960 D. I. Kozlov's sector became subsidiary N°3 of Korolev's OKB.

In 1958 Korolev began working on solid-fuel rockets. In November 1959 a decree ordered the creation of the RT-1 missile. Korolev then absorbed the activities of neighboring OKB-1: institute N°58 (TsNII-58) of constructor V. G. Grabine. After the RT-1, Korolev developed three missiles: the RT-15 in cooperation with TsSKB-7 of the Leningrad arsenal factory, the RT-25 in cooperation with OKB-172 of the Motovilikhinsk factory near Perm, and the RT-2 in cooperation with both companies. It was planned to open a fourth subsidiary to produce this type of rocket at factory N°92 in Gorky, but the project was abandoned.

## DMITRI KOZLOV

Dmitri Ilyich Kozlov was born on October 1, 1919 in Tikhoretske in the Krasnodar region to a family of railway workers. He had two brothers who died at the front during the Second World War. At the outbreak of war, Dmitri was attending courses at the Leningrad military institute of mechanics. He then enlisted as a simple soldier and, after the blockade of Leningrad, shrapnel from a mine near Vyborg severed his left arm on July 12, 1944. He was demobilized and returned to the institute, from which he was evacuated to Perm. In May 1943 he became a member of the party. He finished his studies at the institute towards the end of 1945 with a diploma on a new detonator for artillery shells. He became an engineer, and was a member of the first group of 18 people who did a specialized four month course on rocketry. He was assigned to institute N°88 in Podlipki, which became the institute of rockets of the ministry of armaments on May 13, 1946. He moved near the institute with his wife Zoia Vassilievna, whom he had married in February. Their son Vladimir was born in 1947 (he later directed TsSKB Foton KB) and their daughter Olga was born in 1952. In June 1946 Kozlov went to Nordhausen in Germany, where his first task involved pneumo-hydraulic aspects of the V-2 engine in V. P. Glushko's group. For several months he served as party secretary for the group of specialists led by General Lev Gaidukov, and then for Korolev's group at plant N°3 in Kleinbodungen, which was reconstructing Wernher von Braun's V-2. In August, Kozlov became an engineer for sector N°3 of the OKB of NII-88 (which became OKB-1 in April 1950). He returned to Podlipki in December 1946 and participated in the V-2 flights at Kapustin Yar in

Kozlov.

Recovering Gagarin's capsule, from left to right: Burov (Kuybyshev communist party), Kozlov, Danilov (Saratov region), Voskressenky.

Entrance to TsSKB in Samara.

October 1947. In August 1948 he was group leader and head of the sector of general studies in the design bureau of K. D. Buchuyev. There, he worked on the R-1 which was operational in 1950 (8A11), the R-2 which was operational in 1952 (8J38) and then on the R-3 project in 1949.

In July 1950 Kozlov completed the advanced courses at MVTU (VIK). In June 1951 he became head of general studies and chief constructor for the R-5 (8A62). From 1951-1954 he was secretary of the party for OKB-1. In 1952 Stalin ordered the dismissal of Jews from leadership positions. A directive ordered dismissal of all the Jews from NII-88. Kozlov opposed this order, and three men were spared, including B. E. Chertok. In addition, in 1952 Kozlov and Y. A. Pobedonostsev contributed to

Korolev's joining the party. Pobedonostsev had worked with Korolev in GIRD, at RNII, in Germany, and at NII-88, and in 1950 he had become pro-rector of a special academy. Korolev was finally admitted on July 15, 1953. By that time Kozlov held the positions of constructor and party secretary. He was assisted by Oleg Ivanovsky, who succeeded him as secretary from 1955 to 1957.

During tests of the R-5 in April 1953 at Kapustin Yar, nitric acid splashed onto Kozlov's face. He was wearing glasses and washed his face with cold water, but a few months later he began to suffer impaired vision. After consulting with doctors, he underwent eye surgery. In April 1954 he was chief constructor of the R-5M which became the first Soviet strategic missile in May 1956. For this he received the Order of Lenin on April 20, 1956, along with K. D. Buchuyev, L. A. Voskresensky, S. S. Kryukov, V. P. Makeiev, S. O. Okhapkine, and B. E. Chertok – with Korolev and Mishin receiving the Hero of Socialist Labor medal. On August 14, 1956, OKB-1 became independent of NII-88.

When development of the R-7 was approved on May 20, 1954, Kozlov became the chief constructor. He was involved in R-7 development and flight testing. On December 18, 1957 he received the Lenin Prize for the launch of the first Sputnik. Order N°55ss issued by the ministry on February 21, 1958, appointed him to direct production of the rocket in Kuybyshev. Then he was appointed deputy of the main designer on April 1, 1958. On February 28, 1959 he received the title of candidate of technical science. On March 5, letter N°863ss from Korolev confirmed the existence of the department at Kuybyshev. And on July 23 order N°74 designated it as section N°25 of OKB-1, entrusted to Kozlov. At the time, it employed 100 people. This date is now considered to be the starting point of TsSKB. Order N°1 signed by Kozlov on November 17 designated A. I. Apeksimov as director, and P. A. Gridchin and L. F. Chumny as assistants. Order N°25ss issued by Korolev on July 25, 1960 called for its transformation into a subsidiary. This was confirmed by departmental order N°413ss on August 31. The section became subsidiary N°3 in charge of the production of the R-7. It employed 176 people. On July 29, 1960 Kozlov received a second Order of Lenin, while V. Y. Litvinov received his second Hero of Socialist Labor medal. On January 11, 1961 Kozlov became deputy to Korolev, the chief and main designer of subsidiary N°3. On April 12, the day of Yuri Gagarin's flight, Kozlov was present. Moreover, after the first man in space was recovered near Saratov, he was first taken to Kuybyshev to meet the staff of the company prior to being sent to Moscow for a triumphant reception. On June 17, Kozlov was awarded the Hero of Socialist Labor Medal for Gagarin's flight.

After Korolev's death, Kozlov became V. P. Mishin's first deputy. Decree N°758-316 of September 8, 1966 made him director of TsKBEM's Kuybyshev subsidiary, as of January 1, 1967 (mail box N°G-4213). That year, he began teaching at the Kuybyshev institute of aviation (later Samara Aerospace University). He initially directed the chair of flight dynamics and guidance systems from 1967-1970, then the chair of flying machines from 1980-1989. On January 16, 1970 he received the title of doctor of technical sciences.

On July 30, 1974, ministry order N°252 transformed the subsidiary into TsSKB, which became independent from NPO Energiya. In 1976 Kozlov received his first

Soldatenkov.

Anshakov.

Kozlov and Anshakov at Samara in 1999.

State Prize for his contribution to the Apollo-Soyuz mission. In 1979 he received his second Hero of Socialist Labor Medal. In August 1983 he was promoted from main designer to general designer. That same year he received his second State Prize, this time for the Yantar spy satellite. In 1984 he was elected a corresponding member of the Academy of Sciences. After dissolution of the Soviet Union, Kuybyshev resumed its former name of Samara. In 1994 Kozlov received his third State Prize. On April 12, 1996, TsSKB merged with the Progress factory and became the TsSKB-Progress State Center, also known as "Samara Space Center". Kozlov was director general and general designer. That year, he received the S. P. Korolev gold medal from the Academy of Sciences. Created in 1966, this award was given to V. P. Mishin, M. K. Yangel, V. P. Makeiev, N. A. Pilyugin, V. I. Kuznetsov, O. M. Belozerkovsky, and V. F. Utkin.

Kozlov retired in 2003 and died on March 7, 2009. His deputies included A. M. Soldatenkov, V. M. Saigak, G. P. Anchakov, G. E. Fomin, Y. G. Antonov, A. V. Chechin, V. I. Krainov, V. A. Kapitonov, and M. F. Chum. Soldatenkov graduated from the Kuybyshev aviation institute in 1951 and then worked at plant N°1 before arriving at TsSKB in December 1959. He was initially chief constructor. In 1961 he became deputy, and then first deputy to Kozlov. By 1964 he was technical director for flight testing, and directed the launch of some 1,000 Semyorkas during his career. He received the Order of Lenin in 1966 for the Voskhod program and the State Prize in 1976 for Apollo-Soyuz. From 1983 to 2006 he was deputy general designer and main designer for rockets. In 1987 he received the Hero of Socialist Labor Medal. Saigak graduated from the University of Dnepropetrovsk in 1960 and joined TsSKB, where he worked on satellites. In 1974 he was a main designer for a specific subject. In April 1985 Saigak became deputy chief and deputy general designer at TsSKB. He was a candidate of technical science, Hero of Socialist Labor, and a recipient of the State Prize in 1971. He died in 1996. Anchakov graduated from the

Kuybyshev institute of aviation in 1961 and joined TsSKB, rising to become deputy to Kozlov for satellites in 1979 and then first deputy in 1985. From 2003 to 2005 he had three jobs: first deputy to the director general, general designer, and then director of the company. In January 2006 he was appointed deputy general designer. He became a doctor of technical sciences in 1983 and a professor. He got the Hero of Socialist Labor in 1983, the Lenin Prize in 1988, and the State Prize in 1977. He became a corresponding member of the Academy of Sciences in 1991. Fomin graduated from the Moscow institute of aviation in 1956 and joined plant N°1 prior to arriving at TsSKB in 1960. He participated in the preparation of launches at Baikonur. In 1964 he began designing new versions of the Semyorka, including for Apollo-Soyuz, for which he was awarded the Order of Lenin. He also worked on the creation of the company's satellites. In the 1990s he directed the cooperative program with François Calaque of Aerospatiale and went on to initial the document which created Starsem in August 1996. He retired in 2006.

Today, TsSKB-Progress is led by A. N. Kirilline, who succeeded Kozlov in 2003. He graduated from the Kuybyshev institute of aviation in 1973 and became director in Samara on December 26, 1996. R. N. Akhmetov has been first deputy and general designer since December 2005. The first deputy for economy and finance has been B. N. Melioransky since 1997. S. V. Tiuvelin has been the chief engineer since 2008.

## "PROGRESS" PLANT N°1

The factory which began series production of the Semyorka in 1958 was created at Khodinka in Moscow in 1894. Initially it produced bicycles for the German company Meller. But aviation was in the ascendancy at that time and in 1909 it became the Dux plant, which produced aircraft under license from Farman and Nieuport of France. After the revolution it was renamed State Aviation plant N°1 on February 19, 1919. It subsequently changed names several times, becoming GAZ-1 imeni Aviakhim in 1925, plant N°1 imeni Stalin in 1941, the Progress plant in 1962, and plant imeni Ustinov in 1984.

In 1922 it began to produce aircraft developed by constructor N. N. Polikarpov, and by 1940 it had delivered almost 15,000 flying machines of 23 different types. A special constructions division led by A. Y. Cherbakov began in 1935. He developed stratospheric gliders. On February 28, 1940 he made the first flight of Korolev's RP-318 jet-powered glider from Podlipki airfield (later the site of RKK Energiya). The engine was a nitric acid RDA-1-150 by L. S. Dushkin. Cherbakov then studied the G-14 project for an RDK-1-150 liquid oxygen engine. Likewise, the engineer I. A. Merkulov manufactured his first ramjet at plant N°1 in 1939. Its first flight powered the VR-3 rocket on March 5, 1939. In turn Cherbakov proposed the IVS fighter plane with an M-120 engine and two Merkulov ramjets in 1939-1940. During the war he led an OKB in Orenburg that improved the U-2 aircraft from 1941-1943, then built 550 units of the Che-2 from 1943-1946. After the war he proposed the VSI vertical-landing-and-takeoff airplane using two Rolls-Royce Nene engines, but this was not built. From November 1949 to October 1950 Cherbakov led the OKB of plant N°385 in Zlatoust, which should have produced the R-1, the Soviet version of the V-2, in series. He unsuccessfully attempted to equip the rocket with wooden tanks. Then he worked on a project for an experimental winged rocket for Korolev from 1950-1953 and on the Buria missile for Lavochkin from 1953-1960.

In October 1941, with the Germans advancing on a broad front, the Soviet Union transferred a large part of its defense industry eastward. Plant N°1 imeni Stalin was moved to the site of the Bezymian NKVD camp on the outskirts of Kuybyshev (later Samara) where aircraft plants N°122 and N°295 and engine plant N°337 were under construction. These were replaced by plants N°1 (Progress), N°18 (Aviakor), and N°24 (Motorostroitel). Nearby were Bezymianka airfield and factories N°525 (later Metallist), aggregate N°35, hydraulic automation N°305 (later Gidroautomatika), airport equipment N°454, and aeronautical armor N°207 (later Salyut). During the war, plant N°1 made the Mikoyan-Gurevich MiG-3 (3,122 units) then the legendary Ilyushin Il-2 "Sturmovik" (11,863 units), and finally the Il-10 (1,225 units). It was directed by A. T. Tretyakov from 1941-1944, then V. Y. Litvinov from 1944-1962,[31] who received the Hero of Socialist Labor Medal in 1945 and in 1960, and the Stalin Prize in 1946 and in 1950. As the tide of the war changed in 1942 some plants were returned to Moscow, but plant N°1 remained permanently in Kuybyshev. Plant N°30 was transferred back to its former site in Khodinka. It became in turn Znamya Truda, MAPO imeni Dementiev, MiG-Mapo, and then VPK Mapo. After 1946 aviation meant jet propulsion. Plant N°1 produced the MiG-9 (10 planes built in 40 days), MiG-15 (713 units) and MiG-17 (392 units from 1952-1953), the Il-28 (50 units in 1953), the Tupolev Tu-16 (545 units from 1954-1960), and the Lavochkin La-250 (3 units).

On January 2, 1958 the government decided to transfer series production of the Semyorka from Podlipki to Kuybyshev. The rocket was assigned to "Progress" plant N°1, and the engines of the core and lateral stages to plant N°24 "Motorostroitel". The objective was to build three flight models by the fourth quarter of the year. On February 28, D. I. Kozlov arrived to organize production. The government focused particular attention on this program. D. F. Ustinov visited the plant in April 1958, as did N. S. Khrushchev in August 1958 and L. I. Brezhnev in April 1959. On August 16, 1958, the military industrial commission (VPK) reported upon the delays in execution of the January 2 decree. Measures were taken to maintain the production target. A ministry commission inspected the premises in November. The problems were overcome. The first rocket was assembled and tested in November, then sent to Baikonur. Two other rockets were manufactured in December. The engine of the third model arrived on December 17 and tests of the rocket were completed at 11:00 pm on the last day of the year, just in time to meet the original target. The personnel had worked round the clock without a day off. The first Kuybyshev R-7 arrived at Baikonur on January 3. But while it was outside the assembly building a collision occurred between a locomotive and a railway car. The next day, the damage was evaluated. The chassis to which one of the lateral blocks of the rocket was attached, had been torn off of its bindings. A special commission was formed to make a series of specific checks of this block. The driver of the locomotive was severely punished. The rocket was finally launched on February 17, 1959.

Mounting the boosters.

Mounting the boosters.

Mounting the boosters.

Mounting the boosters.

Mounting the central core.

In 1957 the ministries were transformed into a state committee. The factories were placed under the responsibility of regional councils (sovnarkhoz). Only the design offices remained within the state committees. Plant N°1, which went by the name of "Mail Box N°208", belonged to the 3rd directorate of the Kuybyshev region, directed by I. T. Borisov from 1957-1962. In 1958 it was transferred to the 4th directorate, under the responsibility of L. E. Benkovich. Then from 1962 to 1965 Litvinov was president of the regional council. From 1965-1973 he was a deputy at the ministry of general machines.

On June 10, 1959 the VPK decided to produce the improved version of the R-7A (8K74) in parallel with the R-7 (8K71). In the meantime, on May 13, 1959, a decree

Mounting the third stage.

Mounting the third stage.

Mounting the fairing.

called for the creation of two new-generation intercontinental missiles: the Korolev R-9 and the Yangel R-16. The R-9 was a two-stage 81-ton vehicle. Both stages used liquid-oxygen-kerosene. The first stage was equipped with a 140-ton-thrust Glushko 8D716/RD-111 engine, while the second was equipped with a 30-ton thrust Kosberg RD-0106 engine. Twenty-three units were built in OKB-1's experimental plant from 1961-1962. Using a system of super cooling, the stored missile only lost 2-3% of its liquid oxygen per year. The first launch took place on April 9, 1961 from Baikonur, but the rocket exploded. Four launch facilities were developed at Baikonur: Site N°51, Desna-N, Dolina, and Desna-V. Starting in April 1961 the ground platform of site N°51 was used for tests. Three ground platforms at site N°75 (two Desna-Ns and a Dolina) were used for flight tests, starting in June 1962 for the Desna-Ns and in February 1963 for the Dolina (whose semi-automatic system enabled a launch to be prepared in 10 minutes). The three silos of site N°70 (Desna V) were used from September 1963. On October 24, 1963 an accident in a silo caused the deaths of eight soldiers. Three series of flight tests were conducted. In the first series there were only eight successes out of 20 launches. In the second there were 15 successes out of 25 launches. The third series had eight successes out of nine launches. Thus there was a total of 31 successes out of 54 launches. The rocket was declared operational on July 21, 1965. Series production was to have been in Krasnoyarsk, but it was transferred

Production of the R-9 missile at the Progress plant.

to Kuybyshev in 1962. About 100 rockets were produced in total. The missile was deployed in two regiments in Plesetsk and Kozelsk and a brigade at Bogadinsky near Tyumen. In all, 68 launches were conducted at Baikonur from 1961-1969 and 17 at Plesetsk from 1967-1975. Following the SALT agreements, the launch facilities were dismantled between 1974-1977.

In 1961 Korolev studied a new family of rockets based on the R-9A, including the 8K710 (R-7M) using 8D78/8D79 engines, the 8K711 (launcher), the 8K712 (ICBM), the 8K713 (GR-1 "global rocket"), and the 8K513 (anti-satellite). The 8K711 was to be capable of placing 5.8 tons into low Earth orbit. It used Molniya technologies, but did not meet the specifications because it was too heavy and was abandoned. The GR-1 was to launch an "orbital bomb". It was up against the Chelomei UR-200 and the Yangel R-36Orb. The first and second stages were to be powered by NK-9 and NK-19 engines made by N. D. Kuznetsov. Series production was to be at plant N°1 in Kuybyshev, but the development of the engines took longer than expected and the project was abandoned in 1964.

Plant N°1, in cooperation with plant N°18, also produced the first four stages of the N-1 lunar rocket which were then sent to a subsidiary in Baikonur.[32] Building the N-1 mobilized all of the companies in the region of Kuybyshev: factories N°1, N°18, N°24, N°525, N°207, the "Lenin" metallurgical collective, the Plastik plant, and the Syzran heavy machinery plant. Furthermore, on July 30, 1974 a subsidiary of NPO Energiya was created from a department of subsidiary N°3 of OKB-1 to produce the tanks for the Energiya-Buran system.[33]

Throughout most of its history, TsSKB-Progress employed 27,000-30,000 people. In 1999 it had a workforce of 20,500 but this included those involved in processing other products (machine tools, vodka, caramels, etc.). The staff actually assigned to the Soyuz rocket totaled 5,000 people, with the assembly hall employing 360 people. By 2005, there were no more than 17,500 people including 2,600 working at TsSKB.

## PLANT N°24 "MOTOROSTROITEL"

The plant which began series production of engines for the Semyorka in 1958 was established in Riga in 1895.[34] It was moved to Moscow in 1915, and in 1918 was named GAZ N°4 "Motor". In 1924 it absorbed GAZ-6 Amstro (formerly Salmson), and in 1927 it was merged with GAZ-2 "Ikar" to become plant N°24 "Frunze". In October 1941 the factory was evacuated to Kuybyshev, where it has remained. It has series produced aircraft engines,[35] ramjets (RD-900, RD-012U), and rocket engines (RD-107, RD-108, NK-31, NK-33, NK-39, and NK-43). It has a test bed in Vintai, some 70 km from Samara. In 1999 the factory employed 10,000 people to make 95% of the parts on-site (combustion chambers, turbopumps, control boxes, automation systems, etc.). Other parts were bought. The plant works 50% on rocket engines and

Monument at Motorostroitel plant at Samara.

Production of engines at Motorostroitel.

Production of engines at Motorostroitel.

50% on aeronautical engines. It masters all technologies. Its expertise encompasses metal foundry skills, batteries, mechanical engineering, laser technology, the use of plasma, and other applications.

In 1957, a design bureau (SKBM) opened.[36] In 2008 Motorostroitel, SNTK imeni Kuznetsov, and SKBM – which together employed 11,000 people – were integrated into the OboronProm Engine Corporation, which also included NPO Saturn, the Ufa engine plant, Perm Motors, etc. Due to significant financial difficulties, these firms received aid from the government in late 2008 and early 2009. NPO EnergoMach engines are supervised by specialists of the Samara subsidiary.[37] These include the RD-107, RD-108 (modifications 8D727, 8D728, 8D74K, 8D75K, 8D75M, 8D75F, 11D511, 11D512, 11D511PF, 14D21, and 14D22), and the RD-111/8D716 for the Korolev R-7/R-7A and R-9/R-9A rockets.

## TSSKB SATELLITES

Decree N°569-264 of May 22, 1959 called for the construction of an experimental variant of the Vostok spacecraft to be used in the development of a spy satellite and manned spacecraft. The spy satellite was named Zenit 2. Indeed, the Zenit NPO in Krasnogorsk produced the onboard optical cameras (main designers V. A. Beshenov, Y. V. Ryabushkin, V. V. Nekrasov, amongst others). The images were processed by the space imaging department of the army's GRU of the ministry of defense, headed by General P. T. Kostin.

The first Zenit 2 (11F61 N°1) lifted off from Baikonur on a Semyorka (8K72K) on December 11, 1961. It was equipped with an FTOR-2 photographic system (from the Krasnogorsk factory) which comprised an SA-20 camera with a focal length of 1 m for a ground resolution of 5-6 m, an SA-10 camera of 0.2-m focal length, a Baikal

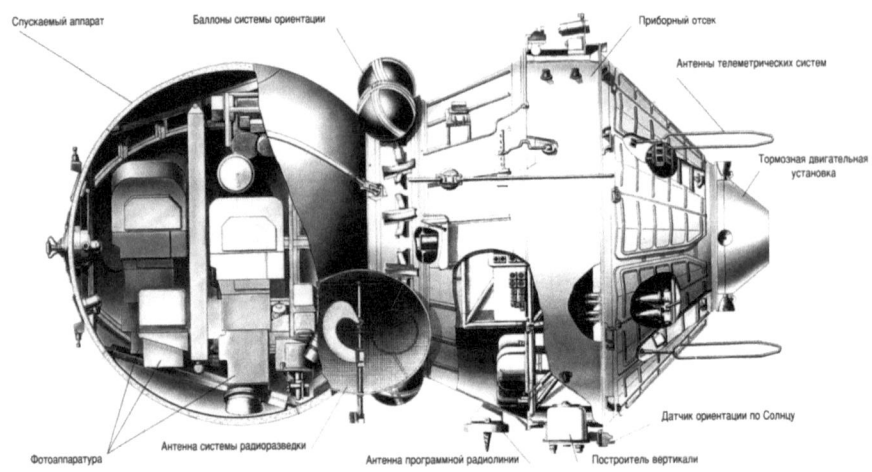

Diagram of the Zenit 2 satellite.

Preparing the Zenit 2 satellite.

television camera (NII-380), and a Kust 12M electronic intelligence device (TsNII-108). The remote control system was provided by NII-648 and NII-10, the telemetry by OKB MEI (Rubine and Tral), onboard timing by NII-33 (Liana), and the guidance system by OKB-1 (Chaika). The state commission was headed by General K. A. Kerimov.[38] However, the launch failed 407 s into the flight due to a fire in the third stage. Tubing burned between the gas generator and the turbopump. The top-secret spacecraft was destroyed in flight and the debris fell in Yakutia. After this failure, it was decided to transfer production of the third stage from plant N°88 in Podlipki to plant N°1 in Kuybyshev. Order N°37s on September 25, 1962 called for the transfer of the technical documentation from OKB-1 to D. I. Kozlov.

Recovery of a Zenit 2 capsule.

On April 26, 1962 Zenit 2 N°2 was put in orbit and named Kosmos 4. It carried out a 3-day mission, but a guidance failure prevented it from fully accomplishing its tasks and it only photographed 10 million sq. km. On June 1 the first Vostok-type Semyorka, 8A92 N°E15000-01, launched Zenit 2 N°3. But an 8D74 engine broke down after 1.8 s of flight and the booster fell back onto the launch platform. On July 28, Zenit 2 N°4 was orbited by 8A92 N°T15000-07 and flew a 4-day mission as Kosmos 7. This was the last Zenit 2 built by Korolev because decree N°470-174 of May 13, 1961, transferred operations concerning the spy satellite to Kuybyshev. Kozlov, a rocket builder, did not want to deal with satellites. General Kostin visited Kuybyshev unofficially to convince Kozlov. He was eventually compelled to endorse the decision, which was confirmed by decree N°963-124 on July 31, 1961. Kozlov had to ensure the production of the complex which included the 8A92 rocket and the Zenit 2 satellite. The first Zenit 2 built in Kuybyshev was launched on September 27, when 8A92 N°T15000-006 placed Zenit 2 N°5 into orbit. It carried out a 4-day mission as Kosmos 9 with the Baikal television system having been replaced by two cameras of 1-m focal length.

On October 17 Zenit 2 N°6 was launched by 8A92 N°T15000-03. It was equipped with four cameras, including three of 1-m focal length (180-km swath of terrain). It carried out a 4-day mission as Kosmos 10. On December 22 Zenit 2 N°7 was placed in orbit by 8A92 N°T15000-10 and flew an 8-day mission as Kosmos 12. On March 21, 1963, Zenit 2 N°8 was launched by 8A92 N°T15000-01 and carried out an 8-day mission as Kosmos 13. On April 22 Zenit 2 N°9 was orbited by 8A92 N°T15000-08 and flew a 5-day mission as Kosmos 15. On April 28 Zenit 2 N°10 was launched by 8A92 N°E15000-02 as Kosmos 16. It carried out a 10-day mission with loss of pitch orientation. On May 24 Zenit 2 N°11 was launched by 8A92 N°E15000-12 and flew a 9-day mission as Kosmos 18. The attempt to launch Zenit 2 N°12 on July 10 using 8A92 N°04-E15000 failed when the launcher exploded on the pad. Finally, Zenit 2 N°13 was placed in orbit on October 18 by 8A92 N°G15001-01, where it performed a 12-day mission as Kosmos 20. It was the last test flight before commissioning into the arsenal by decree N°210-87 of March 10, 1964. Ten launches had been planned for the test flights, but due to three failures there were thirteen in reality.

Of the Zenit 2s that achieved orbit, three had problems in flight: Kosmos 216 (the capsule landed in the Volga River 1 km from the shore, then sank after 42 minutes with the loss of 85% of the data), Kosmos 235 (following a failure of the parachute system, the hard landing resulted in the loss of 30% of the images), and Kosmos 344 (camera SA-10B broke down during the 42nd orbit and the mapping mission could not be completed). A total of 81 Zenit 2s were launched (51 from Baikonur and 30 from Plesetsk) including seven launch failures up to May 1970.

TsSKB developed seven variants by 1982. The first was the Zenit 2M (11F690), or "Gektor", which performed its maiden flight on March 21, 1968. It was launched by the Voskhod version (11A57) of the Semyorka, which enabled the payload to be increased from 4.7 to 6.3 tons. It had a Nauka (1KS) container to study high energy gamma emissions. It completed a 12-day mission as Kosmos 208. Flight tests were conducted through January 21, 1970, when it was declared operational. A total of 94 Zenit 2Ms were launched (41 from Baikonur and 53 from Plesetsk) up to March

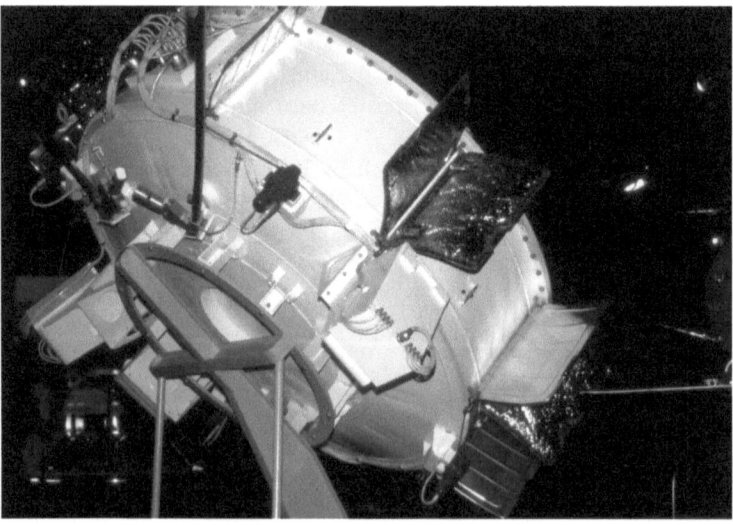

The Nauka container.

1979 with four failures. On October 8, 1970 the launch of a Zenit 2M was witnessed by French President Georges Pompidou as operation "Palma 4". It became Kosmos 368 and flew a 6-day mission. Its payload included a 1KS container and biological experiments (flies, weevils, cells of Syrian rats, human cells, cultures of carrot skin, ginseng, cabbage and crepis capillaris, onions, Chlorella algae, salad and pea seeds, various species of fungi and bacteria). Of 94 Zenit 2M flights between 1968-1979, 44 carried 600-kg Nauka scientific containers. For example Kosmos 243 and 384 carried the 2KS container equipped with a microwave radiometer to study the Earth. Kosmos 309 was equipped with the 3KS container to calibrate the Vega orientation system. Kosmos 410, 443 and 669 carried the 8KS container which had a cryostat Obzor radiometer to measure submillimetric emissions from water vapor in the upper troposphere and the stratosphere. Kosmos 561 and 731 had the 9KS container with gamma ray telescopes for the study of radiation. Kosmos 477 and 555 carried the 14KS container with the Spektr-Obzor-K experiment to study radiation. Kosmos 728 was equipped with the 15KS container to test a magneto-plasma-dynamic (MPD) electric motor using liquid metal (potassium). Kosmos 525 and 552 were equipped with the 16KS container (N°161/1L and N°162/2L) to test elements of the Yantar 2K spy satellite. Kosmos 635 included the 17KS container (N°M15000-171/1L) to test the star-sight and radio altimeter of the Yantar. The other Zenit 2Ms were equipped with a liquid-fuel engine for maneuvers.

A "civilian" version of the Zenit 2M was produced: the Zenit 2M/NKh "Gektor-Priroda". Five satellites were launched by the Soyuz-U launcher (Kosmos 741, 1102, 1106, 1118 and 1122) between May 1975 and August 1979. The images were turned over to the new Priroda state center[39] created by the main directorate of geodesy and cartography (GUGK) in September 1973. In 1974 the hydro-meteorological service (GUGMS) established the center for the study of the Earth's resources

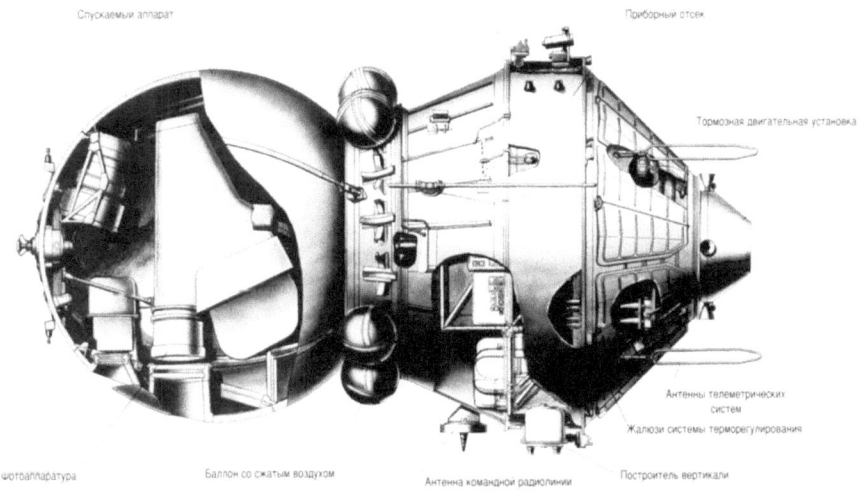

Diagram of the Zenit 4.

(GNITsIPR)[40] which became NPO Planeta in 1990. In April 1975 a commission for the study of the Earth's resources (IPRZ) was created at the Academy of Sciences.[41] Decree N°1087-472 of December 8, 1961, called for development of the Zenit 4 (11F69) by TsSKB, to be launched by the new version of Voskhod (11A57). It was equipped with a long focal length FTOR-4 camera (images of 30 x 30 cm with a resolution of 3-4 m) and a short focal length camera (a spool of 500 images of 18 x 18 cm). The first was placed in orbit on November 16, 1963 by rocket N°G15000-06 and flew a 6-day mission as Kosmos 22. Flight tests lasted through July 12, 1965, when it became operational. A total of 76 Zenit 4s were launched (43 from Baikonur and 33 from Plesetsk) up to August 1970, with four failures.

There were five variants: Zenit 4M (11F691), Zenit 4MK (11F692), Zenit 4MKM (11F692M), Zenit 4MT (11F629), and Zenit 4MKT (11F635). The Zenit 4M, known as "Rotor", had a liquid-propellant engine mounted at the front of the capsule so that it could maneuver. A pair of solar panels were mounted on the engine to increase the duration of the mission. It was launched by Voskhod. The first one was launched on October 31, 1968, and flew an 18-day mission as Kosmos 251. The FTOR-4 camera was replaced by an FTOR-6. Both this and the second mission (Kosmos 264) carried astronomy and gamma emissions experiments. The Zenit 4M became operational in 1971. A total of 61 were launched (29 from Baikonur and 32 from Plesetsk) up to July 1974, with four failures.

Zenit 4MK (11F692) "Germes" carried a camera with a resolution estimated at 1.5-2.0 m. The first was launched by a Voskhod rocket on December 23, 1969, and flew a 13-day mission as Kosmos 317. The Zenit 4MK was declared operational in 1972. Starting in August 1976 it was launched by the Soyuz-U (11A511U). A total of 81 were launched (28 from Baikonur and 53 from Plesetsk) up to June 1977, with two failures. The Zenit 4MKM (11F692M) "Gerakles" was first launched on July 12, 1977 as Kosmos 927. It entered service in 1978 with a camera that had a resolution estimated at 1.0-1.5 m. A total of 38 were launched (11 from Baikonur and 27 from Plesetsk) up to October 1980, with one failure.

The Zenit 4MT (11F629) "Orion" was intended for topographic mapping. It had an SA-106 camera, a laser altimeter, and an NPO Radiopribor Doppler system. The first launches used the new version of the Semyorka called Soyuz-M (8 launches) but in March 1975 it switched to the Soyuz-U (15 launches). The first unit was launched on December 27, 1971, and flew a 10-day mission as Kosmos 470. It was declared operational in 1976. A total of 23 were launched from Plesetsk up to August 1982.

The Zenit 4MKT (11F635) "Fram" was intended to succeed the Zenit 2M/NKh. The 6.1-ton 11F635 was launched by the Soyuz-U and had a multispectral Priroda 3 camera (five channels) with a resolution of 20-30 m in panchromatic mode or 30-50 m in color (180-km swath). The first was launched on September 25, 1975, and flew a 13-day mission as Kosmos 771. A total of 27 were launched from Plesetsk up to September 1985, with one launch failure. In May 1978 the capsule of Kosmos 1010 crashed on landing because the parachute did not open.

The Zenit 6 (11F645) "Argon" was operated in parallel with the Zenit 4MKM. The first was launched by a Soyuz-U on November 23, 1976, and flew a 13-day mission as Kosmos 867. To simultaneously ensure panoramic (survey) and detailed

The Resurs F satellite.

(close-up) missions, TsSKB developed a unified system. The Zenit 6 was integrated into the arsenal in 1978. A total of 96 were launched (28 from Baikonur and 68 from Plesetsk) up to June 1984, with two failures.

The Zenit 8 (17F116) "Oblik" was first launched on June 11, 1984. The Telegoir 12MK lens was replaced by a Rubin 77, T38-T film was introduced, and the control system enabled 40° rotation on the roll axis. It was declared operational in late 1984. A total of 100 were launched (30 from Baikonur and 70 from Plesetsk) up to June 1994, with one failure. The last one (Kosmos 2281) was renamed Resurs T.

The Zenit also led to three types of civilian application satellites: Resurs F for remote sensing, Bion (12KS) for medical-biological experiments, and Гoton (34KS) for microgravity experiments.

Resurs F1 (17F41) weighed 6.3 tons. The first was first launched on September 5, 1979, by Soyuz-U. It had a KFA-1000 camera with 1-m focal length and a resolution of 6-8 m in panchromatic mode (147-km swath), and three KFA-200 cameras with a focal length of 200 mm and a resolution of 20-30 m (225 km swath). It carried out a 13-day mission as Kosmos 1127. A total of 52 satellites were launched up to June 1994, including 28 of the 17F41 type, seven of the 14F40 type (including one failure) and 17 of the 14F43 type (including one failure) from Plesetsk. The Resurs F1M (14F43M) variant with a KFA-1000 camera and a KFA-200 was launched twice, in

The Resurs F2 satellite.

November 1997 and September 1999. Ten of the Resurs F2 (17F42) version with a multispectral MK4-camera that had a focal length of 300 mm and a resolution of 5-8 m were launched between December 1987 and September 1995. The Resurs F2M with a MK-4M and the Resurs F3 equipped with two KFA-3000 cameras that had a focal length of 3 m and a resolution of 2 m never left the drawing board. Finally, a satellite launched in November 1992 as part of the Europe-America-500 cooperative program was renamed Resurs 500.

A total of eleven Bion missions were produced for the institute for medical-biological problems (IMBP) of the ministry of health between 1973 and 1996. The first, Kosmos 605, was launched by a Soyuz-U on October 31, 1973. It carried 45 rats in nine Bios containers, each having five chambers, six turtles, and various flies, weevils, species of fungi and bacteria on a 22-day mission. The second, Kosmos 690, was launched on October 22, 1974. It carried 35 rats in seven Bios containers, each with five chambers, six turtles, flies, weevils, and various species of fungi and bacteria on a 21-day mission. The third, Kosmos 782, was launched on November 25, 1975. It flew a 20-day mission carrying 25 rats in five Bios containers, each with five chambers, a centrifuge with ten samples subjected to 1 g, four turtles subjected to 0.6 g, and other biological samples.[42] The other launches were Kosmos 936 on August 3, 1977, Kosmos 1129 on September 25, 1979, Kosmos 1514 on December

The Bion satellite.

The Foton satellite.

14, 1983 (with two monkeys, Abrek and Bion), Kosmos 1667 on July 10, 1985 (with Verny and Gordy), Kosmos 1887 on September 29, 1987 (with Erocha and Drioma), Kosmos 2044 on September 15, 1989 (with Jakonia and Zabiaka), Kosmos 2229 on December 29, 1992 (with Krosha and Ivashi), and Bion 11 on December 24, 1996 (with Lapik and Moultik). Two of the monkeys, Bion and Moultik, died shortly after landing, provoking a reaction among the animal protection societies. The launch of a Bion-M (6.4-6.8 tons) based on the Yantar-Kometa, is planned for 2012 as Bion 12.

As for Foton, fourteen units were launched between 1985 and 2007.[43] It carried furnaces for the production of alloys in microgravity and for the crystallization of semiconductors, etc. Because the program was open to international cooperation, Foton 11 was equipped with the German capsule Mirka. The flight in October 2002 was a resounding failure: the rocket exploded shortly after takeoff, killing a soldier on the launch team who was located nearby. A team from CNES was present (Denis Thirion, Catherine Ivanov, and others). The last two flights were mainly devoted to experiments provided by the European Space Agency.

In 1965, TsSKB participated in the preparation of manned flights in the context of the Voskhod program. In fact, Korolev had planned to do experiments on generating artificial gravity (IT). This involved linking the spacecraft to the Block I stage using a tether. The schedule was for Voskhod 5 with two dogs to make a 1-month flight in late 1965, Voskhod 6 with the IT experiment for a 2-week flight in late 1965, and a repetition with Voskhod 7 in February 1966. But development was more difficult than expected, the flight was postponed to Voskhod 8 in the second quarter of 1966, then canceled. Tethered experiments were carried out by the US Gemini program in 1966 using Agena target vehicles. In addition, in August 1965, Korolev initiated the Zvezda program with the 7K-VI (11F73) spacecraft, planning for the inaugural flight in 1967. This 6.6-ton spacecraft was too heavy to be launched by the 11A511 rocket, so the improved 11A511M version was developed. The prime contracting of work on the 7K-VI was entrusted to TsSKB in July 1966. In September, cosmonauts began to prepare for this (Popovich, Kolesnikov, Gubarev, Belussov, Artyukhin, Gulaiev). By mid-1967 a wooden model had been assembled in Samara but the first flight had slipped to 1968, and then in February 1968 the Zvezda program was canceled. In November 1967, V. P. Mishin proposed a new project to replace that underway. This Soyuz VI (11F730) complex included an OB-VI orbital space station (11F731) and a 7K-S (11F732) transport spacecraft which would be built in two versions: the 7K-S1 (11F733) for short missions, and the 7K-S2 (11F734) for long missions. There would also be a 7K-G (11F735) cargo craft (in effect the ancestor of the current Progress). In January 1968 TsSKB halted work on the 7K-VI to focus on the OB-VI. The first flight for this station was set for 1969 (later for 1970). Cosmonauts were assigned at Star City to train for this program from 1968-1970 (Popovich, Gubarev, Glazkov, Zudov, Stepanov, Sarafanov, Kramarenko, Kizim, Petrushenko, and Lissun). Three candidates from the anti-aircraft forces (Burdaiev, Alekseyev and Porvatkin) joined this group in 1969. But this program was abandoned in February 1970.

Decree N°715-240 of July 21, 1967 ordered TsSKB to develop the Yantar 2K, a completely new 6.6-ton spacecraft with a conical capsule instead of the spherical one of the Zenit. The spacecraft would remain in orbit for up to a month before returning the capsule containing the camera to Earth. It was also equipped with two small SpK spherical capsules to return exposed film to Earth as appropriate during the mission. With a diameter of 700 mm, the smaller capsules weighed 245 kg and could carry 29 kg of payload. The main engine was an 11D430 from KB KhimMach. It had a thrust of 2.7-3.3 kN and was capable of being fired 50 times over a period of 30 days. The capsules were equipped with an 11D864 solid-fuel engine from MKB Iskra which had a thrust of 34-59 kN. Subcontractors included the institute for instrumentation (NIIP, later RNIIKP), the institute of control apparatus (NIIKP), TsKB Geophyzika for the star camera, NPO Elas for the Salyut 3M computer, the Krasnogorsk plant (KMZ) for the Jemchuk 4 camera (500 kg) using a Mezon 2A 600-mm diameter lens with an estimated resolution of 1.8 m. The small capsule had been tested in flight by five Zenit 2M satellites: Kosmos 596 in October 1973, Kosmos 629 in January 1974, a launch failure in August 1974, Kosmos 692 in November 1974, and Kosmos 769 in September 1975. On the first flight, an electrical short circuit prevented the parachute compartment from opening and the capsule (or functional experimental installation) FEU-170 N°1L was lost. In the second flight

The Yantar 2K satellite.

the FEU-170-13 N°1702/2L capsule was retrieved. But the FEU-170 N°3L capsule was destroyed when the third stage of the rocket failed 338 s after launch. The FEU-170 N°4L and FEU-170 N°5L capsules were recovered successfully.

The first Yantar 2K (11F624 N°1) "Feniks" was launched on May 23, 1974 by a Soyuz-U from Plesetsk, but there was a premature separation of the third stage 439 s into the flight and the safety system destroyed the satellite. The second Yantar 2K was launched on December 13 and flew a 12-day mission as Kosmos 697, but it did not have the two small capsules. On completion of flight testing, the Yantar 2K was declared operational on May 22, 1978. A total of 30 were launched (7 from Baikonur and 23 from Plesetsk) up to June 1983, with two failures. The longest mission was 31 days. In three cases the missions were not completed: the explosions of Kosmos 758 in September 1975 and Kosmos 844 in July 1976, and a sealant failure on Kosmos 1079 in March 1979. Kosmos 758 did not have the small capsules but was equipped with the new Kondor attitude control system which broke down on the second day of flight. Kosmos 805 (the fourth Yantar 2K, in February 1976) was the first to carry small capsules. The main capsule was retrieved successfully, but the small capsules were lost: the first one suffered a parachute failure and crashed to the ground, and the engine of the second one did not ignite at the right time, causing it to come down in an unplanned area and it was not recovered.

Recovery of a Yantar 2K capsule.

Recovery of a Yantar 2K capsule.

The small capsule of the Yantar 2K.

The next version was the Yantar 4K1 (11F693) "Oktan". The Jemchug 18 camera had a resolution estimated at 1.2 m. The mission lasted 60 days. This version was launched on 12 occasions between April 27, 1979 and November 30, 1983 (6 from Baikonur and 6 from Plesetsk) and was declared operational on September 8, 1981. It was replaced by the Yantar 4K2 (11F695) "Kobalt", with a resolution estimated at 90 cm). The first one was launched by a Soyuz U on August 21, 1981, and carried out a 42-day mission as Kosmos 1298. It was integrated into the arsenal on October 21, 1982. Series production was entrusted to the St Petersburg arsenal plant in 1984, which delivered about 50 units. A total of 81 Yantar 4K2s were launched (23 from

Recovery of a small capsule.

Baikonur and 58 from Plesetsk) up to February 2002, with three failures. The longest mission lasted 134 days. Four missions were not completed, in all cases as a result of explosions: Kosmos 1654 in May 1985, Kosmos 1866 in July 1987, Kosmos 1916 in February 1988, and Kosmos 2030 in July 1989. Since September 2004 four Kobalt-M versions of the Yantar have been launched from Plesetsk as Kosmos 2410, 2420, 2427 and 2445.

The fourth version, the Yantar 1KFT (11F660) "Silouet" (later "Kometa") was to supersede the Zenit 4MT in mapping. It was a hybrid composed of a spherical Zenit capsule and a Yantar service module (i.e. the Ikar module equipped with a 17D61 engine). But the development and construction of the service module was protracted, lasting a decade. The Yantar 1KF project began in 1969, was abandoned in 1973 and was replaced by Yantar 1KFT. But the specifications requested by the military made the spacecraft too heavy for the Soyuz launcher. Using a Zenit or Proton rocket was considered, but in 1977 work was halted. The military changed their requirements, and the first Silouet was launched by a Soyuz-U on February 18, 1981. It carried out a 23-day mission as Kosmos 1246. The Jemchuk 104 photographic system included a panoramic Topaz (or KVR-1000) camera with an APO-Oktan 8 lens provided by the Krasnogorsk plant and a Yakhont 1 (or TK-350) topographic camera from BeLOMO (Minsk). The Topaz had a focal length of 1 m

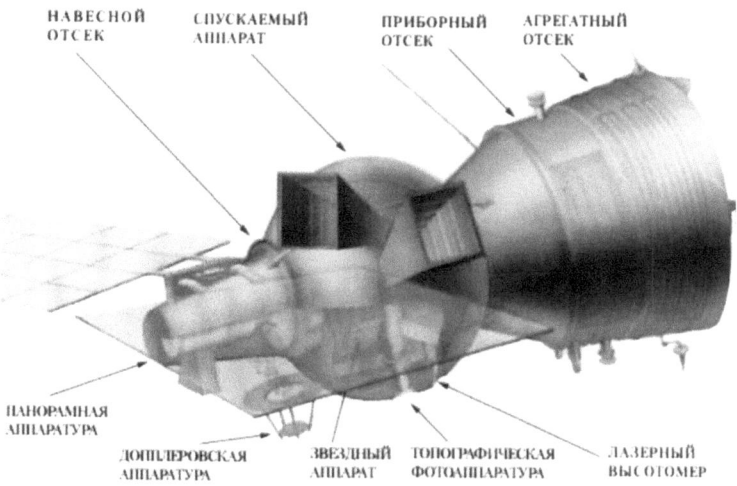

НАВЕСНОЙ ОТСЕК    СПУСКАЕМЫЙ АППАРАТ    ПРИБОРНЫЙ ОТСЕК    АГРЕГАТНЫЙ ОТСЕК

ПАНОРАМНАЯ АППАРАТУРА

ДОППЛЕРОВСКАЯ АППАРАТУРА    ЗВЕЗДНЫЙ АППАРАТ    ТОПОГРАФИЧЕСКАЯ ФОТОАППАРАТУРА    ЛАЗЕРНЫЙ ВЫСОТОМЕР

The Kometa satellite.

and a resolution of 2 m, while the Yakhont 1 had a focal length of 350 mm and a resolution of 10 m. In addition, there was a star camera, a laser altimeter, and a Salyut 3M computer. This spacecraft was integrated into the arsenal on July 22, 1987. A total of 21 Yantar 1KFT type were launched from Baikonur up to September 2005, with one launch failure. The longest mission lasted 45 days. Two missions were not completed: in April 1993 Kosmos 2243 suffered a loss of control, and in September 1994 Kosmos 2284 had a parachute failure that prevented the capsule from being recovered. In addition, the Kometa was used for the US-Russian Spin 2 program. Later, TsSKB proposed employing the Kometa for non-military roles (e.g. the Nika-Kuban for Earth observation and the Nika T for microgravity experiments) but none of these projects were pursued.

The next version was equivalent to the US "Keyhole" reconnaissance satellite which had the first CCD sensors and transmitted its imagery to Earth in digital form instead of returning film in a capsule. The first model was the Yantar 4KS1 (11F694) "Terilen". The first satellite was launched on December 28, 1982, and carried out a 67-day mission as Kosmos 1426. Shortly before, on May 18, the first "Geizer" type satellite-relay (known as "Potok") was launched into a geostationary orbit. A total of ten such satellites were launched up to July 2000. The Yantar 4KS1 was declared operational on January 21, 1986. A total of nine were launched by the Soyuz-U from Baikonur up to December 1990, with two failures. The missions lasted from 49 to 207 days. The Yantar 4KS1 was replaced by the Yantar 4KS1M (17F117) "Neman". The first was launched on February 7, 1986 as Kosmos 1731, and it had a resolution estimated at 30-40 cm. After four test flights, the Yantar 4KS1M was integrated into the arsenal on March 17, 1989. A total of fifteen were launched by the Soyuz-U from Baikonur up to May 2000. The missions lasted from 201 to 418 days. The last ceased functioning on May 4, 2001, and since then there have been no more Geizer satellite-relays available for image transmission. The civil version was the Resurs DK-1 (6.5

The Ikar module of the Kometa.

tons) launched by a Soyuz-U from Baikonur on June 15, 2006. It measured 2.72 m in diameter and 7.93 m in length, and was equipped with solar panels with an area of 36 sq. m. The Geoton L1 camera was provided by the Krasnogorsk plant (NPO Zenit), while the optronics came from Optex in Tselenograd (formerly NPO Elas). It had a resolution of 80 cm in panchromatic or 2-3 m in three-band multispectral mode. The satellite had 768 gigabits of memory and a downlink of 300 megabits per second. The transmission system was provided by the institute for precision instrumentation (NII TP) in Moscow. It also carried the MRI-Pamela Russian-Italian experiment to study the composition of primary cosmic rays and the Arina experiment to develop a methodology for predicting earthquakes. Its expected operational life was 3 years. The next satellite, Resurs P, is scheduled for launch in September 2012.

Development of the next generation started in 1979 and was set against the US Strategic Defense Initiative (SDI) "Star Wars" of the Reagan administration. TsSKB studied several programs, including "Zircon" (Yantar 4KS2), "Orletz" (Yantar 5K), and "Sapfir". A constellation of two or three satellites in a 300-500-km orbit, Sapfir had four objectives: imaging of stationary and mobile civil objects, operational global observation with monitoring of mobile nuclear delivery systems, monitoring of local crises, and global mapping. The satellites were to be launched by Zenit 2 rockets, then serviced and repaired in orbit by the Buran shuttle. But the project did not obtain the necessary financing, and the US-Soviet thaw of 1987 left it dormant (it was 70-75% ready). Sapfir was not officially abandoned until 1992. The Orletz project continued, and involved satellites which periodically released small capsules to return film to Earth. The first Orletz 1 (17F12) "Don" was orbited by a

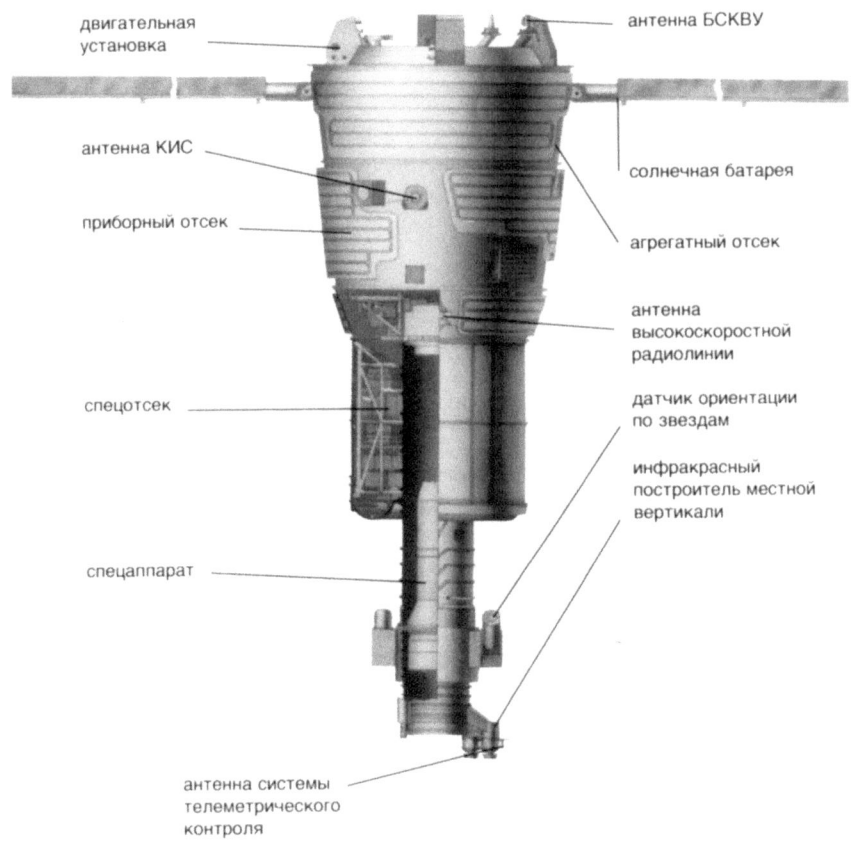

двигательная
установка

антенна БСКВУ

антенна КИС

солнечная батарея

приборный отсек

агрегатный отсек

антенна
высокоскоростной
радиолинии

спецотсек

датчик ориентации
по звездам

инфракрасный
построитель местной
вертикали

спецаппарат

антенна системы
телеметрического
контроля

Diagram of the Resurs DK satellite.

Soyuz-U on July 18, 1989. Its instruments include a mirror 900 mm in diameter. It has been operational since August 25, 1992. In total eight such satellites were launched (Kosmos 2031, 2101, 2163, 2225, 2262, 2343, 2399 and 2423) on missions lasting from 58 to 126 days. The Orletz 2 (17F113) "Yenisey" was launched by the Zenit 2 (11K77) because the device weighed 10.4 tons. Four models of rocket blocks were tested in flight between July 1986 and August 1987 as Kosmos 1767, 1786, 1871 and 1873. Three were in orbits inclined at 65° and the fourth at 97° (mass reduced to 9.6 tons). The missions lasted from 10 to 16 days with the exception of Kosmos 1786, which was placed into an orbit ranging in altitude between 149 and 1,054 km and fell into the atmosphere on March 6, 1988. Orletz 2 N°1L was launched in August 1994 and carried out a mission of 220 days as Kosmos 2290. Yenisey N°2L was launched in September 2000 as Kosmos 2372 and flew a mission of 207 days.

TsSKB's last spy satellite, Kosmos 2441, was launched from Plesetsk on July 26, 2008 by a Soyuz-2.1b. It was an upgrade of the Resurs DK satellite called "Persona". The primary instrument, a telescope with a large LOMO lens, was

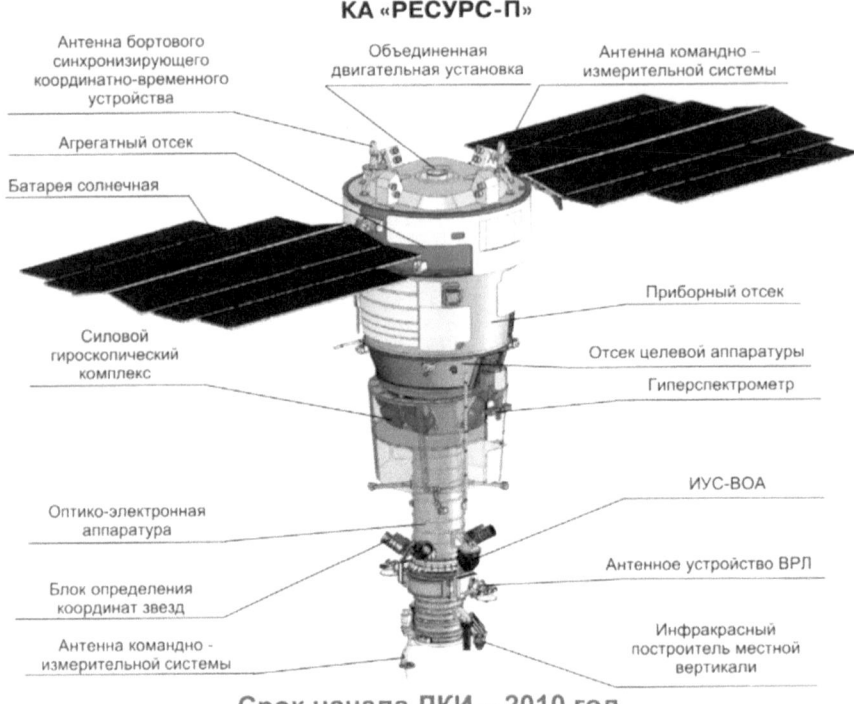

КА «РЕСУРС-П»

Антенна бортового синхронизирующего координатно-временного устройства

Объединенная двигательная установка

Антенна командно – измерительной системы

Агрегатный отсек

Батарея солнечная

Приборный отсек

Силовой гироскопический комплекс

Отсек целевой аппаратуры

Гиперспектрометр

ИУС-ВОА

Оптико-электронная аппаратура

Антенное устройство ВРЛ

Блок определения координат звезд

Антенна командно - измерительной системы

Инфракрасный построитель местной вертикали

Срок начала ЛКИ – 2010 год

The Resurs P satellite project.

derived from the Arkon (or Araks R) from NPO Lavochkin. The resolution was expected to be several dozen centimeters. Unfortunately the satellite failed soon after achieving an orbit at 98.3° that ranged in altitude between 714 and 732 km.

In addition, TsSKB developed the Liana electronic surveillance satellite based on the Kobalt-Resurs DK, intended to replace the Tselina for electronic intelligence and the US-P for monitoring the positions of naval forces. The first launch by a Soyuz-2 was in November 2009.

A total of 772 units of the many versions of the Zenit spacecraft were launched between 1961 and 2007 (including 29 failures), and 173 Yantars of all types have been launched since 1974 (including 8 failures), ten Orletz and one Persona, for an overall total of 956 spy satellites in 47 years. Of this total, 15 were built by Korolev's OKB, 50 by the Arsenal factory, and approximately 890 by Kozlov's OKB.

Today, Russia has five types of spy satellites: Yantar 4K2M "Kobalt-M" (last launch in June 2007), the Orletz 1 "Don" (last launch in September 2006), the Orletz 2 "Yenisey" (last launch in September 2000), the Arkon (last launch in July 2002), and the Persona (launched in July 2008).

# 5

# The various versions

On August 22, 1956, while the R-7 was still in development, V. P. Glushko proposed a version of the R-7 called the R-8 that would use new 100-ton-thrust engines to achieve a total of 1,000 tons of thrust at take-off. On February 12, 1960 he proposed four-stage R-10 and R-20 rockets with take-off masses of 1,500 and 2,000 tons that would use LOX-UDMH engines. In the meantime, Korolev began to study a nuclear-powered version. Glushko's OKB-456 and M. M. Bondariuk's OKB-670 developed nuclear/ammonia rockets. Work on this type of propulsion started as early as 1954 in laboratory V at Obninsk (physics-energy institute) and NII-1 (Keldysh center). It was decided to pursue solid phase reactors (model A) that could be finalized rapidly, as well as a gaseous phase version (model B) that would require longer research.

Korolev studied a modified R-7 of 850-880 tons at launch, a single-stage ICBM of 87-100 tons at launch, and a super-rocket of 2,000 tons at launch. The R-7 known as the YaKhR-2 would have been equipped with a model A engine delivering 140-170 tons of thrust. The four boosters would have 36 N. D. Kuznetsov engines. It could have put 35-40 tons into low Earth orbit. In 1959 OKB-456 developed the RD-401 project (using a water moderator) and RD-402 project (a beryllium moderator) with a thrust of 170 tons and a specific impulse of 428 s. The propulsion fluid was liquid ammonia. Then in 1962-1963 they developed the RD-404 project with 200 tons of thrust and a specific impulse of 950 s (liquid hydrogen as propulsion fluid) and the RD-405 project (zirconium hydride moderator and beryllium reflector) with 40-50 tons of thrust. However, in July 1963 OKB-456 ceased work on the model A project in order to begin work on the model B project.

## THE R-7/R-7A ICBM

The initial version of the R-7 (8K71) was launched 28 times with 19 successes and 9 failures; a 67.8% success rate. The improved R-7A (8K74) was also launched 28 times with 26 successes and 2 failures; a 92.8% success rate.

The rocket was 34 m high in version 8K71, 31 m in version 8K74, and 29.1 m in the Sputnik version (8K71PS). At launch it weighed 278 tons in the 8K71, 275 tons in

The nuclear powered R-7.

the 8K74, and 272.8 tons in the 8K71PS. The structure represented 7% of the mass, leaving 93% for propellant (i.e. 20 tons dry and 258 tons of propellant). The central core was 27.6 m high with a maximum diameter of 2.95 m. The boosters were 19.6 m high with a diameter of 2.68 m, and they were 3,915 kg empty. The entire bundle had a base diameter of 10.3 m. For the launch of the first Sputnik, the 273-ton rocket was reduced to 267 tons and the take-off thrust was 398 tons.

The RD-107 (8D74) engine (1,155 kg) had four combustion chambers supplied by a single group of turbopumps. In one second, it consumed 52 kg of liquid oxygen and 21 kg of kerosene. Oxygen was fed directly into the chamber, but the kerosene first went through the cooling circuit where it was heated to 210°C prior to being fed into the chamber. A total of 337 injectors produced the mixture. Ignition was provided by a pyrotechnic device. The gases in the chamber were at a pressure of 60 bars and a temperature of 3,250°C. The pressure fell to 0.4 bars and the temperature to 1,690°C at the nozzle aperture. The gas reached a speed of 2,950 m/s, with a thrust of 23 tons. The turbopump unit consisted of a 5,200 hp, two-stage turbine which spun at 8,300 rpm, two single-stage pumps for the oxygen supply (91 kg/s at 95 bars) and kerosene (226 kg/s at 80 bars), and two auxiliary single-stage pumps which were driven by a speed-increasing gear for the supply of liquid nitrogen (pressurization) and hydrogen peroxide (gas generator). The gas generator had a catalyzer to

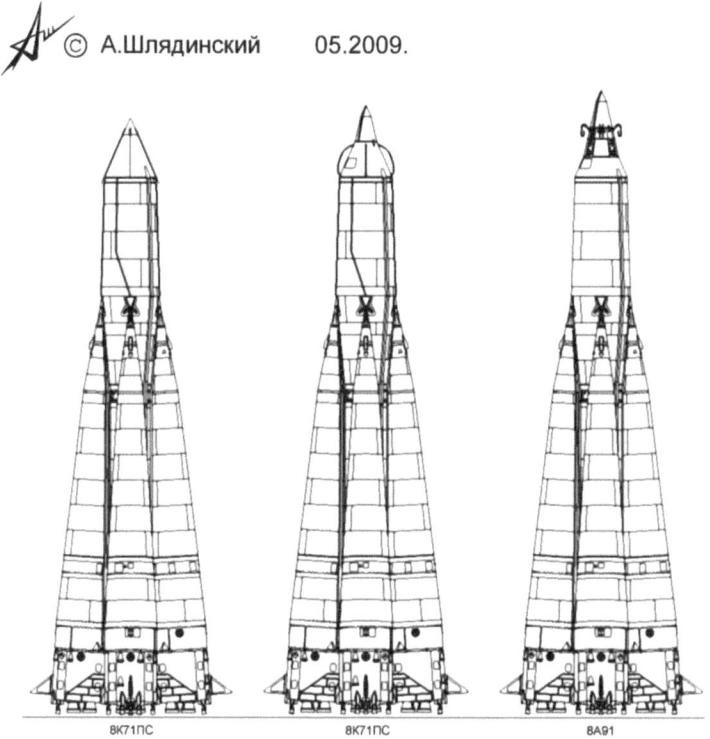

© А.Шлядинский        05.2009.

8K71ПС            8K71ПС            8A91

The 8K71 and 8A91.

dissociate the peroxide into water vapor and gaseous oxygen at 55 bars and 560°C. The 9 kg/s flow drove the turbine prior to entering the exhaust manifold as a gas at a pressure of 1.5 kg and 200°C. The exhaust gas exited at 450 m/s, was heated by the decomposition ejected by the turbine, and served to gasify the nitrogen that pressurized the fuel tanks. The engine was equipped with two vernier engines, with one-sixth the thrust of the main engines. One engine was installed in each of the lateral blocks of the Semyorka, and operated for 120 s. They were mounted at an angle of 3.5° on the longitudinal axis in order to be parallel to the axis of the launcher. The vernier engines could rotate to an angle of 45°. The RD-108 (8D75) engine (1,250 kg) had four vernier engines to steer the core block, and it ran for another 180 s until orbit was achieved.

Between the kerosene tank and the engines, two ring tanks were mounted: the first containing liquid nitrogen to supply the propellant tanks and the second containing hydrogen peroxide or perhydrol (82% concentration) to turn the engine turbine. The auxiliary pumps sent nitrogen to a heat exchanger and peroxide to the gas generators.

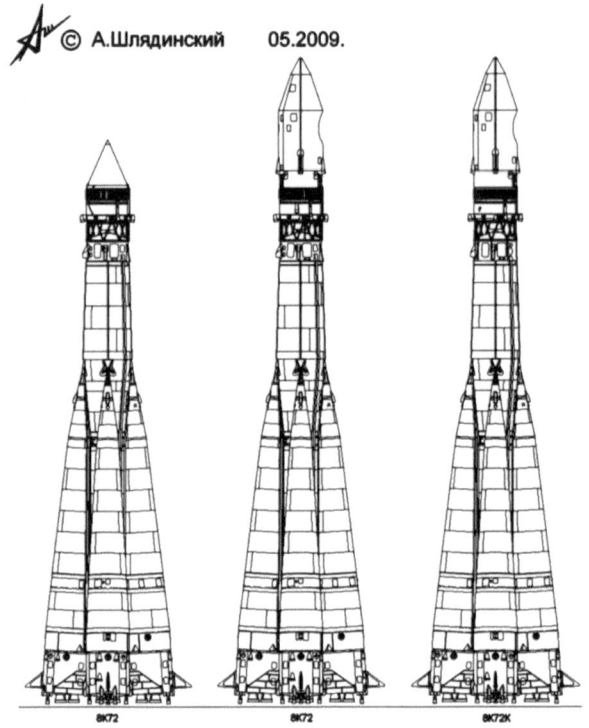

Various models of the 8K72.

## THE THREE-STAGE 8K72

In 1958 Korolev developed a third-stage called Block E. Two versions were planned: the 8K72 using an engine made by M. V. Melnikov's sector of OKB-1 and S. A. Kosberg's OKB-154, and the 8K73 using an engine made by Glushko's OKB-456. Glushko suggested the RD-109 (8D711), with a thrust of 10 tons. This engine had the peculiarity of replacing kerosene with UDMH for the first time. In fact, he had already tested a 7-ton-thrust engine of this type, and wanted to generalize its use. But development would have taken too long, and in 1960 he switched to the RD-119 (8D710) model and in December 1961 this was mounted on the second stage of the Kosmos 2 (11K63) rocket. The Block E was therefore equipped with Kosberg's RD-0105, developed between February and September 1958. This was the first Soviet engine ignited in a vacuum. It delivered a thrust of 5 tons and had a specific impulse of 316 s. The turbopump unit was parallel to the combustion chamber. The gas from the turbine powered the four roll-axis nozzles. The Block E used hot separation (it was affixed to the core using 12 supports, half of which had pyrotechnics), and was steered by a jet deflector.

The Block E third stage.

The RD-0105 engine of the Block E.

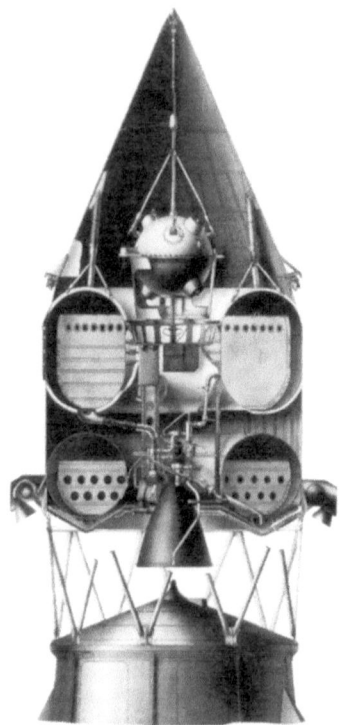

The Block E carrier of the Luna 1 probe.

An R-7 with a Block E launched the first Luna probes to the Moon (8K72L) and the first Vostok spacecraft into Earth orbit (8K72K). In the first case the guidance system comprised the KI-11-29 horizontal gyro, the I-55-11 vertical gyro, and the KI-12-18 and KI-12-19 speed sensors. The second case had the I-11-15, KI-55-16, and KI-12-20. The Block E also had an integrator (IG) N°I-22-8 while the spacecraft had a KI-00-8 orbital gyro and a KI-27-1 free gyro unit. Telemetry was provided by A. F. Bogomolov's Tral P1-1 and Tral P1-2 systems. The 8K72L weighed 279 tons, including 255 tons of propellants. The probe was designated as part of the Block E stage, making it a type E device. The non-steerable E1 which was studied by V. V. Molodtsov led to Luna 1 and 2 and the steerable E2 studied by V. K. Algunov led to Luna 3.

The first Luna (E-1 N°1) was launched on September 23, 1958 by 8K72L N°B1-3 but the rocket failed after 93 s of flight due to oxidant pressure surges at the intake of the engine pumps. A special commission consisting of S. P. Korolev, M. V. Keldysh, V. P. Glushko, N. A. Pilyugin, A. Y. Ichlinsky, B. N. Petrov, V. P. Mishin, N. A. Akkerman, N. D. Prokopov, G. S. Narimanov and V. A. Bokov was established to investigate this phenomenon. The following flight (N°B1-4) on October 12 with the E-1 N°2 probe failed 104 s into the flight. The third flight (N°B1-5) on December 4, with E-1 N°3, failed 245 s into the flight. The first success (N°B1-6) was achieved on

Launching a Luna probe.

January 2, 1959 with E-1 N°4 (Luna 1). This 361.3-kg probe flew within 6,400 km of the Moon and went into a solar orbit. It was equipped with five scientific instruments, and transmitted to a range of 600,000 km from Earth. On September 9, launcher 8K72L N°I1-7A carrying E-1 N°6 was sent back to the factory after five aborted launches. As 8K72L N°I1-7B, it successfully launched E-1 N°7 (Luna 2) on September 12. The 3,902-kg probe hit the Moon the next day in Palus Putredinis, located between the Apennine mountains and Mare Imbrium.

On October 4, 8K72L N°I1-8 successfully launched the E-2A probe (Luna 3). Equipped with a camera, this 278.5-kg satellite took some thirty images of the far side of the Moon from a distance of 60,000 km. As the probe returned towards Earth it transmitted the pictures using a facsimile system to Koshka, a Crimean mountain station. Luna 3 re-entered the atmosphere on May 20, 1960. On April 15, 1960, 8K72 N°I1-9 took off with the E-3 N°1 probe. But the Block E extinguished prematurely and ended up in an elliptical orbit of the Earth whose apogee fell short of the Moon. The probe was equipped with a 750-mm lens to take better photographs than Luna 3. The next day, 8K72 N°L1 9A was launched with the E-3 N°2 probe, but one of the lateral blocks of the bundle detached during launch while the other three remained attached, causing the vehicle to deviate from its trajectory by 150-200 m before the other blocks separated. Two blocks exploded immediately. The third fell towards the assembly building (MIK) and exploded at an altitude of 30-40 m. Unfortunately, the core block fell on the MIK and caused considerable damage. The Block E continued on its path for 800 m and exploded in a salt lake. The lunar program was halted until 1962.

Studies for the OD-1 steerable satellite were conducted by G. Y. Maximov. The return capsule was conical in shape. In 1958 K. P. Feoktistov took over the project and made the capsule spherical (the classic "Sharik"). The 4.7-ton spacecraft existed in an experimental form (1K), a spy satellite (2K), and a manned version (3K). The 8K72K rocket then measured 38 m in height and weighed 287 tons (including 258 tons of fuel) at take-off.

Unit N°L1-11 performed its maiden flight on May 15, 1960 carrying the 4,540-kg Vostok 1KP spacecraft (Sputnik 4) which had neither thermal protection nor a life-support system. After a 4-day flight the guidance system ceased functioning and the retrorocket accelerated the spacecraft rather than slowing it, causing it to move into a higher orbit where the cabin remained until October 15, 1965. The infrared guidance sensor on the local vertical axis had broken down owing to a spontaneous welding of metals in vacuum. The second flight (N°L1-10) on July 28, 1960 carried the Vostok 1K N°1 spacecraft. This had thermal protection, a life-support system, and a payload of two female dogs Lissichka and Chaika. But the rocket exploded at launch. The following flight on August 19 was a complete success. 8K72K N°L1-12 took off with the Vostok 1K N°2 (Sputnik 5) which carried the dogs Belka and Strelka. After the 18th orbit the capsule returned to Earth and landed within 10 km of the target in the area of Orsk. The dogs were ejected in the air and recovered successfully. On December 1, 8K72K N°L1-13 was launched carrying the 4,563-kg Vostok 1K N°5 spacecraft (Sputnik 6) with the female dogs Pchelka and Mouchka. The launch was a success but the recovery was a failure. The two dogs were incinerated during re-entry into the atmosphere. For the December 22 launch (N°13A L1) the Block E stage was equipped with the new RD-0109 (8D719) engine, but this malfunctioned when the Vostok 1K N°6 spacecraft reached an altitude of 214 km. It flew 3,500 km and then dropped near Tura in Yakutia. The separation of the capsule and the service module was not triggered by the thermal relay, but by either overheating or acceleration. Two hatches were successfully jettisoned, but the ejection seat carrying the two female dogs Jemchouzhnaya and Joulka was prevented from exiting the capsule because it had been deformed. A helicopter crew located the capsule and recovered the dogs.

A dress rehearsal for a manned flight went according to plan on March 9, 1961. 8K72K N°E10314 launched the 4,700-kg Vostok 3KA N°1 spacecraft (Sputnik 9) with the female dog Chernushka. After one revolution of the Earth she was retrieved 15 km northwest of the target in the Zainsk Kuybyshev region. A second rehearsal took place on March 25. 8K72K N°E10315 launched the 4,695-kg Vostok 3KA N°2 spacecraft (Sputnik 10), carrying the dog Zvezdochka. It made one revolution before being retrieved safe and sound 45 km southeast of Votkinsk near Izhevsk. With two successes it was decided to attempt a manned mission. That occurred on April 12. 8K72 N°E10316 launched the 4,725-kg Vostok 3KA N°3 spacecraft (Vostok 1) with Y. A. Gagarin on board. After a flight lasting 1 hour 48 minutes he was recovered at

Launch of Gagarin's Vostok.

Smelovka near Engels in the Saratov region. The risk was high, because the rocket and spacecraft had reliabilities of 0.56 and 0.60 respectively, whereas the reliability required for a manned flight was 0.99. The state commission was composed of K. N. Rudnev (president), K. S. Moskalenko, M. V. Keldysh, S. P. Korolev, V. P. Glushko, N. A. Pilyugin, V. I. Kuznetsov, M. S. Riazansky, V. P. Barmine, A. M. Isaiev, S. M. Alekseyev, G. I. Voronin, A. G. Zakharov, F. A. Agaltsov, N. P. Kamanine, V. I. Yazdovsky, and others. 8K72 N°E10317 launched the 4,731-kg Vostok 3KA N°4 spacecraft (Vostok 2) on August 6. After a flight of 25 hours 18 minutes, G. S. Titov was recovered at Krasny-Koul in the Saratov region.

On December 11, 1961, N°E10321 8K72K launched the first Zenit 2 (11F61 N°1) spy satellite, but it was a failure. On April 26, 1962, N°E10320 8K72K successfully launched the second Zenit 2 (11F61 N°2) as Kosmos 4. On August 11, 1962, 8K72K N°E10323 launched the 4,722-kg Vostok 3KA N°5 spacecraft (Vostok 3) carrying A. G. Nikolaiev. The next day, N°E10322 8K72K launched the 4,728-kg Vostok 3KA N°6 spacecraft (Vostok 4) carrying P. R Popovich. The two spacecraft were in similar but not identical orbits, so the separation varied. At one point they were just 5 km apart. They were recovered at the same time in the Karaganda region, Nikolaiev after a flight of 94 hours 22 minutes and Popovich after 70 hours 57 minutes.

On June 14, 1963 an 8K72K launched the Vostok 3KA N°7 spacecraft (Vostok 5) with V. F. Bykovsky. This had been scheduled for June 11, but was postponed due to a powerful solar flare. Two days later an 8K72K launched the Vostok 3KA N°8 spacecraft (Vostok 6) carrying V. V. Tereshkova, the first woman in space. They were recovered at the same time in the Karaganda region, Bykovsky after a record flight of 119 hours 6 minutes and Tereshkova after 70 hours 50 minutes.

Comparison of the 8K71, 8K72 and 8K78.

An 8K72 launched the Elektron 1 (350 kg) and Elektron 2 (460 kg) satellites on January 30, 1964, and then Elektron 3 and Elektron 4 on July 11. The satellites were designed by Korolev to study the radiation belts surrounding the Earth. A total of 26 8K72s were launched, including eight failures; a 69.2% success rate.

Another project studied in 1958 was to launch a satellite which would intercept a target at an altitude in the range 300 and 1,000 km. This anti-satellite (ASAT) project was drawn up jointly by Korolev for the launcher, by aircraft maker A. I. Mikoyan for the ASAT (1 ton), and by G. V. Kissunko (main designer of OKB Almaz missile systems) for the radar. This was why the acronym of the project was KOMIK. It was proposed to Khrushchev on November 16, 1960, but in June he had entrusted a space interceptor program to V. N. Chelomei.

## THE FOUR-STAGE 8K78 "MOLNIYA"

On December 10, 1959, decree N°1386-618 "On the development of the study of cosmic space" ordered the creation of a launcher with four stages for the exploration of the Moon (Object E6) Venus (Object 1V) and Mars (Object 1M). It also created the space committee comprising M. V. Keldysh, S. P. Korolev, A. A. Blagonravov, K. D. Bouchouyev, V. P. Glushko, M. S. Riazansky, N. A. Pilyugin, M. K. Yangel, G. A. Tiouline, and V. P. Barmine.

Korolev designed the 8K78 between January 15 and May 10, 1960, and then on February 28, 1960 he completed the plans for the 1M probe. The decree of June 4, 1960 "On the cosmic space plan" called for creation of the 8K78 to launch the Mars probes in August-September 1960. Then decree N°999-414 of September 9, 1960, defined the composition of the state commission for the 1M program.[44] The 8K78 was 43.4 m in height and weighed 306 tons at take-off. The R-7A was equipped with 8D74K/8D728/RD-107MM and 8D75K/8D727/RD-108MM engines with two levels of intermediate thrust instead of three. The Block I third stage was equipped with S. A. Kosberg's RO-9/8D715K/RD-0107 engine delivering 30 tons of thrust, and the Block L fourth stage with M. V. Melnikov's 11D33/S1-5400 engine. Like the Block E, these two stages used hot separation. The S1-5400 was the first engine to use the integrated flow configuration that had been tested by NII-1 in 1958. It had a thrust of 6.8 tons with a specific impulse of 340 s. The RD-0107 underwent two tests at NII-229's N°2 bench at Zagorsk on August 22 and 31, 1960; the S1-5400 was ignited six times in August-September 1960.

The Block I stage was 6.7 m in height, 2.66 m in diameter, and weighed 25.3 tons

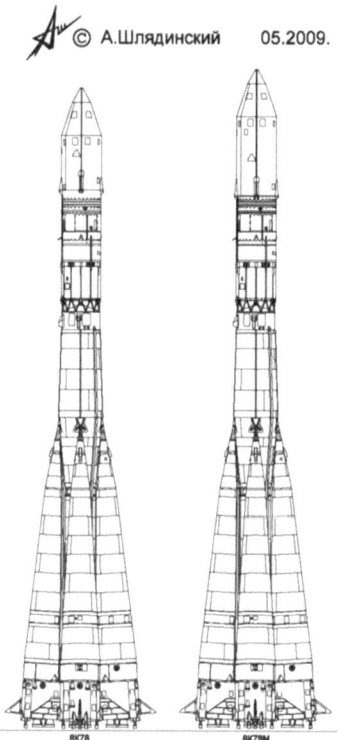

Diagram of the 8K78 Molniya.

Block L.

(including 22.8 tons of propellants). The RD-0107 included a turbopump unit (with a powder starter), a gas generator powered by LOX-kerosene, four main combustion chambers (with pyrotechnic ignition), and four 600-kg-thrust vernier engines. The tanks were supplied by a liquid oxygen gasifier to pressurize the oxygen tank, and a heat exchanger that cooled the gas produced by the turbine in order to pressurize the kerosene tank. The engine operated for 250 s. Flight steering was managed by the Block L guidance system.

The Block L stage was 3.5 m in height, 2.35 m in diameter, and weighed 6.9 tons (including 3.7 tons of propellants). The fuel was carried in ring tanks. The S1-5400 engine was placed in a gimbal to provide pitch and yaw control. For the roll axis a nozzle directed the hot gas from the gas generator. Kerosene tank pressurization was provided by the gas generator. The oxygen tank used a heat exchanger. As with the Block I, the Block L engine operated for 250 s. For ignition in a vacuum, two solid-fuel engines mounted on the intermediate stage produced 572 to 860 N of thrust for an acceleration lasting less than 42.5 s. The period of ballistic flight could last 50-60 minutes, during which stabilization was controlled by pneumo-electrical-valve gas nozzles. Pilyugin's I-100 unit steered the third and fourth stages and the payload. It included a heading and pitch stabilizer (SKT) and a range control automat (AUD). The SKT comprises a set of floating gyroscopes (GIP) and accelerometers (F142J3).

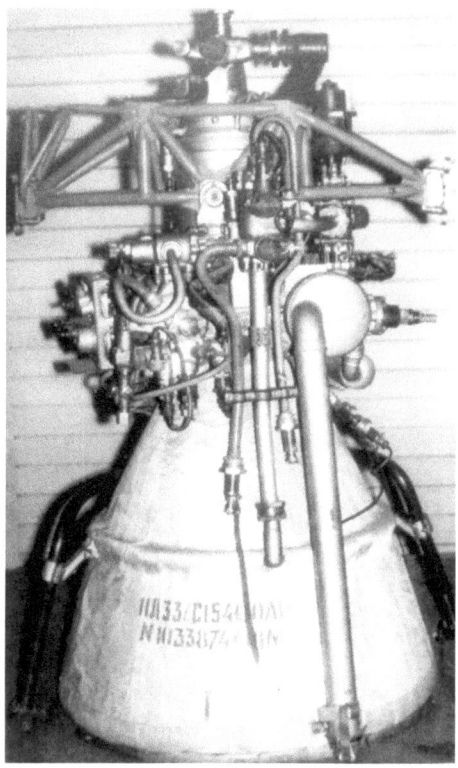

The S1-5400 engine.

Guidance was maintained with an accuracy of $\pm 5°$ on each of the three axes. When the engine had reached 75% of thrust, the I-100 unit separated from the fourth stage in order to increase the performance of the burn. The engine turned off when the programmed speed was reached, and then 8 s later the payload separated at a speed of 1.7 m/s.

The first launch was on October 10, 1960, with 8K78 N°L1-4 carrying the 1M N°1 probe intended for Mars. But the Block I stage malfunctioned due to a defective kerosene supply. The next flight on October 14 was 8K78 N°L1-5 with the 1M N°2 probe, but again the Block I failed. "The Martian probe was too heavy," says Mark Gallai, who was an instructor for the first cosmonauts in 1960. "One of the scientific experiments had to be unloaded. Korolev suggested testing the infrared spectrometer in the steppes of Kazakhstan but the instrument, which was intended to detect life on Mars, detected nothing on Earth. Consequently it was not mounted on the probe."

On February 4, 1961, 8K78 N°L1-7 launched the 1VA N°1 probe (Sputnik 7) that was intended for Venus. The Block I stage achieved "parking orbit" but a fault in the Block L prevented the escape burn. Two days later the 6.48-ton device re-entered the atmosphere. The state commission was M. V. Keldysh (chairman), S. P. Korolev, V. P. Glushko, V. P. Barmine, G. N. Pachkov, S. A. Zverev, A. I. Shokin, A. I. Sokolov, A. I. Semenov, etc. On February 12, 8K78 N°L1-6 successfully launched

The 8K78 Molniya launcher.

1VA N°2 (Sputnik 8-Venera 1) and the 643.5-kg probe flew past Venus on May 19 at a range of 100,000 km. However, the solar guidance system worked poorly and disconnected the control receptors in order to save energy. The onboard programmer established contact with Earth on February 17, at which time the probe was 2 million km away, but there were no further signals. The blackout was due to a design error. The solar orientation sensor was not pressurized and some of its elements were not designed to withstand vacuum. In addition, the probes had arrived at Baikonur without having undergone complex testing at OKB-1, whose staff were overworked. Radio devices which were meant to function for four months in a vacuum had been tested for only a few hours.

On August 25, 1962, 8K78 N°T103-12 launched 2MV-1 N°1 (Venera/capsule) but the fourth stage failed to leave parking orbit. On 1 September, 8K78 N°T103-13 launched 2MV-1 N°2 (Venera/capsule) but again the probe was stranded by a failure of the Block L. On 12 September, 8K78 N°T103-14 launched 2MV-2 N°1 (Venera flyby) and yet again the fourth stage failed. The same happened on October 24, 1962 when 8K78 N°T103-15 launched 2MV-4 N°1 (Mars flyby). The first success came on November 1, when 8K78 N°T103-16 launched 2MV-4 N°2 (Mars 1/flyby). The 893.5-kg probe flew past the Red Planet on June 19, 1963, at a range of 193,000 km. However, by then it was inert because on March 21 the directional system failed (a valve of the micro-motors gas supply system caused leakage of nitrogen reserves). Nevertheless it had functioned perfectly for 140 days and set a new record for radio-connections: 106 million km. The sixth and final launch of the series took place on November 4 when 8K78 N°T103-17 launched 2MV-3 N°1 (Mars/capsule), but yet again the fourth stage failed to leave parking orbit.

The first launch of a lunar probe was on January 4, 1963, with 8K78 N°T103-09 carrying E-6 N°1 but the fourth stage failed and stranded the probe in parking orbit. The same thing occurred on February 3 with 8K78 N°T103-10 carrying E-6 N°2. On April 2, 8K78 N°T103-11 succeeded in placing Luna 4 (E-6 N°3) on its translunar trajectory, but the midcourse correction system failed and the 1,422-kg probe missed its target by 8,500 km. Each malfunction on these three launches was different: non-ignition of the Block L engine supplied by NPO Lavochkin, failure of the SKT from Pilyugin, and failure of the OKB Mars astronavigation system. On November 11, 8K78 N°G15000-17 launched interplanetary probe 3MV-1 N°1 (Zond) but the fourth stage failed and the probe (named Kosmos 21) re-entered the atmosphere after three days. On February 19, 1964, 8K78 N°G15000-26 was launched carrying 3MV-1 N°2 (Zond) intended for Venus. This time the third stage failed. On March 21, 8K78 N°G15000-20 launched the E-6 N°4 probe intended to make a lunar landing, but the third stage failed. Six days later, N°G15000-27 8K78M launched 3MV-1 N°3 (Zond) intended for Venus, but the fourth stage failed and the probe became Kosmos 27. On April 2, 8K78 N°G15000-28 successfully launched 3MV-1 N°4 (Zond) for a Venus flyby, but Zond 1 soon broke down because the glass of the solar orientation sensor dome was not airtight, allowing the orbital module to

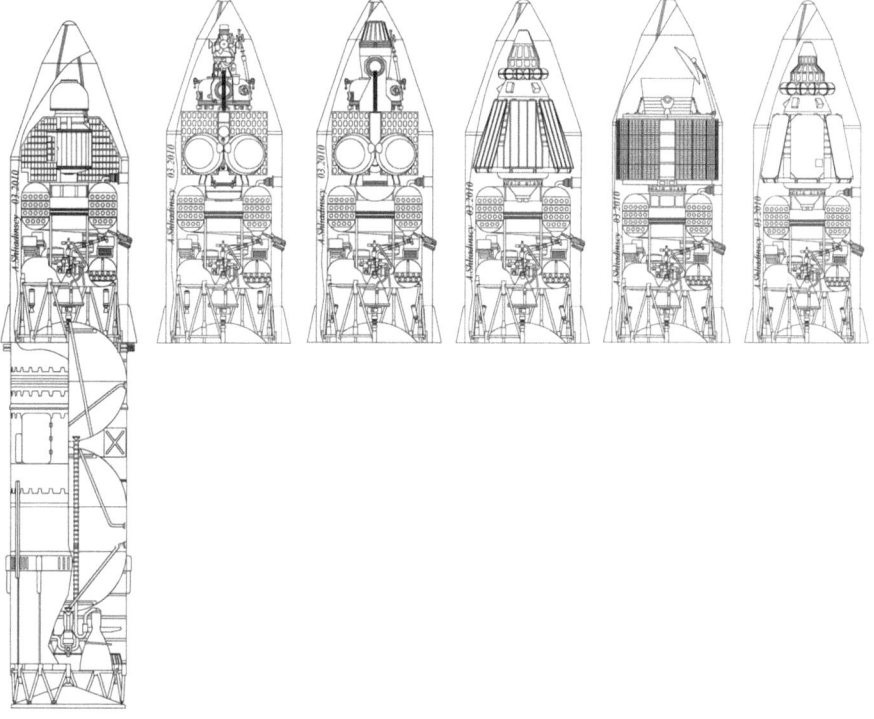

Left to right: 1VA probe Venera 1; 2MV4 probe Mars 1 (1962); 3MV3 probe Venera 3 (1965); 11F67 satellite Molniya 1 (1968); Oko satellite (1975); and 11F637 satellite Molniya 3 (1974).

depressurize. The low pressure in the module caused a short circuit in the transmitter connection, but communication was maintained with the descent capsule, which transmitted for two months, and two trajectory corrections were successfully performed. Then contact was lost. The probe flew past its target on July 18, 1964 at a range of 100,000 km. On April 20, 8K78M N°G15000-21 launched E-6 N°5 intended for a lunar landing, but the third stage failed.

On June 4, 1964, the first Molniya 1 telecommunications satellite (11F67 N°1) was launched by 8K78 N°R103-34, but it was lost 104 s into the flight as a result of a SOB system failure in Block A. The state commission was chaired by General K. A. Kerimov. The 1,600-kg satellite, built by S. P. Korolev, carried an Alfa payload from NII-695 (Y. S. Bykov and I. D. Bogachev). The second Molniya 1 (11F67 N°2) was successfully launched by 8K78 N°R103-36 on August 22, but the satellite failed to deploy its parabolic antenna and therefore could not transmit. It was then written off as Kosmos 41. On November 30, 8K78 N°G15000-29 launched 3MV-4 N°2 (Zond) intended for a Mars flyby. The launch was successful, but one of the solar panels of Zond 2 did not deploy completely because of a failure of that mechanism. Then the onboard programmer failed because the thermoregulation system was not connected between the communication sessions. Only the reserve programmer was designed to connect to thermoregulation, but the probe could not use this. Nevertheless, contact was maintained for a month, during which six plasma engines were tested twice (on December 8 and 18). The inert probe passed within 1,500 km of Mars on August 6, 1965.

On March 12, 1965, 8K78 N°R103-25 launched E-6 N°9, intended for a lunar landing, but the fourth stage failed and it became Kosmos 60. On April 10, 8K78 N°R103-26 launched E-6 N°8, but the third stage failed. On April 23, 8K78 N°103-35 launched Molniya 1 N°3. The deployment was successful, but solar deterioration halted transmissions after only 7.5 months. On May 9, 8K78 N°103-30 successfully launched the 1,476-kg Luna 5 probe (E-6 N°10) intended for a lunar landing, but the gyroscope's oil froze and it was not possible to achieve a midcourse correction. The probe crashed into Mare Nubium. On June 8, 8K78 N°103-31 launched the 1,445-kg Luna 6 probe (E-6 N°7), but the midcourse correction engine did not ignite. It passed the Moon at a range of 160,000 km. On July 18, 8K78 N°103-32 launched the 3,950-kg 3MV-4 N°3 interplanetary probe (Zond 3), which photographed the far side of the Moon from a distance of 9,220 km. For the first time trajectory corrections were able to be made using stellar orientation (based on the bright star Canopus). Communications were maintained for 7.5 months to a record distance of 153 million km. On October 14, 8K78 N°103-37 launched Molniya 1 N°4. The deployment was successful, but deterioration of the solar panels interrupted transmissions on February 18, 1966.

On November 12, 8K78 N°103-42 launched the 963-kg 3MV-4 N°4 intended for a Venus flyby, but Venera 2 broke down after three months and transmissions ceased 17 days before the flyby. The probe had difficulty receiving orders from Earth due to overheating of the receivers and decoders. This high temperature was due to intense solar activity and insufficient layers of protection on the thermoregulation radiators. The silent probe flew past its target on February 27, 1966, at a range of 24,000 km. On November 16, 8K78 N°103-31 launched the 960-kg 3MV-3 N°1 intended to drop a capsule into the atmosphere of Venus, but Venera 3 suffered similar problems to its partner and transmissions ceased 11 days before it reached its target. But for the first time, a man-made object entered the atmosphere of Venus on March 1, 1966. It was just 800 km from the projected target point. On November 23, 8K78 N°103-30 launched 3MV-4 N°6 intended for a Venus flyby. The third stage suffered a problem that left the vehicle tumbling. The fourth stage achieved orbit but was then unable to ignite for the escape burn. The probe was written off as Kosmos 96. It re-entered the atmosphere on December 9.

On December 3, 8K78 N°103-28 launched the 1,552-kg Luna 8 probe (E-6 N°12). The landing in Oceanus Procellarum was nominal but a failure occurred after contact with the surface. Korolev died on January 14, 1966, so did not witness the successful landing of Luna 9 on February 3.

On March 27, 8K78 N°N103-38 launched Molniya 1 N°5, but the third stage failed. On April 25, 8K78 N°N103-39 launched Molniya 1 N°6. The deployment was successful but its operational life was limited to five months. On October 20, in the context of operation "Palma 2", 8K78 N°N103-40 launched Molniya 1 N°7, which functioned for a year. On May 24, 1967, an 8K78 launched Molniya 1 N°8, the first unit built by NPO PM in Krasnoyarsk. On August 31 an 8K78 launched Molniya 1 N°9 (Molniya Yu, named Kosmos 174). It was equipped with a DRS (radio system for long range links) and was to calibrate the tracking system for lunar missions. On October 3, 8K78 N°Ya716-83 launched Molniya 1 N°10. And finally on October 22, 8K78, N°Ya716-82 launched Molniya 1 N°11. A total of 40 Molniya rockets were launched from 1960-1967, with 20 failures (eleven related to the Block I third stage and nine to the Block L fourth stage); a 50% success rate.

## 8K78M

The Molniya-M was equipped with an improved power plant: the 8D75F/8D727K engine of the central core had its thrust increased by 5% (from 75.8 to 79.3 tons) and an extra 2.9 s of specific impulse on the ground. Extinction occurred in the nominal regime. The third stage was equipped with an 11D55/RD-0110 engine that had been developed by Kosberg's OKB-154 for the Soyuz launcher in response to a decree issued in April 1963. The first unit, 8K78M N°103-27, was moved onto the launch platform on September 4, 1965, carrying the 1,506-kg E-6 N°11 lunar probe. But it was removed from the pad because of a failure in the guidance system. It finally took off on October 4. Unfortunately, when Luna 7 reached its destination the braking engine failed to ignite and the probe crashed into the Moon. On November 26, an 8K78M was prepared in order to launch 3MV-3 N°2, intended to drop a capsule into the atmosphere of Venus. But it was withdrawn from the pad when it became evident that it was unable to take off. On January 31, 1966, 8K78M N°103-32 launched the 1,583-kg Luna 9 (E-6-13 N°202). It made a soft landing in Oceanus Procellarum on February 3 and sent back the first panoramic view of the lunar surface. The image was picked up by the Jodrell Bank Observatory in England, which published it in a national newspaper ahead of the official release by the Soviet Union.

The RD-0110 engine.

On March 1, 8K78M N°N103-41 launched E-6S N°204, intended to enter lunar orbit, but the fourth stage failed and the probe, named Kosmos 111, re-entered the atmosphere two days later. On March 31, 8K78M N°N103-42 was launched carrying E-6S N°206. Once in lunar orbit the 583-kg Luna 10 released a 245-kg satellite. This transmitted the "Internationale" during the congress of the CPSU, and continued to provide scientific data until May 30. On August 24, 8K78M N°N103-43 launched E-6LF N°101. Luna 11 achieved orbit around the Moon and transmitted until October 1, but due to a breakdown did not provide any photos of the surface. On October 22, 8K78M N°N103-44 launched E-6LF N°102. Luna 12 sent photographs of the surface to assist the landing of cosmonauts on the Moon that was scheduled for 1968-1969. On December 21, 8K78M N°N103-45 launched E-6 M N°205. Luna 13 landed in Oceanus Procellarum on Christmas Eve, took panoramic pictures, and deployed two mechanical arms equipped with a densitometer and a penetrometer. Then on May 16, 1967, 8K78M N°Ya716-56 launched E-6LS N°11 to test the radio-technical systems for the next generation of lunar probes (E-8, scheduled to go into production on June 14), but a failure of the fourth stage left it in an orbit ranging in altitude between 380 and 60,600 km. Named Kosmos 159, it re-entered the atmosphere on November 11, 1977.

On June 12, 8K78M N°Ya71670 was launched carrying 4V-1/V67 N°310. The

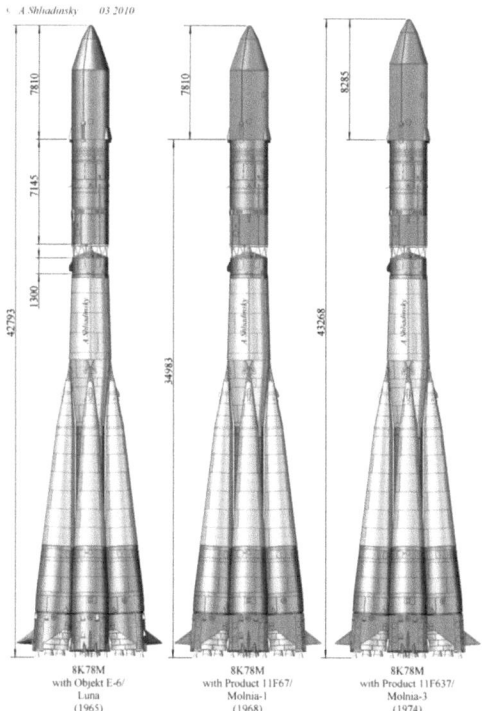

A Shkadovsky    03.2010

| 8K78M with Objekt E-6/ Luna (1965) | 8K78M with Product 11F67/ Molnia-1 (1968) | 8K78M with Product 11F637/ Molnia-3 (1974) |

The versions of the 8K78M: E6 (1965), Molniya 1 (1968), and Molniya 3 (1974).

1,106-kg Venera 4 probe penetrated the atmosphere of Venus on October 18 and its 383-kg capsule transmitted for 93 minutes until falling silent at an altitude of 23 km, where it measured an atmospheric pressure of 17.6 bars and a temperature of 270°C. On June 17, 8K78M N°Ya71671 launched 4V-1/V67 N°311 to perform the same mission as its immediate predecessor, but the failure of the fourth stage stranded it in Earth orbit as Kosmos 167.

On February 7, 1968, 8K78M N°Ya716-57 launched E-6LS N°112 intended to enter lunar orbit, but the third stage failed. Its twin, E-6LS N°113, was successfully launched by 8K78M N°Ya716-58 on April 7. Luna 14 went into lunar orbit (160/870 km) and tested the radio-technical systems of the E-8 probes. It also carried a series of transmissions with gears made of different materials, using different coatings and oils that would require to function in vacuum without sticking or forming cold welds. These transmissions would be necessary for the wheels of the Lunokhod and for the drilling mechanisms of missions sent to bring back samples. In addition, the study of orbital parameters provided a better understanding of the Moon's gravitational field and what would become known as the "mascon" phenomenon. On January 5, 1969, 8K78M N°V/1672 launched V69 N°330. The 1,130-kg Venera 5 spacecraft reached its target on May 16 and dropped a 405-kg capsule that spent 53 minutes descending through the atmosphere before falling silent at an altitude of 16 km. On January 10, 8K78M N°V71673 launched V69 N°331. Venera 6 experienced the same fate as its partner, with its capsule falling silent at an altitude of 16 km where the pressure was 27 bars and the temperature was 320°C.

On August 17, 1970, 8K78M N°Kh15000-62 launched 4V-1/V70 N°630. Venera 7 dropped a capsule into the atmosphere on the night side of Venus on December 15. It descended via parachute for 35 minutes and then functioned on the surface for 23 minutes. On August 22, 8K78M N°Kh15000-61 carried 4V-1/V70 N°631, intended for a similar mission, but the fourth stage stranded it in Earth orbit as Kosmos 359. On March 27, 1972, 8K78M N°S15000-63 launched 4V-1/N°670. Venera 8 dropped a 495-kg capsule into the atmosphere on July 22. It descended via parachute for 53 minutes and transmitted from the surface for 50 minutes, where the pressure was 90 bars and the temperature was 470°C. Instruments were able to detect the presence of free ammonia in the atmosphere, photometers measured luminance as a function of altitude, and a gamma spectrometer studied the chemical composition of the rock. On March 31, 1972, 8K78M N°78014-21513 launched 4V-1/N°671, but the fourth stage failed and it remained in Earth orbit as Kosmos 482.

In April 1968, the Molniya-M began launching Molniya telecommunications satellites made in Krasnoyarsk: four in 1968 (including Molniya Yu/Kosmos 260 to calibrate the lunar mission tracking system), two in 1969, five in 1970 (including the first Molniya launched from Plesetsk on February 19), two Molniya 1s and the first Molniya 2 (11F628) in 1971, two Molniya 1s and three Molniya 2s in 1972, four Molniya 1s and four Molniya 2s in 1973 (including the first Molniya 1M on November 30), then a Molniya 1, three Molniya 2s and the first Molniya 3 (11F637) in 1974. The state commission was headed by General A. A. Maximov until 1978, then by N. F. Shlygov. A total of 99 Molniya 1, 1M and 1T satellites were launched between 1964 and 2004 (with two launch failures and four Kosmos write-offs), 19 Molniya 2s from 1971-1977 (including two Kosmos write-offs) and 56 Molniya 3s and 3Ks in 1974-2005 (including one launch failure and two Kosmos write-offs). On June 21, 2005, rocket N°770466694 with the fourth Molniya 3K failed because of a problem with the second-stage engine, which was not the same type as on the Soyuz launcher. Consequently, the rocket launched 175 Molniya satellites. In 2006, they were superseded by the Meridian satellites.

In 1972 the Molniya-M began launching early-warning satellites of the US-K/Oko (5V95) type built by NPO Lavochkin and NPO Kometa. The 2-ton satellite was equipped with infrared telescopes supplied by the institute of optics imeni Vavilov, the institute of television (VNIIT) and TsKB-589 Geofysika. It was placed in a Molniya-type orbit. The first satellite, Kosmos 520, was launched on September 19, 1972. In total the rocket launched 83 Oko satellites from 1972-2008.

The Molniya-M began launching Prognoz satellites in 1972 in order to study the relationships between the Earth and the sun. NPO Lavochkin developed this 845-kg device which was produced in series by the Vympel Moscow plant. It included 130 kg of scientific experiments for the study of plasma, charged particles in cosmic rays, electromagnetic emissions of the sun, etc. The first (1 SO-M N°501) was launched on April 14, 1972 into an orbit ranging in altitude between 940 and 201,000 km. As a result of its frequent passes through the radiation belts, it stopped working after 172 days. In total, the Molniya-M variant launched 12 Prognoz satellites (including two Interbols) from 1972-1996. In addition, it launched the Indian IRS-1C satellite and the American Skipper satellite on December 28, 1995 (launch price $7 million). By December 31, 2008, the Molniya-M had launched 279 units (53 from Baikonur and 226 from Plesetsk) including two failures (excluding Block L) for a 99.2% success rate. In 1996 there had been a total of 259 launches of Block L, including 13 failures. In 2007, this figure totaled 278 launches with 14 failures; a 96.7% success rate.

## 8A91

The 8A91 variant was an 8K71 with 8D76 and 8D77 engines. On April 27, 1958, 8A91 N°B1-1 carried D-1 N°1 (the first Object D), but vibrations caused a failure 88 s into the flight. It weighed 268.6 tons (5.9 tons less than the R-7). On May 15, 8A91 N°B1-2 successfully put the D-1 N°2 satellite into orbit as Sputnik 3. This 1,327-kg device carried 12 scientific experiments to study the physics of space. In total, there were two 8A91 launches with a 50% success rate.

## 8A92

The Vostok 2 (8A92) began launching Zenit 2 satellites (11F61) on June 1, 1962. In total, 45 of these rockets were launched from 1962-1967 with five failures giving an 88.8% success rate: five in 1962, with a failure on June 1 (a booster engine broke down 1.8 s into the flight and the booster fell onto the launch pad); eight in 1963, with failures on July 10 (the breakdown of a booster engine 1.9 s into the flight caused the vehicle to explode on the pad, totally destroying it) and on November 28 (the 8D719 engine failed to ignite 310 s into the flight); nine in 1964, with the first launch from Zone 31 at Baikonur (Kosmos 28), a launch in the context of operation "Kedr" (Kosmos 46) and a recovery failure (Kosmos 50); eight in 1965, including a recovery failure (Kosmos 66) and a launch failure on July 13 (break down of the first stage, with impact 200 km southwest of Semipalatinsk); ten in 1966, with the first launch from Plesetsk on March 17 (Kosmos 112) and a launch failure on September 16 (failure of the 8D74 first stage engine); and finally five in 1967.

## 8A92M

The Vostok 2M (8A92M) launched the Meteor 1 and 2 series satellites, the Tselina D intelligence satellites, the Meteor-Prirodas, the Resurs OE and O1, Astrofyzika (Kosmos 1066), and Interkosmos-Bulgaria 1300, as well as the Indian satellites IRS-1A and 1B.

On August 28, 1964, 8A92M N°T15000-25 was launched with the first Meteor meteorological satellite (11F614 N°1), but that could not orient itself properly and

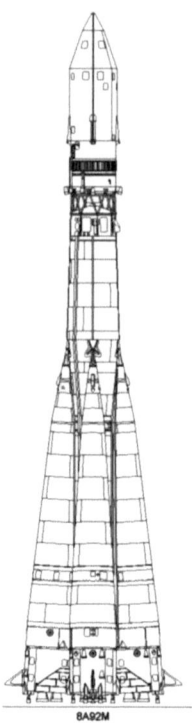

The 8A92M.

was written off as Kosmos 44. The satellite was built by A. G. Iossifian's NII-627 (later VNIIEM) with a TV camera from NII-380 (I. A. Rosselevich) and an infrared camera and a radiometer from OKB Geophysika (V. A. Khrustalev). General K. A. Kerimov headed the state commission. On February 26, 1965, 8A92M N°R15000-09 launched Meteor N°2 but it immediately broke down (Kosmos 58). An investigative commission was appointed to find the origin of the fault. On December 17, 8A92M N°R15000-31 launched Meteor N°3, which also broke down (Kosmos 100). On May 11, 1966, an 8A92M launched Meteor N°4 but it suffered the same fate as its two predecessors (Kosmos 118). On June 25, 8A92M N°N15000-21 launched Meteor N°5 for operation "Palma 1". As Kosmos 122, it was the first of its type to transmit images.

After the explosion of a Soyuz launcher at Baikonur in December 1966, Meteor was transferred to Plesetsk and the orbital inclination increased from 65° to 81.2°. On February 28, 1967, an 8A92M launched Meteor N°6. As Kosmos 144, it became the first operational satellite of the state hydro-meteorological service. On April 27 an 8A92M launched Meteor N°7. As Kosmos 156, it operated simultaneously with Kosmos 144. Kerimov was replaced as president of the state commission by General V. I. Sheulov. On October 25, an 8A92M launched Meteor N°8 as Kosmos 184 but it soon broke down. It was replaced on March 14, 1968 by Meteor N°9 as Kosmos 206.

On June 12 Meteor N°10 (Kosmos 226) joined Kosmos 144, 156 and 206 to form a constellation of four satellites. Then series production was assigned to the YoujMach plant of Dnepropetrovsk in Ukraine. It manufactured 26 Meteor 1 satellites and they were launched between March 1969 and April 1977. VNIIEM acquired a subsidiary at Istra in the suburbs of Moscow to produce its satellites. This began operations in 1973 and was directed by V. I. Adasko. It built two Meteor-Prirodas (11F614ME) between 1974 and 1976 (900 km circular orbit inclined at 81.2°), five Resurs OEs (11F651) from 1977-1983 (circular orbit of 650 km inclined at 98°), four Resurs O1s (11F697) from 1985-1998, 21 Meteor 2s (11F632) from 1975-1993, seven Meteor 3s (17F45) from 1984-1994, a Meteor 3M (17F45M) in December 2001, Interkosmos-Bulgaria 1300 in August 1981, and then Elektro 1 (11F652) in October 1994. These satellites were launched by some fifty Vostok 2Ms (8A92M), twenty Tsyklon 3s (11K68), three Zenit 2s (11K77) and a Proton-K (8K82). On September 17, 2009 the new generation Meteor-M 1 was successfully orbited by a Soyuz-2.1b/Fregat.

The Tselina D N°1 (11F619) electronic surveillance satellite was launched from Plesetsk on December 18 1970. The satellite was built by NPO Youjnoe. It became Kosmos 389, with a TsNII-108 payload. The 1,640-kg satellite was equipped with a system of gravitational orientation that determined angular position in relation to the stars. Using a large reception antenna, it could precisely measure the coordinates of a transmitter (radar). The Vostok 2M launched 40 satellites of this type between 1970 and 1983. After an explosion on the launch platform at Plesetsk killed 48 people on March 18, 1980, these satellites were launched by the Tsyklon 3, the same launcher as was being used for the Meteor satellites. Experimental operation of the system began in 1976.

On December 23, 1978, a strange satellite was launched by a Vostok 2M from Plesetsk. This Kosmos 1066 Astrofyzika (11F653) satellite was equipped with Fakel SPD-50 electric motors. Its unique mission remains a mystery. Finally the Vostok 2M was used to launch two Earth observation satellites for India: IRS-1A on March 17, 1988 and IRS-1B on August 29, 1991. This marked the retirement of this variant. A total of 94 units were launched, with only two failures; a success rate of 97.8%.

## 11A55 AND 11A56

On April 16, 1962, the decree concerning "Development of the Soyuz complex for a fly-by of the Moon by cosmonauts" gave Korolev the green light to start to design the space system composed of the 7K manned spacecraft, the 9K tug and 11K tanker. Assembling this train of vehicles in orbit required several launches and rendezvous in space. The requirement for rendezvous gave the vehicle its name, because Soyuz means "Union" in Russian. Later, this name was given to the new variant of the R-7 launcher. The design was completed on December 24.

The 7K spacecraft was to be launched by the 11A55 (with a reliability level of a manned mission), while the 9K and 11K were to be launched by the 11A56 (the codes 11A52, 11A53 and 11A54 had been attributed to the N-1, N-11 and N-111 launchers in September 1962). The two launchers differed in the type of engines, telemetry system and fairing. On the central core, the RD-108MM/8D75K/8D727 of the Molniya variant was replaced by the 8D75M/8D727P on the 11A55, and by the 8D75F/8D727K on the 11A56. The boosters kept the RD-107 MM/8D74K/8D728 of the Molniya. For the third stage, the RD-0107/8D715K engine of the Molniya was replaced by the RD-0108/8D715P, which was certified for manned flight.

The program was validated on March 20, 1963, and adopted by decree N°1184-435 on December 3. But the start of the Voskhod program slowed down the work. And then on August 3, 1964, the lunar flyby was entrusted to V. N. Chelomei and Korolev began the N1-L3 lunar landing project. The 7K-9K-11K program virtually stopped at that point. But in late 1964 Korolev proposed a rendezvous between two 7Ks in Earth orbit. The 7K-OK (11F615) project was adopted in early 1965. It used a new version of the R-7: the 11A511. The lunar flyby was placed back in Korolev's hands in the spring of 1965. After the success of the first Proton flight in July 1965, a military industrial commission decree on August 26, 1965 directed Korolev and Chelomei to join forces to achieve a circumlunar flight of a spacecraft with two cosmonauts in 1967. This was the 7K-L1 program, in which a stripped down Soyuz spacecraft would be launched by a Proton fitted with an extra upper stage made by Korolev. After two months of study, it was adopted by decree on October 25, 1965.

## 11A57

The Voskhod (11A57) was powered by an 8D75F/8D727K in the core (similar to the 8D75M/8D727P except for extinction, which occurred in the intermediate regime) and 8K74K/8D728 engines in the lateral blocks. The third stage had an 8D715P/RD-0108. The first unit, N°G15000-06, was launched on November 16, 1963. It orbited the first Zenit 4 (11F69) reconnaissance satellite as Kosmos 22. On May 18, 1964, 11A57 N°G15000-12 launched the second Zenit 4 (Kosmos 30). On July 1, 11A57 N°T15000-04 launched the third Zenit 4 (Kosmos 34). On September 9 the failure of the central core caused a launch to be aborted. This resulted in the postponement of the Voskhod 1 mission to October. On September 13, N°R15001-01 launched a Zenit 4 (Kosmos 45). On October 6, 11A57 N°R15000-02 launched the 3KV N°1 prototype of the Voskhod spacecraft as Kosmos 47. This 5.32-ton spacecraft carried two test dummies for a 24 hour flight. The soft-landing system worked perfectly. After this success, 11A57 N°R15000-04 launched 3KV N°2 on October 12 as Voskhod 1 with V. M. Komarov (pilot), K. P. Feoktistov (engineer), and B. B. Egorov (doctor) on board. The flight lasted 24 hours 17 minutes and ended with a recovery 312 km northeast of Kustanay in Kazakhstan. While they were in space, Khrushchev was sacked and replaced by Brezhnev as general secretary of the party and Kosygin as president of the council of ministers.

On February 22, 1965, N°03-R15000 launched the 3KD N°2 spacecraft prototype as Kosmos 57. It carried two test dummies. The inflatable airlock was deployed and jettisoned properly. The return to Earth after 24 hours failed when the return cycle connected spontaneously and fired the retrorocket. A few minutes later, in response to the unexpected change in orbit, an onboard system detonated explosives to prevent the capsule coming down on foreign territory. The failure resulted from two tracking sites simultaneously sending order N°42, which was interpreted as the single order N°5 ("Return"). On March 7, 11A57 N°R15001-05 launched a Zenit 4 (Kosmos 59). Its capsule was fitted with the base ring of the Voskhod 2 airlock to verify that this would not impair re-entry into the atmosphere. The 8-day mission was a success. On March 18, N°R15000-05 launched 3KD N°4 as Voskhod 2 (also known as "Vykhod", meaning "Exit" in Russian). It carried P. I. Belayev and A. A. Leonov, who made the first spacewalk – spending 23 minutes 41 seconds outside of the hermetically sealed cabin with 12 minutes 9 seconds outside the airlock. The flight lasted 26 hours 2 minutes. The automatic return system broke down on the 17th orbit (the solar positioning did not work), so on the next orbit Belayev conducted a manual return using the backup system. The capsule descended 368 km off target, touching down in a snowy forest 100 km from Solikamsk, in the Urals north of Perm. It took 24 hours to recover the crew.

On February 22, 1966, rocket N°R15000-06 launched 3KV N°5 as Kosmos 110. It carried the dogs Veterok and Ugoliok on a 22-day flight through the radiation belts and then landed 210 km southeast of Saratov, 60 km off target. This cleared the way for a 15-20-day manned flight of a 3KV a month later. The prime crew was Volynov and Shonin, with Beregovoy and Shatalov in backup. But the life-support

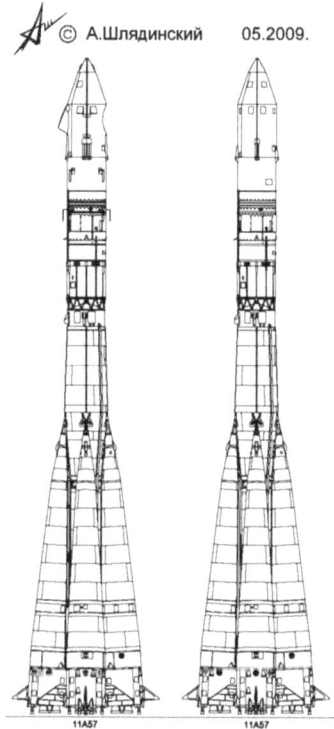

The 11A57 Voskhod launcher.

A Voskhod launcher.

system was not ready and the flight was delayed from month to month until December 1966, at which time it was canceled with the launch only a few weeks away. The plan called for this to be followed by a 3KD flight lasting 15-20 days with an all-female crew of Ponomareva and Solovieva (Sergueichik and Pitzkhelauri in backup). It was intended to make an extra-vehicular excursion with a UPMK chair. Two other flights were to have undertaken spacewalks lasting 3 to 6 hours and artificial gravity experiments using a rigid tether connecting the spacecraft to the upper stage of the rocket. Three doctors and three journalists were selected to participate in these flights, scheduled to take place in the first half of 1966.

On April 6, 1966, the 11A57 was launched for the first time from Plesetsk. The last flight took place on June 29, 1976. In all there were 299 flights of this variant of the R-

7, including 138 from Baikonur (3 failures) and 161 from Plesetsk (11 failures) with a success rate of 95.3%. The payloads were five Voskhods, 293 spy satellites (Zenit 2, 2M, 4, 4M, and 4MK), and the scientific satellite Interkosmos 6 which was derived from the Zenit. The first failure occurred on May 17, 1966, when an 8D728 booster engine extinguished 4 s into the flight due to an oxygen valve failure. The rocket fell to the ground and exploded. On June 20, 1967, the 8D727 engine of the core stage was inadvertently shut off by the steering system (AVD) and the rocket fell to the ground. On July 21, 1967 the 11D55 engine of the third stage extinguished prematurely, 482.6 s after take-off, due to a failure of the oxygen regulator. The payload was destroyed by the automatic system (APO). On September 1, 1967 the third stage failed 296 s into the flight and the payload burned up in the atmosphere. On January 16, 1968, the Kosmos 199 satellite remained attached to the third stage. It was destroyed on falling back into the atmosphere. There were also failures on July 21, 1970, March 5, 1971, June 25, 1971, August 19, 1971, December 3, 1971, September 2, 1972, July 4, 1973, April 12, 1974, and August 30, 1974. The final one was the result of a third stage failure 338 s after launch. The satellite was destroyed in flight.

After the launch of Kosmos 632 from Zone 31 at Baikonur on February 12, 1974, several soldiers (three or four depending on sources) died in a fire which broke out in the underground oxygen storage area, where 500 tons of propellant were stocked. If the oxygen supply had exploded, the launch platform would have been destroyed.

## 11A58

The plan for a Semyorka with a cryogenic upper stage originated in 1958. It was to be able to place 9.5 tons into low orbit or 2.8 tons on an interplanetary trajectory. But putting such a stage on the R-7A required making the rocket itself more powerful. In 1963 Korolev proposed equipping the core with a set of four of N. D. Kuznetsov's NK-9 engines (37 tons of thrust per unit) to increase the take-off thrust from 413 to 470 tons. Liquid oxygen would have to be replaced with super-cooled oxygen, as on the R-9A. Two variants were envisaged: the 11A58 with a cryogenic third stage, and the 8K78 with a cryogenic Block M fourth stage.

At that time, the cryogenic engines in development were Isaiev's 11D56/KVD-1 (7.5 tons of thrust), Liukla's 11D54 and 11D57 (40 tons of thrust), and Kuznetsov's NK-35 (200 tons of thrust). They were to be used on a cryogenic N-1 (i.e. NK-35 on the second stage, 11D54 on the third stage, 11D57 on the fourth stage, and 11D56 on the fifth stage), but none of these engines advanced beyond the ground-test phase. However, the 11D56 was subsequently modified to power the 12KRB upper stage of the GSLV Mk1 Indian launcher.

## 11A59

In March 1962 it was decided that the first IS anti-satellite interceptor and RORSAT radar surveillance satellites produced by V. N. Chelomei would not be launched by his UR-200, but by Korolev's R-7. The 8K74 with the IS became the 11A59, and the 8A92 with the RORSAT became the 11A510. In the case of the IS, the two-stage version of the R-7A required the satellite to use its own engine for insertion into orbit. A solid-fuel engine was used to achieve separation from the rocket stage (a mechanism tested on a Jukovsky test base) and then discarded. The 1.95-ton IS was developed by Chelomei's OKB-52, A. A. Raspletine's KB-1, N. A. Pilyugin's NII AP, A. M. Isaiev's KB-2, S. K. Tumansky's OKB-300, G. I. Voronin's OKB-124, etc.

The first IS, 5V91/I-2B N°102, was launched by an 11A59 on November 1, 1963 from Baikonur and named Polyot 1 (meaning "Flight"). It was equipped with an engine built by KB-2 (400 kg of thrust) and became the first spacecraft to maneuver in space, changing its initial orbit of 339/592 km to 343/1,437 km and reducing its inclination from 60° to 59°. The second unit, 5V91/I-1B N°112, was launched on April 12, 1964 as Polyot 2. This time it used an engine developed by OKB-300 (two chambers, one turbopump, 600 kg of thrust). The 242/485 km orbit inclined at 60° was modified to 310/500 km at 58°. As the UR-200 was abandoned on December 24, 1964, the program continued with M. K. Yangel's 11K67 launcher, which launched I-2BM N°104 on October 27, 1967, as Kosmos 185.

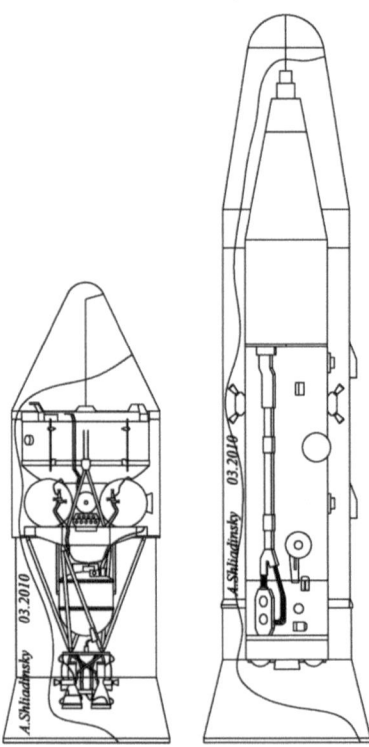

The IS (interceptor) and US-A (RORSAT radar ocean surveillance satellite).

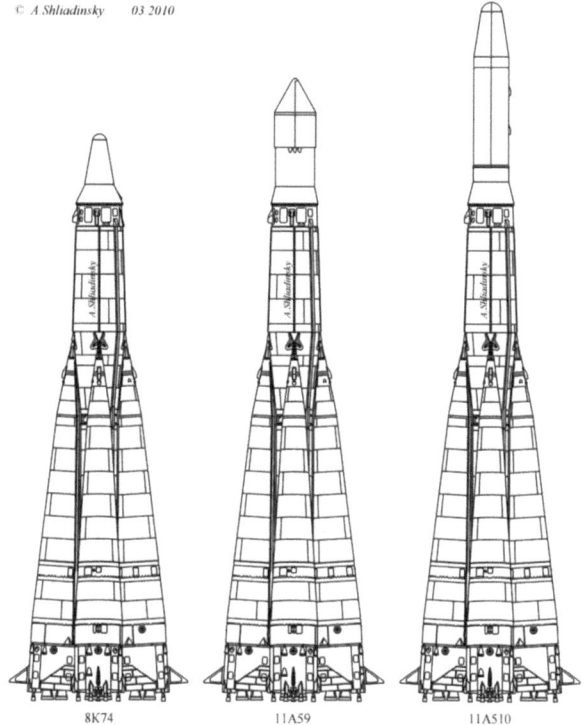

© A Shliadinsky    03 2010

8K74                11A59                11A510

8K74/11A59/11A510.

## 11A510

The prototype of the RORSAT radar ocean surveillance satellite was ready in 1965. It was equipped with two directional and stabilization systems along three axes. The inertial system used a gyroscopic platform from KB-1 which was regulated using the Neptune astronomical system. The orbital system used gyroscopes and an infrared vertical. Once their outputs had been compared, it was decided to adopt the inertial system. The first unit, 4Ya11/US N°EA 0110, was launched on December 28, 1965, by the 11A510 N°G15000-01 (based on the 8A92). It was named Kosmos 102 and stayed in orbit for 16 days at an altitude of 218/278 km inclined at 65°. The second satellite 4Ya11/US was launched on July 20, 1966 as Kosmos 125. It was placed into a circular orbit at 250 km, where it remained for 13 days.

## 11A511

In 1963 the Soyuz rocket was designed to launch the Soyuz spacecraft. Its engines were 8D75M/8D727P (core), 8D74K/8D728 (boosters), and 11D55/RD-0110 (third stage). It stood 51 m tall (with the escape system), weighed 310 tons at take-off, and could place 6.7 tons into low Earth orbit.

The first flight took place on November 28, 1966. Rocket N°U15000-02 launched the prototype spacecraft (11F615 N°02/7K-OK/active) from Zone 31 at Baikonur. It was named Kosmos 133, and the plan was to launch another spacecraft the next day and have them perform a rendezvous and docking. But after the first orbit all the fuel in the guidance system had been inadvertently consumed, making it impossible to steer. The launch of the second spacecraft was canceled. After 33 orbits, Kosmos 133 re-entered the atmosphere over China and self-destructed. On December 14, rocket N°U15000-01 launched the second spacecraft (11F615 N°01/passive) from Zone 31, but the rocket exploded at launch.

On February 7, 1967, rocket N°03-U15000 launched the third Soyuz (11F615 N°03 L/passive), this time from Zone 1. Named Kosmos 140, it quickly lost its fuel reserves and stellar orientation functions. Unable to orient its solar panels toward the sun, its batteries did not recharge. After two days the emergency return went badly. The capsule came down 500 km off target, splashing into the Aral Sea, 3 km from shore. An object in the thermal shield had burned during re-entry of the atmosphere, creating a hole through which water entered the capsule, causing it to sink. Although three test flights of the spacecraft had failed, preparations for the rendezvous of two manned Soyuz capsules in April went ahead.

On April 23, rocket N°U15000-04 launched Soyuz 1 (11F615 N°04 L/active) with V. M. Komarov aboard. The president of the state commission was General K. A. Kerimov. The following day, Soyuz 2 (11F615 N°05 L/passive) was to be launched carrying V. F. Bykovsky, A. S. Elisseiev and E. V. Khrunov. The latter pair were to make a spacewalk from one spacecraft to the other, to rehearse the external transfer planned for the N1-L3 lunar program. But the left-hand solar panel of the 6,558-kg Soyuz 1 did not deploy and the solar directional system did not work. Komarov was obliged to orient the spacecraft manually to generate the required 14 amps. He tried unsuccessfully to use ionic steering to orient the craft for a nominal deorbit burn on the 17th orbit. He then had to use the Vzor optical viewfinder on the daytime side, maintain this orientation using gyroscopes, and then correct it as he came out of the shadow to trigger a return on the 19th orbit. The maneuver was carried out correctly, but unfortunately the main parachute remained stuck in its container, which had been deformed by the difference in pressure. The reserve chute came out, but tangled with the braking chute which had not been jettisoned. The capsule hit the ground at a speed of 50 m/s (180 km/h) some 65 km from Orsk, killing Komarov instantly. NASA had already experienced its first tragedy on January 27, with the deaths of the three Apollo 1 astronauts in a capsule fire during training.

It was therefore decided to proceed with the automatic rendezvous. On October 27, 11F615 N°06/active was launched from Zone 31 as Kosmos 186. It was joined in orbit three days later by 11F615 N°05 L/passive, Kosmos 188, launched from Zone 1. The rendezvous lasted 1 hour 12 minutes. The two vehicles remained docked for 3 hours 30 minutes, then separated and later returned to Earth. Owing to a failure of the stellar sensor, the capsule of Kosmos 186 performed a purely ballistic return. Nevertheless, it landed safely. Kosmos 188 used ionic orientation for the return,

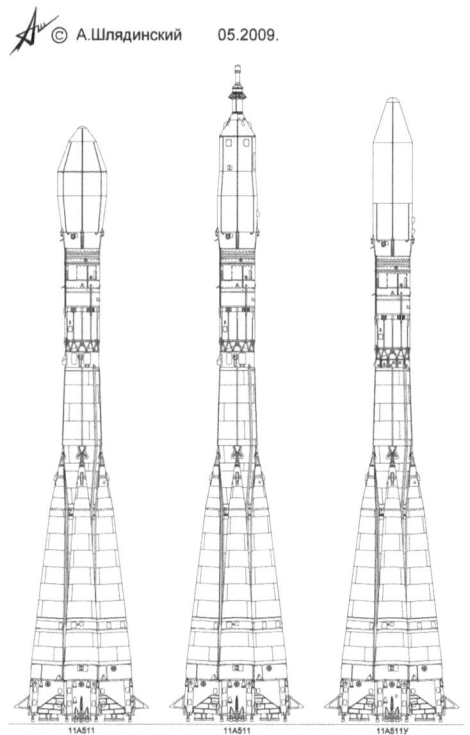

© А.Шлядинский 05.2009.

Versions of the 11A511 Soyuz launcher.

but the trajectory was incorrect and after flying over Irkutsk, heading toward China, the capsule was destroyed in flight over the Mongolian border.

On April 14, 1968, 11F615 N°08 L/active was launched as Kosmos 212. It was joined the next day by 11F615 N°07/passive (Kosmos 213) with which it performed an automatic rendezvous in 46 minutes. The vehicles remained docked for 3 hours 50 minutes and then separated and returned to Earth, each after a flight of 5 days. Both capsules achieved soft landings. The mission was a total success, but it had required eight Soyuz launches to achieve the result initially expected in November 1966. The program was therefore 18 months behind schedule. On August 28, 1968, 11F615 N°09/passive (Kosmos 238) was launched to perform the dress rehearsal for a 4-day manned flight. On October 25, 11F615 N°11/passive was launched unmanned as Soyuz 2. The following day, 11F615 N°10/active set off from Zone 31 as Soyuz 3. It carried G. T. Beregovoy. It was the first time that a manned spacecraft was launched from this area. After the first orbit, Beregovoy took manual control at a separation of 200 m, but was unable to dock. A second attempt was made the next day during one revolution and although he closed to within a few meters of the target there was no docking. The two capsules were safely recovered on October 28 and 30.

On January 13, 1969, for the first time, a Russian cosmonaut, V. A. Shatalov, had to disembark his spacecraft on the launch pad, owing to a 1-day postponement of the

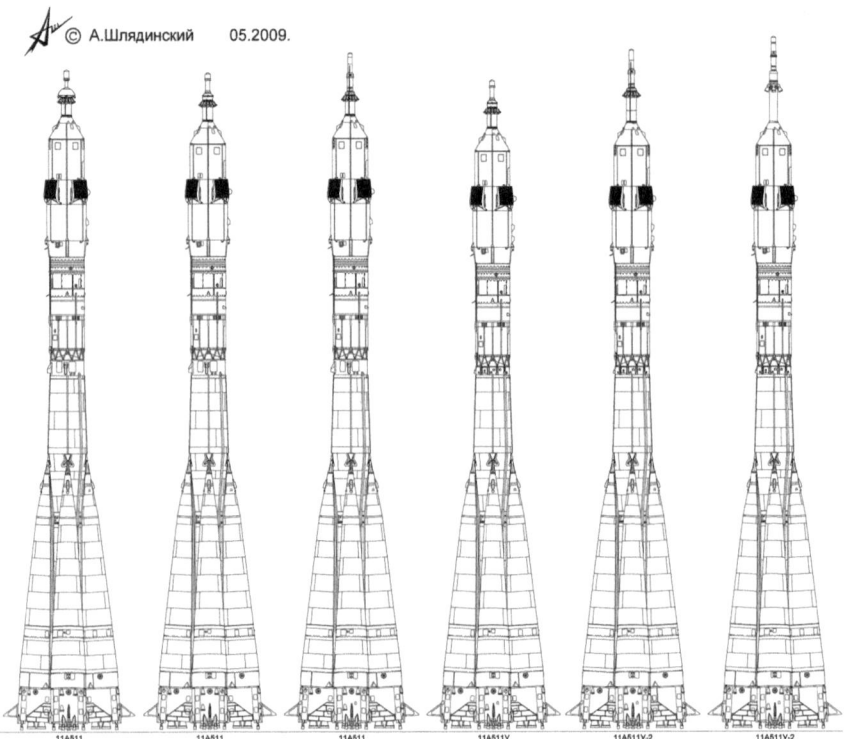

Manned versions of the 11A511.

Launch of Soyuz 3 in 1968.

launch from Zone 31. But 11F615A N°12/active set off the next day as Soyuz 4. On January 15, 11F615A N°13/passive launched from Zone 1 as Soyuz 5, carrying B. V. Volynov, A. S. Elisseiev and E. V. Khrunov. The rendezvous was completed on the 34th orbit of Soyuz 4 and the 18th of Soyuz 5. The two spacecraft remained docked for 4 hours 35 minutes, forming an "experimental orbital station" weighing 13 tons. During this time Elisseiev and Khrunov conducted a 37-minute spacewalk wearing Yastreb suits and transferred over to Soyuz 4, which then landed 40 km northwest of Karaganda. When Volynov attempted to return to Earth, the service module failed to detach from the capsule. Therefore, the heat shield could not protect the capsule as it plunged into the atmosphere in free fall with the service module forward. However, the heat melted through the metal of the service module and the capsule was able to position itself correctly. Volynov endured a deceleration of 9 g. The capsule landed 200 km southwest of Kustanay. It had required 13 spacecraft to achieve this mission, rather than the four originally planned.

On October 11, 1969, 11F615A N°14/active lifted off from Zone 31 as Soyuz 6, carrying G. S. Shonin and V. N. Kubasov. Originally, this week-long mission was to have flown in April-May, to make a weld in space and perform the Svinetz military experiment. The E. O. Paton Electric Welding Institute of Kiev, Ukraine, created the Vulkan unit, and selected two cosmonaut candidates (Fartuchny and Lankin). The Svinetz experiment from NII-2 of the Soviet air defenses (PVO), had been prepared for the canceled Voskhod 3 flight in 1966. It was to film the flight of two ballistic rockets launched from Kapustin Yar. However, the flight of Soyuz 6 was eventually combined with those of Soyuz 7 and Soyuz 8, which were to rendezvous and dock in August-September. This "space troika" mission was to last 4-5 days, with two of the ships remaining docked for 2-3 days. But at that time there was only one simulator at Star City, which did not have the capacity to prepare six crews simultaneously. So there could only be four: the three main crews and a single backup. In September the crew of Soyuz 8 (A. G. Nikolaiev and V. I. Sevastianov) failed their examination and were replaced. On October 12, 11F615A N°15/passive was launched from Zone 1 as Soyuz 7 with A. V. Filipchenko, V. N. Volkov and V. V. Gorbatko. The next day, 11F615A N°16/active lifted off from Zone 31 as Soyuz 8 with V. A. Shatalov and A. S. Elisseiev. But the Igla rendezvous system did not work. On October 14 they were a few kilometers apart. The next day they came within 1,700 m of each other, but the rendezvous failed again. The three capsules returned to Earth in sequence on 16, 17 and 18 October, coming down in the Karaganda region. At that time, the next two Soyuz flights (11F615A N°17 and N°18) were scheduled for January 1970 to test the Kontakt rendezvous system and life-support system for the L-3 lunar spacecraft. Two other flights were scheduled with 11F615A N°19 and N°20 in April-May 1970. Crews were trained on Kontakt from May 1969 to July 1971, but the program was abandoned in October 1971.

Having lost the "race" to the Moon, the Soviet Union redirected its manned space program towards civilian (Salyut) and military (Almaz) orbital stations. To service them, Soyuz 7K-T/11F615A8, equipped with a new docking system, was developed. On June 1, 1970, vehicle N°17 was launched from Zone 31 as Soyuz 9 with A. G. Nikolaiev and V. I. Sevastianov. They returned to Earth after a flight of 424 hours

59 minutes (17 days) that beat the US record of 14 days set by Gemini 7 in December 1965, but were in poor physical shape and experienced great difficulty in readapting to gravity. Since then, cosmonauts in space have performed compulsory gymnastics.

On April 23, 1970, 7K-T N°31 was launched as Soyuz 10 with V. A. Shatalov, A. S. Elisseiev and N. N. Rukavishnikov. It rendezvoused with Salyut 1 but the docking system did not function correctly. The crew had to return to Earth. The film cassettes intended for the OST-1 solar telescope of the station were destroyed when the Soyuz discarded its orbital module to burn up in the atmosphere, and replacement cassettes were not available in time for the next flight. On June 6, 7K-T N°32 was launched as Soyuz 11 carrying G. T. Dobrovolsky, V. N. Volkov and V. I. Patsaiev. This was the backup crew, appointed on the eve of the departure owing to a health issue with a member of the main crew of A. A. Leonov, V. N. Kubasov and P. I. Kolodin. They occupied Salyut 1 for 24 days, but during the return to Earth on June 30 the capsule inadvertently depressurized in space and they were asphyxiated. The follow-up flights by Leonov, Kubasov and Kolodin, and by A. A. Gubarev, V. I. Sevastianov and A. F. Voronov were canceled. On October 11 the station de-orbited itself and burned up over the Pacific. The Soyuz was reconfigured as a two-seater with the cosmonauts wearing pressure suits.

There were 24 flights of the 7K-T over the next ten years: N°33A/Kosmos 496 (June 1972), N°36/Kosmos 573 (June 1973), N°37/Soyuz 12 (September 1973), N°34A/Kosmos 613 (November 1973), N°33/Soyuz 13 (December 1973), N°38/Soyuz 17 (January 1975), N°39/Soyuz 18A (April 1975), N°40/Soyuz 18B (May 1975), N°41/Soyuz 21 (July 1976), and an almost straight run from N°42/Soyuz 25 (October 1977) through to N°56/Soyuz 40 (May 1981). In parallel, a special 7K-T was developed for the Almaz military stations which flew as Salyut 2, 3 and 5. This was the 11F615A9. Seven were launched by the 11A511: N°61/Kosmos 656 (May 1974), N°62/Soyuz 14 (July 1974), N°63/Soyuz 15 (August 1974), N°64/Kosmos 20 (November 1975), N°65/Soyuz 23 (October 1976), N°66/Soyuz 24 (February 1977), and N°67/Soyuz 30 (June 1978, to Salyut 6). Another modification was made for the Apollo-Soyuz program: the 7K-TM. Five of these were launched: N°71/Kosmos 638 (April 1974), N°72/Kosmos 672 (August 1974), N°73/Soyuz 16 (December 1974), N°75/Soyuz 19 (July 1975), and N°74/Soyuz 22 (September 1976).

There was one launch failure: Soyuz 18A with V. G. Lazarev and O. G. Makarov. When the third stage ignited, it failed to fully separate from the core. The escape system was triggered automatically. The capsule separated at an altitude of 150 km. The crew was subjected to a force of 21 g, and returned on a free fall trajectory. The capsule came down on a mountainside to the southwest of Gorno-Altaisk, and rolled on the steep snowy slope towards a precipice, but the parachute, which the crew had not jettisoned, caught in the trees, drawing the capsule to a halt and saving the crew from certain death. The cosmonauts were evacuated by helicopter. Their flight lasted 21 minutes 27 seconds.

The next development resulted from the work carried out on the military Soyuz 7K-VI, starting in November 1967. The plans for the 7K-S/11F732 were adopted in October 1968 and those for the two-seater version in August 1972. Three prototypes were tested in flight. N°1L (Kosmos 670) flew in August 1974 but made a ballistic

Launching Soyuz-TM 15 in 1992.

return. N°2L (Kosmos 772) in September 1975 was a success. For N°3L (Kosmos 869) in November 1976, the initial plan was for it to return after 8 days but a loss of transmission and infrared vertical extended the flight to 18 days. After that, the three-seater 7K-ST was also tested three times in automatic flight. In April 1978, N°4L (Kosmos 1001) suffered a failure of its remote control system. N°5L (Kosmos 1074) in January 1979 encountered the same problem and prematurely returned to Earth. But the flight of N°6L as Soyuz-T 1 in December 1979 was a complete success, including 100 days docked at Salyut 6 during a period when this was unoccupied. This type made 15 manned flights from 1980-1986: N°7L/Soyuz-T 2, N°8L/Soyuz-T 3, N°10L/Soyuz-T 4, N°11L/Soyuz-T 5, N°9L/Soyuz-T 6, N°12L/ Soyuz-T 7, N°13L/Soyuz-T 8, N°14L/Soyuz-T 9, N°16L/Soyuz-T 10A, N°15L/ Soyuz-T 10B, N°17L/Soyuz-T 11, N°18L/Soyuz-T 12, N°19L/Soyuz-T 13, N°20L/ Soyuz-T 14 and N°21L/Soyuz-T 15. To service the Mir orbital station, thirty Soyuz-TMs were used from 1986-2000. Finally, for the International Space Station, four Soyuz-TMs and sixteen Soyuz-TMAs ("anthropometric") flew from 2001-2009. This should continue to be used at the rate of four spacecraft per year until the arrival of a new generation of manned vehicles around 2015. A total of 126 Soyuz spacecraft flew from 1967 to 2009.

In addition, since 1978 orbital stations have been resupplied using the Progress vehicles derived from the Soyuz transporter. Progress (7K-TG/11F615A15) plans were completed in February 1974 and the first unit weighing 7,020 kg was launched on January 6, 1978 to Salyut 6, the first station to have two docking ports. In total, 43 Progress spacecraft (including Kosmos 1669, which was so-named after suffering an early problem from which it was able to recover) were used from 1978-1990. It was replaced by the modified Progress-M/11F615A55, of which 67 units were launched between August 1989 and July 2009. A third variant, the Progress-M1, was used 11 times from February 2000 to January 2004. In addition, two Progress-ENs (N°301 and N°302) were used to deliver small modules (essentially customized Soyuz orbital compartments) to the International Space Station: the DC-1 "Pirs" in September 2001 and the MRM-2 "Poisk" in November 2009. The current "digital" version of the Progress (11F615A60) has been in service since November 2008, and by October 2009 three units had been launched (M-01M N°401, M-02M N°402, M-03M N°403). A total of 124 cargo spacecraft flew from 1978 to 2009.

The 11A511 launcher was modified in 1969 for in-flight tests of the LK lunar lander. This Soyuz-L (11A511L) was used to launch Kosmos 379 on November 24, 1970, Kosmos 398 on February 26, 1971, and Kosmos 434 on August 12, 1971. The success rate was 100%.

The next version was called Soyuz-M (11A511M). It had been developed for the military Soyuz 7K-VI, and after the abandonment of that program was reassigned to the launch of Zenit 4MT "Orion" spy satellites. Eight rockets flew from December 27, 1971 to March 31, 1976, with a 100% success rate.

The Soyuz-U (11A511U) was developed between 1969 and 1971 in order to unify the 11A57, 11A511, 11A511L and 11A511M. It was powered by new versions of the RD-107 and RD-108 engines: the 11D511/RD-117 and 11D512/RD-118, which had a new adjustment program and an extended operation time in a forced regime. The

Launching a Soyuz-L to test the LK spacecraft.

new rocket weighed 308 tons at take-off and could place 6.85 tons in low Earth orbit. The launch platforms were modified in 1972-1973. The first flight, on May 18, 1973, from Plesetsk, placed a Zenit 4MK spy satellite in orbit as Kosmos 559. Then this became the launcher for manned space flights. It was used to launch five spacecraft in the Apollo-Soyuz program from 1974-1976, then the 7K-S from 1974 to 1984 (until Soyuz-T 11). It was also used to launch the automatic Zenit, Bion, Foton, and Yantar satellites. Commissioning of the 11A511U made it possible to retire the 11A57 in June 1976, the 11A511 in October 1976, and the 11A511M in March 1976. During its 10-year career, the 11A511 was launched 32 times with two failures (in December 1966 and April 1975) for a success rate of 93.7%. The 11A511U was launched 744 times, with 20 failures for a success rate of 97.3%. There were similar failures on May 14 and June 21, 1996, respectively 124 s and 49 s into the flight, due to a fault in the new composite fairing. The problem was a lack of binding locks. In fact, during the stamping of a series of locks at the Progress plant the minimum mechanical tolerance applied was insufficient, with the result that when supersonic speed was attained 49 s into the flight the dynamic pressure caused the fairings to collapse. The final failure occurred on October 15, 2002. That day, the Foton M1 satellite was launched by a Soyuz-U from Plesetsk. But 29 s after liftoff, the rocket exploded in the forest a kilometer away. The blast broke the windows of buildings in a radius of 2.5 to 3 km, caused the death of one soldier and injured eight others. During mounting of the RD-107 engine on a lateral block at the Motorostroitel plant, it seems that an object was left in the pipe that delivers hydrogen peroxide to the pump or to the turbopump. Two seconds after the nominal thrust phase, the engine pressure dropped. That booster separated shortly afterwards under its own weight, tearing apart the tank housing. The tanks therefore emptied before impact. The other part of the launcher continued its trajectory to an altitude of 300 m until the other engines were automatically stopped at $+20$ s, whereupon it fell into the forest.

In 1979 it became necessary to increase the payload by 300 kg in order to launch Soyuz and Progress weighing 7.15 tons. To do this, NPO EnergoMach developed the 11D511PF/RD-117PF engine which used syntine, a synthetic kerosene. The pressure in the combustion chamber was increased from 54 to 55 bars. Chamber cooling was improved. The extinction of the engine was delayed by 1.2 seconds. This increased the specific impulse by 6.5 s on the ground. From December 1979 to May 1981, 176 static tests were made using 16 engines. In addition, the vernier engines underwent independent testing. The 11A511U2 rocket weighed 310 tons and its take-off thrust was 416.5 tons. The first flight took place on December 23, 1982 from Baikonur, and placed a Zenit 6 (Kosmos 1425) spy satellite in orbit. Then this version replaced the 11A511U for manned missions, starting with Soyuz-T 12 in July 1984. But after the dissolution of the Soviet Union the syntine plant closed in 1996. After Soyuz-TM 22 in September 1995 the 11A511U2 was abandoned. There had been 70 launches with a 100% success rate. The loss of performance then had to be offset by changing the engine injection head. This gave rise to the 14D21/RD-108A ( $+4.6$ s specific impulse in vacuum) and 14D22/RD-107A ( $+6.2$ s specific impulse in vacuum) to power the 11A511FG (Forsunki Golovka injector head). The 14D21 underwent 69 static tests using 15 engines, while the 14D22 underwent 70 tests using 12 engines. The rocket weighed 308.3 tons and the take-off thrust was 421.3 tons. The maximum payload was 7,270 kg for a Soyuz and 7,450 kg for a Progress. The first four 11A511FG were Progress-M1 6 on May 21, 2001, Progress-M1 7 on November 26, 2001, Progress M1-8 on March 21, 2002, and Progress-M1 9 on September 25, 2002. Then the first manned flight took place with Soyuz-TMA 1 on October 30, 2002. By December 31, 2009, a total of 17 Soyuz-TMAs had been launched successfully.

With the creation of the Franco-Russian company Starsem in 1996, two other Soyuz rocket models were produced. The first was the Soyuz-Ikar, which was used to put first-generation Globalstar satellites in orbit. The Ikar fourth stage was derived from the Yantar-1KFT/Kometa propulsion module. It was 2.59 m in length, 2.72 m in diameter, weighed 3.29 tons, and carried up to 900 kg of UDMH and $N_2O_4$ in spherical tanks. Propulsion was by a 17D61 engine providing 300 kg of thrust and 16 vernier engines for steering. Six Soyuz-Ikar flights occurred in 1999 (100% success rate). They carried a payload of 2.19 tons consisting of four satellites (450 kg each) and the multiple dispenser structure (390 kg).

The latest version is the Soyuz-Fregat. The Fregat fourth stage was designed and developed by NPO Lavochkin, starting in 1992, as a derivative of the propulsion module for the Fobos and Mars-96 deep space probes. It is 1.5 m in height, 3.35 m in diameter, weighs 1.0 ton empty, and can carry up to 5.35 tons of propellants. It is equipped with a KB KhimMash S5-92 engine with 2.0 tons of thrust and a specific impulse of 327 s. It can operate for a total of 877 s, and can be reignited 20 times. There were two test flights, the first on February 9, 2000 with the IRDT-1 inflatable shield demonstrator, and then on March 20, 2000 with a mock-up. Then this version was used to launch the Cluster satellites for ESA (two launches), Mars Express, Amos 2, Galaxy 14, Venus Express, Giove A, Globalstar (two launches), Radarsat 2, and Giove B. A total of 18 Fregat stages flew between 2000 and 2009: four Soyuz-U/ Fregats, nine Soyuz-FG/Fregats, and five Soyuz-2/Fregats. A heavy version is part of the Zenit 3F/Fregat-SB launcher for the Fobos-Grunt probe, Elektro L satellite, Spektr R, and others. The Fregat-SB features a nozzle lengthened by 200 mm and additional tanks – one helium tank, two spherical hydrazine tanks, two hemispherical bulges on each of the four propellant tanks, and a jettisonable ring tank for 3.1 tons of propellants.

In addition to the abovementioned, there are projects that were abandoned. For example, when designer G. E. Lozino-Lozinsky initiated the Spiral space shuttle project, it was envisaged that the 6.8-ton EPOS demonstrator would be launched by a Soyuz (11A511) in 1969-1970. It was supposed to be put into an orbit at an altitude of 150-160 km inclined at 51°. This configuration of the launcher would have been 37 m high, including 10 m for the payload. But the EPOS only made atmospheric flights using the MiG 105-11 aircraft, which is now in the Air Force Museum at Monino near Moscow.


## SOYUZ-2 "RUS"

The Soyuz-2 launcher program (14A14), also known as "Rus", began in 1991 with the intention of replacing the Vostok 2M, the Soyuz-U, and the Molniya-M by about 1996. The evolution of the Soyuz launcher took place in two phases: the introduction of a new digital control system on the Soyuz-2.1a (311.7 tons at take-off) and a new engine on the third stage of the Soyuz-2.1b. The new control system was to eliminate the need for rotating the azimuth of the platform, trim 200 kg from the weight of the vehicle, and reduce the extent of the stage fallback areas by 40%. It was supplied by NPO Automatiki of Sverdlovsk rather than SKB Polisvit of Kharkov in independent Ukraine. The new engine was an RD-0124 (14D23) of KB Khimavtomatiki (CADB) of Voronezh. It differed from the RD-0110 by having an integrated flux engine with four combustion chambers mounted on a two-plane gimbal, eliminating the need for vernier engines. The thrust remained at 30 tons but the pressure in the chamber was raised from 70 bars to 160 bars and the specific impulse extended by 33 s to 359 s. The overall performance gain was 15% (950 kg). The Fregat upper stage enabled the launcher to place payloads in high, elliptical, sun-synchronous, geostationary transfer and other orbits. The new fairing by the Samara Space Center (TsSKB-Progress) was 11.4 m in length and 4.11 m in diameter. It was finished in about three years, and is expected to be produced at the rate of one or two units per year. The launch platform tower was modified to service

The Soyuz-2.1a at Plesetsk.

The RD-0124 engine.

The Soyuz-2.1b at Baikonur with the Corot satellite.

the vehicle. With the Soyuz-2.1a, the performance increase was 250-300 kg to place a payload of 7.4 tons into an orbit at 200 km, and with the Soyuz-2.1b the increase was 1.1-1.2 tons for a payload of 8.6 tons into low orbit.

The inaugural flight of the Soyuz-2.1a took place on November 8, 2004, from Plesetsk. It had been delayed twice: first on October 29, then on November 6 because of issues with the onboard software. The payload (a Zenit 8 Oblik satellite) made a suborbital flight. This version launched the Metop A (4,085 kg) European satellite on October 19, 2006 from Baikonur, then the Meridian 1 military telecommunications satellite from Plesetsk on December 24, 2006. The Meridian 2 satellite followed on May 22, 2009. The Soyuz-2.1b version, with the RD-0124 engine on the third stage, made its first flight on December 27, 2006, from Baikonur. Vehicle N°001 equipped with the Fregat N°1013 stage carried the 600-kg French Corot astronomical satellite. As this was the maiden flight, CNES obtained a promotional price of approximately €15 million. If it had failed, CNES would not have had the resources to construct a second satellite. The second launch of the three-stage version took place on July 26, 2008 and launched the Persona 1 spy satellite from Plesetsk as Kosmos 2441. Then, on September 17, 2009, a four-stage version launched the first Meteor-M satellite along with six microsatellites.

Soyuz launches from the Guiana Space Centre at Kourou began in 2011. From this equatorial launch site, the capacity of the Soyuz-2.1a is 1.18 tons in medium Earth orbit, 4.23 tons in sun-synchronous orbit, or 2.85 tons in geostationary transfer orbit. For the Soyuz-2.1b the corresponding values are 1.57 tons, 4.9 tons, and 3.24 tons. When equipped with a new safety system for self-destruction in flight provided by ETCA of Belgium, a European telemetry system, and adaptations to accommodate the tropical climate it will become the Soyuz/ST.

## YAMAL

This project began in 1996 and involved building a Semyorka capable of launching the Yamal geostationary telecommunications satellites for Gazprom. For this, it was necessary to push performance up to 13 tons in low orbit or 1.8 tons delivered into geostationary orbit. The requirements were studied by RKK Energiya and its subsidiary in Samara, TsSKB-Progress and Motorostroitel. To increase the mass of fuel, the diameter of the central core block was increased from 2,050 to 2,660 mm. The RD-108 engine was replaced by either an NK-33 from the N-1 lunar rocket (177.4 tons of thrust on the ground) or by two NPO EnergoMach RD-120s (each

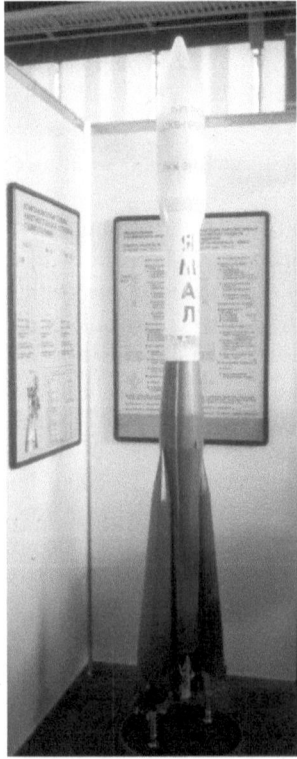

The Yamal launcher project.

delivering 80.7 tons of thrust on the ground). Thanks to the inventory of thirty NK-33s stored in Motorostroitel (they were imported to the United States by Aerojet) three years of development and five years of operation were assured. It was also planned to increase the diameter and mass of the third stage, equipped with the RD-0124 engine. For the fourth stage, the plan was to use a Block LM (or Taymyr) while the Molniya stage was equipped with an 11D58M engine from a Proton Block DM. The first launch was expected in 2003. But when it was realized that the 374-ton rocket would be able to put only 1.36 tons into geostationary orbit the project was abandoned, having already cost 871 billion rubles ($146 million). In 2001 it was renamed "Sotrudichestva" (Cooperation) as a joint Russian-Kazakh venture.

## AURORA

The Aurora version of the R-7 was to be launched from Christmas Island (105.37°E, 10.29°S) under an agreement with Australia. The APSC (Asia-Pacific Space Center) established in Sydney in 1995 had offices in Moscow, San Diego and Singapore. The shareholders were Australian, American and South Korean. The partners were

The Aurora launcher project.

RKA, RKK Energiya, TsSKB-Progress, KBOM, Gutteridge Haskins & Davey, and Saab Ericsson Space AB. The cost of the project was estimated at $408 million (launcher and platform). In June 2001 the Australian government made a loan of $51 million dollars to APSC, who wanted to start work on Christmas Island in September. At that time the first flight was projected for 2003.

The central core was equipped with a Motorostroitel NK-33 engine. Each lateral block had a 14D22 engine. The third stage was a Block I with an RD124E engine, with the chamber pressure increased from 70 to 160 bars. And the Block L fourth stage was replaced by the new Korvet stage derived from the Block DM and fitted with an 11D58MF engine. The digital guidance system was that developed for the Zenit 2 and Sea Launch Zenit 3SL launchers. The payload was to be 11 tons in low orbit, 6 tons in a circular orbit at 1,500 km, and 1.4 tons in geostationary orbit.

In July 2002 a rumor spread that the project was about to be canceled, but APSC officially denied it. Nonetheless, negotiations on building the launch platform were taking longer than expected and the maiden flight was pushed back to 2005. Finally, the competing project at the Guiana Space Centre in cooperation with ESA prevailed and APSC's project was abandoned. Since Energiya was already a partner of Boeing in the Sea Launch program and the International Space Station, Aurora's nickname became "Soyuz by Boeing".

The Onega launcher project.

The launcher project for the Kliper winged spacecraft.

## ONEGA

The Onega project began in August 2002. It was a "heavy" version of the Soyuz, and heir to the Yamal and Aurora. It was intended to use the RD-0126A engine (annular chamber and bell-shaped nozzle) on the third stage, and the SPPS helio-thermal system on the fourth stage. With a mass of 390 tons at takeoff, it would be able to put 15.3 tons in low orbit (in particular the Kliper spacecraft with its escape system) or 2.6 tons in geostationary orbit. The first flight was planned for 2007.

The May 2004 design was less revolutionary. It would have an RD-191 engine of 201 tons of thrust on the core block (with its maximum diameter increased from 2.95 m to 3.55 m), an RD-120.10F engine of 93 tons of thrust on each the lateral blocks, a cryogenic RD-0146 (R-095) engine on the third stage, and a cryogenic RD-0126 (R-097) on the Yastreb fourth stage. The engines for the upper stages would be supplied by KB KhimAvtomatiki in Voronezh: the RD-0146 having a thrust of 10 tons and a specific impulse of 463 s, and the RD-0126 having a thrust of 4 tons and a specific impulse of 467 s.

Although the Onega was to operate from any Soyuz platform (Baikonur, Plesetsk and even Kourou), the facilities would have to be modified to accommodate the 25% increase in take-off weight, an additional 600 mm in the diameters of the cryogenic upper stages, and other contingencies. Development was expected to last three or four years and the cost was projected at 4 to 5 billion rubles (€117-147 million), but the project had to make way for Soyuz-2-3.


## SOYUZ-2-3

In 2005 RKK Energiya proposed replacing the Soyuz-TMA with the Kliper/Parom system by 2015. Kliper was an 11-ton winged spacecraft to be launched by a Soyuz-2-3. This version was powered at take-off by an NK-33-1 with a deployable nozzle on the core block (195 tons of thrust on the ground) and an RD-120.10F engine (85.7 tons of thrust on the ground) on each lateral block. The third stage had four RD-0146 engines with 10 tons of thrust each. It would be able to put 11 tons (or 12.5 tons in an improved version) in a circular orbit at 200 km or 2.75 tons in geostationary transfer orbit from Baikonur. From Kourou the corresponding figures would be 12.7 tons and 4.8 tons (the significant improvement for a geostationary transfer deriving from the fact that there was no need for a plane change). But the Kliper quickly gained weight, increasing first to 13 tons and then to 16 tons. An improved version of

The Soyuz-2-3 launcher project.

The Soyuz-1, Soyuz-2 and Soyuz-2-3 family.

the Soyuz-2-3 was proposed in which the NK-33-1 would also equip the four boosters. Weighing 481 tons, the vehicle would be 46.3 m tall and have a fairing 4.11 m in diameter. For manned flights it would have a third stage equipped with the RD-0110 engine and be able to put 16 tons in low orbit. For unmanned flights the third stage engine would be an RD-0124 capable of placing 17 tons in low orbit, 5.5 tons in geostationary transfer orbit or 11.2 tons in sun-synchronous orbit.

After the Kliper was abandoned in favor of the future PPTS manned spacecraft, a final version of the Soyuz-2-3 was studied which would be able to place 10 tons in low orbit, 2.48 tons in geostationary transfer orbit or 6.2 tons in sun-synchronous orbit from Baikonur. From Plesetsk, the corresponding figures would be 9 7 tons, 2.1 tons, and 6.7 tons (the differences reflecting the plane changes for the two latitudes). In this version, only the core (the lower part of which was of expanded diameter) had an NK-33-1. The upper stage was that of the Soyuz-2. But this plan was abandoned in favor of the Rus-M. In 2009, Roskosmos entrusted the Rus-M launcher to RKK Energiya, TsSKB-Progress, and the Makeiev center. It was to be developed in three versions: light, medium and heavy capable of placing 3 tons, 6-8 tons and 20 tons in low orbit respectively. It would be launched from the new Vostotchny cosmodrome, starting in 2015. However, on October 7, 2011, the entire Rus-M project was canceled due to its high projected cost.

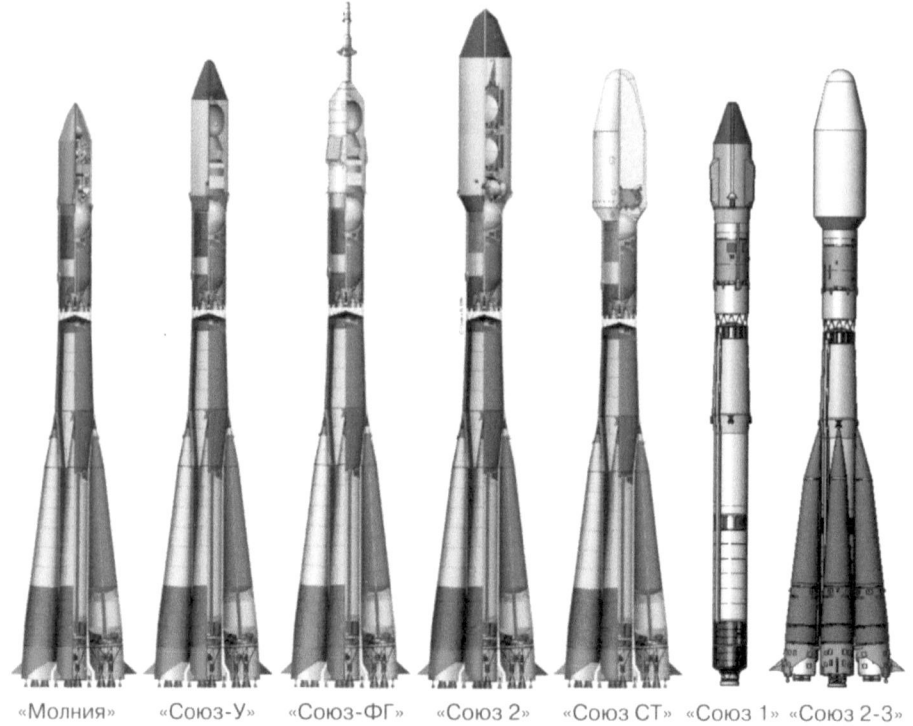

«Молния»   «Союз-У»   «Союз-ФГ»   «Союз 2»   «Союз СТ»  «Союз 1»  «Союз 2-3»

The Soyuz launcher family.

## SOYUZ-1

In 2008 TsSKB proposed a new two-stage version called Soyuz-1 in which the four lateral blocks were omitted. It would be 44 m tall and weigh 136 tons at take-off. The core would have an NK-33-1 instead of the RD-108 engine. The second stage would have an RD-0124. The diameter of the fairing would be 3 m. This small launcher was to be able to put 2.25 tons in low orbit launched from Baikonur (inclined at 51.8°) or 2.4 tons from Plesetsk (at 62.8°). It was to use the same platform as the Soyuz-2. In the fall of 2011 it was renamed Soyuz-2.1v. The maiden flight from Plesetsk was expected in the second half of 2012.

## CONCLUSION

The R-7 gave rise to three types of space launcher: 8A91 (Sputnik 3), 8K72L (Luna), and 8K72K (Vostok), while the R-7A led to development of the 8K78 (Molniya), 8A92 (Vostok 2), 11A57 (Voskhod), 11A59 (Polyot), and Soyuz (11A511). As of January 1, 2010, a total of 1,752 units of the R-7 Semyorka had been launched, with

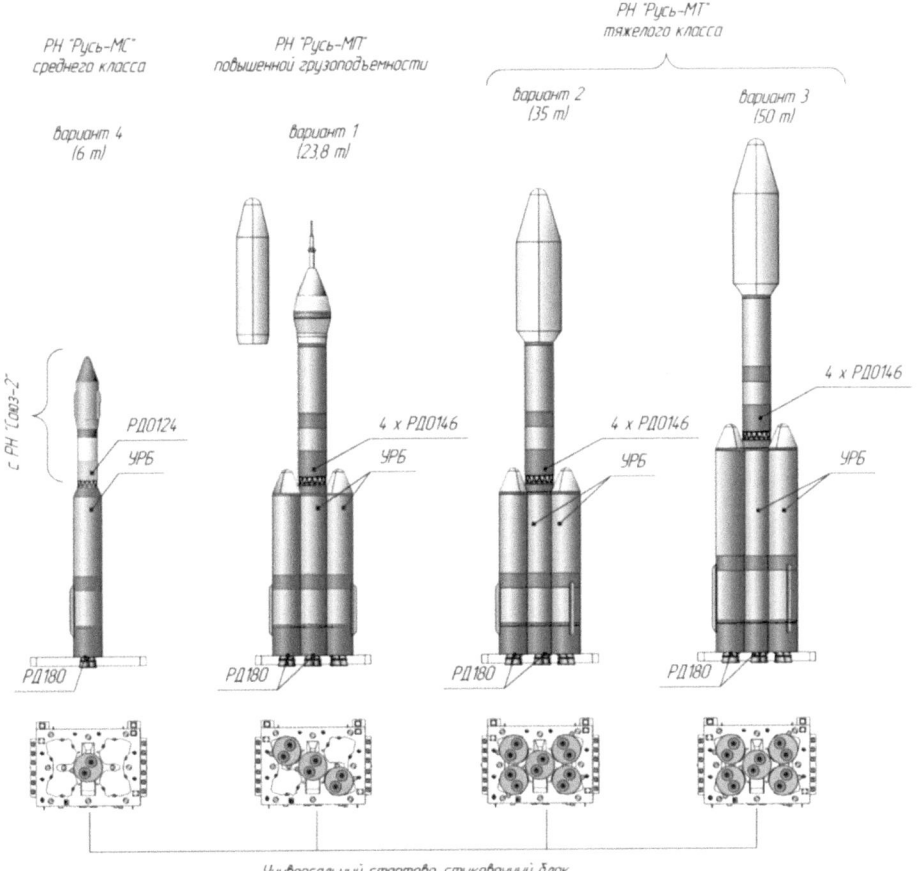

The Rus- M launcher family.

76 failures; a success rate of 95.7%. They were launched from six platforms: two at Baikonur (zones 1 and 31) and four at Plesetsk. The platform was originally planned for 10 launches but some of them have been used in excess of 300 times. A seventh and final platform was built at the Guiana Space Centre in Kourou, and was used for the first time in October 2011. This should be able to be used for several hundred launches.

# 6

# The launch bases

## BAIKONUR

On March 10, 1955, a directive from the chiefs of staff established the organization group from a selection of Kapustin Yar officers. Nine days later, General Nesterenko was appointed base commander, a post that he held until 1958.[45]

On January 1, 2009, Baikonur became a civilian site under Roskosmos. The only military facilities remaining at Baikonur were the 8th IU, which was responsible for the SS-18/15A18/Dnieper, SS-19/15A35/Rokot and 14A036/Strela ballistic rockets, and an aviation regiment. On June 10, 1955, there were 3,000 construction workers on the site. By late 1955 it already had 5,000 personnel. By late 1959 the village of Zarya had 8,000 people. At the end of 1960 Baikonur had 10,000 residents. A year later the population had reached 13,000. In the 1980s there were as many as 100,000 residents. But over the past fifteen years this population has decreased by half.

The first street was named Naberezhnaya. It was soon followed by Pestchanaya, Aviatsionaya, Pioneerskaya, and Oktiabrskaya (including the house at N°7 that was occupied by S. P. Korolev from 1955 to 1957), then Pervomaiskaya, Shkolnaya, and Sapiornaya. In Soldatski Park, 17,000 trees and 9,000 shrubs were planted. The first middle school (School N°30) in Shkolnaya Street was built of wood. It enrolled 136 students in the autumn of 1956. The following summer it transferred to a new four-story building. A 600-seat combined cinema and theatre was opened near Zone 0 (Nulevka) in November 1958. Initially, airplanes landed at the Joussaly airport, 100 km from Tyuratam. In April 1957 the first airfield, called Lastochka, opened with a dirt runway capable of handling Ilyushin Il-12 passenger transporters. The first hotel went up at Place Truda. Another hotel was built in Zone 0 (also called Nulevka) in 1959 for VIPs, including N. S. Khrushchev (October 1964), L. I. Brezhnev (1965), General de Gaulle (June 1966), Georges Pompidou (1970). The main designers had small individual houses of the Finnish type. For example, some of these houses were located in Zone 2, near the MIK. The first was occupied by Marshal Nedeline until his death in the R-16 accident on October 24, 1960. Then it was a residence where cosmonauts (including Gagarin) would spend the night before launches in the period

1961 to 1965. Starting in 1967, they resided at the "Kosmonavt" Hotel. The second house was Korolev's residence from 1956 to 1965. The third was shared by Mishin, Chertok and Voskresensky, while Pilyugin, Glushko and Riazansky occupied the fourth.

The first launch facility, built between July 20, 1955 and May 15, 1957 in Zone 1 would become known as the "Gagarin" platform. The Semyorka rocket arrived by train in seven cars, and was assembled in the launcher integration building (MIK) in Zone 2, some 1.8 km away. The MIK was housed in a building measuring 100 x 50 x 20 m. Nearby (until December 1966) was the storage building for nuclear warheads in Zone 2A. The platform was rebuilt once following the explosion of 8K71 N°B1-4 on July 10, 1958. It was repaired after the failure of a Vostok (8A92 N°E15000-01) launch on June 1, 1962, when an 8D74 engine shut down 1.8 s into the flight. Then it was renovated from June 1970 to February 1971, and again in 1979-1980. A second reconstruction was necessary between September 1983 and June 1984 because of the explosion of the launcher intended for the Soyuz-T 10 mission (V. G. Titov and G. M. Strekalov were lifted clear by the escape system). Finally the platform underwent a third renovation from July to December 1992, and in 1994 it was placed under the responsibility of KBOM. A total of 425 launches took place there from May 15, 1957 to January 2005.

The second Semyorka launch complex was constructed in Zone 31. Construction lasted from December 1958 to August 1960. It was fairly similar to Zone 1, but the MIK was closer to the launch platform. Zone 32 included living quarters for staff of the complex. The nuclear warhead storage building was in Zone 37 (until December 1966). The first R-7A missile flight was on February 14, 1961. Between September

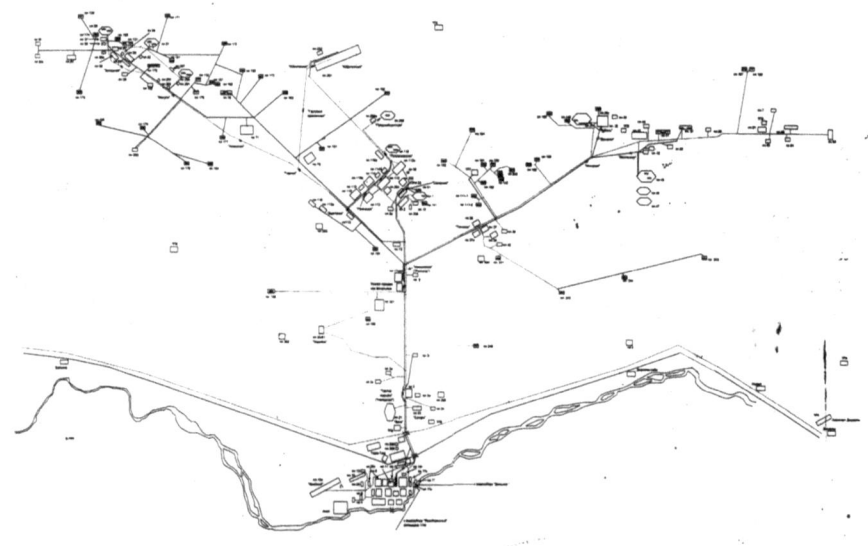

Map of Baikonur.

1963 and March 1964 the rocket was modified to launch satellites, and the first space launch took place from there on April 4, 1964 (Kosmos 28). On December 14, 1966, the first launch of the new manned spacecraft Soyuz (11F615 N°01/passive) was to leave from Zone 31, but one of the boosters could not reach nominal thrust and the rocket (11A511 N°U15000-01) did not take off. The onboard systems did a search for the fault. After 25 minutes the escape system triggered. But fire broke out on the rocket, which exploded, killing an officer who was on-site. The repair was made by transferring a platform from Plesetsk. It was completed in June 1967. From October 1968 to June 1970 some manned flights were carried out (Soyuz 3, 4, 6, 8 and 9). During the renovation of the platform in Zone 1 in 1979-1980, manned flights were again transferred to the Zone 31 platform (Soyuz 32 to 36). This was the case again after the Soyuz-T 10A accident (Soyuz-T 10B, 11 and 12). The platform was rebuilt in 1990-1991. In 1998 the complex came under the responsibility of KBOM. In 2005 the platform was modified to operate the Soyuz-2 with a 4.1 m diameter fairing. The launch of Giove B in April 2008 was the 359th launch from this platform.

The third launch complex was for M. K. Yangel's R-16 missile. It included Zone 41 with two launch platforms built in 1959-1960, Zone 41A with the storage building for nuclear warheads built in 1959-1960, Zone 42 with the MIK built in 1959-1960, Zone 43 with living quarters constructed in 1959-1960, Zone 60 with three silos built from September 1960 to September 1962, and Zone 61 with staff living quarters. During preparation of the first flight on 24 October 1960, the rocket exploded on a platform in Zone 41 killing 92 people – the so-called "Nedeline disaster". It has been turned into a memorial. The second platform was used for the qualification launches of the R-16 until February 1962. Zone 60 had three "Sheksna V" silos. In all, there were 31 R-16 launches from 1961-1962 and 168 launches of the R-16U from 1962-

The entrance to Baikonur in 1966 (sometimes referred to as Zvezdograd, "Star City").

Baikonur's central square.

Platform N°1, the "Gagarin" platform.

1972. The Zone 41 platform which remained intact was transformed to launch the Kosmos 1 rocket (65S3) of OKB-10 in Krasnoyarsk. Eight launches, including one failure, took place in 1964-1965. They put the Strela 1 satellites in orbit in clusters of three or five satellites, and the first Strela 2 (known as "Pteshla"). Then six launches (including two failures) of the Kosmos 3 (11K65) took place from 1966-1968 with the

Launching Soyuz-TM 15 in 1992.

Left: In 1989 Zone N°1 had made 314 launches. Right: By April 2008 Zone N°31 had made 359 launches.

The platform of Zone N°31 in April 2008.

Zone N°31.

Strela 2 satellites and two suborbital probes (VKZ) whose trajectories peaked at an altitude of 4,000 km. When the improved Kosmos 3M (11K65M) was introduced, it operated from Plesetsk.

Between 1978 and 1983 two launch platforms for the Zenit 2 (11K77) were built at Zone 45. The first launch took place on April 13, 1985, but failed due to a defect in the control of the fuel supply system. The ensuing launch on June 21 was also a failure due to an imbalance in the roll-control engines on the second stage. The first success came on October 22. But the launch on December 28 failed because the nose fairing did not separate. Starting in 1986, the Zenit 2 was used to launch Tselina 2 intelligence satellites, as well as Orletz 2, Resurs O1-4, Okean O, and Meteor 3M-1 satellites. A launch accident destroyed one of the platforms on October 4, 1990. The other platform has been managed by KBTM since 1998. In all, there have been 37 launches of the Zenit 2, including 8 failures. The Zenit 3SLB has been marketed by Space International Services and Sea Launch since 2005. It had its first launch on April 28, 2008.

The fourth launch complex was for Korolev's R-9A missile. It included Zone 51 with a pair of Desna N ("Romashka") platforms built in 1960, Zone 52 with a radio guidance station (RUP) built in 1960, Zone 53 with two tracking stations (SM-129 Signal and SM-130 Zarya) built in 1960, Zone 70 with three silos built from 1961 to 1963, Zone 71 with living quarters (where the 8th IU has been located since 1993), and Zone 75 with three platforms (two Desna N, and a semi-automatic Dolina) built from 1961-1963. There were 68 launches of the R-9A from 1961-1969. The complex was to have been used for flight testing the GR-1 (8K713) "Global Rocket", but this was canceled. Later Zone 51 was used to test Soyuz escape systems (EU-58, 1972-1973; EU-97, 1974-1975; EU-81, 1977-1978, and EU-81.4, 1986).

The fifth launch complex was for V. N. Chelomei's UR-200 (8K81). It included Zone 90, built from 1962-1963, and Zone 95 with living quarters built in 1963. Eight flights of the UR-200 took place in 1963-1964 (seven from the left platform and two from the right). It was modified to handle Yangel's Tsyklon 2A (11K67) launcher in 1966-1967. Eight flights (including one failure) took place from 1967-1969 (i.e. five IS anti-satellite interceptors and three RORSAT radar ocean monitoring satellites). In 1969 the platform was modified for the Tsyklon 2 (11K69) launcher. In 1988 fire broke out on the left-hand platform while preparing a launch. Repairs could not be financed and the platform was taken out of service. In 1998 the right-hand platform was turned over to KBTM. A total of 115 launches have occurred to date, especially for the RORSAT (active, nuclear thermal generators) and EORSAT (passive, solar panels) satellites. Zone 94A was used to prepare these nuclear generators from 1968-1988. The first BEU-5/Buk N°31 reactor was launched on the Kosmos 367 (4Ya11 N°310) satellite in October 1970. In all, 31 reactors were launched (29 Buks and two Topaz 1 on Kosmos 1818 and Kosmos 1867 in 1987). OKB Vympel has managed it since then.

The sixth launch complex was for Yangel's R-36 (8K67). It included Zone 67 with two ground platforms built in 1962-1963, Zone 80 with three silos built in 1962-1963, and the silos of Zones 140 (1965), 141 (1965) and 142 (1968). The first launch on September 28, 1963 was a failure. The Zone 67 platform was destroyed and the main designer of KBTM was fired (V. P. Petrov, who had been director since 1948, was replaced by V. N. Soloviev). The silos in Zone 80 were N°16, N°17 and N°18. On January 13, 1965, a launch from N°17 resulted in the rocket falling back into the silo. Then two successful launches took place from silo N°18 on April 27 and May 18, 1965. A total of 139 missiles were launched up until 1975. In August 1968 the first MIRV system (three warheads) was tested using an 8K67P. After five test launches targeting the Koura polygon in Kamtchatka, a campaign was organized in 1969-1970 with the warheads falling in the Pacific to demonstrate the maximum range. Then the missile was declared operational. The Zone 67 platform was modified from January-December 1965 for the 8K69Orb "orbital bomb". Four suborbital launches were made from December 16, 1965 to May 20, 1966. Two orbital launches took place from the silos of Zone 162 on September 18 and November 2, 1966; both of which failed. The first success occurred on January 25, 1967 (No. 7L/Kosmos 139). In all, there were 23 launches from 1965-1971. It was declared operational on November 19, 1969. The Americans described this as a "fractional-orbit bombardment system" FOBS. Three silos were installed in Zones 160 (1966), 161 (1966), 162 (1966), 163 (1969), 164 (1969) and 165 (1969), and were in service until decommissioning began in 1983.

The seventh launch complex was built for Yangel's R-36M (15A14). This heavy missile (referred to in the West as the SS-18 "Satan") was modified to become the R-36MU (15A18) and R-36M2 (15A18M). Its silos were in Zones 101-109. There were 85 launches of the R-36M from 1973-1984 and 49 of the R-36MU from 1977-1997. There have been 33 R-36M2 launches since 1986 (not counting silo ejection tests). Kosmotras marketed the Dnieper space version, with the first launch in 1999. The 11th Dnieper launch was on July 29, 2009. Thus far, this type has launched a total of 45 satellites.

The eighth launch complex was for Chelomei's UR-100 (8K84). It included Zone 130 (two ground platforms), Zone 131 (two OS silos), and Zone 132 (two OS silos) built in 1964-1965. The first flight took place on April 19, 1965 from Zone 130. The first launch from a silo was on July 17. The UR-100 was declared operational in July 1967. It was deployed in silos in Zones 170-180. Two silos were added to Zone 175 between 1975 and 1978. The UR-100M version was tested from December 1970 to November 1971 and integrated into the arsenal in October 1972. In all, 182 UR-100 launches took place until 1975 (170 UR-100s and 12 UR-100Ms). The UR-100K (15A20) version was tested from July 23, 1969 to March 11, 1971, followed by the UR-100U (15A20U) from June 16, 1971 to January 30, 1973. They were integrated into the arsenal on December 28, 1972 and September 26, 1974 respectively. A total of 97 launches of the UR-100K and UR-100U took place until 1977. An improved UR-100N (15A30) was developed from 1970-1973. The UR-100N was tested from June 1973 to December 1974 and declared operational on December 30, 1975. In all there were 63 launches until 1984. The UR-100NU (15A35) was tested from October 1977 to June 1979 and declared operational on December 17, 1980. There were 74 launches until 1998. The UR-100N was used as the first stage of the Rokot (14A01) space launcher which was developed from 1987-1990. It was equipped with a Breeze upper stage and carried the Nariad anti-satellite system (14K11). There were three launches, in November 20, 1990 (Zone 131), December 21, 1991 (Zone 131), and December 26, 1994 (Zone 175). The first two were suborbital and the third put the Radio-Rosto ham radio satellite into orbit. On December 5, 2003 the Strela (14A036) was launched from a silo of Zone 175 and put a dummy satellite into orbit.

The ninth launch complex was for Chelomei's UR-500/Proton (8K82). It included Zone 81 built from 1963-1965 (platforms N°23 and N°24), Zone 92 (integration building), Zone 95 (living quarters), and Zone 200 built from 1971-1976 (platforms N°39 and N°40). The first launch (with only two stages) took place on July 16, 1965 with the Proton 1 satellite which gave the launcher its name. The four-stage version launched Kosmos 146 in March 1967. The three-stage version launched Proton 4 in November 1968. The first geostationary launch was in March 1974 (Kosmos 637). Zone 81 was not used from 1979-1987, and its platforms underwent reconstruction from 1987-1989 and from 1997-1999. Platform N°39 underwent repairs from May to November 1986, while N°40 has been used since 1991 for Kazakhstan's Baiterek program. In all, by the end of December 2009 there had been 351 Proton launches.

The tenth launch complex was for Yangel's MR-UR-100 (15A15). It included the silos in Zones 170, 172, 173, 176, 177 and 181. Flight testing ran from December 26, 1973 to December 17, 1974, and it was declared operational on December 30, 1975. The 15A16 version was tested from April 27, 1977 to February 2, 1981. The 15A11 version (Perimetr program: in-flight command center to control other missiles) was launched from Zones 176 and 181 from 1982-1984. In all there were 55 launches of the 15A15, 26 of the 15A16, and 11 of the 15A11. On September 12, 1978, during a 15A16 launch from Zone 177 the rocket exploded in the silo, resulting in the death of a soldier.

The eleventh launch complex was built for the N-1 lunar rocket (11A52) and later converted for the Energiya-Buran (11K25/11F35) system. It included Zone 110 built from 1964-1969 (with the "Raskat" platforms N°37 and N°38), Zone 112 built from 1966-1967 (integration building), Zone 112A (rocket fuel station, and dynamic test area), Zone 113 (living area), Zone 250 for NII KhimMach universal tests (UKSS) used for the first launch of Energiya on May 15, 1987 (with the 17F19DM/Skif-DM satellite), and Zone 254 (integration building for the Buran space shuttle). Four N-1s were launched from Zone 110 from 1969-1972, all of which failed. This area was not used from 1974-1978 and was then modified between 1978 and 1988. It was used to launch Energiya-Buran on November 15, 1988. This was the only flight of the Buran space shuttle, which cost 20 billion rubles to develop. Zone 110 has belonged to NII KhimMach since 1996. Zone 112 has contained Starsem clean rooms since 1996. In addition, satellites which were integrated in Zone 2 were transferred to MIK areas in Zones 112 and 254. In 2002 the ceiling of the building in Zone 112 collapsed, killing eight people, and destroying an Energiya launcher and the flown Buran. In addition, the Starsem clean rooms required repairs. Buran N°2, whose flight in 1991 had been canceled, was in the fuel filling station (MZK), and the model of the Energiya-M was in the dynamic test bench (SDI).

From 1957 to 1999, 1,173 intercontinental missiles were launched, and 1,127 space launches took place from Baikonur (1,344 domestic satellites and 105 foreign ones). As of January 1, 2005 these figures had increased to 1,195 missiles and 1,230 space launches.

During its long history the Baikonur base has received a number of official visits. The first was that of N. S. Khrushchev and the leaders of the fifteen republics of the USSR on September 27, 1962, during the Cuban missile crisis. They witnessed the launch of an R-16 intercontinental missile and of Kosmos 9 (a Zenit 2 spy satellite launched by 8A92, N°T15000-006). On September 24, 1964, the Soviet government spent two days visiting Baikonur (operation "Kedr"). The delegation included Khrushchev, Malinovsky, Grechko (first deputy), Biriuzov (chief of staff), Epichev (political bureau), Krylov (RVSN-nuclear forces), Verchinin (air force), Sudetz (anti-aircraft defense), Gorchkov (navy), Serbin (central committee), Ustinov (council of ministers), Smirnov (military industrial commission), Pachkov (his assistant), Zverev (defense industry), Kalmykov (radio industry), Keldysh (Academy of Sciences), etc. Korolev, Yangel and Chelomei presented their achievements. They all observed the launch of R-9A, UR-200, R-16U, R-36 missiles and the Kosmos 46 satellite (a Zenit 2 launched by 8A92 N°R15001-05).

On June 25, 1966, General de Gaulle became the first Western chief of state to visit Baikonur (operation "Palma 1"). He observed the launch of two R-16U missiles from Zone 60 and Kosmos 122 (a Meteor satellite launched by 8A92 N°N15000-21). An R-36 had also been scheduled for launch from Zone 80, but it was canceled due to a fire in the silo. On October 20, 1966, operation "Palma 2" saw Brezhnev visit with the leaders of the socialist countries (János Kádár of Hungary, Willi Stoph of GDR, Nicolae Ceaucescu of Romania, Fidel Castro of Cuba, Yumjaagiin Tsedenbal of Mongolia, etc.). A UR-100, two R-16Us (silo and ground) and an R-36 were launched in the presence of guests who visited the N-1 launch platform and observed the launch of a satellite (8K78M N°N103-40 launched a Molniya 1). On October 29, 1969 in the context of operation "Palma 3", Brezhnev, Kosygin and Grechko visited with Czech leaders Gustáv Husák and Ludvik Svoboda, to observe the launch of an R-36 from a silo. On October 8, 1970, French President Georges Pompidou was the guest of operation "Palma 4", and saw R-16U and R-36 launches from silos as well as the launch of Kosmos 368 (a Zenit 2M satellite by an 11A57).

Mikhail Gorbachev visited Baikonur from May 11 to 13, 1987. He witnessed the launch of a Gorizont telecommunications satellite by a Proton and of Kosmos 1844 (a Tselina 2 satellite) by a Zenit 2. At that time, he requested the "disarmament" of the Skif-DM space laser demonstrator satellite (for the Soviet counterpart of the US "Star Wars" program) assigned as the payload of the first Energiya launch later that year.

In November 1988 President François Mitterrand went to Baikonur, landing in a Concorde on the Youbilényi runway, in order to observe the launch of the second Soviet-French manned mission (both of which were flown by Jean-Loup Chrétien). And then French President Jacques Chirac toured TsSKB-Progress in Samara on July 3, 2001.

## PLESETSK

On July 19, 1955, decree N°1313-749 established a commission to select an R-7 launch site in the northern part of the country. It was to be a complex of two or three

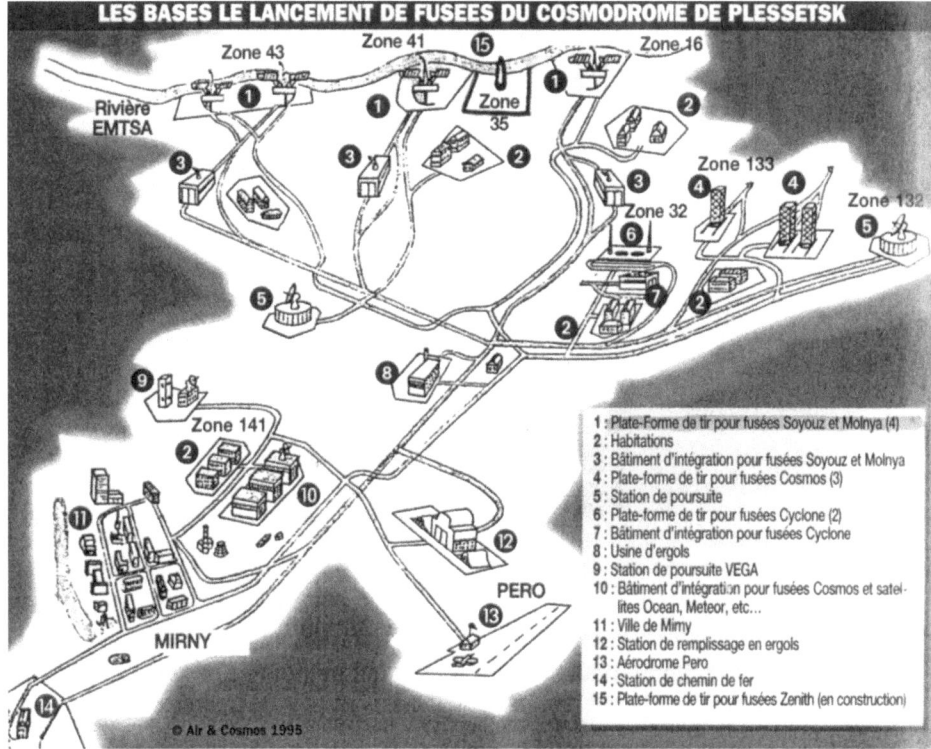

Map of Plesetsk.

launch platforms. General I. F. Dibrov investigated two sites: Project Volga in the Vorkuta region and Project Angara in the Arkhangelsk region. Angara was selected. The initial surveys began in 1956, before the Semyorka had actually flown. Decree N°61-39 concerning Project Angara was signed on January 11, 1957. The mailing address was initially Moscow 400, then Leningrad 300. The 57th regiment UIR of military engineers commenced construction (under the direction of General N. S. Stepanchenko, P. Z. Prestensky and others). The first echelon arrived in March 1957. The first commander was Colonel M. G. Grigoriev, who served as first deputy of the strategic rocket forces (RVSN) from 1968 to 1981. The first order was signed on July 15, which is taken to be the date of creation of the cosmodrome.

In July 1958 construction began of the city of Lesnoe (Mirny since August 1961). On February 7, 1959, the base was named the 3rd artillery training range (UAP) for the reason that it was still under the command of the chief of artillery in the ministry of defense. On February 20, 1959, the cost of the project was estimated at 1.547 billion rubles. Construction of four Semyorka launch platforms began at that time. The first (SK-1, later 317/1 Zone 41) was built between April 1957 and December 1959. On May 19, 1958, unit N°13973 (the 42nd BSS) was created to operate it. The first commander was Colonel G. K. Mikheiev. The unit carried out a test launch from Baikonur on July 30, 1959. The platform was declared operational on

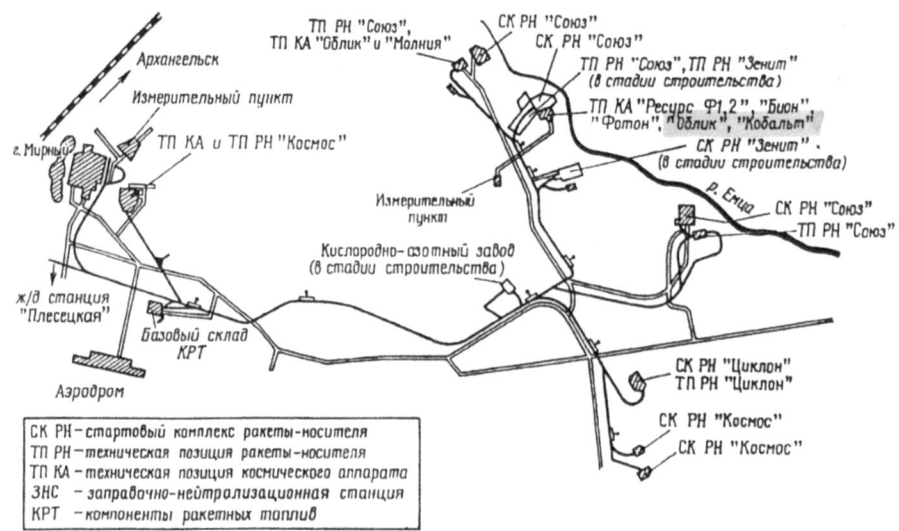

Упрощенная схема космодрома Плесецк

Map of Plesetsk.

December 15, 1959, two days before the formation of the RVSN. Modification of the first platform started on March 19, 1965. The first launch of the R-7A took place on December 14, 1965. The first space launch occurred on March 17, 1966, when a Vostok (8A92) launched Kosmos 112, a Zenit 2 (11F61) spy satellite. Then on April 6 a Voskhod (11A57 N°U15001-02) launched Kosmos 114, a Zenit 4 (11F69). The second platform (SK-2, later 317/2 Zone 16) was built at Gorny. Unit N°14003 (the 48th BSS) operated it. The first commander was Colonel N. I. Tarassov. It made its test launch from Baikonur on November 21, 1959. The platform was declared operational on April 15, 1960. It was transferred to Baikonur in February 1967 as a replacement for one that was damaged. A new platform was built in 1979-1980. Since February 1981 it has conducted 127 launches. The third and fourth platforms (SK-3 and SK-4, later 317/3 "left" and 317/4 "right" of Zone 43) were built in Skipidarny. They were operated by unit N°14056 (the 70th BSS). The first commander was Colonel G. M. Merzliakov. The first training launch at Baikonur took place on June 4, 1960.

The platforms were declared operational on July 15, 1961. The first launch of the R-7A from SK-3 (17P32-3) occurred on December 21, 1965, and from SK-4 (17P32-4) on July 25, 1967. Between March 22, 1968 and June 8, 1970, the two platforms were modified to accommodate the Zenit 2M, 4, 4M, and Meteor satellites. The first space launch from SK-4 was on December 3, 1969, with a Zenit 2M on a Voskhod as Kosmos 313. The first from SK-3 was on February 18, 1971, with a Zenit 4M on a Voskhod as Kosmos 396. Then the platforms were modified to launch Soyuz and Molniya rockets. On March 18, 1980, a Vostok 2M (8K92M) that was to launch a Tselina D satellite exploded on SK-4, killing 48 soldiers and injuring 42 others. The inquiry commission composed of L. V. Smirnov, S. A. Afanaseiev, N. N. Alekseyev,

V. F. Toloubko, A. A. Maximov, F. L. Cherniavsky, D. I. Kozlov, V. P. Glushko, V. P. Barmine, and others, did not find the cause of the accident. But the reason turned out to be a human error that was duly punished. It would take a reoccurrence in July 1981 to finally identify the cause as a lead weld on a hydrogen peroxide filter. When overheated it decomposed, then exploded in the supply pipeline. The platform was rebuilt with new fire protection and anti-explosion measures (PVB). The first launch was on April 8, 1983. Then on March 21, 1984, SK-4 launched the first Soyuz-U-PVB rocket. SK-3 underwent repairs from April 15, 1983 to December 26, 1986. But another accident occurred on SK-3 on June 18, 1987. During the launch of a Resurs F1 satellite, a Soyuz-U (N°77015105) exploded at a height of 100 m because of a foreign object in the oxygen supply to a pump in one of the boosters. The platform was severely damaged and trees were yellowed for a 300 m radius. The repairs lasted until October 1988. In the meantime, on July 27, 1988, another failure occurred on SK-4. During the launch of another Resurs F1 satellite, the Soyuz-U (N°78039130) exploded at a height of 200 m. This time, repairs took six months.

The four platforms were decommissioned from military service on July 20, 1968. On April 22, 1968, the decision was made to modify the cosmodrome to operate the military Soyuz VI in 1969. The first four launches were to occur between January and October 1971 using two spacecraft variants, but this program was canceled.

Today, platform N°1 in Zone 41 has been disassembled and will not be reused. As for platform N°2 in Zone 16, it is currently being used by Soyuz-U and Molniya-M launchers. Platforms N°3 and N°4 in Zone 43 have been modified to accommodate the Soyuz-2.

In August 1960, construction began on three divisions of R-16 missiles. The first launch complex (surface pads N°5 and N°6) was finished on October 27, 1961. The commander was Colonel M. T. Gushi. The second complex was finished on January 15, 1962 (pads N°7 and N°8), and the third (silos N°9, N°10 and N°11) on March 30, 1963. Within the context of the "Groza" maneuvers, an R-16 was launched from silo N°11 on October 22, 1963. In 1962-1963, two regiments of R-9A (two launch pads each) were installed at Maloe Ussovo and Bolshoe Ussovo (Colonel P. D. Goltsov). The first launch of an R-9A took place on May 16, 1967. On July 20, 1970, a decree ordered the conversion of the facilities of the R-9A to accommodate the Tsyklon 3 rocket that was to launch the Meteor and Tselina D satellites. The work was carried out from 1971-1976. The first launch occurred on June 24, 1977. On September 24, 1963, the 348th R-14 rocket division arrived in Plesetsk (Y. E. Suslov). An R-14U launch platform was built at Medvejny-Gory (Zone 131) to test elements of the Aldan missile system. The first flight took place on July 18, 1969, but was a failure. The second launch was a success. In all there were six launches through 1971.

In the period 1964-1966, KBTM in Moscow built a Voskhod platform (11P865) for the Kosmos 3M rocket (11K65M) in Zone 132, and a Raduga platform (11P863) for the Kosmos 2 rocket (11K63) in Zone 133. KBTM completed another Voskhod-type platform in Zone 132 in 1969. They were operated by unit N°63551 (Colonel I. M. Kozhemiako), and were declared operational on January 11, 1968 (Raduga) and December 30, 1971 (Voskhod). From 1964-1968 the Kosmos 3M was launched from platform N°41 at Baikonur, then transferred to Plesetsk.

The two platforms of Zone N°43.

A Soyuz platform at Plesetsk.

Launch of a Soyuz.

Launch of a Soyuz.

The first launch from platform N°1 of Zone 132 occurred on May 15, 1967, with Kosmos 3M N°16LM carrying the Tsyklon navigation satellite 11F617 N°1. But the nose fairing did not separate and the satellite did not function once in orbit. It was written off as Kosmos 158. A notable failure was that of June 26, 1973. A Kosmos 3M was being prepared for the launch of a Tselina O (11K616) intelligence satellite. It was still in the mobile tower when an explosion occurred and fire broke out. Nine soldiers died in this tragedy. As of December 31, 1998, 186 launches had taken place

Launch of a Soyuz.

from platform N°1 (including 12 failures) and 172 from platform N°2 (including 10 failures). On July 21, 2009, the launch of Kosmos 2454 was the 422nd launch of a Kosmos 3M from Plesetsk. A total of 816 units of this rocket were launched, with 30 failures; a 96.3% success rate. There were 462 orbital launches of the 65S3, 11K65 and 11K65M, and 354 suborbital flights of the K65M-R K65UP/Vertikal and K65M-RB/Bor versions.

The first launch from Zone 133 was on March 16, 1967, with Kosmos 148, a radar calibration satellite (DS-P1-I N°2) from the Youjnoé OKB. The last was on June 18, 1977 (Kosmos 919). In all there were 90 launches of the Kosmos 2, including six failures. Next, the Raduga platform was replaced by a Voskhod (11P865P). As of December 31, 1998 there had been 39 launches, including two failures. The platform was modified in 1998 for the Rokot launcher (14A05), and the first launch from this 11P865PR platform was on May 16, 2000.

On May 8, 1962, General S. F. Shtanko was made the commander of the Plesetsk cosmodrome. He was in charge during the Cuban missile crisis. From September 11 to November 21, 1962, the four R-7A missiles were on maximum alert with nuclear warheads ready to be launched at the United States. In late 1964, Plesetsk had four R-7As, seven R-16s, and four R-9As in service. During the occupation of Prague in 1968, the missiles were on maximum alert from August 20 to September 2.

On January 2, 1963, decree N°13-5 transformed the ICBM base into a scientific research (NIIP) range, intended to launch satellites into polar orbit. Shortly after that, General G. E. Alpaidze assumed leadership, and on September 16 the merger of the 3rd UAP and the NIIP led to the creation of the 53rd NIIP of the ministry of defense. In July 1964 three directorates (IU) were formed: the first for military rockets (G. K. Mikheiev, 1964-1974; B. G. Zudin, 1974-1981), the second for space missions (V. M. Yibshitz), and the third for the computing center and the tracking networks (N. N. Borisov).

The first IU included the 48th BSS, the 70th BSS, and the five regiments of R-16 and R-9A missiles. After the destruction of the platform of Zone 31 at Baikonur on December 14, 1966, platform N°2 (48th BSS) was transferred from Plesetsk to Baikonur on February 3, 1967. The third and fourth platforms were withdrawn from military service and transferred to the second IU in 1968. The R-16 and R-9A launch platforms were withdrawn from the arsenal in February 1975. There was one launch of the R-7A, 107 launches of the R-16U, and 17 launches of the R-9A; a total of 125 launches. In 1974 the first IU became unit N°10939 for the light rockets Kosmos 2, Kosmos 3M and Tsyklon 3. The second IU was the space directorate led by Colonel V. M. Yibshitz (1964-1972), M. Y. Kolessov (1972-1974), G. K. Mikheiev (1974-1976), S. V. Essenkov (1976-1984), B. N. Morozov (1984-1986), then V. A. Grin (1986-1989). It included the 42nd BSS and unit N°63551. In 1974 it became unit N°07376, unifying the three Semyorka launch installations of areas N°41 and N°43. From 1979-1980, a fourth launch pad was constructed to replace the platform sent to Baikonur. It was operated by unit N°14003 (V. A. Vorobiev, 1979-1983, V. A. Grin, 1983-1984, etc.). The third IU managed the tracking network, which included the stations at Jeleznogorodny, Naryan Mar and Novaya Zemlya.

As of December 31, 1998, 1,500 launches had taken place (1,433 successes, 67 failures) placing 1,900 satellites in orbit (1,877 primaries and 23 subsatellites). They were carried by Vostok 2 rockets (six launches from March 17, 1966-February 28, 1967), Vostok 2M rockets (79 launches from February 28, 1967-October 28, 1983, including one failure), the Molniya-M (215 from February 19, 1970-September 29, 1998, including seven failures) Voskhod (166 from April 6, 1966-June 16, 1976, including 11 failures), Soyuz-M (eight from December 27, 1971-May 18, 1973), Soyuz-U (420 from May 18, 1973-June 24, 1998, including ten failures), Kosmos 2 (90 from March 16, 1967-June 18, 1977, including six failures), Kosmos 3M (397 from May 15, 1967-December 24, 1998, including 24 failures), Tsyklon 3 (117 from June 24, 1977-June 16, 1998, including six failures) and Start (two launches in 1993 and 1995). Of these, 894 launches involved versions of the Semyorka, 604 involved Kosmos and Tsyklon, and two involved the Start. On July 6, 2009, Plesetsk launched its 2,000th satellite.

The fourth IU (Colonel P. P. Sherbakov 1966-1975, I. I. Dolinov 1975-1986, etc.) was created in March 1966 to handle solid-fuel rockets: SS-13/8K98/RT-2 (1966-1968), SS-15/8K99/RT-20 (1967-1969), SS-16/Temp-2S (1972-1974), SS-24/RT-23 (1982-1988), etc. It included unit N°01349 of Y. A. Yashin (the future first deputy of the RVSN and later deputy minister of defense) which had been formed in February 1965 to flight test Korolev's RT-2. There were 25 flights from November 4, 1966 to October 3, 1968 (including 9 failures) with a salvo of three rockets being fired on August 28, 1968. Then the modernized RT-2P version was tested between January 16, 1970 and 1972 (51 launches). In total, the directorate made 142 launches of this kind of rocket. In addition, 12 of Yangel's RT-20s were launched between October 24, 1967 and October 1969. There were many failures and the missile was canceled. Then in 1972-1974 the directorate carried out 35 launches of Nadiradze's Temp-2S missile. This program was curtailed by the SALT disarmament agreements. Initially, it had been intended to deploy the Temp-2S at Plesetsk. It would have been operated by the fifth IU specially created for this purpose. From 1982 to 1988, two versions of V. F. Utkin's RT-23 were tested: 32 launches of the RS-22V from 1984-1988 and 28 of the RS-22A from 1986-1988. The sixth IU (Colonel M. D. Zholudev) was created in August 1983 for the RT-2PM Topol missile. Fifteen test launches took place from 1983-1985. In all, there were 68 launches through 1988. In 1992 the directorate took charge of the Start launcher derived from the Topol, making two launches in March 1993 and May 1995. In 1994, it began flight testing the Topol-M. A grand total of 456 intercontinental missiles were launched from Plesetsk, including 56 failures.

In 1982, the first and second directorates were merged under the responsibility of the deputy director of the cosmodrome for space. B. G. Zudin held this position from 1981-1985, then V. Y. Riazantsev 1985-1986 and B. N. Morozov from 1986-1989. In December 1989 it became center N°1278 (TsIPKS). In July 1992 the space forces (VKS) separated from the strategic rocket forces (RVSN) and became independent. A year later, TsIPKS became the main center (GTsIPKS). On November 11, 1994, decree N°2077 declared it the leading state cosmodrome (1st GIK). But in December 1997 the VKS was reabsorbed by the RVSN, and the launch range (53rd NIIP) and the cosmodrome (1st GIK) were reorganized as a single entity. As a result, during the period December 1989 to December 1997 Plesetsk was divided into a military base and a cosmodrome.[46]

# Part Two –
# Soyuz in the West

by Stefan Barensky

# 7

# A fantasy launcher

On the morning of October 5, 1957, one of those cleverly terse press releases – a specialty of TASS, the Soviet news agency – introduced the R-7 launch vehicle to the collective imagination of the 20th century:

> For several years, scientific research and experimental design work have been conducted in the Soviet Union on the creation of artificial satellites of the Earth.
>
> As already reported in the press, the first launching of satellites by the USSR were planned for realization in accordance with the scientific research program of the International Geophysical Year.
>
> As a result of very intensive work by scientific research institutes and design bureaus, the first artificial satellite in the world has been created. On October 4, 1957, this first satellite was successfully launched in the USSR. According to preliminary data, the carrier rocket has imparted to the satellite the required orbital velocity of about 8,000 meters per second. At the present time the satellite is describing elliptical trajectories around the Earth, and its flight can be observed in the rays of the rising and setting sun with the aid of very simple optical instruments (binoculars, telescopes, etc.).
>
> According to calculations, which now are being supplemented by direct observations, the satellite will travel at altitudes up to 900 km above the surface of the Earth; the time for a complete revolution of the satellite will be one hour and thirty-five minutes; the angle of inclination of its orbit to the equatorial plane is 65 degrees. On October 5 the satellite will pass over the Moscow area twice, at 1:46 a.m. and at 6:42 a.m. Moscow time. Reports about the subsequent movement of the first artificial satellite launched in the USSR on October 4 will be issued regularly by broadcasting stations.
>
> The satellite has a spherical shape 58 centimeters in diameter and weighs 83.6 kilograms. It is equipped with two radio transmitters continuously emitting signals at frequencies of 20.005 and 40.002 megacycles per second (wavelengths of about 15 and 7.5 meters, respectively). The power of the transmitters ensures reliable reception of the signals by a broad range of radio amateurs. The signals have the form of telegraph pulses of about 0.3 second's duration with a pause

of the same duration. The signal of one frequency is sent during the pause in the signal of the other frequency.

Scientific stations located at various points in the Soviet Union are tracking the satellite and determining the elements of its trajectory. Since the density of the rarefied upper layers of the atmosphere is not accurately known, there are no data at present for the precise determination of the satellite's lifetime and of the point of its entry into the dense layers of the atmosphere. Calculations have shown that owing to the tremendous velocity of the satellite, at the end of its existence it will burn up on reaching the dense layers of the atmosphere at an altitude of several tens of kilometers.

As early as the end of the nineteenth century the possibility of realizing cosmic flights by means of rockets was first scientifically substantiated in Russia by the works of the outstanding Russian scientist Konstantin E. Tsiolkovsky.

The successful launching of the first man-made Earth satellite makes a most important contribution to the treasure-house of world science and culture. The scientific experiment accomplished at such a great height is of tremendous importance for learning the properties of cosmic space and for studying the Earth as a planet of our solar system.

During the International Geophysical Year the Soviet Union proposes launching several more artificial Earth satellites. These subsequent satellites will be larger and heavier and they will be used to carry out programs of scientific research.

Artificial Earth satellites will pave the way to interplanetary travel, and our contemporaries will witness how the freed and conscientious labor of the people of the new socialist society makes the most daring dreams of mankind a reality.

There was no name, no other information. The very concept of a launcher was hardly mentioned. The designation "Sputnik", which the Western press unanimously adopted in the hours following the announcement, is simply the Russian word for "satellite". The word was so rare in bilingual dictionaries that some commentators even translated it as "fellow traveler".

But the thing which surprised observers even more was the fact that the Soviets, who were supposedly trailing the Americans, had managed to get the jump on them with a satellite that weighed 83.6 kg! Dr. Joseph Kaplan, who chaired the American program of the International Geophysical Year, described this as "just fantastic". The satellite of the American Vanguard project, on which the press had published many reports, was just 10 kg, and would only fly if the launcher demonstrated its viability by placing a satellite in orbit weighing 1.4 kg and of a size and shape that inspired its nickname of "the grapefruit".

The first obvious conclusion was that, contrary to the conventional wisdom of the day, the Soviets in fact possessed a powerful launcher – far more powerful than the arsenal of missiles developed by the United States. In addition, as the announcement of upcoming launches implied, in all probability they also had the ability to produce

Thanks to its simplicity, the first Sputnik had everything it took to become an icon, overshadowing the real technical prowess represented by its unseen launcher.

them in large numbers. In reality, this information was not new. Thirty-eight days earlier, on August 27, another Tass press release published in Pravda announced the first successful flight of the R-7 as:

> ...a multi-stage intercontinental ballistic missile with a very long range.
>
> The missile tests have been successful; they helped to fully confirm the accuracy of the calculations and the choice of design. The missile flew to a very high altitude, which has never before been reached. Covering a huge distance in a very short time, the missile reached the selected area. The results show that it is possible to launch missiles to any region of the globe. The solution to the problem of building intercontinental ballistic missiles makes it possible to reach remote areas without the use of a strategic air force, which is currently vulnerable to modern anti-aircraft defenses.

The press release, though exceptional in its wording and its purpose, went totally unnoticed by the Western media. With Sputnik, the potential threat became tangible.

Shrouded in the morning haze of Baikonur, a Soyuz rocket is prepared for launch in 2004. For ten years, its design was concealed from Western eyes.

## WAVE OF PANIC

That October day in 1957 marked a lot more than just the idea that mankind had entered the "space age". The United States of America was jolted awake with the unpleasant realization that it had suddenly become vulnerable, lacking any effective defense against enemies capable of sending a nuclear conflagration wherever they wanted. The world's balance of power was threatened because the centralized Soviet system, with its planned economy, had dealt an overwhelming blow to the capitalist world and the market economy with its "invisible hand". An irony of history is that the wave of panic translated into a sudden interest on Wall Street for shares in the missile industry. American columnists despaired at this demonstration of Soviet scientific superiority over the United States, pointing the finger at an educational system that was deemed deficient. Dr. Elmer Hutchisson, director of the American Institute of Physics, put it bluntly: "...unless we give children more disciplined scientific schooling, our way of life is doomed".

While the launch itself went almost unnoticed as an unexciting "technical" event relegated to an inner page of Pravda on October 5, the panicked reactions of the Western media, especially in the US, were featured in the October 6 issue. The reaction was such that Washington officials immediately raised the pressure a notch to attempt to defuse the crisis. William A. Holaday, special assistant to the secretary of defense for guided missiles, explained in the New York Times that Sputnik's success did not necessarily mean the Soviets had technological superiority in the field of missiles and rockets, and that indeed it must have cost them a great deal of time and money to be the first. And Rear Admiral Rawson Bennett of the Office of Naval Research, in charge of the Vanguard program, stated in an interview with NBC that it was not a question of being in a "race", and dismissed Sputnik as "a hunk of iron that almost anybody could launch". He then speculated about a possible typographical error of transcription concerning its mass. Unfortunately for him, the stated mass was correct and Radio Moscow soon announced that Sputnik was not the only object to become "satellitized": the last stage of the rocket and the protective aerodynamic fairing also went into orbit.

Although the Eisenhower administration had anticipated the possibility of a Soviet launch prior to Vanguard – after all, the Soviets had been announcing for over two years[47] that they intended to launch a satellite for the International Geophysical Year – the White House was overwhelmed by the reaction of the public. As early as October 5, secretary of state John Foster Dulles sent a memo to James Hagerty, White House press attaché, with the intention of putting Sputnik into its proper perspective in the light of the government's official position: "The launching by the Soviet Union of the first Earth satellite is an event of considerable technical and scientific importance. However, that importance should not be exaggerated. What has happened involves no basic discovery and the value of a satellite to mankind will for a long time be highly problematical." He also mentioned the advances already made by the Nazis during the Second World War, and minimized the Soviets' input: "The Germans had made a major advance in this field and the results of their effort were largely taken over by the Russians when they took over the German assets – human and material – at Peenemunde, the principal German base for research and experiment in the use of outer space." This assertion takes on its full significance when we know that at the same time, Wernher von Braun and senior members of his team at the Redstone Arsenal in Huntsville, Alabama, were furious at having been denied permission to launch a satellite using a modified Jupiter-C missile as early as 1956.

John Foster Dulles continued in the same vein: "This encouraged the Soviets to concentrate upon developments in this field with a use of resources and effort not possible in time of peace to societies where the people are free to engage in pursuits of their own choosing and where public monies are limited by representatives of the people. Despotic societies, which can command the activities and resources of all their people, can often produce spectacular accomplishments. These, however, do not prove that freedom is not the best way."

After having congratulated the Soviet scientists for their "peaceful success", he encouraged Moscow to continue peaceful development and to "seek to enrich the spiritual and material welfare" of their people, then called on the Western powers to create a working group so that "outer space shall be used only for peaceful, not military, purposes".[48]

Perhaps the memo from John Foster Dulles seemed too close to misinformation. Whatever the case, it was not used to make an official statement but did serve as a background reference for President Eisenhower in his press conference of October 9, during which he admitted his embarrassment that the Soviets had taken the lead in the production of intercontinental ballistic missiles. But he attempted to minimize the impact of the Sputnik launch by reiterating that there was no race to space, because the United States had never considered launching a satellite to be a

After Sputnik's success, the US "grapefruit" weighing 1.4 kg appeared quite tiny, even if it was a symbolic payload in preparation for a 10-kg satellite.

One of the consequences of the stir caused by Sputnik 1 was the creation of NASA on October 1, 1958, by President Dwight Eisenhower (center), with the first administrator of the agency, Keith Glennan (right), and deputy administrator Hugh Dryden.

priority.[49] The launch of the second Sputnik on November 3 proved to any doubters that the success of the first was not a fluke, and implied that the Soviets had a stock of missiles. On December 6, the spectacular failure of the first Vanguard orbital test launch, which exploded after lifting just a few feet off the launch pad, drove the point home. In the popular imagination this gave rise to the myth that Soviet space capabilities were far ahead of their Western rivals, and triggered a race to space that resulted in men walking on the Moon just 12 years later.

## PHANTOM THREAT

The fantasy of Soviet technological superiority had its military corollary in the form of the "missile gap", the Pentagon's conclusion that the Soviets possessed nuclear ballistic launch capabilities far superior to those of the United States.

On October 7, 1957, the first secretary of the CPSU, N. S. Khrushchev, matter-of-factly announced during an interview lasting 3 hours and 20 minutes that the USSR would need to review its proposals for disarmament talks with the United States, as it was necessary to inspect launch sites rather than military airfields because bombers had become far too vulnerable and were doomed to cede the future to missiles. As missiles gave the USSR an exclusive advantage at that time, he was clearly hinting at Soviet strategic superiority.

Khrushchev had fully understood the value of Sputnik, which enabled the USSR

Broadcast live on television, the first US attempt to put a satellite into orbit turned into a spectacular fiasco, thus reinforcing the American public's feeling of powerlessness.

to demonstrate the capabilities of its ballistic system without the need to reveal the technical details of the missile itself. On the contrary, the mystery surrounding how Sputnik got into orbit added to the psychological effect on civilian populations.

Over the next ten years, the Soviets went to great lengths to maintain the secrecy surrounding the launcher that had won them first place in space, with no scruples about supplying disinformation in order to maintain the illusion of a strong nuclear strike capability which, in reality, the R-7 missile could hardly give them. It was this illusion of a "missile gap" which prompted the American military to accelerate its missile programs in order to counteract the apparent imbalance of terror that Khrushchev skillfully manipulated on the diplomatic scene; for example boasting during a trip to United Nations headquarters in New York on October 11, 1960 to support a proposal for disarmament, that Soviet factories could crank out missiles "like sausages!"

This vaunting of Soviet space achievements to instill in the minds of third-party countries the image of an all-powerful USSR gave rise to the expression "Sputnik diplomacy". The same day that Khrushchev gave his first post-Sputnik interview, the annual congress of the international astronautics federation opened in Barcelona and two Soviet papers were presented on satellites, but they gave no information on

Nikita Khrushchev met with John F. Kennedy in June 1961 in Vienna. By turning space into a zone of technological confrontation between the two superpowers, the Soviets pushed the Americans to launch the "Moon Race".

the launchers. On October 9, Pravda succeeded in the remarkable feat of publishing a long article on the achievement of Sputnik, written anonymously by S. P. Korolev and his team, which managed to mention no names, no places, no institutions, and gave no information whatsoever about the launcher.

In the absence of hard data, civilian analysts were reduced to guesswork. Some imagined that Sputnik was boosted into orbit by an air-launched vehicle which was released from a giant bomber at high altitude. On October 18, the British magazine Flight estimated that the Soviet launcher must weigh 80 to 100 tons at lift-off and consist of three stages and the launch site would be in Central Russia, 50-60°N and 50-60°E, perhaps in the Ural mountains. In its October 19 issue, the French magazine Paris Match had a diagram of Sputnik on a three-stage launcher taking off from the shores of the Caspian Sea. The article went on to extrapolate a future of lunar bases serviced by rockets of a configuration evoking the V-2 or the Moon rocket from the popular Tintin comic books.

On November 3 came the announcement of a second satellite launch, this time carrying the female dog Laika, giving the Soviets another first: the shock of the first Sputnik had not yet worn off, and the Vanguard program with its "grapefruit" had yet to attempt to leave the ground. To make matters more painful for the Americans, at 508 kg this new Sputnik was six times heavier than the first, and was orbiting at an even higher altitude. On November 13, Pravda announced that in fact the satellite equipment was mounted directly on the top stage of the launcher, but did not disclose the total mass that had been placed in orbit. It would have certainly caused a further stir if it had been revealed that this exceeded 6 tons.

With this new success, Western scientists wondered what the Soviets would do next. Dr. Fred L. Whipple of the Smithsonian Astrophysical Observatory located in Cambridge, Massachusetts, speculated that when they launched the second Sputnik into orbit they might have sent another rocket to the Moon to detonate a hydrogen bomb at the time of the lunar eclipse on November 7, which would be visible from California to Iran, to the Pacific, and the Far East, as well as most of the USSR. Dr. S. Fred Singer of the University of Maryland claimed in US News & World Report magazine that the flash produced by such an explosion would appear several times brighter than the 'full' Moon. This magnificent fireworks display would mark the 40th anniversary of the revolution. Of course this did not happen, but the rumor says a great deal about Western anxieties about Soviet launch capabilities at that time.[50] The launch of the 1,327-kg Sputnik 3 on May 15, 1958, plunged Westerners into perplexity. How many different models of space launchers did the Soviets have? Was it possible that they were using combinations of fuels unknown to the West?

## MISLEADING DISPLAYS

The first clues about the Russian launchers were released in 1959, in relation to three Moon probes – promptly dubbed "Luniks" in the West. On January 2, 1959, the mass of the first probe was said to be 361.3 kg, but it was accompanied by an upper stage weighing 1,472 kg that reached escape velocity. After passing 6,000 km from the Moon, the two objects flew on into orbit around the sun to become the first artificial planets. The masses of the other two probes were essentially the same. The first lunar impact was achieved on September 13 with a 390-kg probe accompanied by an upper stage of 1,511 kg. Launched on October 4, the third probe transmitted pictures of the Moon's mysterious far side. Once again, the Soviets had sent more than 1,830 kg to the vicinity of the Moon – a mass equivalent to that announced for the future US manned Mercury capsule that would be placed in low orbit around the Earth! In comparison, Pioneer 4, the first American probe, which flew close to the Moon on March 4, 1959, weighed a mere 6 kg. In their presentations, the Soviets displayed a system composed of each probe, the upper stage of the launcher, and its protective fairing. The size of these assemblies was impressive and left analysts struggling to imagine the dimensions of the launcher capable of sending it all into interplanetary space. On February 4, 1960, the announcement of the latest orbit of a "Sputnik 7" weighing 6,483 kg again shook the observers.[51] The Czech magazine Letectvi + Kosmonautika revealed that the launcher could include several lateral boosters, but only had two, which would be triggered at an altitude of 2 km!

Disinformation efforts reached a peak with the Vostok program, and the flight of Y. A. Gagarin on April 12, 1961. The sport aviation commission of the USSR central aero club sent a telegraph on April 18 to the Fédération Aéronautique Internationale in Paris to certify records established during the first manned space flight. The launch site was given as Baikonur, a mining town some 350 km north-east

By presenting the entire upper portion of the 8K72 launcher as a lunar probe, the Soviets impressed Western imagination with the power of their space capabilities.

As with the first lunar probes, the Vostok appears to be a much larger spacecraft – with a quite different design – when attached to the Block E stage and the protective fairing.

of the true site of Tyuratam. The launcher was said to be "powered by six engines" (without specifying whether this referred to the engines on the first stage or the total number of engines of all the stages) which produced "20 million shaft horsepower". Analysts were at a loss to translate this unit into thrust. A Rolls-Royce engineer calculated it as 1,238 tons of thrust, a value almost double anything previously imagined.

The first photo of the Vostok 1 launch was published on April 14 in Moscow. It supposedly showed the tip of the launcher rising above the cloud of smoke from the lift-off. But it was retouched to such an extent that it only showed a black oblong form with a rounded top, similar to a sugar loaf, emerging from a cloud. No part of the launcher was actually visible. It was not until July that the Vostok first appeared in a film "First Journey to the Stars" shown at the Moscow international film festival. In fact the image showed the (Block E) upper stage and the fairing that concealed the spacecraft itself. The assembly bore the reference "USSR Vostok" in Cyrillic. The scale was implied by technicians working on the spacecraft, and gave the impression that it was not just 2.5, but 7 times larger than its American counterpart; once again fueling the obsessive fantasies about how powerful the launcher must be to place it in orbit.

This propaganda poster proclaims "Glory to the first Soviet Cosmonaut" and shows a much more fanciful launcher than the real Semyorka.

A retouched photo of the boarding for Gagarin's historic Vostok flight: The launcher has been airbrushed out, along with S. P. Korolev's otherwise anonymous silhouette.

Images of the launcher in flight appeared in September in the film "With Gagarin to the Stars", but they were taken at such a distance that only the smoke plume was discernible. Some analysts nevertheless concluded that it was possible to infer the presence of several engines. A photo taken under the launch platform at the time of ignition confirmed this, but plunged the same analysts into even deeper perplexity. They were far from imagining that what they perceived was the flames produced by 32 combustion chambers operating together.

The design of the Vostok spacecraft was finally revealed on April 29, 1965, when a life-size mock-up was displayed at the exhibition of "Economic Successes" held in Moscow. Once again, the upper stage was presented as if it were an integral part of the vessel. This deception was also used to give the impression that the design of the Voskhod spacecraft – announced as a maneuverable vehicle capable of carrying a crew – was quite different, which was not the case.

The first images of details from the Semyorka launcher appeared – skillfully selected to avoid revealing the exact overall design – in September 1965 in the film "First Steps into Space". They nourished the latest deception on the exact identity of the vehicle that had launched the Sputniks and the Vostoks.


## MISSILES ON RED SQUARE

At the height of the Cold War, May 9 was an important date for all military analysts, because in the Soviet tradition new equipment – or supposedly so – was revealed during the WWII Victory Day parade. In reality, it was also an opportunity for the Soviets to engage in a skillful display of misinformation by including supposedly operational equipment that in fact had not yet entered production, in order to mislead Western observers.

On this particular May 9, in 1965, two large missiles were paraded in Moscow's Red Square. These were the largest rockets ever revealed by the Soviets, which they presented as a semi-orbital bombing system (or fractional-orbit bombardment system – FOBS – in US terminology), and suggested were the military variant of their space launchers. Observers immediately made the rapprochement between these missile stages, linked together by a network of straps and rods, and the little that was known about the vehicle that had launched Vostok, thus believing they had finally identified the mysterious launcher which had enabled the Soviets to get the jump on the Americans four years earlier.

They could not guess that the only point in common was that these missiles and the R-7 had come out of the same design bureau: that of Korolev. In fact, these were the only two flight models of the GR-1 (Global Rocket), designated Object 8K713, developed since 1962 by Korolev's OKB-1 and finally abandoned in 1964 in favor of Yangel's R-36 (which led to the Tsyklon launchers). The GR-1 never flew, but its skillfully staged appearance in Moscow was sufficient for NATO to assign it the reporting name SS-X-10 "Scrag", and for it to be considered in the West for several years to be an operational system.

In the aerospace press a multitude of diagrams sprang into existence, extrapolating on the space version of the missile that was imagined to have helped launch the first probes to the Moon, Venus and Mars, as well as the Vostok spacecraft. In the French weekly Air & Cosmos, young Alain Dupas – future expert for CNES – assigned it a weight of 280 tons, producing 5,000 kN of thrust at take-

The two prototypes of the GR-1 missile on parade in Red Square on May 9, 1965.

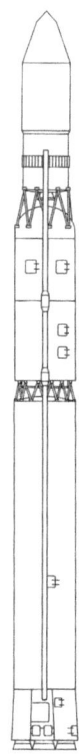

The GR-1 imagined as a "Vostok-launcher" in the French Air & Cosmos magazine.

off. In fact, the size of the missile had been overestimated: 40 m in height by 3.2 m in diameter, whereas in reality the GR-1 measured 35 m by 2.85 m. Without more complete information to go on, he attempted to identify the different versions of the "Scrag" through the test campaigns which the Soviets carried out over the Pacific between January 1960 and December 1964 (some were actually R-7A missiles, as well as R-9s and R-36s).

In early 1967, an engineer of the British Aircraft Corporation concluded that the "Scrag" was too small to have launched the Vostok and the Voskhod spacecraft, but probably launched the Elektron, Kosmos and Polyot satellites.

In fact, if it had flown, the GR-1 would have been a far less powerful rocket than the R-7. With a take-off mass of just 117 tons, it would have produced no more than 1,676 kN at launch for an orbital capacity of barely 1.5 tons at an altitude of 150 km, well below the capacity required to put a 4,750-kg Vostok in orbit.

## CIA SEEKING INTELLIGENCE

Although civilian specialists got bogged down in the incoherence concocted out of Soviet disinformation, CIA analysts succeeded in deducing a fairly representative picture of the R-7 missile and the space launch vehicles derived from it. Since its creation in 1947, the US Central Intelligence Agency had endeavored to expand its sources of information to try to determine what was actually happening in the Soviet Union. From 1949 to 1954, undercover agents were parachuted into the territory of the USSR, but most were quickly captured. Aerial reconnaissance was also utilized systematically by the end of the 1940s. From 1950, specially lightened stratospheric bombers, equipped with cameras, crisscrossed Soviet airspace at an altitude where Soviet fighter planes could not reach them. But after 1953, they proved vulnerable to anti-aircraft missiles.

That same year, regime change after Stalin's death led to more flexible exchanges between the Soviet Union and the West. The CIA then recruited tourists, diplomats, athletes and students to collect information such as the sites of antennas, factories, warehouses, or even photographs of equipment, including prototypes of aircraft or submarines. However, because a large part of the country remained inaccessible to foreigners, "human intelligence" was limited to the information transmitted by a few, very rare "moles".

In fact, the essence of what the CIA knew in the early 1950s on the development of the Soviet missiles came from veterans of Peenemünde who had collaborated with the Soviet teams between 1947 and 1950 (some until 1954) before being returned to East Germany once their expertise had been transferred over to the Soviet teams. By extrapolating on this basis, as early as 1954 the CIA analysts believed that the USSR would seek to develop intercontinental ballistic nuclear capability in the near future because US targets were too well defended against aerial bombing. This ability was predicted to be attained by 1963, or as early as 1960 if given developmental priority. For the CIA analysts, it was clear that if the US had not adequate countermeasures in place by then, it would have ceded a major military advantage to the Soviets that would pose an extremely grave threat.

In late 1955, new information must have been transmitted concerning the start-up of the R-7 missile program, because the CIA refined its forecasts to indicate that an

Stratospheric B-36 bomber, converted into a spy plane, at high altitude, early 1950s.

A U2 spy plane with fictitious NASA markings.

intercontinental missile was to be a priority and might be ready for series production by 1960-1961. But its performance was underestimated as being able to fly a range of 9,000 km with a payload of 1.4 tons. At the same time, a simple artificial satellite could be launched as early as 1958, as a "permanent Interdepartmental Commission on interplanetary communications" had been implemented in November 1954, with the development of a satellite supposedly in its mission orders.[52] No connection was made between the missile and the satellite, which it was assumed would be launched by a space variant of a medium range missile.

From January 1956 to September 1958, in the guise of "meteorological research", the CIA launched stratospheric balloons equipped with cameras from Norway and Turkey. Carried by the jet stream, they took 8 to 10 days to cross the USSR prior to being recovered in Japan or Alaska. Despite a loss rate of over 90% of the more than 500 balloons released, this program, dubbed "Moby Dick", covered 8% of Soviet and Chinese territories.[53] The Lockheed U2 spy plane made its appearance in Soviet skies on July 4, 1956, but after protests from Moscow, for several months it was restricted to intelligence missions along the borders.

In March 1957, two months before the first test flight of the R-7 missile, the CIA had to admit that it had no actual evidence of the development of an intercontinental missile in the USSR. It could only reiterate its estimates dating back to 1955, because the analysis of the strategic situation of the USSR indicated that such a development was indispensable and must have received top priority. The Pentagon even referred to it as the SS-6 "Sapwood" (meaning the sixth surface-to-surface missile inferred to exist in the Soviet arsenal). The imagined design (a two-stage missile with 150 tons of thrust at liftoff, powered by two engines of 1,000 kN in the first stage and one of 350 kN in the second stage) was very far from reality, but a rapid economic analysis convinced the CIA that 1,000 of these missiles might be deployed between 1960 and 1965 at a total cost of 33.8 billion rubles at 1951 value, including development. The forecast for the ability to launch a simple satellite was pushed up to 1957, in the context of the International Geophysical Year, with spy satellite capabilities by 1963-1965.

The intensive use of the U2 for reconnaissance overflights did not begin until late 1956. Its tasks included identifying production and launch sites for intercontinental ballistic missiles and mapping the rail network which linked them together. It was precisely during a mission to photograph launch sites near Sverdlovsk, and what later proved to be the site of the future Plesetsk cosmodrome, that American pilot Francis Gary Powers was shot down over the Urals on May 1, 1960. His capture and trial triggered a diplomatic crisis. In the meantime, the R-7 launch pad was photographed in the summer of 1957 at a site in Kazakhstan which the CIA chose to refer to by the name of the nearest train station: Tyuratam.[54] Prior to that, only one launch site was known: the Kapustin Yar range, near the Caspian Sea, where captured German teams had carried out test launches. The new facilities were of considerably greater size.

## PARANOIA IN WASHINGTON

In December 1957, after the launch of the first two Sputniks, the CIA drafted a secret report on the Soviet ICBM program which summarized available knowledge and estimates,[55] revealing in passing that the first failures of the R-7 went completely unnoticed. In this document – some portions of which are still classified more than fifty years later – the Tyuratam complex is correctly identified as the launch site of the Sputniks, and analysts acknowledged that the intercontinental missile and space launcher programs might well be very similar.[56] But the proposed design remained highly hypothetical, and different versions were even imagined for the two Sputnik launchers. It was believed to be a two-stage launcher, equipped with one or two 1,000 kN thrust engines (or one 2,000 kN) on the first stage, with a payload capacity of 1 ton and a range of 7 to 10,000 km depending on the version. A test program of 20 to 50 flights was expected, with 3 to 4 flights per month, leading to operational deployment. The CIA estimated then that the Soviets could have 10 intercontinental missiles available by 1959, 100 in 1960, and 500 in 1961 or 1962, at over 50 sites across the country. Later 500 more missiles could be added each year. The US Air Force had even more alarming estimates: with 1,000 missiles deployable by 1961! The paranoia of the "missile gap" was born.

The following summer, the absence of the predicted test flights unsettled some CIA analysts, who wondered how the Soviets could deploy their missiles without determining their accuracy and reliability in the course of intensive campaigns like those in the United States. In addition, performance estimates were revised upward, with intercontinental ballistic capability now projected not at 1 ton but at 2.5 tons, which implied a launcher far more powerful than previously imagined. A two-stage configuration was still thought likely, but the photographic observation of stages put into orbit with Sputnik 1, 2 and 3 indicated that they would measure approximately 25 m, which was considerable for an upper stage. Consequently, the idea of a design using one and a half stages – like the US Atlas missile[57] – or the use of stages "in bundles" was suggested. This was in mid-1958 and the predicted performance levels were still far from precise, but apparently someone in the CIA was already on the right track.

The Soviet advance impressed CIA analysts so much that they predicted probes being sent to the Moon, Venus and Mars by 1958-1959, men in Earth orbit by 1960-

Launch pad No.1 of Tyuratam and its impressive flame trench photographed by a U2 in 1959.

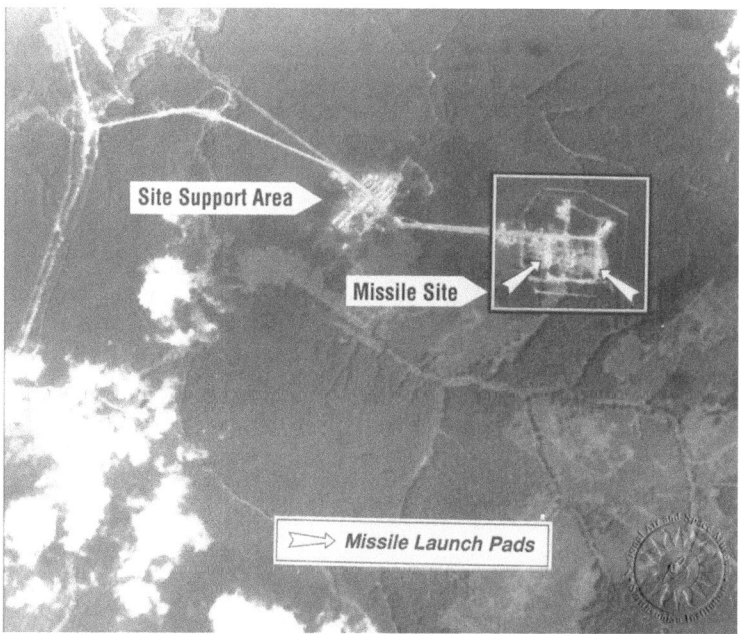

The Yuria test site, near Kirov, photographed by an orbiting Corona satellite.

Railway tracks leading to launch pad No.1 at Baikonur. To identify Soviet missile sites, US spy aircraft and satellites followed the tracks.

1961, followed by manned flight around the Moon in 1961-1962 and a lunar landing "after 1965" – although it pointed out that these objectives would not necessarily be met simultaneously.

However, U2 surveillance did not detect the movements and missile deployments corresponding to those forecasts. And it did not distinguish formally identifiable sites such as silos or operational launch complexes. A squabble amongst experts ensued, pitting the Air Force's adamantly alarmist predictions that perceived launching sites everywhere,[58] against the Army and Navy assessments which revised their estimates downward, and the CIA which tried to adopt an intermediate position by arguing that the deployment of the SS-6 was "imminent" if it had not already begun in 1959, and that the system might involve missiles transported by train to launch sites prepared in advance. In April 1961, the CIA considered that some SS-6s had been operational since early 1960, whilst the Navy did not believe that they had been deployed yet. In the absence of formal evidence, Western experts were reduced to pure conjecture.

In early 1961, a new evaluation of SS-6 performance – with a range that would be over 13,000 km compared to the 9,000 km range previously envisaged – forced the experts to revise their estimates on the possible location of operational launch sites,

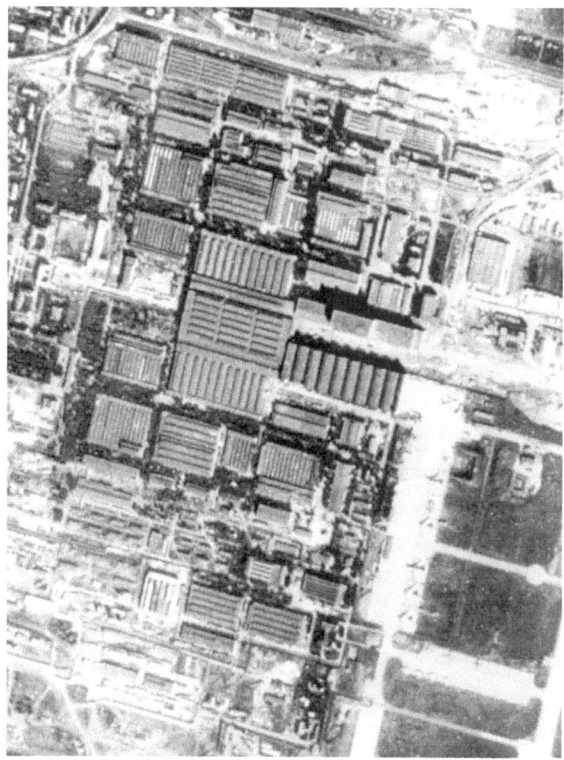

The R-7 production site at Kuybyshev, photographed by a Corona satellite.

which could be virtually anywhere in Soviet territory, even at Tyuratam where the construction of a second and then a third launch platform was detected. It was not until the summer of 1961 that the Soviets' strategy of bluff was revealed by the first real success of the US Corona spy-satellite program (in the guise of the Discoverer technology program). After three years of intensive research, on September 1, 1961 the Discoverer 29 satellite returned the first images of a true R-7 missile launch site at Plesetsk, with two launch pads, two deployed missiles, and two in reserve. Fifteen days later, all suspect sites were photographed from orbit and only Plesetsk proved to be an operational launch base. The Soviets had only two R-7 missiles deployed at a time that the US arsenal possessed 78 Atlas and Titan ICBMs. There really was a "missile gap", but it was in America's favor. As for the imagined design of the SS-6, the idea of two stages in tandem was abandoned as early as 1959. The propellants – liquid oxygen and kerosene – were correctly identified. However, the CIA imagined that two space variants existed: a 160-ton launcher for the first two Sputniks and another of 225 tons for Sputnik 3. As intercontinental missiles, these two launchers would be able to deliver warheads of 1.3 and 2.5 tons respectively. In late 1959 the reliability of the SS-6 was estimated to be 60%.[59]

In April 1961, the accumulated data on SS-6 flight testing enabled the CIA to refine its estimates, and presume a lift-off mass of 200-225 tons and a lift-off thrust

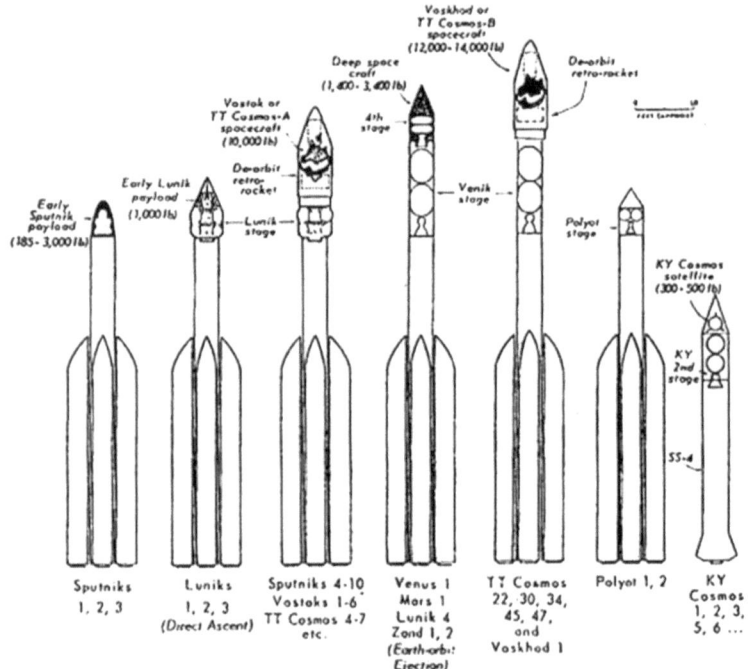

FAMILY OF SOVIET SPACE VEHICLES USED, 1957-1964

The CIA's interpretation of Soviet launchers in 1965: The general design of the Semyorka is clearly identified here.

of 3,500 kN provided by five engines, of which four were jettisoned at altitude "and probably their propellant tanks as well".[60] By that time, it had been established that there had been 14 space launches for the first Sputniks, "Luniks", "Veniks" and the prototypes and first model of the Vostok spacecraft, all of which used the same basic launcher derived from the SS-6, possibly with upper stages. In the aftermath of the Gagarin flight, the CIA began to have a fairly good impression of the reality of the Semyorka launcher, even if the capacity was still slightly inflated at 6,350 kg in low orbit due to having been extrapolated from data disseminated by the Soviets on the launch of "Sputnik 7".

In January 1965, the CIA published sketches detailing the different designs of the Semyorka used to launch the Sputniks, "Luniks", "Veniks", Vostoks, Voskhods and Polyots. The proportions were inaccurate and the launcher appeared much slimmer than it was in reality, probably due to an underestimation of the mass (assessed to be 230 tons), but the basic configuration was correct.

## THE GREAT REVELATION

The first Westerners to officially approach a Semyorka launcher were French – and VIPs at that – General Charles de Gaulle, then president of the Republic, his aide de camp Admiral François Flohic, and Foreign Minister Maurice Couve de Murville. On June 25, 1966, on the sixth day of his trip to the Soviet Union, de Gaulle had the privilege of becoming the first foreign visitor admitted to the site of Baikonur, in the context of a Soviet operation code named "Palma". At that time the exact location of the cosmodrome was a state secret, although Westerners had possessed very accurate maps for several years. To preserve the illusion, dummy signs were planted along the itinerary that was to be used by the visitors. In the space of a few hours, the French delegation observed the launch of a Vostok-M (8A92M) carrying the Kosmos 122 meteorological satellite, and the launch of an R-16U intercontinental missile that was known in the West as the SS-7 "Saddler". De Gaulle was very impressed by Soviet mastery of the technology.

The debriefing of the delegation on its return to Paris was never revealed, but the design of the Semyorka was now secret only to the general public. Another "Palma" operation was organized on October 8, 1970, with de Gaulle's successor, Georges Pompidou, witnessing the liftoff of a Voskhod launcher carrying the Kosmos 368 spy satellite. By then, the Semyorka had been revealed to the world for a long time.

By 1967, the secrecy around the launcher of Sputnik and Vostok was no longer necessary. As of 1965, thanks to Vladimir Chelomei's Proton launcher, the USSR had a launch capacity which exceeded that of Wernher Von Braun's Saturn I, and the three-stage version was expected to soon match the capacity of the Saturn IB. Development of Sergei Korolev and Vasily Mishin's N-1 lunar launcher was underway to compete with the American Saturn V. Whilst the Semyorka remained the workhorse of Soviet cosmonautics, with close to 40 flights per year, it was no longer the flagship.

General de Gaulle was the first Westerner to visit the Baikonur cosmodrome, but contrary to this newspaper headline, he was not alone.

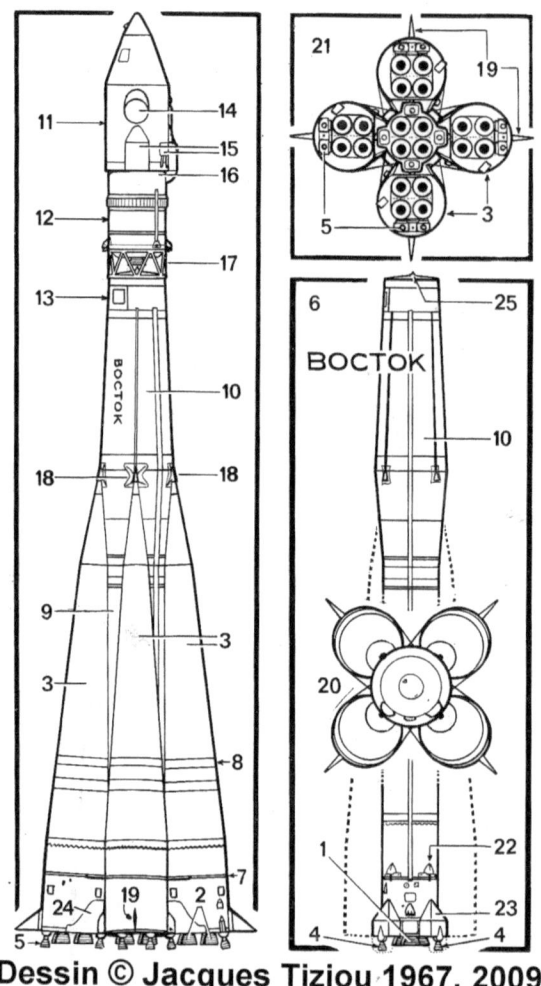

**Dessin © Jacques Tiziou 1967, 2009**

LEGENDES : 1: Moteur à quatre tuyère du corps central (deuxième étage, mis en marche en même temps que les accélérateurs. 2: Moteurs à quatre tuyères des accélérateurs (4). 3: Accélérateurs accolés (4). 4: Moteurs verniers (pilotage) du corps central (4). 5: Moteurs verniers (pilotage) des accélérateurs (4). 6: Corps central dépourvu des accélérateurs. 7: Fixation inférieure principale des accélérateurs. 8: Fixation supérieure principale des accélérateurs. 9: Réservoir inférieur de l'étage de propulsion. 10: Réservoir supérieur de l'étage de propulsion. 11: Coiffe de protection du vaisseau "Vostok". 12: Troisième étage mono-tuyère, également à propergols liquides. 13: Case d'équipements des étages d'accélération et de propulsion. 14: Ecoutille du "Vostok". 15: Carénage des liaisons électriques vaisseau - 3e étage.et fusées de séparation de la coiffe. 16: Jupe de liaison Vaisseau - 3e étage. 17: Structure tubulaire de liaison 2e-3e étages. 18: Réceptacles pour l'emboîtement des pointes avant des accélérateurs. 19: Empennages stabilisateurs mobiles participant au pilotage. 20: vue de dessus. 21: Vue de dessous, montrant les trente-deux (32) tuyères utilisées au départ (20 tuyères principales et 12 verniers). 22: Fixations du bloc moteur du corps central. 23: Carénages des verniers (alimentés en propergols par la même turbo-pompe que les tuyères principales. 24: Plaques de protection thermique des accélérateurs. 25: Noter la forme particulière du fond avant en cone très aplati et "à pointe" du réservoir supérieur.

The first three-view diagram of the Semyorka made by Jacques Tiziou on the eve of the 1967 Paris Air Show. It was widely circulated, including by the Soviets.

The horizontal integration of the Vostok launcher presented at the 1967 Paris Air Show impressed observers, particularly since the Soviet technicians casually walked on the body of the rocket.

As an intercontinental ballistic missile, the R-7 had played its role as bugaboo against the Americans, but its long, complex operational deployment, requiring over 20 hours of preparation on the launch pad, did not correspond to the needs of the new nuclear balance of power. The deployment of R-9 missiles since 1964, and especially that of the silo-based R-36 since 1967, made the old Semyorka obsolete in this role and its removal was scheduled for 1968.

In such a situation, secrecy could be lifted, and it was planned to do so in October, 1967, for the 10th anniversary of Sputnik, and as a prelude to the celebrations of the 50th anniversary of the October revolution.

In fact, the calendar was somewhat accelerated. For several years, two French journalists at Aviation Magazine, Jacques Marmain, a specialist on the USSR, and Jacques Tiziou, its space correspondent, had been prodding their Soviet contacts to obtain photos of the Sputnik and Vostok launches, and to convince them to create an event at the 1967 Paris Air Show, which would open its doors in June, by exhibiting more equipment than they had in 1965. By leveraging his political contacts, Jacques Marmain managed to bend them to his purpose just a few weeks before the opening of the show, with an agreement to present the mock-up of the Vostok launch vehicle – which was planned to be unveiled in October – at the highly publicized biennial international aerospace trade show. Consequently, a Soviet cargo ship docked in Rouen in late May 1967 and unloaded a series of crates that were transported to the Le Bourget aerodrome, north of Paris. In the week running up to the show, the Soviet technicians were occupied with the impressive spectacle of assembling a full scale Vostok launcher, with its four conical boosters and the nose fairing typically used for manned flights. It became the high point of the 1967 Paris Air Show, where several thousand visitors were able to gawk at it. The great secrecy was history.

## WHAT'S THAT ROCKET CALLED?

Although the Soviets unveiled the design of the Semyorka in 1967, it would take another 20 to 25 years, together with Gorbachev's *glasnost* and the collapse of the USSR, to finally understand the details of the different versions and their names. For lack of anything better, the Western press and analysts simply referred to them as the "Sputnik launcher", "Vostok launcher", and "Soyuz launcher". But more discerning analyses named the variants according to their most typical payload: Sputnik, Luna, Vostok, Venera, Voskhod, Meteor, Polyot, and Soyuz.

In an attempt to rationalize the situation, the American intelligence community developed two systems of designation that co-existed for nearly three decades. The first came from the Pentagon, which continued to refer to Soviet military hardware by a practice introduced during the Second World War for Japanese materiel. A code name was given to enemy equipment for purposes of taxonomy and identification in the field. Thus surface-surface missiles, without any real distinction in range, were designated "SS", while the space launchers were designated "SL". Consequently, as early as 1957, even before having flown or been properly identified,

| Equivalent designations | | | | | |
|---|---|---|---|---|---|
| | GRAU | Western | Sheldon | DOD | OTAN |
| R-7 | 8K71 | - | - | SS-6 | Sapwood |
| R-7A | 8K74 | - | - | SS-6 | Sapwood |
| Sputnik | 8K71PS | Sputnik | A | SL-1 | - |
| Sputnik 2 | 8A91 | Sputnik | A | SL-2 | - |
| Vostok L | 8K72 | Luna | A-1 (*) | SL-3 | - |
| Vostok | 8K72K | Vostok | A-1 | SL-3 | - |
| Vostok 2 | 8A92 | Vostok | A-1 | SL-3 | - |
| Vostok M | 8A92M | Meteor | A-1 | SL-3 | - |
| Voskhod | 11A57 | Voskhod | A-2 | SL-4 | - |
| Poliot | 11A59 | Sputnik, Poliot | A-m | SL-10 | - |
| Voskhod 2 | 11A510 | - | A-1-m | SL-5 | - |
| Soyuz | 11A511 | Soyuz | A-2 | SL-4 | - |
| Soyuz M | 11A511M | Soyuz | A-2 | SL-4 | - |
| Soyuz U | 11A511U | Soyuz | A-2 | SL-4 | - |
| Soyuz U2 | 11A511U2 | Soyuz | A-2 | SL-4 | - |
| Molniya | 8K78 | Venera, Molniya | A-2-e, A-2-m | SL-6 | - |
| Molniya M | 8K78M | Molniya M | A-2-e, A-2-m | SL-6 | - |
| * Should have been A-1-e. | | | | | |

the Soviet R-7 missile became the SS-6 in the West. This scheme made no distinction between the R-7 (8K71) and R-7A (8K74) versions.

For the space versions, however, a slight blur in the designations illustrated the uncertainties that existed concerning the actual configuration of the different models. Thus, SL-1 designated the launcher of Sputnik 1 and SL-2 of Sputnik 3 because they were believed to be different, although the precise manner in which they differed was not known. The launcher of the "Luniks" and the Vostok spacecraft was designated as SL-3, while that of the Voskhod and Soyuz spacecraft was the SL-4. The scheme was not necessarily linear. For example the SL-6 that launched the Molniya satellites came before SL-4. Subsequently, SL-7 and SL-8 designated Kosmos launchers, then SL-9, SL-12, SL-13 referred to Proton launchers, and finally SL-11 and SL-14 were Tsyklon launchers. SL-5 and SL-10 were attributed to two intermediate versions of the Vostok and Voskhod launchers, with two units of each being launched between 1963 and 1966 carrying the Polyot test vehicles and prototypes of the RORSAT radar ocean surveillance satellites.

The other nomenclature was developed starting in 1962 and published in 1968 by an analyst of the Library of Congress, Charles S. Sheldon. This assigned a letter to each family of launchers. "A" for the Semyorka family, "B" and "C" for the two families of Kosmos launchers, "D" for the Proton, and "F" for the Tsyklon. And if applicable, a digit was added to designate an upper stage version. If needed, a letter was added to clarify the specific capabilities of the system: "e" for an escape stage, "h" for a high-performance stage, "m" for a maneuvering stage, "r" for a re-entry system, and "s" for a satellite station-keeping system. In Sheldon's nomenclature, the launcher for Sputnik was "A", for Vostok was "A-1", for Soyuz was "A-2", and for Polyot was "A-m". Furthermore, depending on whether it was used for a mission in deep space or for a mission to Earth orbit, the Molniya became "A-2-e" or "A-2-m".

After enjoying their heyday in space literature devoted to Soviet astronautics,

With the Apollo-Soyuz mission on July 15, 1975, the world observed the first live lift-off of a Soyuz launcher, carrying the Soyuz 19 spacecraft.

these designations fell completely into disuse in the 1990s when information on the different versions of the Semyorka and their Soviet designations filtered out, thereby highlighting the limitations of these derived classifications.

## FROM FANTASY TO ICON

After having been a well-kept secret, the Semyorka became a symbol. In the absence of more precise data, the characteristic image of the launcher with its "bundles", as displayed at the Paris Air Show in 1967, went on to embody Soviet astronautics in the collective imagination of the West for two decades, particularly since the Soyuz "manned flight" version was regularly presented to the Western media in the context of international cooperation from the Apollo-Soyuz flight to Interkosmos missions with the Salyut 6 and Salyut 7 stations. It was natural, therefore, to suppose that the more powerful Proton would be a similar design on a larger scale.

For more than fifteen years, the Soviets kept the design of this launcher a secret in the same manner as they had for ten years with the Semyorka. Rather humorously, an encyclopedia of Soviet cosmonautics, published in 1970 at the direction of Valentin Glushko, a leading figure in the space industry, described the Proton simply as a launcher "of several stages and many engines". And the image of the Semyorka was so dominant that when experts finally saw the Proton launcher in its entirety in 1984 they misunderstood its first stage structure. They inferred it to be flanked by liquid-fueled boosters, whereas it was a distinct, but ingenious, concept in which six tanks of fuel were strapped around a central tank of oxidizer in a "revolver" configuration, with an engine at the base of each peripheral tank.

In contrast to the simple lines of another iconic launcher of that generation, the Saturn V, the stunning and very graphic design of the R-7 led it to enjoy a discreet career in popular culture. Its image appeared on monuments, posters and stamps of the countries of the East as a symbol of the Soviets' ability to access space, while in the West it represented the Soviet obsession for secrecy. But the lack of good quality images kept it off the big screen, where it was quickly eclipsed by the Saturn rockets and later by NASA's space shuttle. Nevertheless, the R-7 did make a few remarkable ethereal appearances in the film "The Right Stuff", by Philip Kauffman, released in cinemas in 1983, where it embodied the inexorable Soviet lead that bold American space pioneers had to overcome.

One of the finest tributes to the Semyorka came from Japan in the animated film "Oneamisu no Tsubasa" (The Wings of Honneamise), by Hiroyu Yamaga in 1987. This "anime" narrates the personal, psychological, and philosophical journey of a young officer designated to be the first man to fly in space. The story takes place on an alternative Earth, in a complex geopolitical context, and the launcher intended to take the young man into space, an enlarged version of Korolev's R-7, is used as a pretext by his chiefs of staff to draw a rival power into war. Like its real counterpart, this launcher designed for warlike purposes succeeds in its true mission of showing humankind the path to space.

# 8

## East meets West

When Mikhail Gorbachev came to power on March 11, 1985, a certain image of the Soviet Union was about to disappear. Unlike his predecessor, Konstantin Chernenko, who was the textbook incarnation of waning Brezhnevian conservatism, the new first secretary of the CPSU, remarkably young having just celebrated his 54th birthday, would attempt to salvage whatever could be saved of a Soviet regime that was losing momentum and at risk of imploding, even though this would involve questioning the dogma of the state since the days of Stalin.

The priority of the new regime was to establish an economic restructuring policy, *perestroika* as soon as April for the purpose of achieving a "relative" liberalization of internal trade and the operations of large state enterprises. In return, new accountability for results was demanded from the leaders of those state companies. This reduction of the party's omnipresence meant the beginning of the end of an opaque culture which, up until then, had concealed key economic players behind the state apparatus and a forest of committees. The new watchword was *glasnost* – a policy of transparency, which would progressively extend to many areas, allowing outside observers to finally discover the inner workings of the complex bureaucratic-industrial mechanism of the USSR.

In the space field, one of the first illustrations of this restructuring was the creation in October 1985 of Glavkosmos,[61] which was presented to the outside world as an inter-ministerial agency responsible for administering the Soviet civil space program. Aleksandr Dunaev was appointed to lead it, with the rank of minister delegate to the first vice-minister in charge of the space program at MOM,[62] of which Glavkosmos was in reality division No. 13.[63] From the outset, stormy relations developed with the minister, who was supposed to entrust it with a budget. Consequently, Glavkosmos never managed to impose itself as a true space agency like NASA or CNES. Under pressure from MOM, which retained full control of the programs, Glavkosmos found its real role limited to that of single contact point for all international cooperation.[64] It duly attempted to market Soviet launch services abroad.

Glavkosmos chalked up its first export success in September 1986 by signing a semi-commercial agreement with ISRO, the Indian Space Research Organization, to place the first Indian Earth-observation satellite, IRS-1A, in orbit, using a Vostok-M launcher. The contract was worth 75 million rupees ($7.34 million at 1988 value). Although officially non-aligned, India had a long history of relations with the USSR for obvious diplomatic and strategic reasons, since Moscow was the natural enemy of India's strategic rival, China, as well as of the United States, which supported India's other traditional enemy, Pakistan. The consequence of this diplomatic proximity, i.e., Soviet-Indian cooperation in space, was not a recent development. In April 1975, the first Indian satellite, Aryabhata, was launched by a Kosmos 3M rocket, followed by two Bhaskara satellites in 1979 and 1981, while India's first cosmonaut, Rakesh Sharma, flew aboard a Soyuz spacecraft to the Salyut 7 space station in 1984.

Nonetheless, putting IRS-1A into orbit aboard the next-to-the-last Vostok-M, on March 17, 1988, was the Soviet Union's first commercial launch in history. A second contract, valued at 220 million rupees ($15.8 million), followed in November 1988 to launch IRS-1A's twin, and on August 29, 1991 IRS-1B rode on the final Vostok-M. A long-term agreement was signed in February 1991. This included a firm launch of IRS-1C on a Molniya launcher for $12.5 million on December 28, 1995, with two

Mikhail Gorbachev, on the right, poses with US President Ronald Reagan at the Geneva Summit in November 1985, which marked the beginning of the end of the "Cold War".

The IRS-1B satellite, one of the Semyorka's first commercial payloads.

The last Vostok launcher – featuring "Glavkosmos" markings – being put in place for the launch of the Indian satellite IRS-1B.

options for IRS-1D on a Molniya and the Insat-2A telecommunications satellite on a Proton. The launch of IRS-1D at a price of $14 million was canceled in April 1996 when ISRO decided to assign the satellite to its national launch vehicle, known as the PSLV. As for Insat-2A, it was eventually the subject of a contract with Arianespace. In total, three Indian satellite launches earned the Soviet Union and then Russia over $30 million. However, this was minor, because at the same time Glavkosmos signed technology transfer contracts with ISRO worth ten times as much to help India to develop its own launchers.[65]

## COMMERCIAL EFFORTS IN THE WEST

Of course, India was not Glavkosmos's primary market target. The second half of the 1980s saw the sector for geostationary telecommunications satellites gain momentum with the emergence of new national, regional and even international operators, due in part to the rising power of Arianespace, which ended the US launch monopoly and the restrictions resulting from it.[66]

In the West, this period was marked by the return in force of expendable launch vehicles. Indeed, despite its brilliant technical success, the space shuttle failed to live up to its operational promises – as evidenced by the US Air Force's 1984 decision to return to conventional launchers. After the Challenger tragedy in January 1986, the White House directed NASA in June to cease carrying commercial satellites on the shuttle. The US launcher industry regrouped with the objective of breaking into the growing geostationary market, recently surrendered to Arianespace. In September 1986, Martin Marietta announced that its Titan 3 Commercial would be available as early as 1989. In October 1986, the Delta, whose production had ceased in 1982, was restarted by McDonnell Douglas. And General Dynamics resumed making the Atlas Centaur in June 1987.

This growing market also sharpened appetites in the USSR. In addition to the fact that *perestroika* was proving to be bitter medicine for the sluggish economy of the Soviet giant, it was also hit by the collapsing price of oil, its primary natural resource and major source of foreign currency. In 1988 a barrel dropped to $15; less than half its price in 1985. Marketing space launches looked like a good way to raise foreign currency. To penetrate this market, Glavkosmos joined forces with V/O Licensintorg, an import-export organization, and Ingostrakh, which provided insurance services. The goal was to offer satellite operators the range of Soviet launchers currently in production: Vostok, Soyuz, Molniya, Kosmos, Proton, and Tsyklon. Consequently, at the 1987 Paris Air Show, exactly twenty years after the unveiling of the Semyorka launcher to the world, the first commercial brochures appeared (signed Licensintorg) for Soviet launchers, and primarily the Proton.

This was not the first time that the USSR had offered its space launch services to the West. From 1972 to 1977, CNES entrusted three of its satellites to the Molniya and Kosmos launchers.[67] And in 1979 the Soviets even offered the European Space Agency the use of the Proton launcher – still largely secret – to put one of its Marecs maritime communications satellites into orbit, but ESA declined. In 1983, an offer was made to Inmarsat, which operated European Marecs and US Marisat satellites, and to which the USSR was the fifth largest financial contributor. This was to launch satellites in the Inmarsat 2 series. The first offer of $24 million was rejected. In 1985 the price was halved, still in vain. Adding the commercial dimension of Glavkosmos to this effort after 1986 still brought no success.

The Soviets were aware of the lack of trust in a world still split into two opposing blocs, and the word *glasnost* still raised suspicions. So in 1987 the Soviets attempted to play by the rules of the competitive economy and offered an attractive price "20 to 50%" less than Ariane or the Shuttle ($18-20 million). In addition, they promised that satellites entrusted to them would not be inspected. This was to allay Western fears that once satellites were handed over, the Soviets would try to appropriate the advanced technologies, possibly even going so far as to partially disassemble them, jeopardizing their operation. Dmitri Poletaev, in charge of launchers at Glavkosmos, tried to placate potential customers with the assurance that "representatives of the constructor can accompany their satellite during its transfer across the USSR and until it lifts off from the launch pad". It was even jokingly suggested that security officers could be "handcuffed" to satellites shipped to the USSR.

The cover of the sales brochure for the Proton launcher distributed at the 1987 Paris Air Show.

A poster issued by Glavkosmos and the Space Commerce Corporation in 1989. It offered all of the launchers of the Soviet inventory except the Zenit.

Integration of the first stages of the Proton launcher at the Khrunichev plant, in 1990.

In order to "Westernize" its image, Glavkosmos partnered with Space Commerce Corporation (SCC), a firm created in 1986 by Texas corporate lawyer Art Dula.[68] SCC was assigned to represent Glavkosmos internationally, and initially handled the Proton market. After January 1989, it dealt with the whole range of services offered by the Soviets – launches, data, equipment, technologies, as well as conferences, training, and even organized tours of facilities. The Proton was offered to Eutelsat to launch a Eutelsat 2 in 1986, and to Indonesia in 1987 for its Palapa B2R. But despite very competitive rates, both offers were turned down. The USSR even offered to fly an Indonesian cosmonaut to the Mir space station as a bonus in a contract to launch Palapa B2R on a Proton! Finally, a contract was announced in 1987 with satellite manufacturer Hughes[69] to launch an HS-393 satellite,[70] then another with Energetics Satellite Corp in 1989 for its Sat/Track system of eight satellites to be launched as an auxiliary payload. But none of these launches took place, particularly since the US department of state refused to permit the export of "Made in USA" satellites or their components to Soviet territory.

After 1990, major Soviet space companies such as NPO Energiya and the satellite manufacturer NPO Prikladnoy Mekhaniki (NPO PM) were bypassing Glavkosmos and attempting to negotiate directly with Western partners. This was also the case for TsSKB in Samara (formerly Kuybyshev),[71] manufacturer of the Semyorka family, and its KB Foton division which specialized in satellite observation and microgravity research. Taking responsibility for their own marketing, they attended international trade fairs.

Glavkosmos escaped MOM's grip in September 1991 to become the marketing agency of various Russian industrialists. But by that time most of its activities had already been taken on directly by the industrial firms and the design bureaus such as NPO Energiya and KB Salyut. Upon the creation in February 1992 of RKA, the new Russian space agency, the bulk of Glavkosmos' staff transferred over.

Meanwhile, the world in which this powerful military industrial complex had emerged, was crumbling. Weakened at its foundations by the dismemberment of the Eastern European bloc and the consequences of *perestroika*, the Soviet Union's *coup de grace* came from the putsch of conservatives who tried to remove Gorbachev on August 19, 1991, but in less than 72 hours merely undermined the credibility of the country's institutions. In the ensuing days and weeks, power wilted away, republics proclaimed their independence, and in just four months the Soviet giant was history. On Christmas Day 1991 Gorbachev resigned as president of the USSR, declaring the office extinct. He turned the powers which had been vested in the presidency over to Boris Yeltsin, the president of Russia, and the Russian flag replaced the Hammer & Sickle on the towers of the Kremlin.

## ONE WORLD ENDS, ANOTHER BEGINS

For the space industry, which lost some of its political protectors in the putsch,[72] the best approach was to join forces. The final months of the USSR and the first months of the post-Soviet era were a time of confrontation between the inertia of a system with little aptitude for change and the anarchy of a political situation in which new contacts abounded but no one was clearly identifiable as responsible. As early as September 1991 the major players in the military, industrial, space, and ballistics fields decided to establish their own organization to market their products. Their Rosobshchemash association presented itself as a "private" and Russian version of MOM. Rosobshchemash had a space division, "Kontsern Kosmos", which brought together thirty of the top names in the space industry[73] and generated a multitude of sales representation agreements in Europe and the United States before fading out in the face of commercial initiatives pursued directly by its members.

Officially, Russian space budgets, which had been frozen since 1990, underwent a reduction of 40% in current rubles, whereas inflation had increased prices eight-fold between 1990 and early 1992, and would increase five-fold again during that year. In reality, therefore, the resources allocated to the space industry shrank by 95% in two years. Some industrial teams, or even entire factories, received no income for several months and were left wondering what the formerly planned future would bring. The imperative to find ways to survive, coupled with ignorance of Western trade rules, resulted in an anarchical froth of offers that sometimes took the form of clandestine dumping, in which Westerners, themselves often poorly informed about the internal structures of the Russian industry, were struggling to see clearly. For example, it was not uncommon to see three or four representatives simultaneously attempting to sell global exclusivity for the same product. Glavkosmos, design bureaus, a production plant, or several, and even individual engineers or technicians, were all making rival offers to sell items such as launchers, engines, or any other kind of equipment which was in a warehouse for which they happened to have the keys!

One of the highest-profile examples was that of the Proton launcher, marketed "exclusively" by Glavkosmos and its partner SCC, but also by the KB Salyut design bureau that designed it, the MZ Khrunichev plant[74] which manufactured it and NPO Mashinostroeniya to which it reported in the organizational chart, not to mention the all-new Russian space agency, RKA. In November 1992 KB Salyut was selected to sign the first contract for a commercial launch of the Proton with Inmarsat to deliver a satellite of the Inmarsat 3 series into geostationary orbit. The contract's value was announced as $36 million. Immediately, MZ Khrunichev claimed that it would not deliver the launcher for the mission, contesting KB Salyut's right to sign contracts for the Proton launcher. The case went back to the Kremlin and was resolved in part thanks to the introduction of Western capital into the equation, specifically by the US aerospace giant Lockheed.

In the early 1990s, the Lockheed Missile & Space Company produced ballistic missiles, mainly the Trident missile for British and US nuclear submarines, having withdrawn from the space launcher sector upon ceasing production of the Agena upper stages used by Thor, Atlas and Titan launchers for almost thirty years.[75] In February 1992, the company's efforts at rapprochement with the former Soviet industry bore fruit. On getting the go-ahead from Viktor Chernomyrdin, the Russian prime minister, on December 17, and from the US state department on December 23, Lockheed and MZ Khrunichev created Lockheed Khrunichev International (LKI) on December 28 as an equally owned joint company for exclusive marketing of the Proton launcher outside the former USSR. LKI became LKEI in May 1993 when NPO Energiya, producer of the fourth stage of the Proton, joined in with one third of the shares. Lockheed made no secret of its intention to go hunting in Arianespace's territory. To close the deal, the American company agreed to invest 2 billion rubles over several years (approximately $5 million in 1992 terms). In accordance with a decree by Russian President Boris Yeltsin, KB Salyut merged with MZ Khrunichev in June 1993 to form GKNPTs Khrunichev[76] under the direction of Anatoli Kiselev, former director of MZ Khrunichev. This became known in the West as Khrunichev Enterprise.

The contract with Inmarsat was finally signed on April 27, 1993, and then had to await the lifting of the embargo that prevented Inmarsat satellites, built by GE Astro Space[77] in New Jersey, being exported to the Baikonur cosmodrome in the former-Soviet republic of Kazakhstan. The ban was lifted on September 2 of that year, when Russian Prime Minister Viktor Chernomyrdin and US Vice President Al Gore signed the first version of an agreement that placed a limit of eight contracts for commercial launches for geostationary satellites carried on a Proton which could be signed by December 31, 2000. This limitation was considered essential to prevent the Proton, manufactured under conditions quite different from those of the market economy, from posing unfair competition for Western launchers, especially US ones. Moreover, this agreement also prohibited the Russians from offering their launchers at a price lower than 7.5 percent below their least expensive Western competitor.

The Proton was therefore ready to begin its international career, but it was not the only launcher from the Soviet arsenal potentially capable of winning a share of the commercial market. Similar maneuvers took place involving the Ukrainian Zenit and Tsyklon, and the Russian Kosmos 3M as well as some ballistic missiles

Al Gore examining a nozzle of a Proton first stage engine during a tour of the Khrunichev factory in June 1995: Facing him are Yuri Koptev and Anatoli Kiselev.

destined for conversion. The motivation was all the greater, in that the outlook for the commercial launch market had just begun to diversify with the advent of constellations in low Earth orbit. The first of these was Iridium.

## BLOSSOMING OF THE CONSTELLATIONS MARKET

At the end of the 1980s, mobile telephony was still in a "prehistoric" phase. The first cellular phones appeared in Japan in 1979, then in Scandinavia in 1981. The United States had to wait until 1984 before a model developed by Motorola received FCC (Federal Communications Commission) approval. In Europe, adoption of the French GSM standard in 1987 facilitated the emergence of modern digital telephony starting in 1991 and leading to its generalization and subsequent democratization in the mid-nineties.

In 1990 mobile phones were analog, heavy, with low battery life, offering very limited coverage, and no roaming compatibility from one network to another. The mobile phone was reserved for the elite, in exclusively urban settings. But Motorola wanted to change that. On June 26, during a press conference held simultaneously in London, Melbourne, New York and Beijing, the US manufacturer struck a stunning blow by unveiling a project that its teams had been working on for over two years: a network of 77 satellites in low Earth orbit which would enable a mobile phone user to make a call from absolutely anywhere on the planet. Whether in the middle of the Sahara, the heart of the Amazon, on an oil platform, or in the Himalayas, there would always be at least one satellite in line of sight to receive a call and transmit it from satellite to satellite, directly to another subscriber, or to a gateway station that would reconnect to the international switched telephone network to reach a non-subscriber. With its 77 satellites orbiting the Earth like a myriad of electrons around the nucleus of an atom, the project was named Iridium in reference to the 77th member of the periodic table of elements. In 1992 the number of satellites was reduced to 66 but the name remained.[78] The cost of developing and putting the constellation into orbit was estimated at $3.45 billion, with an additional $2.8 billion for the first five years of operation, but the market was promising: a market study conducted by Motorola pointed towards 100 million subscribers by the year 2,000! Lockheed was selected to supply the satellites. In all, 120 were ordered in August 1993, enough to set up the system and ensure its maintenance.

However, raising the funds was more difficult than expected, partly because of an error in the initial presentation of Iridium. Its inter-satellite linking capability was seen by traditional landline operators – enjoying significant financial strength – as an attempt to bypass them, and pushed them into considering investment in competing systems that would soon appear. The most powerful of the competitors to Iridium was a program from Ford Aerospace, which made satellites, and its parent company, Ford Motors. In 1989 they began to study a satellite system to provide assistance to automobile drivers – and this included telephone services. With Loral's acquisition of Ford Aerospace in 1990, the concept was updated in an alliance with Qualcomm, which was already developing satellite messaging systems. The new project, dubbed

Early experience of an assembly-line approach to manufacturing satellites: preparation of structural panels for Globalstar satellites, at Aerospatiale in Cannes in 1993.

The main competitor to Soyuz in the constellations market was the Delta 2, in this case launching its tenth cluster of Iridium satellites on September 2, 1998.

A cluster of seven Iridium satellites mounted on the Block DM stage of a Proton launcher.

"Globalstar", was presented in March 1991 and called for establishing a constellation of 48 satellites in low orbit.[79] Since these satellites would be required only to connect a user to a gateway station in order to link in to the international switched network, they would be considerably simpler than the Iridium satellites. The investment was estimated at $2 billion. Loral's European partners – Aerospatiale, Alcatel, Alenia, and DASA – joined the project, as did France Telecom in 1994.

Iridium and Globalstar were not the only players in a market which analysts and consultants described as stunning.[80] The military satellite manufacturer TRW[81] was offering Odyssey requiring 12 satellites in medium orbit for $1.8 billion. Consortia formed by investors and manufacturers of small satellites and equipment were also making an appearance; for example, Constellation Communications with its Aries system composed of 48 satellites at $292 million, or Ellipsat with 24 satellites for $214 million. Inmarsat, faced with attacks on the marine telecommunications market where it enjoyed a monopoly, also announced its intention to put its own system in place, which could combine geostationary satellites and a constellation of 35 to 40 satellites in low orbit. Finally, at the start of 1993, the California holding company Calling Communications filed an application with the FCC to launch and operate a constellation of 840 satellites in low orbit, later known as Teledesic and intended not only for mobile telephony but also for digital broadband connections. The investment was estimated at $9 billion, but its backers included several Saudi princes and three well-known American billionaires: Bill Gates and Paul Allen, joint co-founders of Microsoft, and mobile-phone mogul Craig McCaw.[82]

The prospect of a market of dozens, even hundreds of satellites – especially taking into account the maintenance of the constellations after initial deployment – quickly led to the emergence of a bloated offering of small and medium launchers, based on existing missiles or engines as well as entirely new concepts. To receive their share of this boon, investors and manufacturers eyed the ballistic missiles of the former Soviet Union and even the United States. When Russian President Boris Yeltsin and his American counterpart George H. W. Bush signed the START 2 disarmament treaty on January 3, 1993, it meant that almost 2,000 missiles which would otherwise have to be dismantled could be transformed into space launchers by 2005.

Many of the prospective operators were seeking the ability to launch one or two satellites at once, but Iridium wanted to deploy its satellites in "clusters". In February 1993 an agreement was reached with LKI to launch three clusters of seven satellites using the Proton. In exchange, Khrunichev committed to investing $40 million in the project. In April 1994 another contract was signed with McDonnell Douglas for nine launches of five-satellite clusters using the Delta 2. As a result, by mid-1994 Iridium had enough launchers to orbit its constellation and had even signed with the CGWIC (China Great Wall Industry Corporation) for 10 launches of satellite pairs aboard the CZ-2C[83] as necessary to maintain the constellation.

The Iridium launch market totally bypassed Arianespace, because the European Ariane 4 launcher was sized for placing satellites into geostationary orbit rather than into low orbit. The Ariane 5, which was planned for 1995 and optimized to deliver pairs of satellites into geostationary orbit, might be adapted for this type of mission, but it appeared oversized and still had not proved its capabilities. Quite simply, the Europeans lacked a launcher capable of getting a toehold in this new market.

## LAUNCHERS TO COMPLEMENT ARIANE

The question of a launcher to complement Ariane was almost as old as Ariane itself. In 1973 the European partners had just barely come to an agreement to develop a launcher suitable for geostationary orbits with the Ariane 1, when the French military began wondering how to derive a smaller launcher from it to satisfy their potential needs. The concept of the "Z launcher" based on the two upper stages of the Ariane never got off the drawing board. In the seventies France had no military satellites and therefore no need for a launcher to put them into orbit.

A dozen years later, the concept of Marianne (or "mini-Ariane") proposed using a PAL (French acronym for liquid strap-on booster) from the Ariane 4 as a main stage, with PAPs (solid strap-ons) as boosters and upper stage to launch small scientific satellites. As such payloads did not yet exist, the project went no further.

By the second half of the 1980s the dynamic was changing, with on the one hand the success of the Spot optical observation satellite launched in 1986[84] that led to the development of a true family of satellites, and on the other hand the implementation by the European Space Agency of the Ariane 5 in 1988, on which CNES had been working for ten years, regularly increasing its power to enable it to carry the Hermes space plane.

As presented in 1978, the Ariane 5 pre-project used the first stage of the Ariane 4 flanked by its PAP strap-ons, and topped with a cryogenic H45 stage, loaded with 45 tons of hydrogen and liquid oxygen. By reusing the H10 third stage of the Ariane 4 it was possible to increase the payload for geostationary transfer orbit to 5,500 kg. But the big innovation was that the two-stage version would have been able to put into orbit a Hermes space plane weighing 10 tons. This would be the first European foray into the very exclusive club of manned flights. But the technology of the Ariane 4's Viking engines did not guarantee the level of reliability required for manned flights, and this configuration was abandoned in 1984.[85] A vehicle that combined an H120 cryogenic central core and P170 solid-fuel boosters (170 tons of propellant) was then introduced. The target capacity was now 5,200 kg in transfer orbit using an L4 upper stage with 4 tons of storable fuel.[86] Over a period of three years, this capacity rose to 5,950 kg while the total mass of the launcher increased from 500 tons to over 700 tons as the mass of the Hermes space plane, intended for low orbit, increased first to 16 tons and then to 22 tons.

Many voices were raised criticizing an Ariane 5 that was now too powerful for the market owing to the Hermes requirements. Although that criticism was questionable for geostationary transfer orbit,[87] it was undeniably true for remote sensing satellites derived from Spot, such as the Helios military satellites whose development was

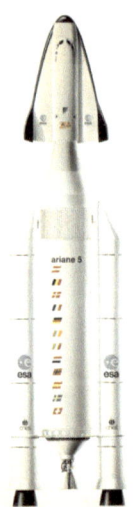

Left: Maturation of the Ariane 5 concept was very largely influenced by the desire to be able to carry the Hermes space plane. Right: A model of the Hermes space plane in a subsonic wind tunnel test.

decided on in December 1985. With a mass of 2-2.5 tons in sun-synchronous orbit at 800 km, these satellites already seemed much too light for even the least powerful version of the Ariane 4. The size of the Ariane 5 would theoretically have enabled it to lift four satellites at a time, but the strategy of these programs and the constraints of celestial mechanics required launching the satellites one at a time. The Ariane 4 was to be withdrawn from active service as soon as the Ariane 5 was qualified for commercial service. As a result, the remote sensing programs would quickly run into a shortage of launchers.

CNES studied one solution as early as 1988 under the leadership of Pierre Marx, one of Ariane's original designers. Since Spot or Helios would only take up a portion of the launcher, why not develop a modular payload capability in order to optimize performance? This led to the concept of the Cariane capsule. Starting with the shape of an Apollo capsule at a scale of 0.9 (which made it possible to benefit from a consistent database on its aerodynamic performance), this automatic reusable vehicle could carry the equivalent of four payload bays of the Columbus laboratory module intended for the Freedom space station (up to 900 kg of experiments) for 7-to-21-day missions in microgravity. An agreement was even reached with MAN

Work on Cariane was reused in the development of ESA's experimental ARD capsule (Advanced Reentry Demonstrator), launched in 1998 on Ariane 5's third qualification flight and retrieved successfully in the Pacific.

Technology AG in Germany for its development and industrialization, but the concept was not selected.

The development of a "Spot launcher" seemed unavoidable. However, the budget situation was not conducive. To start with, all resources were already mobilized for the Ariane 5, whose cost was revised upward by 5% in 1991. In that same year, the portion of ESA's annual budget dedicated to "launchers" (primarily the Ariane 5) reached the record level of 47.23%. In addition, Europe's financial resources for space were directly and permanently impacted by the costs of German reunification, as Germany was the second largest contributor to launchers after France. As long as the development of the Ariane 5 lasted, there would simply be no budget available to develop a complementary launcher.

And it was not just one complementary launcher that would have to be designed, but two.

Early in 1992, in the United States, the transition from the Bush administration to the Clinton administration brought changes at the top levels of NASA. Astronaut Richard H. Truly was replaced by engineer Daniel S. Goldin, who had spent twenty-five years at the TRW Space & Technology Group, becoming its chief executive. He arrived at NASA with radical ideas on how to restore the vitality of NASA programs. He proposed giving priority to missions which could be developed more quickly, and which offered better results at a reduced cost. The famous "Faster-Better-Cheaper" approach was born, which would generalize the implementation of smaller missions and so require smaller launchers.[88] The Europeans observed this policy with interest, because ESA's decision-making structure and financing system had given rise to large, long-term missions that were directly threatened by budget shrinkage.

The ESA council met at the ministerial level in November 1992 in Granada. The fact that just twelve months had elapsed since the session in Munich indicated how quickly the situation had changed. To many observers, this meeting would feel like a failure, due to the cancellation of Hermes – euphemistically termed "a reorientation" – and the reduction of funding allocated to the Columbus laboratory. While declining budgets and enhanced cooperation with the new Russia were on everyone's mind, the closing resolution nevertheless contained a footnote in which the council stressed "the need to search, in conjunction with the Member States, means by which the construction and launch of small satellites could contribute to achieving the objectives set out in the long term space plan". In other words, ESA would like to acquire its own "Faster-Better-Cheaper" approach and the necessary launchers; i.e., a medium launcher and a small one.

## POWDER FIZZLES OUT

As the designer of Ariane and operator of Spot, CNES was very interested in this type of development, particularly since it was studying its own Taos constellation for low orbit, which was to succeed the Argos localization and data collection system.[89] The French space agency was also involved in the military project of developing the

Zénon eavesdropping satellites[90] for which DGA (the French ministry of defense's procurement agency) had just initiated a feasibility study. The mass of these satellites corresponded to the targeted capability of the small launcher.

In mid-1991, at CNES launcher headquarters in Evry, the idea was put forward to derive these new launchers from current developments for the Ariane 5, primarily the solid boosters (EAP) loaded with 237 tons of propellant, each generating a greater take-off thrust than that of the most powerful Ariane 4. The first of these motors was successfully test fired on the bench in Kourou on February 16, 1993, in a reinforced "battleship" structure. This gave the go-ahead for the first ignition test with a normal flight structure on June 25, which was also successful.

In parallel with these initial tests, CNES and Aerospatiale co-financed a study on using the EAP technologies to create a modular family of launchers to be known as DLAs (light Ariane derivatives[91]). In addition to a DLA-S "Spot launcher" capable of placing 3 to 4 tons in sun-synchronous orbit at 800 km, there would be a DLA-P for satellites weighing less than 1,000 kg. These DLAs required the development of two new motors from shortened versions of EAP segments: an 85-ton motor (P85) and a 30-tons motor (P30). The DLA-P would combine the P85, the P30, and the last stage (L5) of the Ariane 5,[92] and the DLA-S would be built by mounting the DLA-P stages on top of an EAP.

The study clearly showed that the sun-synchronous market was still too limited to gain European political and economic support. Only Italy – very involved in the EAP via the Europropulsion consortium which built the solid-propellant motor for it – was interested in the DLA concept, but the DLA-P version came dangerously close to competing with Italy's own small Vega launcher projects.[93] A different design was proposed instead of the DLA-P which more closely matched Italian developments. This ESL (European small launcher) had two solid motors loaded with 50 tons of propellant (P50) as the first and second stages for a 7-ton stage (P7)

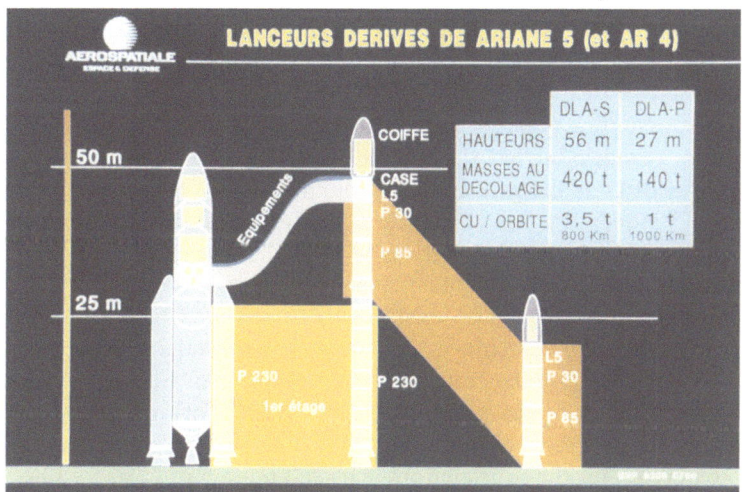

The DLA concept was developed for maximum reuse of Ariane 5 technologies.

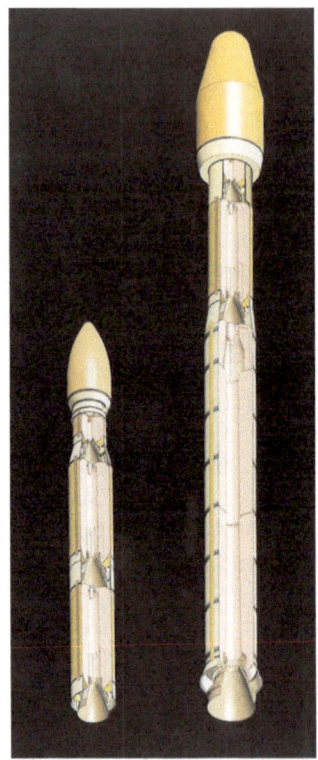

With the ESL and ACL concepts, the use of a P50 engine reduced commonality with
Ariane 5, but made it possible to optimize the designs.

and a hydrazine injection module. Unlike the P85, the P50s of the ESL would use a
wound composite structure like that of Vega's Zefiro motors. Except for a few
avionics components, the ESL had nothing in common with the Ariane 5. The
concept of the two identical stages represented a considerable economy of scale in
production... assuming that a development budget was forthcoming. But the project
faced a major obstacle in that the Italian Space Agency (ASI) wanted its national·
industrial firm BPD as the prime contractor, which CNES vehemently opposed.

François Calaque, director of launchers at Aerospatiale, had little patience with
political-industrial dithering, because he believed that the market would not wait for
the Europeans to come to some sort of agreement among themselves. In his opinion,
by the time they came to a compromise – which everyone would find distasteful – to
develop their own complementary launchers, the constellations would have been
launched by someone else. In June, at the 1993 Paris Air Show, he called together
journalists from the specialist press to talk about the work-in-progress. He did not
mince words. He told them that the DLA would be too expensive and would take
too long to develop for the constellation market, and in particular for Globalstar. He
then evoked a different track using existing components available in Russia. "If we
fail to finalize an industrial organization by the end of the year, with the objective of

a first flight in 1995, we will lose the battle for the small-application satellite market," he insisted. Five months later, in November, he revealed that he had made satisfactory progress on the medium launcher, and the definition would be completed before the end of the year with the objective of being able to orbit clusters of 5 to 6 satellites weighing 500-kg each for $35-40 million.

It was about time, since on January 24, 1994, Globalstar in turn initiated a call for tenders for the launch of its 48 satellites during 1997 and 1998. As for Iridium, it was clear that the Europeans, lacking an appropriate launcher and both the time and the budget to develop one, would not be able to respond effectively. François Calaque's "plan B" therefore arrived at the right time. On February 14, he finally revealed his plan to CNES and Arianespace. It involved modifying the Russian Soyuz launcher to meet Globalstar's needs, and turning it into the complementary launcher that Europe sorely needed. This was project "Irene".

# 9

# The genesis of Starsem

At the first signs of an opening in the Soviet market, French aerospace companies managed to build networks of contacts in the Russian and Ukrainian space industry. By cooperation initiated by General de Gaulle in 1966, Aerospatiale had worked with the Soviets, in particular to provide laser reflectors to equip the Lunokhod lunar rovers from 1970-1973 and to fly the deployable ERA structure aboard Mir in 1988. At the Moscow air and space show in July 1990, CNES organized meetings between the French and Soviet industries. Aerospatiale representatives participated, including François Calaque, director of launchers, and Michel Delaye, head of the space and defense branch. Arianespace was also represented, as were SEP (Société Européenne de Propulsion), Alcatel Espace, Matra Espace, and a multitude of suppliers. French manufacturers had discussions with their counterparts at TsNII Mash, NPO Energiya, NPO Lavochkin, NPO Molniya, NPO Tekhnomash, NPO Kompozit, NPO PM, NII-KP, NPO Yuzhnoe, TsSKB, and MZ Khrunichev. It was the starting point of many cooperative projects in the years that followed.[94]

Aerospatiale's role in the Hermes space plane project, for which ESA wanted to reduce the costs by cooperation with the USSR and then Russia, opened doors for the French industrial company throughout most of the industry, enabling it to evaluate capabilities through preliminary study contracts. Aerospatiale also enjoyed a positive image with its former-Soviet counterparts because of its role as prime contractor for France's strategic nuclear force ballistic missiles. As Hermes was losing momentum, the study topics were diversified. In particular, the technologies and systems for an "Assured Crew Return Vehicle" for what would become the International Space Station were evaluated. Another study explored the concept of an automatic supply vessel, LOVE (Logistics Vehicle) – the predecessor of the current ATV (Automated Transfer Vehicle).

After 1992, competition to get established became tougher because European manufacturers were seeking cooperative projects whilst their US counterparts were more willing to pay for appropriate equipment and technologies with financial backing from the Pentagon and NASA, along with political support from the White House.

Patrick Eymar was one of the top Aerospatiale people in charge of prospecting in

François Calaque and Yuri Koptev, during the 30th encounter on Franco-Russian space cooperation in Toulouse, April 1994. An agreement was signed on cooperation in satellite remote sensing.

Russia for draft projects and future systems. He was also very involved in the issue of small launchers because he had participated in various studies conducted by the industry on the topic since the end of the 1980s. After having tried to define designs based on motors from the French strategic nuclear force's missiles (with the possible addition of motors developed in the United States) these studies established that the main difficulty was the absence of either a large solid-propellant motor or a liquid-propellant stage that was powerful enough and cheap enough to constitute the first stage. Three solutions were therefore possible. The first was simply to purchase this engine in the US, where Thiokol had developed the 50-ton Castor 120 derived from the first stage of the MX intercontinental "Peacekeeper" missile. The second solution was to develop such a motor in Europe, either derived from the Ariane 5's EAP (the DLA concept) or using composite technologies (the ESL P50). An alternative option was to find an "off-the-shelf" engine or stage in the arsenal of the former USSR.

The RT-23U solid-fuel missile (SS-24 "Scalpel") with its 53.7-ton first stage, presented promising characteristics. Ukrainian manufacturer NPO Yuzhnoe offered an airborne version dropped with a parachute from an Antonov 124 "Ruslan" cargo plane.[95] However, many factors worked against this solution: the uncertain future of Russian-Ukrainian relations and therefore that of the 46 missiles based in Ukraine, in addition to the risk of having to maintain a production line which disarmament treaties listed for closure, plus the difficulties related to the use of a nitrocellulose-nitroglycerin double base propellant, with more energy yield than the polybutadiene-based propellant used for space applications.

Systems available in the former Soviet Union were the Rokot from KB Salyut derived from a UR-100UTTKh missile (SS-19 "Stiletto"), the MIT Start derived

from the RT-2PM Topol missile (SS-25 "Sickle"), the small Kosmos 3M launcher from PO Polyot, and the Tsyklon medium launcher from NPO Yuzhnoe. The last two offered advantageous performance and had been operational since the 1960s, so their reliability was proven. In addition, they were no longer related to any ballistic missile in service and were not affected by disarmament issues. So, based on this analysis, in early 1993 Patrick Eymar wrote a proposal to CNES to investigate the possibility of operating one of these launchers from the Guiana Space Centre, which would require introducing some changes in their power lines, as well as in their self-destruct systems (which in fact did not exist on the Soviet launchers; they merely shut down their engines if they strayed off trajectory). Eymar's proposal got a very cool reception from CNES, which then hired Aerospatiale for its feasibility studies on the DLA.

After the first contract was signed for the launch of Iridium satellites aboard the Proton, and the announcement that the rest of the constellation would be placed in orbit by the Delta, Arianespace realized that the "all-Ariane" offer did not correspond to market demand. This called into question the marketing drive that had been touted for two years and which had already pushed the management of Arianespace away from any interest in the DLA concept. In its defense, from 1989 to 1991 Arianespace had an exclusive agreement with Orbital Sciences Corp (OSC) in the US to market the Pegasus, a small airborne launcher, in Europe, and even though the creation of a joint venture had been considered at one point, nothing came of it.[96] The urgent need to get back on course was confirmed when Globalstar published its first deployment scenarios, and neither the Ariane 4 nor the future Ariane 5 appeared to represent a comprehensive and competitive response. Hence on February 8, 1993, Arianespace called for proposals from potential partners capable of providing a launch capacity complementary to the Ariane 4. The organizations consulted included PO Polyot for its Kosmos 3M, NPO Yuzhnoe for its Tsyklon, and ISRO, the Indian space agency, whose medium launch vehicle, the PSLV, was scheduled to make its first flight a few months later, as well as EER Systems in the US which offered its family of small Conestoga launchers composed of bundles of Thiokol Castor 4

The RT-23U (SS-24) missile nourished the hopes of some European industrialists for a while, but in the end there was no space derivative.

solid boosters.[97] The Italian solid rocket motor manufacturer BPD, which was working on its San Marco Scout launcher at the time, was also consulted, as was Aerospatiale, which decided to explore with TsSKB and the MZ Progress factory in Samara the possibilities offered by the Soyuz rocket.

## NEW DEAL IN SAMARA

After the Soviet collapse, TsSKB employed approximately 5,000 people, half of them engineers. Its divisions were specialized in the production of launchers, civilian and military observation satellites, microgravity research satellites, and agricultural equipment through a program of diversification already under way. TsSKB was well managed and had a sustained level of production in proportion to its workforce, all of which enabled it to better withstand this phase than many other industrial clusters of the former USSR. "We have no problem at the moment, but you must understand that we manufacture satellites, launchers, chocolate, grain sorters and sausages," explained Professor Gennady P. Anchakov, first deputy general designer of TsSKB. Nevertheless, he added, "We can't say that the current situation is satisfactory, and we are concerned for the future."

The MZ Progress factory of Samara had approximately 30,000 employees, half of whom worked on TsSKB programs. Others worked for NPO Energiya, in particular manufacturing the central core of the giant Energiya launcher. Samara also had many subcontractors, including the MZ Frunze factory that produced engines for NPO Energomash, and a number of metal-working companies. Unlike KB Salyut and MZ Khrunichev, which were rivals at that time, TsSKB and MZ Progress worked in close collaboration and a merger was even considered after 1991.

In June 1992, TsSKB hired a commercial agent in Sweden, Technology Trade

The stage integration hall at the Samara factory, about 1995.

A Tsyklon 3 launcher is installed on its launch pad at Plesetsk.

A model of the Soyuz launcher at the entrance to the Samara Space Center.

International AB, to represent it in the West. But the results were hardly conclusive, because in the months that followed only a handful of contracts were signed to carry experiments on Resurs F satellites, and in November 1992 an agreement was signed with Kayser-Threde in Germany to market launches of small piggyback payloads on Vostok or Soyuz launchers.[98]

The pace of launches of the Semyorka family, which averaged over one flight per week from 1973 to 1988 (with a record of 61 missions in 1980), decreased by half in 1992 and continued to drop. Production of the Vostok launcher was interrupted that year and the final vehicles – including two for Indian satellites – were mothballed. Production of the Soyuz-U2, a variant of the Soyuz-U using the synthetic fuel syntin instead of kerosene, ceased in 1994 owing to the prohibitive cost of this fuel. Export prospects were limited since the remaining launchers, the Soyuz-U and Molniya-M, were sized to serve orbits that had no real commercial potential. At best, the Molniya could reach a sun-synchronous orbit at an altitude of 800 km with its non-restartable Block L stage, but for a launcher of that size its capacity was poor at only 1,500 kg.

Towards the end of the Soviet era, it had been assumed that sometime in the near future the Semyorka family would be withdrawn and replaced with the new Zenit launchers from NPO Yuzhnoe. But this constructor had become Ukrainian by then, making that particular plan void. This allowed the TsSKB launchers the reprieve they needed to consolidate. Moreover, Kazakhstan's independence put the Baikonur cosmodrome in foreign territory, and negotiations to lease it from the government in Alma-Ata[99] proved so difficult that consideration was given to transferring Russian priority missions, including manned flights, to the Plesetsk cosmodrome north of Moscow. Construction of a new cosmodrome at Svobodniy, not far from the Chinese border on the Amur River, was also under study. The Semyorka launchers had to be adapted to these new geopolitical constraints.

In partnership with NPO Energiya, TsSKB then studied a modernized version of the Soyuz launcher, with its low orbit capacity increased to approximately 7.9 tons. The project was called "Rus" in reference to the founding people of medieval Russia, and it was adopted by RKA and the Russian ministry of defense in early 1993. The launcher – designated "Soyuz-2"[100] – was to be operational by 1996 to replace the existing launchers and possibly service a Mir 2 space station in an orbit inclined at 65° from Plesetsk in order to circumvent using Baikonur. The Soyuz-2 would use the design of the Semyorka family with improved engines and new avionics developed by NPO Avtomatiki i Priborostroeniya (NPO AP) replacing that of NPO Khartron, which was now Ukrainian.

To replace the Block L of the Molniya version for missions to high orbits, NPO Energiya proposed using a variant of Block DM, which already formed the fourth stage of the Proton and had the advantage of being restartable. Ultimately, the more agile Fregat upper stage was selected. This had been developed by NPO Lavochkin from the propulsion module of the Fobos deep space probes launched by Protons in 1988.

With the Russian economy stagnating, the stubborn issue of financing these developments stood out with no visible solution, just at the moment when the French from Aerospatiale arrived on the scene.

## A FRENCH PROPOSAL

Aerospatiale had no intention of funding the development of the Soyuz-2; its idea was to offer an industrial partnership to develop a new maneuvering stage to replace not only the Block L fourth stage but also the Block I third stage. The French called this new stage "Irene", borrowing the first name of the interpreter who accompanied them to Samara.

The Soyuz-Irene launcher would therefore use the Soyuz-U base, with its central core and its lateral boosters directly from TsSKB without any modifications, while Aerospatiale would provide the upper part with the Irene stage. The structure would be manufactured by Aerospatiale Espace & Défence at its Aquitaine site in Saint-Médard-en-Jalles, near Bordeaux, where the workload had considerably decreased due to the decision to halt Hermes and the French government's delay in launching the M5 submarine-launched strategic ballistic missile program.[101] After transfer to Russia, the stage would be integrated at the Samara factory with its two S5.92 engines[102] from KB KhimMash, each developing 2 kN of thrust. Aerospatiale would also provide new fairings and dispensers designed to deploy clusters of satellites.

Adopting a design similar to that of the Ariane 5 upper stage, the Irene stage, equipped with four pressurized helium tanks, would be a compact 3.7 m high, 2.66 m

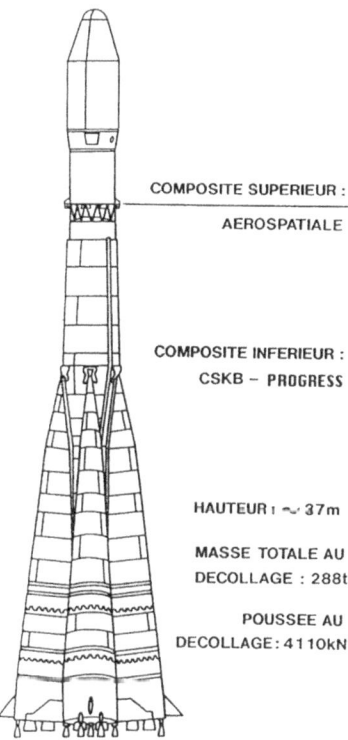

COMPOSITE SUPERIEUR :

AEROSPATIALE

COMPOSITE INFERIEUR :

CSKB – PROGRESS

HAUTEUR : ~ 37m

MASSE TOTALE AU DECOLLAGE : 288t

POUSSEE AU DECOLLAGE : 4110kN

The Soyuz-Irene concept with its Franco-Russian work share.

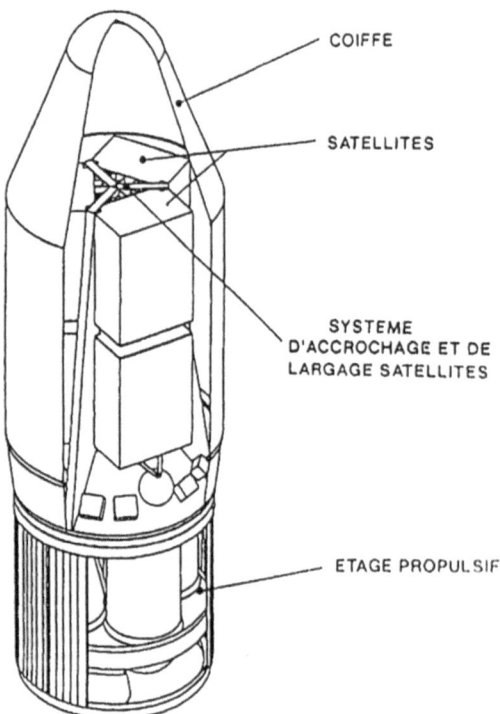

COIFFE

SATELLITES

SYSTEME
D'ACCROCHAGE ET DE
LARGAGE SATELLITES

ETAGE PROPULSIF

The maneuverable Irene stage topped by its payload, as imagined by Aerospatiale.

in diameter, and have 9 tons of storable propellants (UDMH and nitrogen peroxide). It would have its own independent avionics, separate from that of the lower part of the launcher, and an off-the-shelf attitude control system selected from equipment already developed for satellites. With its multiple restart capability, it could perform complex deployments and be removed from orbit at the end of the mission in order to limit orbital debris.

For a launch from the existing facilities in Baikonur, the calculated capacity to the Globalstar orbit (circular at 1,400 km, inclined at 52°) was around 3,200 kg, making it possible to launch up to six satellites simultaneously. The sun-synchronous orbit capacity at an altitude of 800 km would reach 2,500 kg, corresponding to the mass then envisaged for the Spot 4 satellite.[103] Moreover, Aerospatiale was exploring the feasibility of installing a "short term" launch complex adapted to Soyuz vehicles at the Guiana Space Centre, in Kourou, which would increase the payload capacity, while protecting the system from any Russian-Kazakh crisis concerning the use of Baikonur.

To bring the project to the phase of a qualifying flight planned for early 1997, Aerospatiale estimated the necessary investment at 350 million francs on the French side and 100 million francs on the Russian side. This investment could be offset by accepting paying passengers on the qualification flight (a total of 60 to 100 million francs) and by selling five flights at $36 million (or 10% less than launches on the Delta 2 made by McDonnell Douglas).

On February 14, 1994, François Calaque presented this concept to CNES and Arianespace in the context of a broader strategy of developing a range of launchers complementary to the Ariane 5, with ESL for small satellites and a possible future adaptation of the Irene stage to the DLA concept, using a new medium launcher architecture consisting of an Ariane 5 EAP, an ESL P50 and the Irene stage. The Soyuz-Irene concept would become an intermediate solution, allowing Europeans to immediately ensure themselves a place in the constellations market – starting with Globalstar – whilst limiting the penetration of other launchers into this market.

## FROM IRENE TO IKAR

Irene's debut came at an awkward moment. Three weeks earlier, on January 24, the Ariane 4 suffered its first failure in four years. An overheating turbopump bearing in the third stage engine on flight 63 caused the loss of two satellites, Eutelsat F5 and Turksat 1A, both built by Aerospatiale in Cannes. The commission investigating this failure announced its findings on that same February 14 and recommended changes to the bearing in question. Arianespace had to reorganize in order to resume launches as soon as possible, ideally by the end of May, to ensure it met its workload for the year.[104]

In the meantime, CNES was completely occupied with the final development of the Ariane 5, and could not afford such an investment. Nevertheless, the principle of

François Calaque (left), during a meeting in Samara around 1995.

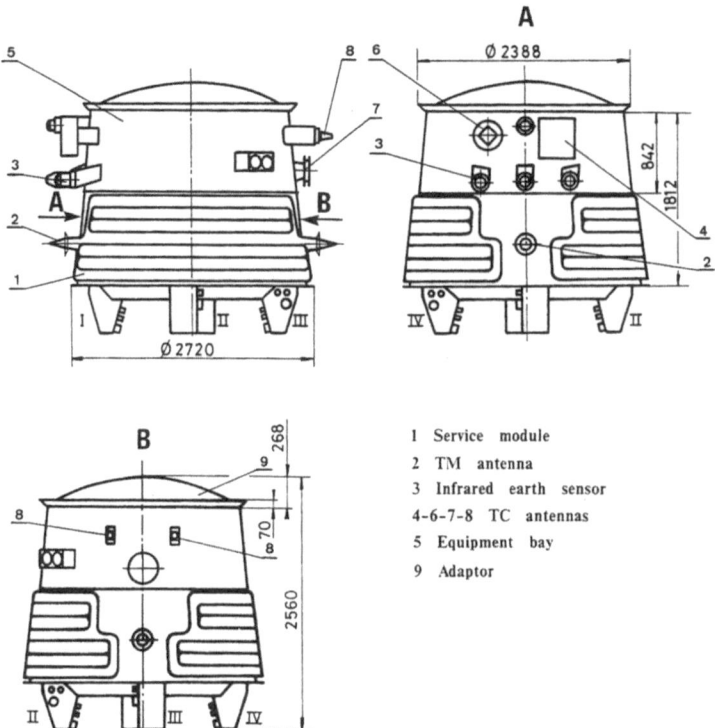

The Ikar stage used the propulsion block of the Yantar satellites.

Franco-Russian cooperation was not in jeopardy. René Pellat, president of CNES, even reaffirmed it as the way forward when he presented the French space agency's budget in March. On the same occasion, chief executive, Jean-Daniel Levi, indicated that no decision would be made concerning a small launcher before 1996 owing to a lack of market visibility. On the Russian side, the project lacked anything resembling unanimous support. To start with, there was no budget to finance the Russian share of the investment. Secondly, many at RKA and Samara were lukewarm to the idea of producing a part of the launcher in France when it was Russia that intensely felt the lack of currency.

So Irene was stuck on the drawing board, and another round of negotiations began between the French and Russians to find a solution that would enable them to launch Globalstar satellites.

Despite language and cultural barriers, contact with the TsSKB teams went well, since both the French and Russians spoke the language of engineers above all, and had learned to appreciate each other for their respective qualities.

If no European funding was available, and if the Russian partners were reluctant to relocate part of the industrial work, a technical solution would have to be found using pre-existing Russian components that had already been qualified. The Fregat stage that NPO Lavochkin was developing from the propulsion module of its Fobos probes would be ideal, but would not be available in time. However, it transpired that TsSKB had also built satellites equipped with a maneuverable propulsion module that could be made autonomous. Better yet, a year earlier, in the context of a contract with Aerospatiale unrelated to the Soyuz project, a study had been conducted on the structural and dynamic behavior of such satellites, so these were well known. This involved the Yantar spy satellites, of which over 60 units had been launched since 1974 in the framework of the Feniks, Oktan, Terilen, Neman and Kobalt military programs,[105] from which TsSKB had been considering making Resurs-Spektr civilian derivatives since 1989 for remote sensing. The Ikar upper stage would be developed based on this propulsion module, built in series in Samara, while keeping most of the equipment unchanged and modifying just a portion of its avionics and its mechanical and electrical payload interfaces. Powered by a 2.94-kN-thrust 17D61 engine made by KB KhimMash, the Ikar stage would be able to perform up to 50 restarts. Sixteen vernier engines would provide steering. In addition, like the Yantar satellites, the Ikar would be able to operate in either a standalone or a ground-controlled mode. Loaded with 2,344 kg of UDMH and nitrogen peroxide and installed under the fairing of the Soyuz, it would be able to put a payload of 3,300 kg in a circular orbit at an altitude of 1,400 km at 51.8° when launched from Baikonur, which was the orbit selected for the Globalstar satellites.

This solution was certainly elegant from an engineering standpoint, but it raised a question of industrial relevance. Aside from the marketing aspects of the launches, which only mobilized a small team, Aerospatiale's role in the Soyuz-Ikar launcher would be limited to producing multi-payload dispensers for Globalstar satellites. In other words, it would generate minimal industrial activity at the Aquitaine plant. Did this joint venture really make sense? For Michel Delaye, head of Aerospatiale Espace & Défense, the answer was an unequivocal affirmative.

## BREAKING OUT OF THE CIRCLE

François Calaque sought to establish cooperation with TsSKB, above all to enable Europeans to gain a foothold in the constellations market. But Michel Delaye gave him carte blanche to pursue this project with a broader strategic vision in mind, and an equally urgent goal – to break the encirclement that was being drawn around the Ariane system.

As the world leader in commercial launches, with 60% of the accessible market, Arianespace had everything to lose from an increase in the number of competitors. Even though Martin Marietta had decided to abandon its Titan 3 Commercial, and General Dynamics had still to make a profit with its Atlas,[106] the McDonnell Douglas Delta 2 was doing well out of contracts with the Pentagon. Commercial contracts began to elude the Europeans. The Chinese launchers had made their breakthrough in 1990. And in 1994 the Japanese would have their first fully indigenous launcher, the H-2, ending their reliance on licensed US technology. It was therefore necessary to stem the rise of the Russian launchers if Arianespace was to preserve the viability of the Ariane system, especially because the cost of labor in

Russia – like in China – might lead to a price collapse that would be fatal to Europeans. To avoid confrontation with the Russians, it was best to make them allies.

This idea was not new. In the fall of 1992 Michel Delaye had organized a dinner with Claude Goumy, CEO of Matra Marconi Space,[107] and Jean Sollier, president and CEO of SEP,[108] to talk about Ariane's situation in the face of the eminent arrival of former Soviet launchers in the market. At that time, the greatest danger seemed to come from the Proton in a trade agreement with Lockheed. As Lockheed was not present in the launcher market other than by way of its Russian initiative, an alliance between Khrunichev and Arianespace therefore seemed possible that would prevent the emergence of a potential competitor capable of causing long term damage to the European launcher industry. The three men who at that time represented most of Arianespace's French industrial shareholders, brought out the heavy ammunition to convince Charles Bigot, the Arianespace CEO, to go to Russia to negotiate with the Khrunichev management.

Charles Bigot was one of the last pioneers of the French space industry still in command of a major company in the sector in the early 1990s. Like Michel Delaye, he was a graduate of the elite Institut National Polytechnique, but while Delaye had

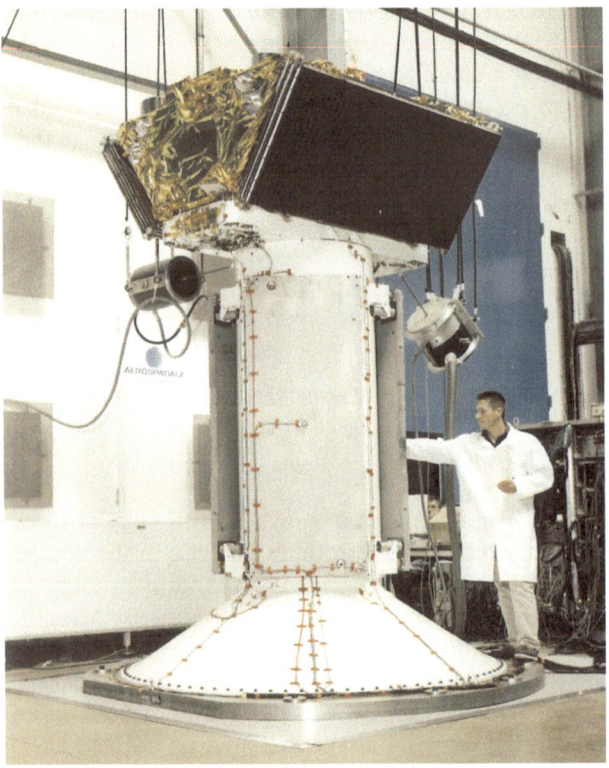

A Globalstar dispenser tested at Aerospatiale's Aquitaine facility in October 1998.

Charles Bigot of Arianespace (on the right) in discussion with Yuri Koptev of RKA (center) and Dmitri Kozlov from the Samara Space Center (left).

come from the nuclear ballistic missile sector,[109] Bigot had participated in the birth of the European launcher industry. He worked on the Véronique sounding rockets at Hammaguir in the Sahara for CNRS's Aeronomy Service. He was the first director of launchers at CNES during the development of the Diamant B and the Europa 2. After some time in the automotive industry and air transport, he was the commercial director of Aerospatiale Systèmes Balistiques et Spatiaux, for which he won the first international contract to build the satellites of the Arabsat 1 generation. In 1982 he became chief executive of Arianespace, and then CEO in 1990. With his predecessor Frédéric d'Allest, Bigot was a key architect of Ariane's success in the international market, and hence was determined to protect the European launcher and its industrial sector, whatever the price.

Consequently, Bigot also assessed the Russian threat with his secretary general, Roland Deschamps, another veteran of SEREB and the CNES launchers division who had been an initiator of Ariane and Arianespace, and noted that Ariane's market share would be the first to suffer from an uncontrolled rise by the Proton. At a press conference on October 1, 1992, he did not hesitate to stigmatize the "maneuvers" of the US government to thwart Arianespace's international penetration by pitting it against competition from the Russian and Chinese launchers. According to Bigot, the Americans, concerned about the loss of momentum in their own launcher industry, were seeking to support it by trying to "shove their opponent's head under water".

Bigot therefore undertook a series of trips to Moscow to meet with key officials, including Yuri Koptev, director general of RKA, and Anatoli Kiselev, then director of MZ Khrunichev. After several months, a foundation for commercial cooperation was laid. This could even involve technical cooperation on upper stages which could be used interchangeably on Proton or Ariane launchers. The bylaws were drafted to form a joint company called API (Ariane Proton International). The formal signature of the agreement was scheduled for June 1993, at the Paris Air Show. But it never took place.

Officially, the reason invoked by the Russian partners was the inadequate number of launches reserved for the Proton in the first years. In the initial agreement, there would be only one per year. In fact, that matched more or less the quota agreement to be signed on September 2 by Viktor Chernomyrdin and Al Gore giving Khrunichev the right to sign eight commercial contracts for geostationary satellites in the period 1993-2000; that is, one per year.[110]

## NEW RULES OF THE GAME

For Charles Bigot, to a certain extent this failure resulted from naivety on the part of the Europeans, in contrast to US aggressiveness. This was especially noticeable in the Europeans' surprisingly disorderly dispersion. For example, at the hotel where he was staying during one of his trips, Bigot happened to meet two other European delegations – one from an industrial firm and the other from an agency – who had come to meet the same interlocutors as Bigot to discuss cooperation on launchers. None of the three delegations knew anything about the presence of the other two! In fact, as Arianespace was negotiating a commercial alliance for the Proton, the French of Aerospatiale were interested to varying degrees in Soyuz, Tsyklon, Kosmos and Proton launchers. As for the Germans, DASA was preparing an agreement on the small Rokot launcher and was also interested in the Tsyklon. The Italians of BPD Difesa e Spazio also had projects for the Tsyklon that were totally incompatible with those of the Germans. Even the Brits of British Aerospace made the trip in an effort to adapt the Spelda dual launch structure that they manufactured for the Ariane 4 for use with the Proton! In parallel, ESA and CNES were studying cooperation projects that might lead to using Russian launchers to orbit some of their satellites.

All these European partners were elbowing each other to get their share of the pie without realizing that each one of them was eroding European credibility in the eyes of their mutual contacts. Naturally, the Russians were quick to play the Europeans against each other in order to obtain the best negotiating positions. Representatives of the space agencies even openly questioned the legitimacy of Arianespace to negotiate agreements which, they insisted, ought to be the responsibility of political entities.

Coordination was needed, particularly since at this pivotal moment Russia was not the only field of exploration for Arianespace and its shareholders: contacts were also made in Japan and the United States. In the months that followed, the Arianespace board of directors set out certain rules of conduct for its shareholders. First, they had to keep Arianespace informed of their initiatives for the sake of coordination, and to ensure no one lost sight of the main objective, which was to save Ariane's markets. Then, if an agreement was reached, it was established that the commercial role would revert to Arianespace. Lastly, in the case of the creation of a

A Rokot in its launch container is installed on a modified Kosmos pad at Plesetsk.

joint company, the non-Russian share of the capital should involve several industrial partners, to prevent any single entity from setting up its own project to the detriment of the others, and, above all, Arianespace would hold the largest share among these industrial shareholders.

One of the first projects to be subjected to these rules was submitted by DASA. It involved the small launcher, Rokot, developed by Khrunichev based on the UR-100 intercontinental ballistic missile, all of which were otherwise to be dismantled under the terms of disarmament agreements. When Boris Yeltsin visited Germany in May 1994, the Erno Raumfahrttechnik division of DASA signed a cooperation agreement with GKNPTs Khrunichev to prepare for the joint operation of the small launcher. Initially, it was even envisaged that the DASA division would develop a small upper stage for the launcher.

For DASA, the Rokot was regarded as a solution to the issue of a small European launcher – and available at a discount price. This was obviously not to the liking of shareholders involved in other projects, whether BPD Difesa e Spazio and its small Vega launcher, or Aerospatiale, SEP and CNES who were working together on the competing ESL concept. To implement this cooperation, a total investment of $30 million was required, part of which would be provided by the German government in the framework of its program to aid in the conversion of the Soviet Union's military industrial complex, which in turn was financed by Russia in the context of paying off the Soviet debts owed to Germany. Many points remained uncertain, in particular concerning Khrunichev's capacity to recondition missiles withdrawn from service in order to extend their use beyond 1997.[111] In addition, the Rokot was originally a KB Salyut project and the leadership of Khrunichev was not in favor of it, preferring to focus on marketing the Proton. Under these circumstances the board of directors of Arianespace refused to participate in DASA's program. Nevertheless, the Germans pushed on, and in March 1995 they established Eurockot GmbH as a joint venture between DASA (51%) and Khrunichev (49%). The new company based in Bremen, had a difficult start with the collapse of its financing plan by way of compensation of Russian debt. By finding other sources of financing, DASA succeeded in keeping it afloat long enough for it to sign its first launch contract in June 1997.[112]

Other projects that were evaluated, included adapting the Ukrainian Tsyklon 3 launcher. This was proposed independently by two teams, one led by BPD and the other involving DASA and MMS, but neither was successful.[113]


## OBJECTIVE GLOBALSTAR

During this time, at Aerospatiale, Michel Delaye was questioning the procedure to be pursued after the failure of the alliance, which could have led Arianespace to market the Proton. It was clear from his analysis that rather than running the gauntlet of an Arianespace board that would be largely opposed, it would be better for Aerospatiale to take the lead alone in negotiations on the Soyuz. This would give Aerospatiale the opportunity to prepare a strong case to demonstrate to the other shareholders that the Soyuz was complementary to the Ariane launcher, rather than a competitor, with a possibility for Russian and European teams to work together for mutual benefit.

Louis Gallois, then CEO of Aerospatiale, agreed to give him free rein to explore the possibility of creating a joint company owned equally between Aerospatiale and its Russian partners, provided that he was discreet. To facilitate negotiations, Delaye relied heavily on another alliance that he had formed some time earlier in the field of satellites.

After a failed merger attempt in the late 1980s between Aerospatiale Satellites, the bulk of whose commercial activity was producing Spacebus platforms, and Alcatel Espace, which supplied telecommunications payloads, a technical and commercial partnership agreement was concluded between the two manufacturers in November 1990. It was expanded to the Italian Alenia in March 1991. The following month, the three partners became a 49% shareholder in Space Systems/Loral (SS/L), a satellite manufacturer based in Palo Alto, near San Francisco, and formed what was called the "Space Systems Alliance". When DASA joined in December 1992, the European venture became a quartet.

This investment in one of the largest US satellite prime contractors had been very beneficial for the European partners. For example, to decongest its own cleanrooms SS/L assigned the integration work on several Intelsat 7A satellites to the Cannes facility of Aerospatiale. More importantly, it gave Europeans a central role in the development and manufacture of the Globalstar constellation. Aerospatiale therefore had leverage favoring the selection of the Soyuz to participate in the deployment of the constellation.

While technical discussions between engineers always went well, commercial

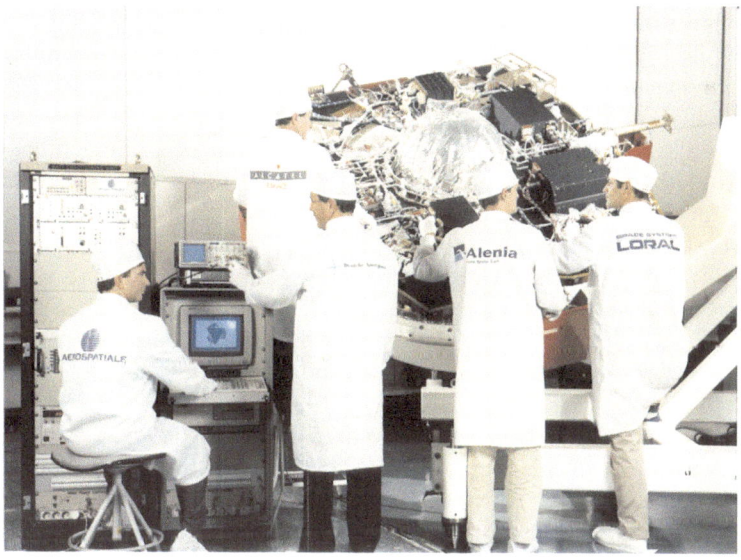

The "Space Systems Alliance" in which Aerospatiale, Alcatel Espace, DASA and Alenia in Europe were shareholders of the US company Space Systems/Loral.

negotiations between the French and the Russians proved to be much more difficult. To avoid being submerged in bureaucratic chicanery, and to cut short any ideas of obstruction from one of the many subcontractors at TsSKB or MZ Progress, the decision was made to bring RKA into the circle. All of the former Soviet industrial partners concerned were still state-owned enterprises, and would have to bend to the will of their leading prime contractor. In addition, since RKA had been omitted from all the other Russian-Western partnerships at the time, it was clear to everyone that it would be in RKA's interest to promote a Franco-Russian partnership in future trade negotiations.

The next step was to associate Arianespace with the project, because it was clearly agreed that the offer to be presented to Globalstar would comprise launches by the Ariane 4 and Soyuz-Ikar. This meant confronting the opposition of many members of the Ariane family. Aerospatiale would soon need to make a headcount of its allies, and there weren't very many.

CNES was paralyzed by a crisis of leadership. Its president, André Lebeau, who looked favorably on the Soyuz, engaged in a daily clash of opinions with Jean-Daniel Levi, the director general, who held an opposite view on almost every topic. Within the agency's teams, people were suspicious of a project that had not originated with CNES, which since 1973 had led new developments of launchers in Europe. The Soyuz proposal was also criticized for creating internal competition with the ACLV (Ariane Complementary Launch Vehicle) concept that superseded the rejected DLA concept.

Similarly, SEP was going through a difficult period because the transition from the Ariane 4 to Ariane 5 would trigger painful production cutbacks at its Vernon site in Normandy, dedicated to liquid propulsion. While an Ariane 4 was equipped with 6 to 10 engines built in Vernon,[114] with up to 12 launches each year, the Ariane 5 had just one Vulcan engine and its launch rate would not exceed 6 or 7 annual missions. In these circumstances, any project that might compete with the final Ariane 4s or reduce the market for the Ariane 5 was very negatively perceived. The reaction was the same at Matra Marconi Space, where production of equipment bays for the Ariane would be halved once the Ariane 5 was introduced. DASA and BPD, which did not appreciate the way their Rokot and Vega projects were received, and moreover were each preparing to establish their own Tsyklon launcher projects in Kourou, saw no advantage in supporting Aerospatiale's proposal to use the Soyuz.

And even within Aerospatiale, some engineering teams questioned the benefit to be gained from helping a Russian company while they themselves could develop a medium launcher that would keep the industrial activity in France – if only they were financed to do so.

Regrettably, the Europeans waited too long. In March 1995, SS/L signed the first contract with McDonnell Douglas, worth $50 million, to launch a cluster of four satellites using a Delta 2. An option to purchase a second such launch was soon confirmed. In May, another contract was signed with NPO Yuzhnoe in Ukraine for three Zenit 2 launchers. The proposed price of around 88 million dollars for three launches beat out any competition. Each rocket was to carry a cluster of 12 satellites in a dispenser derived from a platform developed to deploy multiple warheads from

A Zenit 2 on its launch pad at Baikonur. Prior to launching the Globalstar satellites, it had already suffered seven failures in 30 flights.

a ballistic missile. Neither Aerospatiale's ownership position as a shareholder of SS/L nor the presence of RKA in the partnership was able to prevent it. Yuri Koptev was quick to denounce the agreement for which RKA was not even consulted, although Russia, not Ukraine, was responsible for Zenit launches from the Baikonur site. The spokesman for SS/L countered that RKA was not the only player with authority over the cosmodrome and that all necessary permissions had been granted prior to issuing the contract.

The *coup de grace* came in July, when a contract was signed with the CGWIC to place the 12 final Globalstar satellites in orbit using a Chinese CZ-2E launcher with a TS (Top Stage) steering module. Unofficially, it was explained that this contract was required by the license granted to Globalstar to operate on Chinese territory. The 56 satellites for the initial deployment of the Globalstar constellation (48 operational and 8 backup satellites), were now booked for launch – the Soyuz-Ikar venture had completely lost out on this contract.

And as misfortune never strikes just once, in April Boeing joined forces with NPO Yuzhnoe of Ukraine, the Russian RKK Energiya, and Kvaerner of Norway, to form the Sea Launch consortium to make Zenit launches from an offshore platform on the equator with satellites heading for geostationary orbit, and then at the Paris Air

Show in June 1995 Lockheed Martin Commercial Launch Services and LKEI announced the creation of International Launch Services (ILS) to jointly market the Atlas and Proton launchers with the ambition, as Vance Coffman, CEO of Lockheed Martin Space & Strategic Missiles, bluntly expressed it, of winning more than 50% of the commercial market and claim Arianespace's leading position in the sector for itself.

It was official: Ariane was surrounded.


## THE POLITICIANS COME ON STAGE

In France, 1995 was an election year. The first round of the presidential elections, on April 23, was the scene of an unprecedented confrontation between two candidates, Jacques Chirac and Édouard Balladur, from the same party, the RPR (Rally for the Republic). Balladur was the outgoing prime minister of a "cohabitation" government with François Mitterrand, the Socialist president who did not run for a third term. Two years earlier Balladur had accepted this position – deemed to be a thankless one under such circumstances – while the leader of his party, Jacques Chirac, prepared for the presidential deadline. Against all expectations, Balladur's popularity rating remained steadily above 50% for the duration of the term of cohabitation, so that he looked like a more electable presidential candidate than Jacques Chirac in the eyes of many RPR elected officials, and since the RPR did not use a primary election process for candidate selection he declared himself a candidate. In this venture, he had the support of several members of his outgoing government, including that of his young minister of advanced education and research, François Fillon. At the age of 41, Fillon had already been a member of three legislatures and had gained a reputation as an expert on defense issues.

In the first round of voting Édouard Balladur was eliminated, ending up in third place, 800,000 votes behind Chirac. In the second round on May 7, Chirac beat the Socialist candidate to become president. Despite Fillon having supported Balladur, Chirac entrusted him with a portfolio. On May 18, Fillon was appointed minister of postal services and information technologies in the first government of the new prime minister, Alain Juppé. His main task was to end France Telecom's monopoly on telephony and prepare its change of status in the context of European deregulation of the telecommunications market. At the instigation of Jérôme Paolini, his advisor for international affairs and space in his previous position, Fillon requested that space affairs be added to his portfolio, and this was done.[115]

At the ministry of research, François Fillon had exercised oversight of CNES's scientific activities, while industrial activities – including the Ariane program – were under the responsibility of the ministry of industry, headed by Gérard Longuet and then by José Rossi over the previous six months. Nevertheless, since Fillon also had oversight of the CNRS (French national scientific research center), he had many contacts with the Russian space community, including the Academy of Sciences. He was able to observe the growth of US-Russian cooperation in the field of launchers and readily recognized the strategy of encirclement that threatened Ariane.

At the 2009 Paris Air Show, François Fillon, the new prime minister, paid a visit to Jean-Yves Le Gall, head of Arianespace by that time marketing the Soyuz.

Fillon's new responsibilities included full oversight of CNES activities and so he had a say on the issue of launchers. Paolini, once again his advisor for space affairs, then began to meet with the leaders of all major entities in the sector in France. When he met with Michel Delaye, the head of Aerospatiale Espace & Défense, the two men took the opportunity to discuss the strategic implications of the latest developments involving US-Russian cooperation on launchers and its impact on Europe's industry. At Paolini's request, Delaye drafted a brief for the minister. When Fillon read it, he observed the success of the encirclement strategy he had suspected a year earlier. By buying into systems whose development had already been funded and amortized by the military industrial complex of the ex-USSR, and manufactured in conditions that still had nothing to do with real market economy constraints, the US industry would be able to compete with Arianespace by offering prices that did not reflect economic reality, and do so at a time when Ariane's entire production chain was in a position of weakness owing to the transition from the Ariane 4 to Ariane 5.

A few days later, Jérôme Paolini again received Michel Delaye, who described the progress of François Calaque's team in preparing for commercial operation of the Soyuz launcher, capable of breaching the encirclement. A meeting was then set with the minister, and Delaye explained to him in detail all the technical advantages of the concept – reliability of a proven architecture, robustness of the industrial equipment, and the capacity for high use rates.

On the strategic side, although the Soyuz and its variants were not advantageous for geostationary orbit, they could place a variety of payloads in low Earth orbit. In addition, the Soyuz was qualified for manned flight, which could be advantageous in preparing for ESA's ministerial conference in Toulouse in October. For the first time since the 1992 Granada conference that had buried the Hermes space plane program, ministers in charge of space affairs from all member states of ESA would gather to approve new programs. In fact, France wanted to reenergize the dynamics of manned space flight through programs such as the Columbus laboratory and the ATV (Automated Transfer Vehicle) supply freighter. There was even a cherry on the cake, because the Soyuz brought very strong symbolic capital on the political level. François Fillon was excited by the project, which he regarded as "politically enabling", and which promised to boost industrial cooperation with Russia. To move forward, he therefore asked Paolini to organize the details between Aerospatiale and Arianespace.

## ARIANESPACE FACED WITH A DILEMMA

In the autumn of 1995, Charles Bigot found himself in an uncomfortable situation, caught between a French government asking him to join a cooperative project which did not respect the criteria set out by the board of directors of Arianespace, and the majority of its shareholders, who were hostile to the project. Even CNES, which then held 32.22% of Arianespace shares, was still in the throes of a management crisis, with its president agreeing with the project and its director general fiercely opposed. The CEO of Arianespace made no secret of his disapproval of an industrial plan that was largely established without his knowledge, and he clearly expressed his doubts about the strategy to pursue after the biggest constellation market had gone to the competition. In the absence of a commercial market in low Earth orbit, there was a serious risk that the Soyuz joint venture would complicate the task of selling the final Ariane 4s at a profitable price for an industry which was already being subjected to incessant demands to lower costs. Arianespace had imposed very lean conditions so that the Ariane 4 would remain competitive long enough for it to be replaced by the Ariane 5.[116] On the other hand, abandoning Soyuz at this point would mean allowing someone else to market it later, and running the risk of eventually having to confront it in less favorable conditions. McDonnell Douglas and the Italian company Alenia Spazio[117] had also approached TsSKB to express their interest in the Soyuz.

For two months in late 1995, Charles Bigot and Jérôme Paolini attended a series of meetings to try to smooth out the differences between the partners. In December, Aerospatiale signed both a shareholders' pact with its Russian partners TsSKB, MZ Progress and the RKA, and a memorandum of understanding with Arianespace.

In the wake of the Franco-German summit between President Jacques Chirac and Chancellor Helmut Kohl in Baden-Baden on December 6, the lengthy process of rapprochement between Aerospatiale and DASA to merge their satellite businesses resulted in a memorandum of understanding between the two manufacturers which was signed on December 18 by Louis Gallois, the president and CEO of Aerospatiale and Manfred Bischoff, the director general of DASA.[118]

And the resignation of the president of CNES, André Lebeau on January 30, 1996, immediately followed by that of the director general, Jean-Daniel Lévi,

The version 40 (no boosters) Ariane 4 was used to launch Earth observation satellites, in this case the military satellite Helios 1B on December 3, 1999.

removed the deadlock at the top of the agency. The new president, appointed by François Fillon, was Alain Bensoussan, former president of INRIA.[119] This renowned mathematician had already had significant experience of cooperation with Russia due to his previous work, and he supported the ministry on the Soyuz proposal. He left for Moscow to attend the 31st Franco-Russian space cooperation summit, where he met Alexander Medvedchikov, the second in command at RKA. No director general was appointed for the moment, but the position of director responsible for the implementation of a strategic plan was created and assigned to Jean-Yves Le Gall. This engineer in optics had been technical advisor to Paul Quilès,[120] the minister in charge of space affairs. This gave Le Gall the opportunity to participate in the negotiation of an agreement for three Franco-Russian manned flights. He had also had experience at Novespace, the CNES subsidiary in charge of activities in microgravity and technology transfer, where he worked with RKA and TsSKB on Foton satellite missions.

The balance of forces had evolved and the project was presented to the board of directors of Arianespace, where it was hotly debated but ultimately adopted. The Soyuz would be integrated into the range of launchers offered commercially by the Europeans. However, the shareholders' pact had to be revised to make it possible for other European partners to invest in the share capital, in accordance with the rules of conduct previously enacted by the ESA council. A bilateral agreement was signed between Aerospatiale and Arianespace on May 6, 1996.

In Samara, not everyone appreciated the French political-industrial combinations. TsSKB and MZ Progress were merged on April 12 by special decree of the Kremlin to form GNPRKTs,[121] which would become known in the West as the "Samara Space Center". Whilst the leaders of this new entity were happy for their launchers to be marketed by Aerospatiale, they worried about Arianespace's new involvement. The Ariane operator had a reputation for toughness in business, and they could imagine that it would sacrifice the interests of the Soyuz to those of the Ariane. Once again an ordeal of explanation and negotiation had to be undertaken before everyone could be brought to the same table to sign.


## STARSEM IS BORN

On July 17, 1996, in Moscow, the agreement that led to the creation of the Starsem company[122] was signed in the presence of François Fillon and Alain Bensoussan. The four shareholders were RKA and the Samara Space Center, each contributing 25% of the capital to form the Russian 50% share, and Aerospatiale and Arianespace, which respectively provided 35% and 15% to form the European 50% share.

This division, imposed by Aerospatiale, did not satisfy the rule which required that Arianespace control the largest share of the Western capital, but it reassured the Russian partners, so that they did not feel that they were surrendering control to the Europeans. Nevertheless, in the agreement signed between Western partners, it was

A technician working on an RD-107 engine prior to its integration on a Semyorka booster.

Installation of a hydrogen peroxide ring tank at the base of a booster.

The booster integration line at Samara.

stipulated that this share division was provisional and would have to be changed at a later date in order to allow other manufacturers to become shareholders.

Starsem was incorporated as a "société anonyme" under French law, and its bylaws were ratified on August 8 in Paris, in the presence of Yuri Koptev, director general of RKA, and Dmitri Kozlov, general designer at the Samara Space Center. The share capital was initially set at 500,000 francs. It was headquartered in Suresnes, a Paris suburb. The president was François Calaque of Aerospatiale. The executive director was Viktor Kuznetsov, a veteran of NPO Lavochkin who had been with Glavkosmos and then RKA. The director of commercial operations, Jean-Charles Vincent, was seconded from Arianespace, with Guy Dubau in charge of operations. Vincent had participated in the development of the Ariane 1 launcher at Aerospatiale and joined Arianespace in 1990 as director of the establishment at

Third-stage electrical integration.

A team prepares to attach the nose section of a booster. Behind them, two aft protective skirts for boosters are awaiting integration.

Kourou. Dubau was a veteran of the ballistic missile programs, as well as of the Diamant and Ariane programs. Patrick Millet, from François Calaque's Aero-spatiale team, was in charge of communications. On the Russian side, Gueorgi Fomin, deputy to Dmitri Kozlov, was in charge of the industrial aspects, while Viktor Kozlov, head of unmanned missions at RKA, directed other space businesses, primarily the marketing of Foton capsules for microgravity work.[123] The team was in place and it would soon be time to release funds for the commercial operation of Soyuz, Molniya, and Soyuz-Ikar launchers. To facilitate the implementation of funds, the ideal was to win a few contracts. But this was not simple, since the main market, the Globalstar constellation, had been picked clean by the competition.

In the space sector, there is a saying that any failure is harmful to the sector as a whole, because it discredits the principle of using space solutions rather than ground

An integrated booster is moved via an overhead gantry crane, while other units undergo preparation.

solutions. In addition, it pushes up insurance rates, thus undermining the profitability of all programs. Nonetheless, Starsem's story turned out to be the exception to the rule: twice.

For the Chinese space program, 1996 was a catastrophic year. On February 14, at the Xichang space center in Sichuan, the inaugural flight of the new CZ-3B launcher, a counterpart to the most powerful version of the Ariane 4, turned into a disaster. As soon as it left the ground, the vehicle pitched over and went crashing into a nearby hill, destroying a rural village which had not been evacuated despite its proximity to the launch complex. The official toll listed six dead and 57 injured, but unofficially it was higher. The hotel that housed the SS/L team that had supervised the Intelsat 708 satellite on the rocket, was destroyed. Fortunately, the entire team was sheltered in the bunkers of the space center at the time and only suffered material damage.

On August 18, ill luck struck the payload of another US manufacturer. Chinasat 7, supplied by Hughes, was inserted into the wrong orbit by a CZ-3, the predecessor of the CZ-3B. The company had already lost two satellites on Chinese rockets in 1992 and 1995. On average, over two years, the Chinese had lost 50% of the commercial satellites entrusted to them. The insurance market was boiling over and premiums soared. In rapid succession, CGWIC customers canceled their contracts. As a result, in September, SS/L switched to Starsem to launch the 12 Globalstar satellites which it had booked on the CZ-2E/TS. At 450 kg each, the satellites were to be launched in clusters of four by Soyuz launchers equipped with the new Ikar upper stage. The first flight was set for late 1998. The contract was officially signed on December 3, 1996. Its exact amount was not disclosed but it reputedly cost SS/L $68 million more than it had agreed to pay the Chinese. For the US manufacturer "it was the price to pay for reassurance that the satellites would be deployed correctly and on time". Ariane was still surrounded, but a breach had been opened.

A Soyuz launcher stands as a monument in the center of the city of Samara.

# 10

## Europeans on the steppe

Once Starsem was created, its bylaws ratified, and the first contracts signed, a great deal remained to be done before the most historic of the Soviet launchers, topped by the propulsion module of a Soviet spy satellite, could take off from Baikonur with a cluster of Western satellites as its payload.

The industrial environment was not encouraging, as although the French minister in charge of space was promoting Starsem, not everyone accepted it yet. The ink of the bylaws was not yet dry when Matra Marconi Space suggested the development of a complementary Ariane launcher (LCA) that was in the same class as the Soyuz but comprised of elements already developed in Europe. The idea was to use the first two stages of the Ariane 4 in conjunction with the upper stage of the Ariane 5. According to Armand Carlier, CEO of MMS, this purely European option – nicknamed "Ariane 4 Lite" – would supersede Starsem's Soyuz from 2000 until a truly complementary launcher to the Ariane 5 could be developed solely by Europeans.[124] Furthermore, at Aerospatiale, studies of an ACLV (Ariane Complementary Launch Vehicle) – which was the new incarnation of the DLA and the ESL – continued with a small ACLV-1 launcher (of which the P85/P16/P10 architecture prefigured the final version of the Vega) and a medium ACLV-3 launcher (an EAP of the Ariane 5 as the booster for an ACLV-1) envisaged as a successor to the Soyuz for deployment of second generation of telecommunications satellite constellations. Clearly for some Europeans, whilst it was too late to prevent the Soyuz from entering the market, there was no question of retaining it in the long term.

Nonetheless, Starsem had many ways to make the operation of the Soyuz benefit the Europeans, rather than exist at their cost. The tough negotiations with the Russian side had resulted in an agreement containing a dual exclusivity clause. The first gave Starsem responsibility for all commercial launches performed from Baikonur using the Soyuz or Molniya launchers for clients from outside the former-USSR. Russian government launches remained the sole responsibility of RKA and the armed forces. But Starsem had exclusive rights to market future versions of the Semyorka launcher family. The alliance with Samara was not an agreement of circumstance, but rather a partnership designed to last.

In the meantime, Europeans had to invest nearly $35 million on upgrading the

Russian infrastructure to provide launch services compliant with international quality standards.

In September 1996, Starsem ordered its first Soyuz launcher from Samara, as well as the elements necessary for the production of four more, in order to have a launch capacity available by late 1997. At that time, a Soyuz launcher cost Starsem $15-20 million and it was sold for $30-35 million. The Russian partners were paid in either dollars in New York or deutsche marks in Frankfurt.

It was also necessary to ensure the quality of the product. In 1996, Russia made 27 space launches. In 20 years, launch activity had fallen by 60%. For the Semyorka family, the pace – which had already dropped from 61 flights in 1980 to 32 in 1992 – was only 12 flights in 1996, including two failures.[125] As a result, it was difficult to maintain the level of involvement required for flawless production. The introduction of "European style" quality control management would make it possible to keep the Soyuz from repeating the misadventures of the Proton (six major failures from 1996 to 1999) and Zenit (three major failures from 1997 to 1999). This effort would not be painless, since overcoming the culture gap on the Russian side represented a big leap. As a representative of Starsem explained, "the first time that I requested a quality control survey, I was given a one-page certificate of conformity!"

These investments quickly led to the need to increase Starsem's capital. To ensure the support of Aerospatiale, Michel Delaye brought in François Auque, director of economic and financial affairs of the French industrial group,[126] to join the Starsem board of directors; he became an ardent supporter. Charles Bigot was in a difficult position, since his shareholders had not given him a mandate to engage more than 10 million francs, and the needs greatly exceeded that amount. His successor Jean-Marie Luton, former director general of ESA appointed to head Arianespace in June 1997, completed the next round. The capital was increased to 377 million francs in June 1999.

## ACCOMMODATING SATELLITES AND TEAMS

In early 1997, work began at Baikonur on the construction of dedicated clean rooms where Western satellites could be encapsulated under the fairing with their upper stage before integrating the assembly on the launcher. During a preliminary study, it became apparent to the Europeans that instead of trying to bring an existing Soviet installation[127] up to Western standards, it would be easier to build a new one. It was decided to do so in an annex of MIK-112, one of the largest hangars available for the reason that it had been used to integrate the Buran shuttle with its Energiya launcher. The first and only flight model of the Soviet shuttle, which flew for 206 minutes on November 15, 1988, was stored there, mounted on a test model of the launcher. This had been gathering dust since President Boris Yeltsin announced the cancellation of the program on June 30, 1993.

In an adjacent hall, the Europeans built their own payload preparation complex. This comprised three clean rooms totaling over 1,150 sq. m of veritable "buildings inside the building", with all the facilities necessary for the maintenance of a class

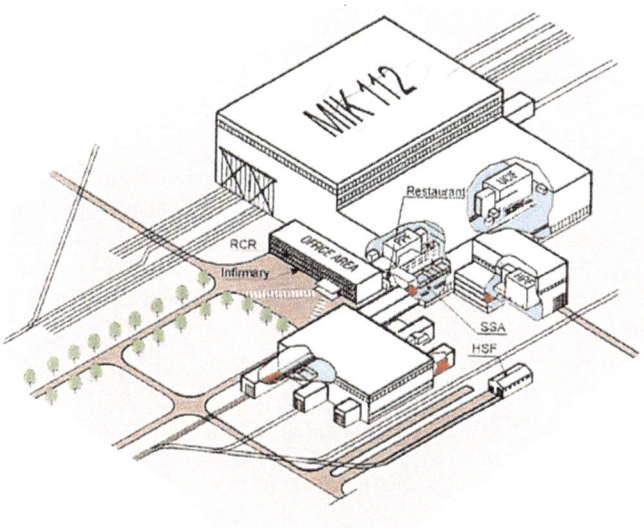

The Starsem clean rooms are located inside the enormous integration hall of MIK-112 at Baikonur.

A prestigious neighbor for Starsem: the only Buran shuttle to go into space.

100,000 clean room. To build this installation, the air conditioning equipment and all the control systems were ordered from Aerospatiale's Aquitaine site. Transporting them via sea and rail gave the French engineers assigned to Kazakhstan insights into the inner workings of Yeltsin's Russia in the midst of economic crisis. It was openly understood that to get the trains to arrive on time, the railway men needed to receive an extra monetary "sweetener".

The SPPF (Starsem Payload Processing Facility) included the PPF (Payload Processing Facility), a room with 286 sq. m of floor space and a 12.8-m ceiling for the preparation of payloads; the HPF (Hazardous Processing Facility), 285 sq. m and 16.8-m ceiling for propellant filling and other hazardous operations; and finally the UCIF (Upper Composite Integration Facility), 587 sq. m and 18.8-m ceiling for the integration of the satellites on the upper part of the launcher and encapsulation under the fairing. Each room had precision cranes, fluid connections and secure power supplies, as well as voice and data communications with offices located in another wing of MIK-112 where a satellite link enabled teams preparing payloads to communicate directly with their offices in Europe or elsewhere.

In order to ensure service quality comparable with that which Arianespace clients were accustomed to in Kourou, the Starsem team also had to arrange any logistical support needed to take full responsibility for satellites sent to Kazakhstan, as well as for the accompanying teams and equipment. This involved agreements with aircraft cargo companies to bring the equipment directly to Yubleniy airport, located within the bounds of the cosmodrome, and the preparation of customs documents, the latter greatly facilitated by support from RKA. Fortunately, the acceptance of equipment by Russian customs also applied in Kazakhstan.

Starsem also gained some unexpected kudos in the realm of tourism by tackling a crucial problem which became apparent as soon as teams arrived to work at the cosmodrome. The town of Leninsk, long absent from official maps, had grown up on the southern side of the Moscow-Tashkent railway, between the Tyuratam station and a lazy meander of the Syr-Darya River. It was a garrison town and a base for engineers and technicians, built in the Soviet mode: streets laid out at right angles, collective housing blocks, and power supplied by a thermal plant located in the town center. Temperatures soared to 45°C in the summer and plunged to minus 40°C in winter. In this city of nearly 100,000 people, temporary accommodations for those on assignment were almost non-existent. The 30 rooms at Hotel Kosmonavt, famous for hosting the crews of space missions the night prior to their launch, and the 80 rooms at Hotel Baikonur, although the best by Soviet standards, did not provide Starsem's teams and clients with acceptable conditions. According to Americans who had lived there on assignment, "Sometimes the water was not good enough to even shower with." As for the Tsentralnaya Hotel, in the mid-1990s it was renowned for its lack of air-conditioning! At the same time that Starsem was installing clean rooms, the company also began construction of the first hotel at Baikonur designed to meet international standards: 120 four-star rooms and suites, autonomous electrical power plant, purified water supply, swimming pool, sports facilities, sauna, satellite TV, and medical center. The Sputnik Hotel opened its doors in 1999 at the city's northern exit, on the road leading to the cosmodrome, and immediately became the meeting point for all Westerners spending any time in Baikonur, including clients and teams from ILS, the main competitor of Ariane-space.

## SEEKING NEW CUSTOMERS

For François Calaque's team, there was also no question of resting on their laurels after the initial success with Globalstar. Studies projected a large potential market for the years 1999 and 2000, and since it took nineteen and a half months to produce a launcher on the Samara assembly line, it was urgent to prepare a new order beyond the three units being made for the Globalstar missions. To streamline production, a "reservation" system prior to ordering was even considered.

To meet the constraints of each type of orbital deployment requested by potential customers, Starsem and RKA had to study the launch azimuths required to reach the target orbits. These azimuths determined the areas in which the rocket stages would fall, and therefore affected the arrangements to be negotiated with local authorities. From Kourou of course, all of the Ariane stages fell into the sea, while from Baikonur the lateral boosters and central core of the Soyuz fell on land. These drop zones had to be in uninhabited or very sparsely populated areas, and RKA had to pay compensation to the local authorities. It depended on the accuracy of the trajectory and drop of each stage (which determined the size of the "reserved area") and the type of propellant used (in particular its degree of toxicity).

In early 1997, as often occurs in the commercial space transport sector, one of the promising markets for Starsem was a direct competitor of the first customer, namely Globalstar. TRW's Odyssey project was to include 14 satellites (i.e. 12 plus two in-orbit spares) of 2,200 kg deployed at 10,354 km of altitude and 50° inclination. The Soyuz was therefore well sized to launch them individually. In 1995, TRW received an FCC license authorizing it to operate this system. An impressive international industrial team was associated with the project, and the Canadian operator Teleglobe joined TRW to market the service. In mid-1997, it all seemed settled and Starsem was sure to participate in the deployment of the constellation. But TRW spent more time defending its patents and its licenses in court than in effectively raising funds to finance the $2.8 billion cost of implementing this system. In September, one of its main investors, PT Satelit Palapa Indonesia (Satelindo), withdrew from the project, and in November TRW failed to attract a sufficient pool of investors to initiate the construction of its first two satellites before the deadline set by the FCC license. It tried to negotiate an extension, but finally threw in the towel on Christmas Eve. No firm contract had been signed with Starsem and therefore no money was paid out.

Two other constellation projects were also extending promises of contracts to Starsem.

The first was in the same industrial vein as Globalstar. It involved a constellation of 64 satellites weighing 800 kg, orbiting at 1,469 km to provide Ku-band broadband data transmission. It was proposed in October 1995 by Alcatel Space under the name of Sativod (Satellite Interactive Video On Demand) and then renamed SkyBridge in February 1997. The target market was no longer mobile telephones but multimedia, and particularly the Internet, whose exponential growth world-wide had just begun.[128] The industrial team included the US company Loral and the French company Aerospatiale Satellites. The announced budget was close to $4 billion.

Launches were planned to start in 2001 but bidders would have to wait until 1999 for the first contracts to be signed.

The second was Bill Gates and Craig McCaw's fabulous Teledesic, which Boeing had just bought into. The US aerospace giant was to take charge of an industrial team which was yet to be organized. The size of the constellation had been reduced from 840 satellites (924 including in-orbit spares) to 288, but the mass of each satellite had been increased from 800 to 1,300 kg. Teledesic had included the Starsem Soyuz-Ikar in its list of potential launch vehicles, together with the Indian PSLV, the Ukrainian Dnepr, and more exotic projects such as the airborne Eclipse launcher proposed by the California firm Kelly Space & Technology (KST).[129] This project was budgeted at $9 billion but, again, launch service providers would have to wait several years for potential contracts to be signed.

To address these potential demands, Starsem planned to order 6 to 14 launchers in early 1998.

Yet, despite all these efforts, Starsem's second client would prove to be neither commercial, nor US-based; it was a public European organization.


## A RESURRECTION AND A DEATH

The day after the in-flight failure of the inaugural Ariane 5 on June 4, 1996, and the loss of its payload – a cluster of four ESA scientific satellites – the space agency and Dornier Satellitensysteme, the prime contractor for the payload, began to investigate solutions to revive the mission. A satellite could be assembled from spare parts still available to Dornier, and three new satellites could be made for a fraction of the initial budget. The question remained of how to launch them. An initial Arianespace proposal for sharing two Ariane launches in 1999 and 2000 was refused by the ESA science program committee (SPC). At an estimated $108 million, it exceeded the $72 million budget made available by ESA member states. Arianespace proposed a joint bid with Starsem for two dedicated Soyuz launches. The announced price was 390 million francs ($69 million). Exceptionally, the SPC took four weeks to consider this offer before accepting it in February 1997. Several weeks of negotiations with Starsem led to measures to keep the cost of the two launches below 60 million ECU (ESA's currency unit based on a basket of currencies, replaced by the Euro in 1999).

On April 3, the SPC approved a Cluster 2 mission with a budget envelope limited to 214 million ECU. The launch of the four identical satellites (1,185 kg each) was be by two Soyuz launches in mid-2000. Since the satellites would be too large for the fairing of the Soyuz using the Ikar stage, Starsem signed an agreement with RKA to augment its offer with a Fregat stage from NPO Lavochkin. The development of the Fregat had been stagnating for years due to lack of funding. The announcement was made on June 18 at the Paris Air Show and a proposal incorporating this solution was transmitted to ESA at the end of the month. It was stipulated that prior to launching the ESA satellites, the Fregat would be qualified in flight. In parallel, an evolution of the Ikar stage in which certain military avionics elements would be

Cluster 2 satellites – replacements for those lost during Ariane 5's inaugural flight – undergoing dynamic testing at IABG in Ottobrunn, Germany.

replaced with less expensive civilian equivalents was abandoned. Thus two new combinations entered the catalogue of launch services offered by Starsem: Soyuz-Fregat using the standard Soyuz launcher and the Soyuz-2-Fregat which incorporated the improved version of the Soyuz studied for the "Rus" program. The introduction of digital avionics would enable the Soyuz-2 to fly in more unstable aerodynamic configurations, in particular carrying 4-m fairings similar to those of the Ariane 4 to accommodate bulkier payloads. This configuration quickly received the trade name of "Soyuz/ST".

The Cluster 2 prime contract was signed with Dornier on November 28, 1997. The same day, a preliminary contract was signed with Starsem to study the complete mission of launching pairs of satellites on the Soyuz-Fregat. For reasons of celestial mechanics and mission geometry (the satellites had to orbit through the geomagnetic field radiation belts flying in a tetrahedral formation) the two launches had to occur during a six-week period between mid-June and early August 2000.

A definitive contract would follow on August 18, 1998, but the founder and CEO of Starsem, François Calaque, would not be there to sign it.

In the years shortly before that time, Calaque had pushed himself over the limit, and his health, damaged by his heavy smoking, had declined. In September 1997 he had had to turn over his "space transportation system" responsibilities at Aerospatiale to Philippe Couillard so that he could completely focus on Starsem. He nevertheless remained an adviser to Michel Delaye. In March 1998 his declining health required treatment, forcing him to give up his leadership role within the company. He died on June 14. Finding a successor to François Calaque proved problematic. Starsem is a Euro-Russian company. Its first CEO was French: the question arose of whether or not the successor CEO should be Russian. All the diplomatic talent of Michel Delaye and other European members of the board of

The Soyuz launcher with the Fregat upper stage shown in red.

directors[130] was needed to convince their counterparts at RKA and Samara to put another Frenchman in the CEO's seat.

Their preferences turned toward Jean-Yves Le Gall. Having established CNES's strategic plan, published on February 1, 1997, Le Gall became deputy director of the French space agency, in charge of ministerial trusteeships, Europe, and partnerships, then strategy and international relations. He was not a newcomer to cooperation with Russia, having gained experience during the preparation of the VeGa missions at the laboratory for Space Astronomy in Marseille in the early 1980s. He had negotiated Franco-Russian manned space missions as advisor to his minister, Paul Quilès,[131] and then as deputy director general of CNES.[132] In the meantime, while at Novespace, he worked with TsSKB on the use of Foton capsules for microgravity research missions.

At the end of March, while he was on an assignment in South Africa for CNES, Jean-Yves Le Gall received two telephone calls from France; one from Jean-Marie

Luton, the former director general of ESA who replaced Charles Bigot at the head of Arianespace in July 1997, and another from Yves Michot, who succeeded Louis Gallois at Aerospatiale in August 1996. Both invited him to become chief executive of Starsem. Le Gall's first reaction was that at 39 years of age he was too young for the job, but this concern was soon swept away when he realized that Sergei Kirienko, the new Russian prime minister who had succeeded Viktor Chernomyrdin on March 23, was only 36.[133] On April 6, 1998, Le Gall became the new CEO of Starsem and rejuvenated the management by bringing with him a part of his strategic affairs team from CNES: Michel Doubovick (31) for administration and finance, Mathieu Weiss (27) for marketing, Bernard Luciani (36) for international development, and Claire Coulbeaux for communications.

Starsem's general management was assigned to Victor Nikolaev. This Francophile Russian, formerly with Khrunichev and Glavkosmos, worked at RKA, where he had helped to establish strong links with CNES and ESA prior to taking charge of the project with Starsem, which he officially joined in February 1997. He played a key role in maintaining excellent working relationships with Alexander Medvedchikov, deputy to Yuri Koptev, who remained the principal contact with RKA, and then from June 1999 at the Rosaviakosmos agency that succeeded it.

In the absence of the contracts initially expected (early forecasts aimed for five launches per year) a plan was enacted to reestablish the company's financial balance. Amongst the actions taken to capitalize on assets, in the summer of 1998 the SPPF facilities in Baikonur were leased to NPO Yuzhnoe to prepare its three clusters of 12 Globalstar satellites.

## "ONE MAN'S LOSS IS ANOTHER MAN'S GAIN"

On September 9, 1998, at 20:29 UT, a Zenit 2 rocket lifted off from complex 45/1 at Baikonur with an impressive payload of 12 Globalstar satellites valued at $16 million each. At the same time, the top marketing moguls of the space industry were meeting in Paris for a major symposium organized at "La Maison de la Chimie" by Satel Conseil, a French consultancy firm. All of the major business leaders of the satellite and launch vehicle industry, along with their operator-customers came in person to discuss the global economy. A grand gala dinner was held that Wednesday evening. At the end of the evening, an announcement came that the first commercial mission of the Ukrainian heavy launcher had successfully placed its satellites in orbit. Glasses were raised to celebrate the good news. In 1998, the Internet was not yet the instant information tool that it would become a decade later, and most of the participants did not learn the sad reality until the following morning.

The launcher had lifted off as planned and the flight went nominally during the boost of the first stage, but 4 minutes 32 seconds after liftoff two electronic failures were detected in the second stage. Former Soviet launchers did not have self-destruct systems, the onboard computer merely shut down the RD-120 engine and the entire unit, consisting of the second stage and its expensive payload crashed in southern Siberia, on the border between the Russian Federation republics of Tuva, Altai and

A cluster of Globalstar satellites on the "dispenser" being integrated with an Ikar stage for launch on a Semyorka.

Khakassia, about 2,000 km from Baikonur. It was the eighth failure of this rocket out of just 31 flights. Leonid Kuchma, the Ukrainian president and once a senior official of NPO Yuzhnoe and MZ YuzhMash, insisted that the failure would not jeopardize the other two launches of Globalstar clusters on Zenit 2 launchers identical to the one that had just failed. But a few days later Robert Berry, director general of SS/L, the system's industrial prime contractor, announced that options on the contracts signed with Boeing and Starsem would become firm orders. The December 1996 contract with Starsem left a lot of leeway: there were eight such options!

At that time Starsem had six launchers in production in Samara, including the first three which were in the completion phase, and it had three dispensers on order from Aerospatiale in Aquitaine. The first unit was scheduled for launch in November. Sixteen satellites intended to fly on the Zenit 2 were in its clean rooms at Baikonur, and 14 more were expected to arrive soon. On October 5, SS/L officially transformed three of its options with Starsem into firm orders.[134] As a result, Starsem, which had not yet carried out a single launch, found itself with six urgent missions to launch 24 satellites within a year! The integration of the last three launchers on order from Samara would have to be accelerated. Three new dispensers were ordered from Aerospatiale on October 28.

A final step remained, but it was a giant one. This was the signature of a trilateral agreement between the United States, Russia and Kazakhstan for the protection of sensitive technologies, to authorize Starsem to mount the Globalstar satellites on its launchers.[135] Intense lobbying was undertaken in Washington to prevent the signing of this agreement. Boeing had just received a contract from SS/L for seven launches of Globalstar satellites in the event that the Starsem contract fell through. The tension was intense. The European correspondent of a US-based aerospace magazine, who had followed Starsem's progress, even received strange letters from a law office in Washington telling him that "Starsem would never launch a US satellite" and firmly enjoining him to "stop peddling false information" by stating the opposite. Despite these maneuvers, the Technology Safeguard Agreement was signed on January 26, 1999.

Two weeks later, in the early hours of February 9, at precisely 3:53:59 UT, the first Soyuz launcher sporting the Starsem livery lifted off from pad N°1 at Baikonur. The Ikar stage, making its first flight, perfectly deployed Globalstar satellites N°23, N°36, N°38 and N°40 into the targeted orbits. From now on, Starsem would have to be considered a fully-fledged player in the international satellite launch market.

The following launches took place with the steady predictability of a metronome.

A historic Soyuz launch on February 9, 1999: the first bearing the "Starsem" logo.

Flights on March 15 and April 15 completed the first series of three launches. The second series was deployed on September 22, October 18, and November 22. In ten months, Starsem, the outsider, had placed half of the operational constellation into orbit with "no errors". As early as June 29, SS/L had signed a contract for a launch to ensure the maintenance of the Globalstar constellation starting in 2001.

## FREGAT MAKES ITS APPEARANCE

At the Paris Air Show on June 14, 1999, Starsem signed its third launch contract with ESA to launch the Mars Express probe on a flight to the Red Planet in June 2003. The mission had several outstanding aspects. First, it represented another resurrection since the probe was designed to fly scientific instrumentation originally developed by Europeans for the Mars-96 heavy orbiter. This mission had survived the economic crisis in Russia that followed the collapse of the Soviet Union only due to Western support and a delay of two years. Unfortunately, at the time of its launch on November 16, 1996, a failure of unknown origin[136] on the Block D2 upper stage of the Proton-K launcher impeded the escape into interplanetary space of what was

Mars-96, the last major probe developed in Soviet times, included a significant international contribution but failed to escape from Earth orbit.

The Fregat stage.

supposed to be the largest probe ever sent to Mars. Instead, it re-entered the atmosphere and disintegrated over the Pacific off the coast of South America.[137]

To salvage something from their work, European scientists turned to ESA, which was seeking precisely to establish a new concept of more reactive scientific missions. Like "Faster-Better-Cheaper" at NASA, these "Flexi"[138] missions were to be decided upon and developed more rapidly than was traditional for ESA. The Mars Express proposal submitted in 1997 was approved in principle in June 1998 and the Starsem Soyuz proved to be the ideal launcher for this class of mission.

However, as with the Cluster 2 launch contract, this one depended on the Fregat stage from NPO Lavochkin being qualified. The ground qualification program was completed in August and the stage passed an ESA engineering review in November. The qualifying flight in early 2000 was to be followed by a rehearsal of the Cluster mission. If these two flights failed, the launch of the four Cluster 2 satellites would be given to the Ariane 4. While scientific missions were never insured, in this case an exception was made and a policy was taken out to cover the additional cost induced by such a transfer.

The qualification flight occurred on the night of February 8-9, 2000. The Soyuz-U launcher injected the Fregat stage and its payload into a suborbital trajectory, with the new stage from NPO Lavochkin working perfectly. The first ignition enabled it to achieve orbital velocity and then a second ignition circularized its orbit at 600 km. After ejecting an inert mass of 1 ton, simulating a satellite, and completing six orbits, it performed a third and fourth ignition to deorbit itself and re-enter the atmosphere. On its way in, it released a 110-kg technological payload: IRDT (Inflatable Reentry & Descent Technology). This was a semi-rigid deployable heat shield developed by NPO Lavochkin, measuring 8 m in diameter after deployment. This technology had been developed for the Mars-96 probe landers, but had never been test-flown due to the loss of the mission. Since 1998, DASA and NPO Lavochkin had cooperated on adapting this type of application to other missions, such as returning equipment from the International Space Station. The experiment on the Soyuz-Fregat test flight was the first of its kind,[139] and was facilitated by ESA, the European Commission, DASA, and Khrunichev each contributing $600,000. Eight hours after take-off, the IRDT landed in the Orenburg Oblast, near the Kazakh border. A beacon failure and terrible weather prevented immediate recovery, but it was found on February 15 and hoisted out via helicopter. An analysis of the recorded data showed that a tear that occurred in the shield during descent resulted in a more violent impact with the ground than expected. A similar 14-m diameter shield was mounted on the Fregat stage, but that was never found.

For Starsem, the Fregat's flight qualification paved the way for a full-scale Cluster 2 rehearsal. On the evening of March 20 at 23:28 local time, the second Soyuz-Fregat took off from Baikonur with a mock-up made by Aerospatiale Matra Lanceurs and NPO Lavochkin. This 2,384-kg "dumsat", loaded with sensors, simulated the dynamic behavior of a pair of Cluster satellites during launch. Once

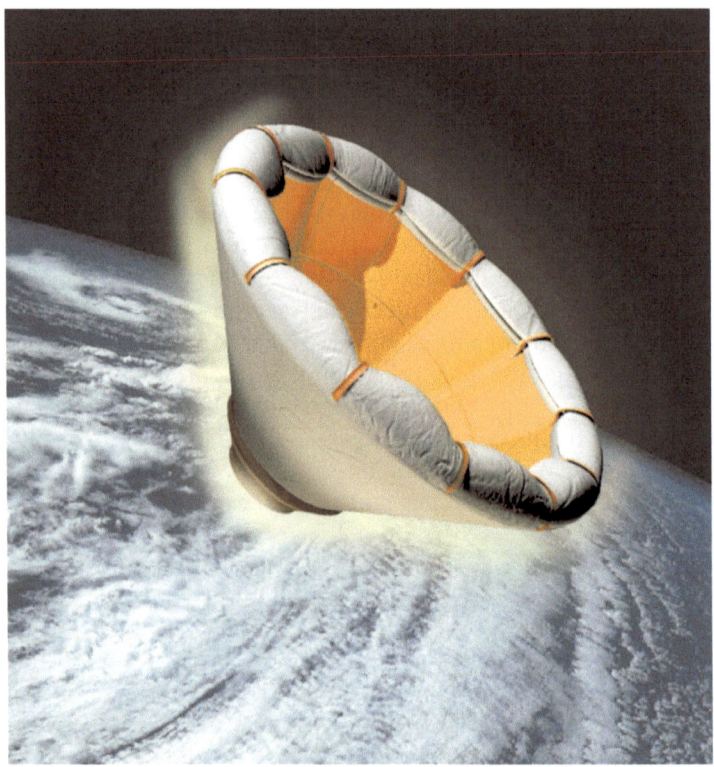

Artist's view of the IRDT demonstrator, with its inflatable shield deployed for re-entry into the atmosphere.

The IRDT demonstrator on the Fregat stage about to be encapsulated by the fairing.

The second Soyuz-Fregat launcher with the "dumsat" payload.

The launcher for the first pair of Cluster 2 satellites being integrated in the MIK.

On July 16, 2000, Starsem's first launch for ESA carrying the first pair of Cluster 2 satellites, "Salsa" and "Samba".

again, the Fregat stage fulfilled its mission. The first ignition, lasting 4 minutes, put the vehicle into a parking orbit at an altitude of 200 km. One orbit later, the Fregat ignited again for about 10 minutes to achieve an elliptical orbit of 242 x 18,000 km inclined at 64.8° for the deployment of the Cluster pair.

Integrating the second pair of Cluster 2 satellites, "Rumba" and "Tango", with the Fregat stage on July 31, 2000.

Encapsulating the second pair of Cluster 2 satellites in the fairing.

Joe Pereira of the Cluster team saying goodbye to "Rumba" and "Tango" during withdrawal of the purge lines from the fairing on the launch pad.

Artist's view of the injection of a pair of Cluster 2 satellites by the Fregat stage.

After a 1-month delay due to an anomaly with the thrusters on the satellites and another 24 hours because of a defective sequencer which interrupted the launch less than a minute before the take-off sequence, flight models N°6 and N°7 of the Cluster program (nicknamed "Samba" and "Salsa") were deployed in orbit on July 16 by the third Soyuz-Fregat. Flight models N°5 and N°8 ("Rumba" and "Tango")

"Tango" moving away after its release in space.

followed on August 9. On this flight, the Block I stage of the Soyuz cut off prematurely due to a propellant-filling error. The mission was saved by the versatility of the Fregat stage, which compensated for the diminished performance by providing the additional 70 to 80 m/s needed for injection.

## MARKET COLLAPSE

Efforts to progress in the constellation market continued in parallel, particularly the SkyBridge project which had increased from 64 to 80 satellites, with the unit mass growing from 800 to 1,400 kg. The budget also jumped from $4 to 6 billion. Once again, Starsem seemed well positioned, but no commitment was made. December 1999 brought the unpleasant surprise of the deployment of half of the constellation being entrusted to Boeing, which would launch them in clusters of four satellites on Delta 3 launchers and clusters of eight on Delta 4 launchers; with the US industrial giant making an investment in the share capital. For Starsem it was a case of déjà vu, because, as previously occurred with the Zenit for Globalstar, these two operationally unproven launchers were offered at a knockdown price specifically to "capture" the market.[140]

SkyBridge also announced that it was negotiating with Arianespace and ILS for deployment of the other half of the constellation. But Arianespace, which proposed to launch two clusters of 10 satellites on the Ariane 5, withdrew from the competition due to the condition imposed by the operator which required the supplier to become a "strategic partner" by investing in the project. Arianespace thus stepped back, giving way to Starsem which signed a "risk-sharing partnership" agreement with

SkyBridge on April 11. This agreement provided for the placement in orbit of 32 satellites via 11 launches (10 triplets and one pair) in exchange for an investment in the operator's capital. Launches were planned for 2002 and, for the first time, the Soyuz/ST model would be used.

The celebration after the success of the first two flights of the Soyuz-Fregat was short-lived. The previous month, the speculative Internet "bubble" had burst and the market collapsed. In a matter of months, 4,300 NASDAQ technology companies lost all of the fabulous profits made since 1995 and were forced into bankruptcy or had to regroup by mergers and acquisitions. The SkyBridge project was frozen indefinitely at the end of November 2000.[141]

Finally, the only firm contract signed by Starsem in 2000, on December 18, was again with ESA, to launch the first two MetOp satellites in a sun-synchronous orbit for Eumetsat, the European meteorological organization, with an option to launch MetOp 3. Again the Soyuz/ST was selected as the launcher. After that, a difficult period of economic stagnation began. The fact that Iridium and Globalstar filed for Chapter 11 bankruptcy protection under US law in August 1999 and February 2002 seriously eroded investors' confidence in the space sector. The $7 billion sunk into these two projects simply evaporated. There were no more constellations in need of launch, and no commercial satellite projects matching the performance of the Soyuz.

Only a few rare payloads were considered, such as the Japanese Muses-C probe, grounded by the failure of the M-5 launcher in February 2000, and which needed to set off by the end of 2002 to have a chance of reaching the near-Earth asteroid 1998 SF36. A Japanese delegation even came to attend the launch of the second pair of Cluster 2 satellites to assess the possibility of launching the probe on a Soyuz-Fregat if the M-5 remained unavailable – which did not prove to be the case.[142] Mitsubishi Electric was also studying the possibility of launching two technological SERVIS (Space Environment Reliability Verification Integrated Systems) platforms by Soyuz to test components, but it eventually signed with Eurockot. In France, consideration was given to the possibility of using the Soyuz for a new flight of the ARD capsule – which had already flown in October 1998 on the third Ariane 5 – in order to test an air-capture maneuver for the future mission to bring back Martian samples for study at CNES and NASA,[143] but this was not pursued.[144]

At ESA and the European Commission, work began on a Galileo constellation of satellite navigation, but its launch remained far in the future, with a suggested date of 2008.

Consequently, for 30 months, Starsem did not sign a single contract or perform a single launch.

## TIME FOR A CHANGE

Those two and a half years nevertheless saw major shifts in the realm of European space transportation, along with Starsem's role within it.

The European industrial sector went through a profound reorganization. On

Integration of the upper part of the launcher with the lower assembly, for the launch of the Mars Express probe on June 2, 2003.

Inside its fairing, the Mars Express probe awaits launch. The two service gangways are pulled back 30 minutes before take-off.

The launch of the first MetOp satellite on October 19, 2006, introduced the 4-m fairing on the Soyuz.

February 15, 1999 – six days after Starsem's first launch – Aerospatiale merged with Matra Hautes Technologies to create Aerospatiale Matra.[145] Hence Matra Marconi Space, which had campaigned for a study of an "Ariane 4 Lite" to compete with the Soyuz, was absorbed by Starsem's main shareholder. The concept was nevertheless presented to the board of Arianespace on April 20, 1999, but was set aside and then permanently abandoned. On July 10, 2000, Aerospatiale Matra merged with DASA – another European opponent of the Soyuz – and with the Spanish company CASA to create EADS (the European Aeronautic Defence and Space company).[146] With the exception of Snecma (which had absorbed its subsidiary, SEP, in 1997) and Fiat Avio (which had absorbed BPD in 1996), the major suppliers of the Ariane family were now within the industrial group that supported Starsem.

With the disappearance of the constellations and the contraction of the market – which also had the effect of reducing appetites throughout the sector – a necessary rationalization of European launcher policy could take place. When the "Proposal for a European strategy in the launchers sector" initiated by ESA's ministerial council in May 1999 was published in June 2000, it was already out of date by proposing a variety of architectures to replace the Soyuz with a purely European

At the 2003 Paris Air Show David Southwood, science director of ESA and Jean-Marie Luton, CEO of Starsem, sign the launch contract for the Venus Express probe.

The Israeli Amos 2 satellite on its Fregat stage being prepared for Starsem's first geostationary mission.

Left: Three Semyorkas at varying phases of integration at Baikonur in September 2007, with the Foton M3 launcher in the foreground. Right: The Foton M3 satellite carrying the European YES-2 experiment, with the Mirka mini-capsule being placed inside the fairing.

A final glimpse of the Venus Express probe through a hatch in the fairing as the launcher is installed on the launch pad.

The junction between the two sections of the launcher requires particularly careful attention.

The Venus Express launcher approaches the "pit" of pad Nº31 at Baikonur. The retaining brackets of the four boosters can be seen.

launcher by around 2005,[147] since there was no budget for such development. At the same time, Arianespace remained realistic, and presented a strategy leveraging the idea of a line of launch vehicles in which the Ariane 5 was offered along with the Soyuz and even the Rokot, which had made its first commercial demonstrator flight on May 16.[148] For Arianespace, it was simply a question of "putting in place a coherent trade policy in the face of renewed and growing competition".

A rapprochement between the executives of Arianespace and Starsem began in August 2001 when Jean-Yves Le Gall, still CEO of Starsem, was appointed director general of Arianespace. Nine months later, in June 2002, he was promoted to CEO of Arianespace, succeeding Jean-Marie Luton, who remained chairman of Arianespace and became CEO of Starsem.

Starsem headquarters, which had relocated in 2000, from Suresnes to the Maine Montparnasse Tower in Paris, where it shared premises with Arianespace, was then moved to Evry, across from Arianespace's headquarters. For reasons of "classified defense information" related to military payloads in Ariane's manifest,[149] a merger of the two headquarters, which would bring Russian personnel to Arianespace, was still impossible.

Meanwhile, Samara was able to fill out its workload thanks to International Space Station operations, which required regular launches of Soyuz and Progress vehicles, with the Soyuz-U being superseded in May 2001 by the Soyuz-FG whose upgraded propulsion prefigured that of the Soyuz-2.

However, a dramatic event nearly compromised Starsem's return to business. On May 12, 2002, roofing repairs on the assembly hall of MIK-112 at Baikonur, where Starsem's clean rooms were located, turned into a catastrophe. The structure could not support the materials piled on the roof and three of the five sections collapsed from a height of 70 m, killing seven workers, destroying the Buran shuttle that flew in 1988, the dynamic test model of the Energiya launcher on which it was mounted, and the stages for two complete Energiya launchers that were stored there. The last vestiges of a bygone Soviet era were reduced to twisted scrap, buried under tons of debris.

Fortunately, Starsem's facilities were located in a section of the building with a lower ceiling where the roof remained intact and they were not damaged. They were still quite capable of receiving the Mars Express probe from ESA in April 2003 for the launch campaign.

Therefore a transformed Starsem, in a modified industrial landscape, resumed its activities in mid-2003. On June 2, Mars Express was launched for a historic flight to the Red Planet. Two weeks later, ESA hired Starsem to launch the Venus Express probe, scheduled for November 2005.

At roughly the same time, Arianespace decided to entrust Starsem with the launch of a satellite from its own manifest, the Israeli Amos 2. This had been booked on an Ariane 5 in late 2003 along with the Eutelsat W3A satellite, but following the failure of the maiden flight of the Ariane 5ECA on December 11, 2002, Eutelsat decided to transfer its satellite to a Proton-M from ILS in April 2003. With no launch slot in the short term available for the small Israeli satellite, Arianespace decided for the first time to use the Starsem Soyuz as a complement to the Ariane 5; the 1,374-kg Amos 2

Top: The Venus Express launcher is placed on the four retention hooks. Bottom: Viewed from the service structure, the Kazakh steppe stretches out beyond the flame trench of pad Nº31.

Frost covers the base of the Venus Express launcher during filling of its liquid oxygen tanks.

The Soyuz-FG/Fregat launcher carrying the Venus Express probe lifts off on November 9, 2005.

was compatible with the geostationary performance of a Soyuz-Fregat launched from Baikonur. On December 27, 2003, for the first time in history, a Semyorka launcher – the 1,684th member of the family – placed a payload into geostationary transfer orbit.[150] This was just the start, because the Soyuz would soon take another step in its "Europeanization".

During the summer of 2003, clearing began at the Malmanoury site of the Guiana Space Centre, some 11,550 km from Baikonur, where a new Soyuz launch complex, recently approved by the ESA council, would be constructed. Three new launchers from the European product line – the Ariane 5, Soyuz and Vega – would soon be operating from the same site in French Guiana.

# 11

## Russians in the jungle

As we have seen, by the end of the 1980s, even before the fall of the Berlin Wall had triggered the impetus that would lead to the collapse of the Soviet bloc, the USSR's military industrial complex had attempted to open up to Western markets to obtain the currencies essential to its survival. Space launchers, as quasi-legendary showcases of know-how for over three decades, were the clear choice, offering the prospect of capturing a share of a satellite launch market that was in full expansion at that time.

Unfortunately, it quickly became apparent that the launchers of the Soviet arsenal suffered from a major handicap due to the geographical location of their launch sites. From Baikonur, at 45.6°N, it is not possible to launch a satellite into an orbit inclined at less than 45.6° without making complex maneuvers that would be costly in terms of energy. In fact, to keep stages from falling into Chinese territory it was necessary to fly an azimuth a little more to the north, making it impossible to achieve orbits inclined at less than 50.5° or 51.6° depending on the launcher. Unfortunately the bulk of the market is in geostationary orbit directly above the equator. Trigonometry added another factor. While from Kourou (5.2°N), a launcher can benefit from an additional boost of 460 m/s merely from the rotation of the Earth, from Baikonur this boost is only 325 m/s. In terms of performance, this can represent a considerable difference in payload capacity.

Moreover, until 1993, the United States maintained an embargo on the export of satellites and their components to the USSR, and later to the republics derived from it.

For all of these reasons, very early on the Russian and Ukrainian design bureaus began to investigate the possibility of finding a hospitable country in which to install a new cosmodrome near the equator so that their launchers could be operated without restrictions and maximize the boost to be gained from the rotation of the Earth. There was no need to solicit investors, because those willing to finance feasibility studies were legion. For the newcomers to the market economy, who lacked their own funds, the motto was simply: "if you've got the site, we've got the launchers".

## GLUT OF LAUNCH SITES

The first project to generate serious talk was at Cape York Peninsula, Queensland. Since October 1986, the Australian government had been exploring the feasibility of establishing a space center for launches out over the Pacific. Two private consortia studied this possibility. The Australian Spaceport Group, led by US aerospace giant Martin Marietta,[151] and Aussat,[152] the Australian satellite operator, was interested in a site near the mining town of Weipa (12.6°S, 142°E), but concluded in 1989 that there was not enough demand to justify such an investment. The Cape York Space Agency (CYSA), a competitor, did not agree and set its sights on a 607 sq. km site at Temple Bay (12.25°S, 143°E) to operate Zenit launchers which would be purchased from the Soviet industry. This was just before the fall of the Berlin Wall, and the Ukrainians had unveiled their powerful vehicle to the West just 3 months earlier. An investment estimated at $470 million would construct an airfield capable of receiving a Boeing 747F cargo plane or its equivalent, living accommodation and facilities for the 700 people making up the teams, and an automated launch pad. Ground-breaking was planned for 1992, with a first flight in 1996. Unfortunately, the US contractors for the project could not obtain export licenses because of Soviet involvement. CYSA failed to raise the money, and had to throw in the towel in October 1991. Following that, another private consortium, Space Transportation

Starting in 1999, the Zenit was operated commercially from an oil platform converted by the Sea Launch consortium led by Boeing.

Systems (STS) Ltd., took up the project, then withdrew in late 1992, giving way to a third group of investors, the Cooksey consortium, which fared no better.

Attempts to find a launch site for the Zenit took a more political turn in 1994, when Leonid Kuchma, former head of the MZ Yuzhmash plant of Dniepropetrovsk where the Zenit launchers were manufactured, was elected president of Ukraine. The uncertainties weighing on relations between Ukraine and Russia, where construction of a Zenit pad at Plesetsk had been placed on hold, led the Ukrainians to try anything in order to find a launch site that would make them independent of the Russian "big brother" as soon as possible.

India was approached about its site in Sriharikota (13.7°N, 80.2°E), as were the Europeans for possible access to Kourou in Guiana. In October 1995, Brazil was approached about its space center at Alcântara, Maranhão (2.3°S, 44.4°W), and then in May 1996, Indonesia, which offered the island of Biak (1°S, 136°E)[153] from which Ukraine proposed launching the Indonesian Palapa B5 satellite.[154] In October 1997 a memorandum of understanding was even signed with Gary Filmon, the Canadian prime minister, to use the Churchill space center (58.7°N, 93.8°W) in Manitoba, on the shores of Hudson Bay. This was used until 1984 for sounding rockets, and was famous for its relatively tame polar bears which earned the nearby town of Churchill the title "Polar Bear Capital of the World". It was far from the equator, but Ukraine was so concerned about being denied access to Baikonur that any back-up site was welcome.

In the meantime STS had decided to construct a Proton launch site in Papua New Guinea, which would increase the launcher's capacity for geostationary transfer orbit to 9 tons. Two sites were envisaged in the Bismarck Archipelago: Emirau, one of the Matthias islands (1.6°S, 150°E) or Manus in the Admiralty Islands (2°S, 147°E). The required investment was estimated at $900 million to achieve operational service by 1998, but a lack of investors led to the abandonment of the project in 1994.

After having taken over marketing responsibilities for the Proton in June 1995, ILS in turn began looking for an equatorial site, focusing on Brazil's Alcântara space center, the Cape Canaveral Air Force Station, Florida (28.5°N, 80.6°W), and the site of Gunn Point near Darwin in the Northern Territories of Australia (12.1°S, 131°E) called the Asia Pacific Space Launch Center. The latter was proposed by Transfield Defence Systems, STS, and Thai Satellite Telecommunications, a shareholder in the Iridium project. The necessary investment was estimated at $590 million (some 800 million Australian dollars) for a first launch in 2000. If the project had continued, the Australian federal government was even ready to provide $93 million.

The ratification of the Baikonur rental agreement by the Duma and the Kazakh parliament on April 27, 1996, took the pressure off the search for a new site for the Proton, which could now make commercial flights from Kazakhstan.[155] Moreover, the launch quotas imposed by the United States in 1993 were increased in January 1996 and the quota system expired on December 31, 2000 without being renewed.

## KOUROU IN SIGHT

With the commercial geostationary market booming at the end of the 1980s, the Soviets, and their Russian and Ukrainian successors, were not the only ones seeking to move closer to the equator to benefit from better launch conditions. The existing facilities at the Guiana Space Centre (CSG) near Kourou in French Guiana were not exempt from this prospecting.

As early as 1989, General Dynamics made overtures to study the possibility of operating the Atlas 2AS launcher from French Guiana. Over the years, discussions took place with McDonnell Douglas for its Delta 2, with Khrunichev for the Proton, with NPO Yuzhnoe for the Tsyklon, and even with NTTs Kompleks for its small Start launcher derived from the RT-2PM Topol missile. Israel Aircraft Industries (IAI) also made an approach about Next, the commercial version of its small Shavit launcher, which was otherwise obliged to launch towards retrograde orbits in relation to the rotation of the Earth from the Israeli site of Palmachim, due to the geopolitical impossibility of launching over the Arab countries that surround the Hebrew State.[156] To guide these discussions, on November 23, 1993, the French government, which owned CSG, and ESA, which owned the ELA facilities, signed an agreement on "the possibility of opening CSG to launchers other than Ariane". The arrival of new players in Kourou would be a way for France to offset the fixed operating costs of its space center. The first version of this agreement was to be renewed before the end of 2000.

The first mention of a possible Soyuz launch from the Guiana site came roughly at the time that Aerospatiale was studying the Irene project. The French then explored the feasibility of such implementation "in the short term". The objectives were not given in detail, but there were probably several. Circumventing the US embargo was a non-issue because it had been lifted in September 1993. Operating the Soyuz from Guiana would provide security against a crisis between Russia and Kazakhstan over the use of Baikonur. And for François Calaque's team, the Soyuz could have a role to play in launching French military payloads such as the Helios 1B optical observation satellite,[157] the next generation Helios 2, the Osiris radar satellite,[158] and the Zenon eavesdropping satellites,[159] all of which were too small for the Ariane 5 and much too sensitive to be exported to either Russia or Kazakhstan.

But the abandonment of the Irene concept and the program's redirection toward the use of an almost unmodified Russian Soyuz launcher consigned these studies to the archive. The difficulties Starsem proponents encountered from the start in getting their project accepted, and especially the overt hostility displayed by some industrial partners, led certain Aerospatiale executives to make sure they never mentioned the option of launches from Kourou: "There was already so much resentment towards us, I didn't want to have to watch my back every time I got home!"

After February 1997, the CNES strategic plan, drafted under the direction of Jean-Yves Le Gall, stipulated that only launchers not in competition with Ariane could be given access to CSG. The following month, a team composed of DASA, MMS and German industrial firm MAN Technologie, proposed that after the

Ariane 4 ceased operating, the ELA-2 facilities (Ariane launch complex N°2) be modified to operate the Tsyklon 3K, a form of the Ukrainian Tsyklon 3 launcher upgraded to the safety standards of the European space port, but this project went no further.

It took until January 6, 1999, during the annual new-year press conference of the Arianespace CEO, for Jean-Marie Luton to announce that a study was underway on the possibility of building a launch site for the Soyuz in Guiana. The situation had evolved since Starsem's creation. The company had won its first eight contracts and was preparing to initiate its launches, but negotiations were still underway with US authorities for the signature of the Technology Safeguard Agreement which would authorize the company to install US manufactured satellites on its Russian launchers. By admitting that studies had been conducted concerning the possibility of operating the Soyuz from Guiana, the Europeans clearly signaled to Washington their desire not to forgo their strategic alliance, even if others created obstacles. The agreement was signed less than three weeks later and Starsem flights could start from Baikonur in the following days.

## A TROPICAL, CRYOGENIC SOYUZ

In reality, it was in late July 1998 that the Starsem board of directors decided to set up a working group to study the possibility of constructing a Soyuz launch complex at CSG. Ever since the idea was first raised in 1993, it had appealed to the Russian partners, although they were reserved about it and never failed to point out that it was a European proposal. One thing was certain: a Soyuz launched from Guiana could access new markets, including geostationary, and both the Samara Space Center and RKA were ready to consider anything that would ramp up production of launchers on the Semyorka lines.

On the European side, people were also conscious of the fact that by bringing the Soyuz to Guiana, Starsem could extend its offer to the broader market of flights to Mir and later the International Space Station, as a prelude to manned flights under European responsibility, without having to seek a significant government investment to achieve this capability.

As a result, in September representatives of RKA, CNES, Samara, Aerospatiale, Arianespace and KBOM began to explore the idea. Two versions were envisaged for operations in Guiana: Soyuz K1 and Soyuz K2. In fact, the first was the Soyuz-2.1a (whose development in Russia was suffering from budgetary difficulties) equipped with a Fregat upper stage. Its first qualification flight was expected in late 2001 and it could be available for flights from Kourou by 2003. Whereas from Baikonur it was only capable of placing 1.6 tons in geostationary transfer orbit, from Kourou it could loft 2.7 tons. The second version, also known as "Soyuz/HO", was equipped with the H10-3 cryogenic upper stage of the Ariane 4.[160] Its development was regarded by the Europeans as "inseparable" from flying the Soyuz from CSG. Estimating its actual performance was complicated by the fact that neither the Russians nor the Europeans had sufficiently precise data on the components

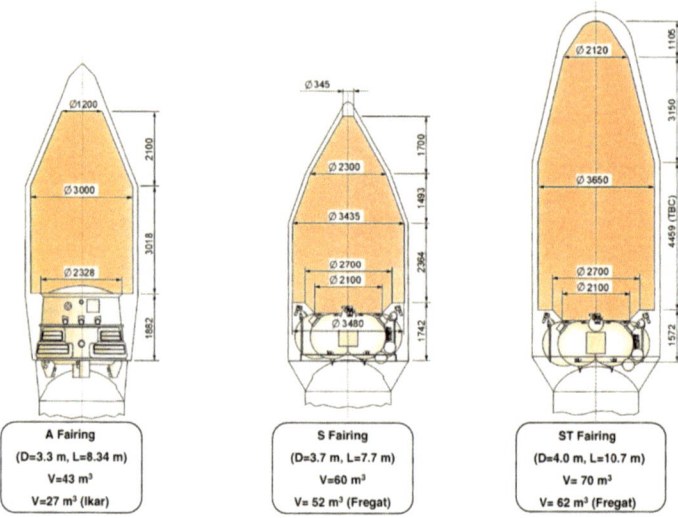

Detail of the fairings proposed by Starsem for commercial flights.

produced by the other party, or their injection procedures, but the figure of 4 to 4.5 tons in geostationary transfer orbit was deduced. This promptly raised the issue of how such a vehicle would be positioned in marketing terms relative to the Ariane 5 launcher.

Moreover, Soyuz rockets launched from Guiana would have to be upgraded to comply with safety rules in force at CSG, which involved the installation of in-flight neutralization systems and related telemetry.

A previously proposed modification, which involved adapting the fairing of the Ariane 4 to the Soyuz-2 to make the Soyuz/ST, proved to be a problem. Although the digital avionics of the Soyuz-2 enabled it to compensate for the loss of aerodynamic stability caused by a fairing 4 m in diameter and 10.7 m tall, European technology was not compatible with the horizontal vehicle integration practiced by the Russians, so a fairing of similar size would have to be built using Russian methods.

Three options were studied for the Soyuz launch complex at CSG. The first would modify ELA-2 after the Ariane 4 was retired. Though the idea looked good initially, it proved to be unworkable due to the very different designs of the launchers. Apart from anything else, the fact that they used different propellants would have required everything to be rebuilt, including the concrete base of the pad. The second option was a site close to ELA-2 in order to benefit from certain facilities, in particular for nitrogen, helium and compressed air. The third option was to seek a completely new site. The first surveys revealed the presence of a granitic subsoil, suitable for digging a vast flame trench, about 10 to 12 km from the existing ELA in the direction of the village of Sinnamary, near Malmanoury Creek.[161] The launch pad that could be built on this site would be a true copy of the Baikonur facility, except for the

The Malmanoury savannah in 2004, on the eve of development work for the Soyuz launch complex.

addition of a gantry to mount the H10-3 stage on top of the Soyuz/HO and the cryogenic umbilical for supplying its propellant.

The total investment to install the Soyuz in Guiana was estimated at $400-500 million dollars, including $250-300 million for infrastructure. Financing possibilities included the European Reconstruction and Development Bank, which had helped to finance Sea Launch, European Union grants for aid to Russia, or even exploitation of the tax exemption enjoyed by investments in Guiana.

Whilst it appeared that the technical issues could be resolved, the question of the commercial market remained a tricky one. Since the launch of the second-generation constellations (Iridium 2, Globalstar 2, SkyBridge, and Teledesic) posed no problem in terms of azimuth from Baikonur, they would be mostly, or even entirely, launched from the Kazakh complex. But it quickly became apparent that in order to achieve the seven to eight launches per year needed to make a Guiana complex profitable, the geostationary market, limited to two or three flights per year, would be insufficient. There was still the hope that the introduction of a complementary launcher would "stimulate the emergence of new programs within ESA".

## POLITICAL GAMES AND STAKES

At the end of January 1999, the working group concluded that it could not decide on the merits of the operation until a clear strategic relationship between the Soyuz and the Ariane 5 had been thrashed out. A decision on the policy relating to a complementary launcher would be made in June, after a meeting of ESA council ministers on May 11-12 held in the Egmont Palace in Brussels. The French delegation to the council was led by Claude Allègre, minister of national education, research and technology. As a fierce opponent of manned space flight in general, and the European contribution to the International Space Station in particular, he was also opposed to several other programs which were to be evaluated in this council. These included the Vega small launcher supported by Italy, for which the ESA council – at the delegate level – had agreed to begin preliminary studies in April 1998.

The Europeanization of the Vega launcher was debated by the members of ESA during the Brussels and Edinburgh ministerial council meetings.

Also, in April, Aerospatiale Matra and Fiat Avio had established a cooperation framework by creating the Vegaspazio joint venture.

To avoid having to negotiate on these subjects, Claude Allègre left immediately after his speech at the opening ceremony, leaving his delegation, led by his advisor Joël Hamelin, with non-negotiable instructions, one of which was firm opposition to pursuing the Vega launcher program. This decision took the Italian delegation by surprise, and they decided to block negotiations on the Ariane 5 Plus program for the development of improved versions of the Ariane 5 with cryogenic upper stages: the Ariane 5ECA and 5ECB. They also transferred the remaining €80 million needed for Vega from this program, as well as from ARTA (Ariane Research and Technology Accompaniment) and from FLTP (Future Launcher Technology Program) – whose roles were respectively to ensure technological follow-up of industrial production of the Ariane and prepare the technologies needed for developing future generations of launchers. It took all of the delegates' diplomatic skills to prevent the meeting from turning into a disaster. A figure of €533 million (10% less than expected[162]) emerged for the Ariane 5, and a budget of €317 million remained allocated for the Vega, but it would not be available until the end of the next council meeting in October – again at the level of delegates. Officially, the purpose of this delay was to complete technical studies but in reality it would above all enable the French delegation to reassess its position.[163]

This Franco-Italian confrontation prevented the specific issue of a complementary launcher to Ariane 5 from being raised, but the ministers requested that a "proposal for a European strategy" on launchers be developed by ESA and submitted to the council before the end of 2000.

In June, the axe fell on the Soyuz/HO, which had to be abandoned for lack of a strategic position in relation to the Ariane 5 within Arianespace's offer of services. The study concerning establishment of the Soyuz-2 in Kourou was quietly extended for another year.

At the request of the ministry, CNES examined the legal aspects of a possible renewal of the agreement to open CSG to launchers other than Ariane. The French government declared that it was ready to consider proposals on a case-by-case basis, if they were consistent with its industrial and commercial strategy on launchers, if they raised no problems of security, and if they respected intergovernmental treaties. Even if the policy regarding CSG should be considered as a sovereign act of France, it had to be unimpeachable in terms of the free competition policy established by the European Union and the World Trade Organization.

A summary report, dated November 15, 1999, recognized the need for a policy concerning a family of launchers and Franco-Russian cooperation in a spirit of "trade complementarity". It emphasized France's priority on the development of the Ariane industrial chain and its unifying role in the industry, but also noted that "the budget situation at the moment makes it impossible to consider the development of a range of exclusively European launchers". The report said there was no insurmountable obstacle to the implementation of the Soyuz in French Guiana in terms of legality, international relations, security or sovereignty, but pointed out that a refusal could be interpreted as protectionism and could ultimately push the Russian partners into the arms of the competition.

On January 12, 2000, the French ministry therefore presented a positive opinion on the implementation of the Soyuz at CSG. However, this opinion was far from unanimous. In particular, the ESA launcher directorate did not share this opinion, and on June 7 published its proposal for a strategy, according to which, "The launch of Soyuz from Kourou would make it a competitor to Ariane 5 on the geostationary transfer orbit market, especially considering its potential for development, thanks to European technologies (from 2,500 to 4,500 kg in geostationary transfer orbit)." In fact, as was already so with the Ariane 4, each Soyuz launch to geostationary transfer orbit would increase the cost of the Ariane 5. In addition, considerable investments for ground infrastructure in French Guiana would be necessary. The report concluded that, "Due to the unique characteristics of Europe's space port, any launcher which would be operated to geostationary transfer orbit from [there] would represent competition with Ariane 5."

Accordingly, on June 20 the ESA council adopted a resolution which called for the development of "small and medium launchers complementary to Ariane 5 and produced in Europe, responding to the diverse needs of users and based on common elements, such as stages, subsystems, technology, production lines and operational infrastructures, thereby improving the competitiveness of the European launcher industry".

This strategy held no place for the Soyuz at Kourou, and Arianespace therefore gave up the project.

But this did not take into account the tenacity of the Russian partners, whose need remained, and they would not throw in the towel, even if it meant introducing new players.

## AUSSIE POKER

The Australian entrepreneur David Kwon had already attracted attention at the Paris Air Show on June 19, 1997, by announcing that Starsem had taken a 5% stake in his company, Asia Pacific Space Center Pty Ltd (APSC), and that he now had a license to market Soyuz launches from a site he proposed to build on Australia's Cape York Peninsula. An impact study was even announced for the following October. François Calaque, CEO of Starsem, had immediately denied the information relayed by the Australian press.

When Europe looked like it was ready to give up on basing the Soyuz in Guiana, David Kwon and APSC were opportunely brought back to take center stage, enabling Rosaviakosmos[164] to shake things up through an intense round of bluff poker with the Europeans. On March 12, 2001, Mikhail Kasyanov, the Russian prime minister, gave his agreement for Russian industry to participate in the development of a commercial launch site project proposed by APSC on Christmas Island, an Australian territory in the Indian Ocean (10.5°N, 105.6°E).[165] An environmental impact study had been presented a year earlier, and the Australian government gave its approval on May 11, 2000. An area of 85 hectares would be devoted to the complex in the southern part of the island, called South Point, at a guano extraction site surrounded by the Christmas Island national park.

The Russians announced the creation of an industrial consortium. It included the Samara Space Center, which was to develop and build the Aurora launchers, derived from the Semyorka using re-engined stages. RKK Energiya was to provide the Korvet upper stage, a variant of Block DM from the Proton and Zenit using liquid oxygen and naphthyl. KBOM would be in charge of constructing the launch complex. All systems were to be operated by Russian technicians and protected by a technological safeguard agreement to prevent any unintended technology transfer. But the Russian government did not plan to invest in the project. Theoretically private investors were to supply the funding, which was initially estimated at $390 million.

APSC and its Aurora launcher, offering a capacity of 4.35 tons in geostationary transfer orbit, looked like a serious future competitor for Arianespace. The issue of the marketing exclusivity granted to Starsem for Soyuz derivatives was swept aside by project proponents who felt that the Aurora, with its new engines and redesigned central core, was sufficiently different from the Soyuz launcher to make this clause inapplicable. That exclusivity was in the marketing agreements signed at the creation of Starsem, which were due to expire in July 2001, and renewal was uncertain.

With this very much in mind, Roger-Gerard Schwartzenberg, who had succeeded Claude Allègre as minister, went to Moscow on April 24 to meet with Yuri Koptev, still in charge of Rosaviakosmos, who asked him to push the Europeans to reconsider their position. A working group was established, headed by Gilles Le Chatelier, the French minister's chief of staff, and Alexander Medvedchikov, still Koptev's deputy, for the purpose of assessing the options and reporting before the end of June, so that a new proposal could be submitted to the ESA ministerial council, which was due to meet in November in Edinburgh.

Two days later, in Paris, the parliamentary office for evaluation of scientific and technical choices (OPECST) issued a report "French space policy: assessment and prospects". This had been prepared at the direction of Christian Cabal, a member of the parliament, and Henri Revol, a senator. Revol was worried about the risk of the Soyuz falling under US domination – in other words becoming "Soyuz by Boeing"[166] – and insisted on the need to maintain strong links with Russian industry to prevent a return to the encirclement previously broken by Starsem's creation.

At ESA, the new director of launchers, Jean-Jacques Dordain, also took up the affair, which France had not yet formally submitted to the agency. Dordain, formerly at Onera (the national office of aerospace research and studies), had just spent eight years dealing with ESA strategy and international relations. He was therefore ready to grasp the complexity of the issues and was convinced of the project's advantages for Arianespace, for ESA as a user of the Soyuz, and for the future of cooperation with Russia in general. He therefore suggested a roadmap for the French minister to follow for the possible "Europeanization" of the project. First it would be necessary to officially open the CSG to other launchers – renewing the 1993 agreement. Only then would it be possible to prepare a program proposal. A revision of the agreement binding ESA and Arianespace would also be necessary, to ensure that operating the Soyuz could never be detrimental to Ariane 5's market share.

Rosaviakosmos raised the pressure on May 23 by signing an intergovernmental agreement on space cooperation with the Australian ministry of industry, science and resources. This covered in particular the operation of Russian-made launchers from Australian territory, including the Aurora-Korvet on Christmas Island.[167]

On June 20, Moscow gave Paris another nudge through Ilya Klebanov, the deputy prime minister, who said "Russia was ready to participate in investments" necessary for the installation of Soyuz in Guiana.

The Australian government counter-attacked on June 25 by announcing that it was prepared to invest up to 100 million Australian dollars (US $52 million) to support the APSC project, whose cost was now estimated at 800 million Australian dollars. In fact, this was an exaggeration, because these funds had already been allocated to support employment and upgrade the infrastructure of the island.

## FROM ELYSÉE TO EDINBURGH

During an official trip to Russia on July 3, French President Jacques Chirac toured the Samara factory with Jean-Yves Le Gall. He spoke with Dmitry Kozlov and took a position in favor of the Soyuz in Guiana, and confirmed that France was ready to support the project "[a]s long as Soyuz is not in competition with Ariane 5". Many observers noted at the time that this support might be related to an expected order of 36 Airbus planes by Aeroflot, worth an estimated $1.5 billion.

The advisors accompanying Chirac on this trip insisted, "Soyuz will not compete with Ariane 5 as the two launchers cover different market segments." In fact, it had recently been estimated that the introduction of Ariane 5ECA in 2002 and the Ariane 5ECB in 2006, with respective capacities of 10 and 12 tons in geostationary transfer orbit, would move the Ariane 5 away from the lower portion of the market (meaning satellites of less than 2 tons) which represented approximately 20% of Arianespace's order book.

The next day, July 4, in Paris, Jean-Michel Boucheron, a member of the defense commission in the national assembly, introduced his report on "image intelligence". This estimated that the Ariane 5 "is not economically suited to the Helios 2 class of satellites. As there was no question of exporting the flagship of French military space observation outside the national territory, the introduction of a new launcher in Kourou would be wise." The Soyuz was the obvious choice.

As happened 5 years earlier with the creation of Starsem, the industry, including some Arianespace shareholders, disagreed about the need to support the project. For example, Joël Barre, director of Snecma's space division, asked about the absence of justification for the project in terms of economic profitability. He claimed that the cooperation which his group had forged with the Russian engine builders proved that alliances could be established without introducing the Soyuz to French Guiana. His counterpart in EADS, François Auque, felt that Ariane and Soyuz would capture a larger market share if they were operated together in Guiana rather than separately at different sites.

It was in this heated atmosphere that Jean-Yves Le Gall, CEO of Starsem, was promoted to chief executive of Arianespace in August. His predecessor, Jacques Rossignol, former CEO of SEP, had occupied the position since February 1999 and made no secret of his opposition to introducing the Soyuz to Guiana. On October 23, Roger-Gérard Schwartzenberg, who went to Baikonur for the launch of the French astronaut Claudie Haigneré to the International Space Station, had joined Lionel Jospin, the French prime minister, on a trip to Moscow during which he announced that progress had been made in negotiations with Russian partners. They had agreed that Arianespace should assume full responsibility for marketing and operating the Soyuz launcher in Guiana. Three other conditions imposed by the French were still to be resolved. These included a contribution of one third of the cost of construction and operation of the launch complex, confirmation of the exclusivity held by Starsem on all derivatives of the Semyorka, including the Aurora, and reducing the purchase price of the Soyuz from $20 to 12 million each, because that would make the project economically viable at only three or four flights per year. At the end of the trip, Ilya Klebanov, the Russian deputy prime minister, expressed his confident opinion that the ESA council in Edinburgh would be in favor of a rapid decision on operating the Soyuz in Guiana.

He was mistaken. The European ministers in charge of space met in the Scottish capital on November 14-15, 2001. The conditions put on the Russians had yet to be formally accepted. So, when the question of the Soyuz came up, the ministers merely asked the ESA executive to define a plan of action to prepare the opening of CSG to foreign launchers on the basis of assessments by the Ariane program committee. This plan, which would include a legal component, would serve as the basis for discussion for a decision concerning the Soyuz, which would be handed down by the board at a date to be determined.

The Russian reaction came the following month. On December 14, a cooperation agreement was signed with APSC. It called for work to begin on Christmas Island in February 2002, with the first Aurora launch in late 2004.

## THE TABLES ARE TURNED

The weeks that followed saw a steady flow of announcements on the progress of negotiations, the signing of agreements and contracts, and even work at Christmas Island. The Russians' objective was to put pressure on the Europeans to accept their conditions before a final decision, which had by now been postponed until the ESA council meeting in June 2002 in Montréal.

A few analysts questioned the ability of Rosaviakosmos – with its well-known budget difficulties – to finance the major investment needed for the development of a new launcher and its installation on an isolated island 1,600 km from the Australian coast, while it was already difficult for them to finance one third of the Soyuz project in Guiana with an overall price tag less than half of that announced by APSC. In this war of press releases, those few skeptical voices went almost unheard.

For example, on January 31, 2002, RKK Energiya announced having ordered 13 cryogenic tanks from OAO UralKrioMash for the transport of liquid oxygen to the APSC site for the Aurora launcher. The first deliveries were expected to be made in October 2003.

On February 28, Yuri Koptev, while visiting CSG for the launch of ESA's Envisat environmental monitoring satellite aboard an Ariane 5, announced that the Russian proposal for the Soyuz in Guiana was already prepared. He took the opportunity to reiterate that the Aurora was not a derivative of the Soyuz and that APSC did not call into question the existing exclusivity clauses. This statement meant another rejection of the French conditions.

Then in April it was APSC's turn to face difficulties. The Indonesian government expressed its diplomatic concern about the construction of a launch site just 360 km south of Java, the country's most populous island, and the risk of stages or debris falling on its territory. Representatives of LAPAN, the Indonesian aerospace agency, claimed that launches into geostationary orbit from Christmas Island could pass over the islands of Java, Bali and Lombok. In reality, none of the azimuths studied by APSC passed anywhere near Indonesia, but for the first time press coverage was critical of the project.

In the days that followed, it was learned that the Australian government had plans to construct a detention center on Christmas Island to hold asylum seekers, whose numbers had increased after the September 11, 2001 attacks, the American military intervention in Afghanistan in October 2001 and again in the Philippines in January 2002.[168] This decision cooled the fervor of some Russian partners concerned about the security of the launch facility and their teams. While Canberra issued reassuring statements, rioting broke out in another detention camp, ironically located near the former Woomera rocket base in Southern Australia. Fourteen guards were injured in the clashes. Moreover, a disagreement between APSC and its Russian partners over ownership of the technologies developed for Aurora prevented any work from really starting.

Finally, the ESA council delegates in Montréal adopted a resolution authorizing the startup of a Soyuz installation in Guiana in exchange for industrial cooperation between Russian industry and European research centers in the fields of propulsion and future space transportation systems, especially hydrocarbon propulsion (liquid oxygen-kerosene and liquid oxygen-methane) and reusable engines or stages, or even the definition of possible future launchers to be developed jointly. A framework agreement was to be negotiated by the end of 2002, but no budget was formally assigned to the program, which was estimated to cost €275 million.[169]

Despite Rosaviakosmos's satisfaction with the European decision, on July 2 it announced that cooperation would continue with APSC. However, as negotiations advanced with the Europeans, the Australian bugaboo became less interesting and eventually disappeared from the news. The CEO of APSC, David Kwon, who had diversified his activities by operating a casino on Christmas Island, attempted to solicit Vladimir Putin, the Russian president, to restart the project during the APEC (Asia-Pacific Economic Cooperation) forum in Sydney, in September 2007, but with no apparent success.

## MEETING IN PARIS

On the eve of the Montréal council, France had changed governments again, and the astronaut Claudie Haigneré was appointed to the ministry of research, responsible for the space sector. She devoted a great deal of her attention to the project. On July 8, she went to Moscow to meet with Yuri Koptev. The first phase of the program – called the "pre-project study" – began in September under the responsibility of the ground segment sub-directorate of CNES, which undertook a technical feasibility study. On November 18, in a Franco-Russian intergovernmental seminar, Jean-Pierre Raffarin, the French prime minister, said that a final agreement would soon be ready and a funding scheme should be finalized in early 2003.

To complete this funding, the negotiators demonstrated their inventiveness. There was talk of a possible financial contribution from the European Commission by way of the budgets for the development of Guiana, for cooperation with Russia, and for

An official visit to the Malmanoury site in 2003: Jean-Yves Le Gall (second from left), Jean-Marie Luton (arms crossed, jacket in hand), Yannick d'Escatha (with sunglasses), Jean-Jacques Dordain, Igor Barmine (striped shirt) and Marcel Agasse (in profile).

technology programs. A French proposal would fund the Russian portion of the investment via bank loans, which would be repaid by profits made on the launches. The regional council of French Guiana was even asked to finance 20% of the project – approximately €60 million – but refused because its infrastructure budget for 2000-2006 was only $1.14 billion whilst its needs were estimated at $3.2 billion. Another attempt was made with the regional fund for employment for a total of €4 million.

In December, a delegation of Russian engineers visited the Malmanoury site, where an ELS (Soyuz launch complex) was to be built, while the first phase of the study by the SDS was completed with a "Review of system design". Project costs were estimated at €270 million, to which €30 million should be added to allow for unforeseen costs. A "detailed pre-project study" was launched in January, targeting a "preliminary design review" just before the next ESA ministerial council meeting in Paris in May.

Although the technical aspects were progressing well, negotiations continued to founder on the finances, particularly the Russian contribution. European funding could also be problematic due to the need for a budget to ensure the return to flight and qualification of the Ariane 5ECA, which had suffered a resounding failure on December 11 during its maiden launch.

Nevertheless, on February 11, 2003, ESA went back to the idea of financing the Russian contribution by bank loans. A reimbursement of €17 million was envisaged between 2006 and 2015. Moreover, because the program was linked to technological and experimental exchanges with Russia on the theme of future space transportation systems, its budget could be integrated into the FLPP (Future

Launcher Preparatory Program), which had succeeded FLTP, thus allowing more flexible management.

On May 27 the space ministers of the ESA member countries met at the agency's headquarters in Paris. The main theme of this council was restructuring European space transport. The return to flight of the Ariane 5ECA was at the top of the list, but establishing the Soyuz in Guiana came right behind it. In contrast to the situation in Edinburgh, the ministers now had a solid plan with clear, well-supported figures, and a detailed financing plan.

They now understood that complementarity between the Ariane 5 and the Soyuz would enable Arianespace to make an unbeatable offer for geostationary satellites of less than 3 tons. This segment of the market had been neglected by both ILS and Sea Launch.[170] It would enable Arianespace to better manage its manifest with "small" satellites riding along with the largest satellites in the market using the Ariane 5, and by launching the surplus on the Soyuz.

The overall investment was estimated at €314 million. ESA member countries would be responsible for €193 million (70% at the start, 30% after 2006), with €121 million the responsibility of Arianespace, which would purchase equipment from the Russian partners, while covering operating costs and reimbursing itself with revenue from Soyuz launches. France agreed to take on 50% of the ESA share, Germany 6%, Belgium 2.9% and Switzerland 2%. This represented only 60.9% of the funding required, but Austria, Spain and Italy also committed to contribute at levels which would be set later. Other states had until December 11 to decide. It was sufficient for the principle of the program to be officially accepted.

In order for the first flight to take place in 2006 as planned, there was no time to lose, and land clearing on the Malmanoury site began in July 2003.

## MONEY MAKES THE WORLD GO ROUND

Jean-Pierre Raffarin, the French prime minister, traveled to Moscow in early October with his minister for research, Claudie Haigneré, to meet with Mikhail Kasyanov, the Russian prime minister, to negotiate the intergovernmental agreement which would provide a legal and regulatory framework for operating the Soyuz from CSG. The signing of this agreement was announced a little prematurely, and set for October 9 during a Franco-Russian intergovernmental seminar. In fact, Rosaviakosmos had not had time to examine the details of the text and the signing ceremony between Jean-Pierre Raffarin and Boris Alioshin, Russia's deputy prime minister, did not actually occur until November 7, while President Vladimir Putin was on a visit to Paris.

At ESA in the meantime, a new fund raising campaign was initiated at the council meeting of December 18. With a reserve of €30 million assigned to cover unforeseen costs, a total of €223 million had to be found. Once again negotiations were difficult because each contribution had strings attached. Germany wanted more investment from other governments in the FLPP program, while Italy was seeking contributions for the Vega small launcher program.

The European space ministers surround Antonio Rodota, the director general of ESA in Paris in 2003. Claudie Haigneré, the French minister, is on the right (red blazer, dark blue skirt).

Delegates found themselves deadlocked. To preclude the risk of a failure, Jean-Jacques Dordain, who left his position as director of launchers to become director general in July, was obliged to put an administrative spin on the situation, saying that the council did not end without achieving its goal – rather, it was "suspended" for a few weeks in order to conclude the negotiations. The session resumed on February 4, 2004 after everyone had come to an agreement. France increased its contribution to 58%, Italy to 8% and then 12% starting in mid-2004, Germany remained at 6% and was joined by Belgium, also 6%, Spain agreed to provide 3% and Switzerland 2.5%. In addition, the European Commission would have to contribute a total of 6.7%. This represented more than 80% and therefore justified a "program statement" authorizing monies to be released and work commitments to be made. In April, ESA appointed CNES as the program's "system architect", in other words its prime contractor. This would farm out all procurement of Russian equipment to Arianespace, which would place orders with Roskosmos,[171] which would then subcontract with the appropriate industrial supplier.

To finance its €121 million contribution, Arianespace negotiated a loan with the European Investment Bank (EIB). However, this was refused despite the guarantee given by ESA. The EIB wanted the European Union to provide the guarantee, but EU regulations ruled out such a commitment before the next budgetary period, which was scheduled for 2007. Alternatively, the EIB would accept one or more European governments as guarantors. On December 10, 2004, an amendment in this sense was therefore introduced in the project budget of the French government for 2005. It was unanimously adopted by the assembly. France would be Arianespace's guarantor for the loan.

The first Soyuz 2.1a being prepared for its suborbital test from Plesetsk, on November 8, 2004.

On December 15, 2004, the ESA council opened another fund raising campaign in order to assure its own share of the financing. Most governments increased their contributions: France upped its share to 63.13%, Belgium to 6.53%, Spain to 3.26% and Switzerland to 2.72%. The killjoys were Italy and Germany, which respectively reduced their contributions to 8.71% and 5.65%. The missing 9% would be provided by the European Union. This financing plan was definitively endorsed on March 21, 2005, by four agreements signed by Jean-Pierre Raffarin along with his ministers of research and economy François d'Aubert and Thierry Breton, Jean-Jacques Dordain director general of ESA, Jean-Yves Le Gall, CEO of Arianespace, and Philippe de Fontaine Vive Curtaz, vice president of the EIB.

In the meantime, the inaugural flight of the Soyuz-2 took place successfully from Plesetsk on November 8, 2004. This Soyuz-2.1a, which sent a mock-up of the Oblik observation satellite on a suborbital trajectory, prefigured the Soyuz-2.1b that would become Arianespace's workhorse for launches from French Guiana.

The first Soyuz-2.1b being prepared at Baikonur. It was launched on December 27, 2006, with the French satellite Corot satellite.

On April 12, 2005, Arianespace was thus able to sign its contract with Roskosmos for the equipment intended for the ELS and to start manufacturing the Soyuz-2.1b. Its maiden flight occurred on December 27, 2006 from Baikonur, and put into orbit the Corot small astronomy satellite built by CNES to detect exoplanets.

## A CONSTRUCTION SITE IN GUIANA

Once the political and financial issues were settled, construction started at the site itself. As is often the case in this phase, the ELS also had its own surprises to offer, particularly since French Guiana is not an easy place to work, and from December to July – with a break in March – the rainy season makes such work impossible at many sites.

After clearing the area of its trees, earthworks began in April 2004. Ground was broken in the presence of Jean-Yves Le Gall and Jean-Jacques Dordain, as well as Yannick d'Escatha, president of CNES, and Igor V. Barmine, general manager of KBOM. In a year, over a million cubic meters of soil were moved on the 120-hectare site to lay out roads and prepare locations for the various facilities.

Located north of "Space Road" (the former RN1 highway that links Kourou to Sinnamary), the site was divided into two sections; a preparation zone (ZP) and a launch zone (ZL). Access to the site would be via ZP, which would include a guard post and a fire station, as well as the launch center (CDL) – the roof of which was a slab of concrete 2 m thick – from where the final operations and launches would be controlled. It was also in this area that the utilities would be installed (electricity, air

The Malmanoury site in February 2005, after clearing and initial earthworks, with the Atlantic Ocean in the background.

Excavation of the ELS flame trench starts in March 2006.

conditioning, compressed gases, etc.), along with a storage area where the launcher components would be delivered. The ZL was about 1 km north of the ZP. Halfway between them, the launcher integration building would be built. Retaining its Russian designation of MIK, this hall would have a footprint of 92 m by 41 m, and be 22 m in height. The launchers would be assembled horizontally and the erector car would be driven 700 m along a rail track to the launch facilities.

The pad itself was designed according to the model of the existing facilities in Baikonur and Plesetsk. It would include a massive concrete base below ground, and a launch table overlooking a flame trench dug into the granite 123 m long, 143 m wide, and 25 m deep. The only visible differences from the traditional Russian complex would be the four 90-m masts to serve as lightning rods, and a mobile gantry, 53 m high and weighing 1,200 tons. This MBO (Mobilnaya Bashnia Obsluzhivania) was built by Russia's Mir company for KBOM, under contract for Rheinmetall Italia (the former Contraves SpA),[172] to protect the launcher from the rain (Kourou has ten times the precipitation of Baikonur) and to facilitate vertical integration of the upper stage and the satellite after its erection in order to be perfectly compatible with Ariane for fitting payloads.

A subtler difference between other Soyuz launch pads and this one would be the absence of the rotating ring used in Russia to adjust the launcher azimuth. The digital avionics equipping all the launchers operated from French Guiana made this feature unnecessary.

On July 19, 2005, Jean-Jacques Dordain and Yannick d'Escatha signed the development contract through which ESA authorized CNES to start building the infrastructures. The following October, CNES awarded a €135-million contract to Vinci Construction Grands Projets, as authorized agent of Soyouz Infrastructure,[173] the industrial consortium placed in charge of ground infrastructure work and mechanical construction. The contract, covering the ZP, launch facilities and CDL totaled 20,000 sq. m of buildings and 86,000 sq. m of roads. This work was to start on December 9 and last 33 months. The inaugural launch was expected to take place at the end of 2008. The first step was to dig the flame trench, which would require the excavation of 200,000 cu. m of rock, and prepare the foundations of the buildings. Excavated rock would be crushed into gravel for concrete and earthworks. This

The "Gagarin Stone", extracted from pad N°1 of Baikonur, was laid at the ELS on February 26, 2007 during the site's inauguration ceremony.

Top: Inside the base of the pad structure, the different utility conduits (propellants, ventilation, purge lines, air conditioning, electricity, remote controls, etc.) are visible along with the "service drawer" which is retracted before launch. Bottom: The ELS with the platform above the flame trench, the mobile gantry in its retracted position, and the four lightning-rod towers.

The launcher is maintained by the four petals of the "Tiulpan" and the four brackets that support the boosters. This system, and the umbilical arm, is attached to a 300-ton ring.

Top: The MIK under construction in early 2008. Bottom: An aerial view of the ELS. In the foreground, the massive base overlooking the flame trench connected by tracks to MIK and, in the background, the Preparation Zone and the control center.

Top: The interior of the MIK, being fitted out. Bottom: Starting assembly of the mobile gantry in February 2010.

Beneath the launch table, the maintenance hooks can be seen in the pit along with the rails on which the "drawer" rides, and the large deflector to protect it during the rocket's launch.

Top: In February 2008, the massive concrete base of the launch area takes shape. Bottom: Another view of the "drawer", partially retracted and leaving part of the "pit" visible.

In June, 2009, tests are carried out to close the "Tiulpan". The umbilical mast is visible on the left.

Top: The vast space needed to retract the "drawer" prior to launch. Bottom: The "pit" of the ELS, with the four hooks (one under each petal of the "Tiulpan") and the two umbilical sockets.

An aerial view of the "Tiulpan".

The first Soyuz to launch from the Guiana Space Centre carried a pair of test satellites for the European Galileo navigation system on October 21, 2011.

single phase would take a year. Digging the flame trench revealed the unpleasant surprise that the rock structure did not match that indicated by the surveys.

On February 13, 2006, during a visit by Dominique de Villepin, the French prime minister, to Moscow, Jean-Yves Le Gall and Anatoli Perminov, director general of Roskosmos, signed an €80-million contract for the first four Soyuz rockets for the Guiana Space Centre. An order for another ten launchers would follow in September 2008.

Once the flame trench was dug, the roads laid out, and the foundations poured, the above-ground construction officially began on February 26, 2007, with the laying of the cornerstone. This was donated by Anatoli Perminov, and had the highly symbolic distinction of having been extracted from launch pad N°1 at Baikonur.

For two years, some 300 to 450 people worked full time on the site to construct the above-ground buildings. They included 100 Russians, housed in Sinnamary, the nearest village. They did not mingle with the Westerners, who resided in Kourou.[174]

In July 2008, a Dutch cargo vessel, the *Flinterland*, began a series of round trips between the port of Vyborg near St Petersburg, and "Dégrad des Cannes" (Sugarcane Harbor), near Cayenne, to deliver Russian equipment for the ELS to French Guiana. The launch complex really began to take shape in early 2009, with the mounting of the "Tiulpan" (Tulip) with its four petals to hold the launcher.

Development of the BMO gantry, which was the only radically new component, proved to be more difficult than expected, primarily due to the extremes of climate in Guiana. KBOM's contractor, Mir, was specialized in large metallic assemblies for amusement parks. Unknown to the Europeans, in 2007 it was selling its facilities to stave off bankruptcy. But Roskosmos and KBOM refused to cancel the contract. This led to many delays in the development of the structure, which was not assembled for testing at NITs RKP, 45 km north of Moscow, until March 2009. It was disassembled in late June and given an anti-rust coating for shipment to Guiana.

At last, in November 2009 the first two Soyuz launchers sailed from St Petersburg on the MN *Colibri* cargo ship (which was more accustomed to carrying the Ariane 5) heading for Dégrad Pariacabo, Kourou's harbor. These two rockets would be used for qualification tests of the facilities. After over half a century of loyal service in Kazakhstan and Russia, the Semyorka prepared to start a new life.

# 12

## Soyuz, launcher of the future

More than 52 years after the launch of Sputnik, the latest versions of the Semyorka were about to open a new chapter in the history of cosmonautics in a very different landscape from the steppes of Baikonur or the forests of Plesetsk – this time on the coastal savannah of French Guiana. The two Soyuz-2.1a launchers unloaded from the MN *Colibri* cargo ship on November 25, 2009, were just the first in a series which Europeans and Russians hope will be a long one. The Soyuz-2.1b followed in June 2011, since even before the first launchers reached South America, Starsem's launch manifest was already quite long.

On May 12, 2004, in other words just 3 months after ESA issued the "program declaration", Arianespace signed its first contract for a Soyuz launch from Kourou during the ILA exhibition in Berlin. As a final slight to the Aurora project, the client was the Australian operator Singtel Optus, with the Optus D2 satellite. The contract included the option of a launch on the Ariane 5 in the event of a delay – and in fact Optus D2 was launched on October 5, 2007 on an Ariane 5GS. The delays resulting from the budgetary negotiation at the ESA council, obtaining a loan from the EIB, management of the construction site in Guiana, and then excavating the flame trench and building the BMO gantry, gave several potential payloads to the first Guianese Soyuz. These included Corot, the small French astronomical satellite that was finally launched from Baikonur on December 27, 2006 on the first Soyuz-2.1b, and Comsat Bw-2, the second German military telecommunications satellite that was launched by an Ariane 5 on May 21, 2010. A very iconic candidate was envisaged during much of the year 2010 with Hylas 1 (Highly Adaptable Satellite) for British operator Avanti Screenmedia.[175] This 2,100-kg satellite for high-speed Internet connections and high-definition television was funded by ESA for €120 million because its purpose was to validate new technologies for the dynamic allocation of bandwidth. It was built by Astrium on an Indian platform provided by Antrix, the commercial branch of ISRO, the Indian space agency. It was originally to be placed into orbit by a Falcon 9 rocket developed privately by the American firm SpaceX (Space Exploration Technologies) set up by Elon Musk, a young billionaire of South African origin who made a fortune on the Internet. This provided an ideal illustration the space sector's globalization, which had developed in leaps and bounds since the birth of the Semyorka launcher family – and indeed since the creation of Starsem.

Unfortunately, assembly of the MBO gantry turned out to be a real nightmare for the teams because the assembly instructions were incorrect and 3,000 struts had to be examined and identified one by one. Assembly did not begin until February 2010. As a result, the scheduled date for the first flight was pushed back to 2011. But Avanti wanted its satellite in orbit before the end of 2010, so it was launched on November 26 by an Ariane 5.

Meanwhile, in October 2010 the technical qualification campaign for the ELS site began. Upon its completion on March 31, 2011, CNES turned over the keys to ESA, which in turn gave them to Arianespace. Operational qualification of the site began, with a "dress rehearsal" launch campaign, whose high point was the erection of a Soyuz-2.1a launcher on the platform on April 29 and the use of the MBO gantry the next night to integrate the upper assembly with its payload. Two launch countdowns were performed in early May to verify the proper functioning of the systems before the launcher was returned to the MIK for disassembly.

Finally, the first launch from the ELS was scheduled for late October 2011, with a Soyuz-2.1b carrying the first pair of European Galileo navigation satellites, the IOV (in-orbit validation) phase of which was entrusted to Arianespace by ESA on June 15, 2009, at the Paris Air Show.[176]

The pace was expected to increase quickly, with the launch the Pleiades 1 and 2 French observation satellites – the first of which would be accompanied by four demonstrators of the French ELISA (ELINT satellite) eavesdropping satellites, and the Chilean SSOT (Sistema Satelital para Observación de la Tierra) miniature

Loading elements of the first two Soyuz launchers bound for French Guiana aboard the MN *Colibri* roll-on/roll-off cargo ship in the port of St Petersburg.

Signing the contract to launch the four Galileo IOV satellites at the Paris Air Show on June 15, 2009.

Elements of the two first Soyuz launchers for French Guiana disembark at Degrad Pariacabo, near Kourou.

remote-sensing satellite. Next would be the second pair of Galileo IOV satellites. The deployment of the first 10 satellites in the final constellation has also been assigned to Soyuz launchers under the terms of a contract awarded by ESA in January 2010. Eventually, the Guianese Soyuz should also put into orbit the satellites of the O3B (the "Other 3 Billion" people) constellation for Internet infrastructure in developing

The Poisk research module docked with the ISS, with its Progress-M-SO-2 maneuvering module attached.

countries, and ESA payloads such as the Gaia astrometry observatory and the first Sentinel environmental satellite of the GMES (Global Monitoring for Environment and Security) program carried out in conjunction with the European Commission, as well as geostationary satellites weighing less than 3 tons left over from the Ariane 5 manifest. In order to lighten the load on the ELS arising from delays, Arianespace is continuing commercial operation of Soyuz from Baikonur in parallel. The Kazakh site will therefore serve as the launch base for a new generation of satellites for the Globalstar constellation in continuation of the missions which enabled Starsem to get going, or Eumetsat's next sun-synchronous meteorological satellite, MetOp-B.[177] The facilities for payload preparation in MIK-112 have been renovated, and can be used for many years to come.

## FROM THE SOYUZ LAUNCHER TO THE SOYUZ SPACECRAFT IN GUIANA

The 1,750th launcher of the Semyorka family lifted off on November 10, 2009, for a mission to deliver a new research module – the MRM-2 (Mini Research Module), known as "Poisk" – for the Russian segment of the International Space Station. Since the beginning of the permanent occupation of the orbiting complex in October 2000, the Soyuz launcher has provided the "life line" for astronauts on board, contributing to provisioning in conjunction with the Progress cargo and maintaining at least one Soyuz vessel on permanent standby at the station for return of the crew

A Soyuz-TMA spacecraft disappears into the sky over Kazakhstan. Is such a sight possible from Guiana someday?

An impressive locomotive tows Soyuz-TMA 15 to the pad at Baikonur. Promoters are looking forward to an equivalent picture from Guiana.

Soyuz-TMA 15 emerging from the MIK. In the event of a problem during the ascent, the escape tower will pull the spacecraft clear. The large square panels on the fairing are deployable aerodynamic stabilizers.

to Earth in case of an emergency. During the downtime of the US space shuttle after the loss of the Columbia orbiter during re-entry on February 1, 2003, and the actual return to flight in July, 2006 (after an initial attempt in July 2005), the Soyuz launcher was the only operational manned launcher in the world and hence the only means of access to the station.[178]

From August 2000 to mid-June 2011, the Semyorka launched 27 Soyuz-TM and Soyuz-TMA ferries, 43 Progress-M and Progress-M1 cargo ships, and two docking modules (Pirs and Poisk). Since 2009, and the increase of the permanent crew of the International Space Station from three to six astronauts, the pace has been ramped up to maintain a minimum of two Soyuz-TMA ships moored at any given time.

Retracting the service tower during preparations to launch Soyuz-TMA 3.

This mission of the Soyuz – of both the launcher and the spacecraft – became even more important for the operation of the station upon the retirement of the space shuttle in July 2011. NASA will have no other way to launch its astronauts, pending commissioning of a private capsule developed in the context of the CCDev program (Commercial Crew Development) in 2014-2015 or that of the Orion capsule, initially studied for the Constellation lunar program, halted in 2010, then redesigned as a new multi-purpose crew vehicle (MPCV) that is currently expected to make its inaugural flight in 2016.[179]

The fact that the Soyuz is "man-rated" has always been an exciting feature for the Europeans, who had planned a similar qualification for the Ariane 5 at the time of the Hermes program and regularly propose to achieve an equivalency for the launch of a hypothetical CTV (Crew Transport Vehicle) that is just as regularly rejected by the politicians.[180] When the Russian partners asked to launch the Soyuz from CSG, they were aiming above all at extending their market. When their European counterparts began to study the feasibility of the project, they kept in mind the idea that Soyuz could carry out manned flights from Guiana. The Russian side usually remained very reserved on the issue, with hostility increasing the higher it climbed into the political sphere. Thanks to its capacity, its reliability, its regularity, and its reduced unit cost, the Soyuz has become an economic and political lever, especially as far as the United States is concerned. The US eventually had to agree to lift the sanctions imposed by Congress on some Russian manufacturers accused of working too closely with Iran, and authorize NASA to order missions from Roskosmos to transport US astronauts during the grounding of the shuttle from 2003 to 2006 and for the post-shuttle period starting in mid-2011. Such a lever is not easily shared.

This has not prevented the Europeans from exploring the issue. Consequently, in September 2004 a report was issued at the conclusion of a detailed study undertaken between late 2002 and early 2003 by Astrium Space Transportation and Dassault Aviation[181] on the technical and operational implications of a launch from Guiana of a Soyuz-TMA spacecraft to the International Space Station, with a particular interest in the downstream ground and tracking facilities.

Such a mission would require the development of a new non-rail transport container for the spacecraft, which could be delivered to Guiana by cargo aircraft (Airbus A300-600 ST "Beluga", or an Antonov 124) or a RO-RO ship (MN *Toucan* or MN *Colibri* operated by Arianespace) during periods when the port of St Petersburg is free of ice (mid-April to mid-December). Most of the launch campaign could be carried out using existing facilities: preparation assemblies for the S3B or S5 payloads for normal operations and S2 or S4 for the escape tower and pyrotechnic items. But airtightness testing would require construction of a vacuum chamber at CSG some 10 m in length and 5 m in diameter. Furthermore, complete deforestation would be necessary within a radius of 3 to 5 km around the launch pad to ensure the safety of the crew in the event of the escape tower being triggered prior to lift-off. If all the necessary equipment were added for the battery of tests which the Soyuz spacecraft would have to undergo in Guiana before its mission (electric docking simulators with the International Space Station, domes for testing airtightness of the docking collars, etc.) then the bill would come to €40 million.

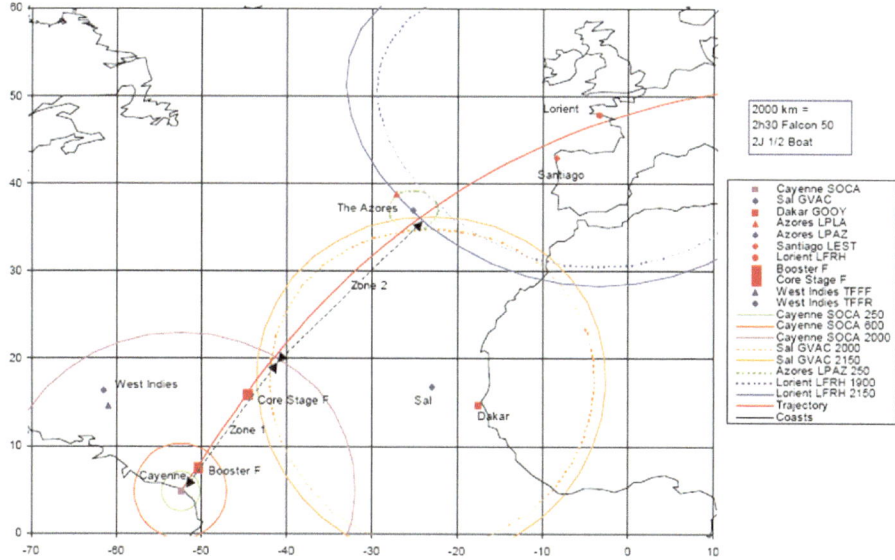

Diagram of the trajectory from the CSG of a manned Soyuz flight with potential intervention zones for safety mechanisms.

Evacuating a Soyuz after a water landing is a clumsy business, as the Italian ESA astronaut Roberto Vittori demonstrated in a training session in the Black Sea in November 2001.

The Semyorka for GIOVE-B is adjusted over the "pit" of the ELS, while the petals of the "Tiulpan" close over it. The umbilical connectors are visible at the base of the fairing.

At the top of the launcher, the covers that help to control the payload's temperature during the transfer to the launch pad are removed.

A view taken from the top of the service structure: The tip of the fairing of the launcher is visible on the left.

A close-up of the intermediate space between the stages of the GIOVE-B launcher. The four nozzles of the third stage RD-0110 engine are visible, while frost has formed on the liquid oxygen tank of the core stage.

The GIOVE-B launcher being filled with propellants.

Whilst it appears that the ground facilities could be adapted to the launch of a Soyuz spacecraft, the launch itself presents some problems. To reach the International Space Station, the Soyuz would use the same azimuth as that used by the Ariane 5ES for ATV missions, which requires the deployment of a tracking ship in the Atlantic Ocean to compensate for the absence of a station – or island to build one on – in the area. This launch trajectory would take the Soyuz over the Azores, and then along the Iberian Coast before beginning to fly over the European continent directly above the city of Nantes in France. In case of failure, it would be along this trajectory that the search and rescue would take place for the capsule and its crew. This path can be covered by two Falcon maritime patrol aircraft based at the Cayenne-Rochambeau airport in French Guiana and the naval air base in Lann Bihoué, near Lorient on the Brittany peninsula of France, and a third deployed from Amílcar Cabral International Airport on the island of Sal in the Cape Verde Islands. A small portion is also located within range of sea-rescue helicopters (600-700 km for a Super Puma RESCO) based either in the Azores or Guiana. However, the most likely drop points in case of failure remain within 1,350 km of the Guianese coast.

Training for Russian cosmonauts and other passengers of the Soyuz spacecraft includes a session in the Black Sea, off the coast of Sochi, in preparation for a water landing. In the general opinion of astronauts, the Soyuz capsule was not intended to navigate on water, and evacuating it at sea – which requires donning a special suit designed to ensure survival for 24 hours in polar waters – is particularly arduous.[182] When landing outside the radius of action of the helicopters, assistance to a crew in

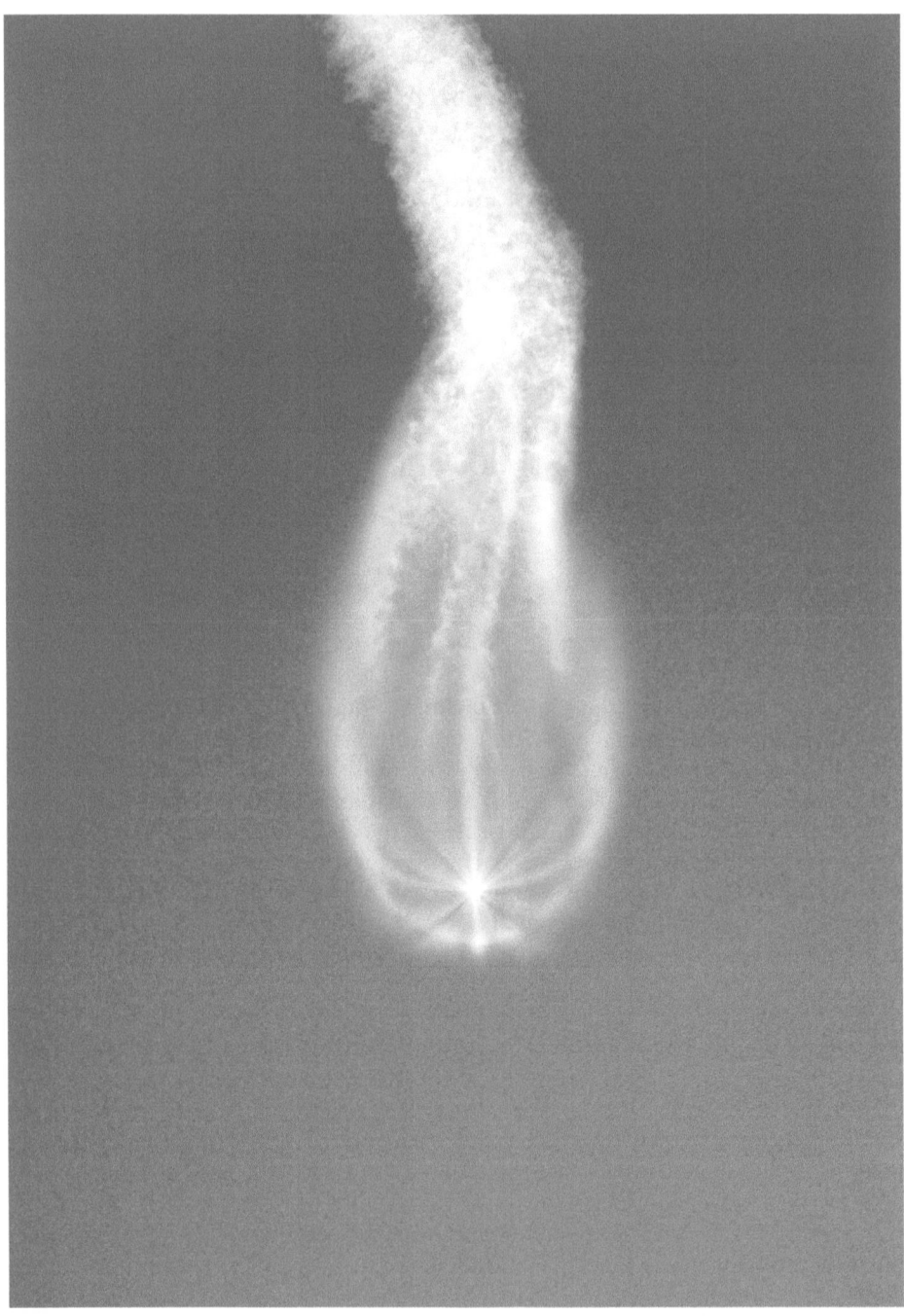

Soyuz-TMA 6 in flight in the Kazakh sky, at the precise moment that its four lateral boosters separate.

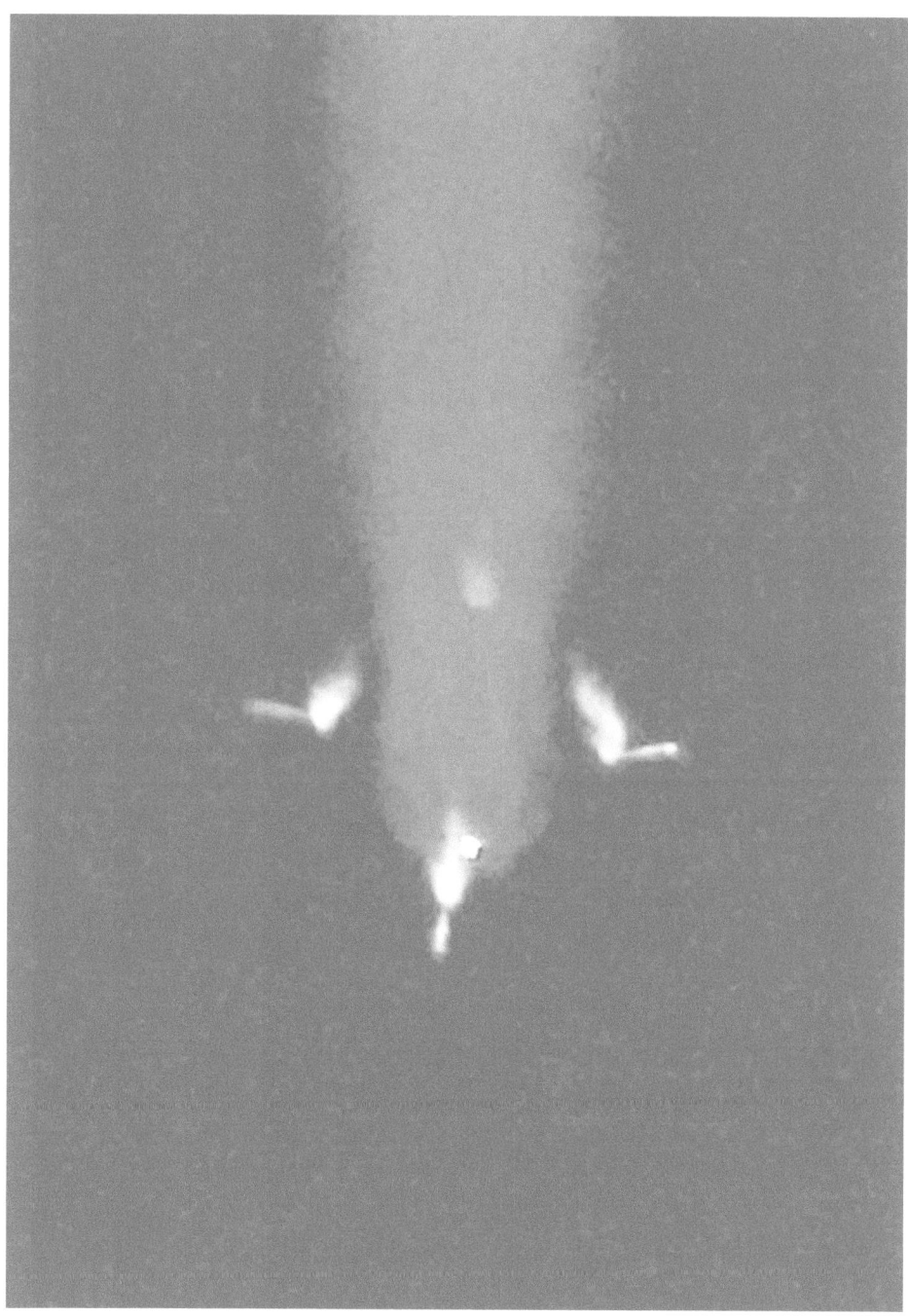

The four boosters of the launcher carrying the Foton M3 satellite fall clear.

Soyuz-TMA 6 approaches its launch pad, with the petals of the "Tiulpan" open and the umbilical mast in its standby position.

Soyuz-TMA 15 ready for installation on pad N°1 at Baikonur. The petals of the "Tiulpan" are open, one arm of two-part service structure is splayed horizontally (facing the camera) and the umbilical mast is in its standby position.

As Soyuz-TMA 6 is transferred to the pad, it passes the disused facilities of the Energiya/Buran program and the MZK integration and filling complex.

distress would imply fixed-wing aircraft dumping containers of survival equipment and parachuting in combat swimmers.

Given these constraints, it seems unlikely that a Soyuz launched from Guiana will carry astronauts any day soon. However, to keep all options open, during definition of the ELS, ESA and CNES and the ground-control branch took care not to make any decisions which would make it impossible to adapt the complex for manned flights later.

## TOWARDS THE 2,000TH FLIGHT?

Counting commercial flights from Baikonur and Guiana, Russian government flights from Baikonur and Plesetsk, and flights to the International Space Station, the Soyuz family began the decade of the 2010s at a rate of about a dozen flights per year with the prospect of increasing to 20 per annum. At this rate, the 1,800th flight could take place in late 2012 and we could imagine a 2,000th launch in the 2020s.

It seems likely that the Semyorka will give at least sixty years of service, and its history will probably not stop with the Soyuz-2 version which will be operated from Guiana and replace all other versions in Russia around 2015. Studies conducted for more than 15 years in Russia – after the cancellation of its retirement in favor of the Zenit – have shown that despite its great age, the concept retains significant growth potential, as evidenced by studies of the Soyuz/HO version and the marketing effort for the Aurora launcher.

Paradoxically, the Samara Space Center is not the initiator of the more ambitious projects; rather, it is RKK Energiya, which is seeking launchers for its own projects and opportunities for its upper stage technology. The Yamal concept was unveiled in 1996. Named after a peninsula in northwestern Siberia, it was to carry the Yamal satellites which RKK Energiya developed for AO Gascom, the telecommunications subsidiary of the Gazprom oil conglomerate. Energiya's objective is to launch these satellites into a geostationary orbit from the existing facilities at Plesetsk, but doing so involves making maneuvers that are extremely expensive in energy terms.

The technical details of this project are discussed in Chapter 5 of this book, but, in brief, this involves increasing the diameter of the central core stage of the Semyorka and using an NK-33 engine adapted by SNTK Kuznetsov from Korolev's N-1 lunar program.[183] It is a staged combustion engine that uses oxygen-rich pre-combustion to power the turbopumps. It provides 1,750 kN of thrust and is extremely lightweight (1,235 kg), giving it the best weight/thrust ratio ever achieved. The lateral boosters remain unchanged. On top of this would be mounted a 30-ton upper stage powered by a KB KhimAvtomatiki RD-0124 engine (later used on the Soyuz-2.1b) and then a 9,840-kg fourth stage, named Taymyr after another Siberian peninsula, developed by Energiya from its Block DM for the Proton and Zenit launchers but powered by a new RD-161 engine from Energomash. As the mass of this launcher would be 374 tons, it would be compatible with the complexes at Baikonur and Plesetsk which can handle vehicles of up to 400 tons. It would be able to launch a payload of 1,360 kg directly into geostationary orbit from Plesetsk, or a payload of 11.8 tons into low orbit from Baikonur – sufficient to launch the new "heavy" Progress-M2 cargo ship or even scientific modules for the International Space Station.

The project stalled for lack of funds, and the Yamal satellites were launched by Protons from Baikonur, starting in 1999.[184] In 2000 a commercial version of Yamal was used as the starting point of the Aurora project on Christmas Island. The NK-33 was not the only power plant available to re-engine the basic configuration of the Semyorka. As of 2000, concepts were developed based on the use of Energomash's RD-120K engine on both the central core and the boosters. This was a version of the engine used by the second stage of the Zenit, adapted to sea level. The use of a 5-ton cryogenic upper stage was also proposed. This concept was introduced in 2000 as the "Rus" or "Soyuz-M", although there was no direct relationship with the 11A511M Soyuz-M version, of which eight units were launched from December 1971 to March 1976.

A new concept was derived from the designs of the Yamal, Aurora, and Rus, and studied from 2002-2004 under the leadership of RKK Energiya. The Onega concept, named after a river, a lake, and a peninsula in northern Russia, was a new launcher able to carry the Kliper winged spacecraft which the company wished to develop to replace its Soyuz spacecraft. The boosters would use the RD-120M or RD-120.10E engine (a slightly reduced version of the RD-120K), and the core would be powered by an RD-191 delivering 1,976 kN of thrust. This engine, designed by Energomash, is the single chamber derivative of the American dual chamber Atlas RD-180 and the RD-170 four-chamber engine of the first stage of the Zenit. It is designed to propel Khrunichev's Angara modular launcher. Unlike the Yamal, the Onega would retain a main propulsion unit from NPO Energomash. The third stage would have been a

Block IE cryogenic stage with increased diameter and four RD-0146E engines from KB KhimAvtomatiki.[185] Finally, the Yastreb cryogenic fourth stage would have used an RD-0126 engine delivering 39 kN of thrust.[186] However, like the others, the Onega was not pursued.

## 3, 2-3, 1: THE FINAL COUNTDOWN?

The latest descendant of this feverish development was presented in August 2005 at the Moscow Air Show as a possible launcher of the Kliper spacecraft. The Soyuz-3 used the expanded core of the Yamal with its NK-33 engine, flanked by four boosters from the Onega with RD-120.10F engines, and its cryogenic third stage powered by four RD-0146E engines. It would have a launch weight of 392 tons and be able to put 14 tons into a 200-km orbit at 51.6° inclination.

An intermediate version, dubbed the Soyuz-2-3, was soon proposed as part of a review of the Kliper, when this put on so much weight that it had to be divided into two distinct elements: an 11-ton Kliper "Lite" and a 6.8-ton Parom tug. This launcher would have a central core redesigned from the Soyuz-2 to use an NK-33 engine and carry 40 additional tons of propellant. The four boosters would be powered by the new RD-0155 engine developed in 1998 by KB KhimAvtomatiki and equipped with two combustion chambers – one stationary and the other maneuverable for steering. This new engine, which had already been proposed in studies for the Yamal, Aurora, and Onega, would provide 900 kN of thrust. Three-phase development was decided by Energiya and Samara in early 2006, beginning with the introduction of the central core using the Soyuz-2 booster, then the boosters propelled by RD-0155 engines, and finally increasing the performance to the capacity offered by the Soyuz-3.

In 2007, on the basis of this work, Samara began studying a Semyorka capable of taking over the market of the Tsyklon 3 launchers, the last of which was launched on January 30, 2009, and even of Khrunichev's Rokot and of Polyot's Kosmos 3M, as their production was interrupted and their inventories were diminished. The Soyuz-1 was unveiled in May 2008 in Berlin at the ILA show. It used the standard core of the Soyuz-2, powered by an NK-33 engine, but the four boosters had been deleted. From Plesetsk, the Soyuz-1 could put 2.8 tons in low orbit, and with the Volga upper stage 1.92 tons in a sun-synchronous orbit. It drew great interest from the Russian military, in particular for navigation satellites. According to Roskosmos, the Soyuz-1 was also offered to Arianespace, which reportedly declined because it would represent internal competition with the Vega, its own small launcher.

In order to demonstrate the capabilities of the NK-33 more than 30 years after its production was halted, a static test was successfully conducted on June 2, 2008. Two tests were also performed in October 2009, but the campaign had to be interrupted in the wake of a fire at the test facility. Nevertheless, to cope with the demand which could exceed the 40 units still available in Russia, the OPK Oboronprom firm, which acquired the manufacturer SNTK Kuznetsov in 2008, began preparations to open a new production line for the NK-33 in 2014.

In the fall of 2011 the vehicle was redesignated Soyuz-2.1v to integrate it into the current Soyuz-2 family, whose launch facilities it would use, and its first flight from Plesetsk was expected in October 2012, prior to the entry into service of the smaller versions of Khrunichev's Angara family which could compete with it.

In fact, the prospects of developing a Soyuz-2-3 have grown increasingly dim as a launcher is taking shape in Samara that may well succeed the Soyuz.

## THE MIRAGE OF A EURO-RUSSIAN SPACECRAFT

In 2004, ESA began to study options for a vehicle capable of bringing material back from the International Space Station after the retirement of the space shuttle. To be precise, while the shuttle can bring back several tons of equipment and experiments with each flight, once it is withdrawn, the return capacity from orbit will be limited to 60 kg aboard a Soyuz capsule.

A number of CARV (Cargo Return Vehicle) concepts based on different types of capsules have therefore been evaluated, including the Energiya Kliper spacecraft, which obviously did not offer enough internal volume to meet the needs of the Europeans. In December 2005, the ESA council in Berlin endorsed the decision taken several months earlier to decline the Russian offer of cooperation based on the Kliper.

Following this failure, in January 2006, Roskosmos launched a tender for another spacecraft to succeed RKK Energiya's Soyuz, which was then celebrating its fortieth year of service. Energiya again proposed its Kliper Parom concept, which could be launched by either the Soyuz-2-3 or the Soyuz-3, or even the Ukrainian Zenit 2 or Khrunichev's Angara 3. In April, the Soyuz-2-3 became the launcher of reference for the Kliper, whose wing area had been significantly reduced in order to be compatible with re-entries from translunar or even interplanetary trajectories, as stipulated by the new specifications imposed by Roskosmos, making the Kliper more of a lifting-body than a winged capsule or space plane. Roskosmos did not like the Kliper Parom, and decided to suspend the invitation to tender in July 2006, asking RKK Energiya to assess the modernization options for Soyuz. For more than a year Energiya tried to save the Kliper, insisting that studies were continuing. It again showed off the Kliper at the 2007 Paris Air Show, but in reality, Roskosmos had already moved on to another concept, and was renewing its efforts to get ESA to take a share in the cost of development.[187]

In the meantime, the Europeans were about to complete the development of the ATV automatic cargo transport vehicle with which they hoped to qualify the skills and technologies that would give them a major role in any future cooperation in this area.

For strategic reasons the US refused any foreign contribution to its Constellation program intended to succeed the space shuttle (a project that would be halted in early 2010 by the Obama administration). Consequently, Europe turned back to Russia for the joint study of the ACTS (Advanced Crew Transportation System) concept which could be launched aboard Soyuz-2-3 from Baikonur or CSG. At the ESA directorate of human space flight there was even talk of the vehicles launched from Kazakhstan being manufactured in Russia and those launched from Guiana being built in Europe. Studies moved quite quickly towards a three-part system in which the Soyuz would have a European orbital module, a Russian return module, and a Russian or European propulsion module. Between August and October 2007, the project was discussed on several occasions among representatives of Roskosmos, ESA, and the participating Russian and European manufacturers. On October 24, 2007, Roskosmos and ESA called upon their respective industries to prepare a preliminary study of the ACTS concept, and this rapidly evolved towards a two-part system consisting of a capsule and a propulsion module. The question of work sharing got sticky, and on February 22, 2008, the Bremen establishment of Astrium Space Transportation, the principal European partner on the project, received instructions to cease all work on ACTS. Negotiations resumed with difficulty the following June, but the European side was now benefiting from the success of the first flight of the ATV and favored the option of a purely European development as part of a concept called "ATV Evolution".

In fact, ESA had become very critical of Roskosmos's reliability as a result of the setbacks encountered on another project: the ERA (European Robotic Arm). This device, in storage since 2005, was to be sent to the International Space Station along with the Russian MLM (Multi-Purpose Laboratory Module) by a Proton in 2007. But development of the module ran into considerable delays, which Roskosmos carefully concealed from its partners. So the Europeans discovered in 2008 that the launch had been postponed to December 2011 without their knowledge![188] Such a

The ARV (Advanced Re-entry Vehicle), essentially an ATV with a capsule for the return of payloads, could have been a first step towards a European manned spacecraft.

precedent made it unreasonable to partner with Roskosmos on a project as strategic as ACTS, especially since, from the Russian point of view, work sharing meant building the capsule in Russia and the service module in Europe, thus depriving the Europeans of all access to the specific technologies of manned flight.

At the ESA ministerial council meeting in The Hague, Netherlands, in November 2008, cooperation with Russia on ACTS was abandoned and preliminary studies were set in motion for an unmanned capsule dubbed the ARV (Advanced Re-entry Vehicle).

## POST-SOYUZ IN RUSSIA

After the withdrawal of the Europeans, Roskosmos chose to resume work on a fully Russian basis and on January 29, 2009 invited tenders for the study of a vessel called PPTS (Perspektivnaya Pilotruyemaya Transportnaya System) that was to be available in two versions to serve a station in low Earth orbit or in lunar orbit. RKK Energiya won the contract in April with a concept very close to that of ACTS that would be able to carry six cosmonauts into low Earth orbit (12-ton version) or four into lunar orbit (16.5 tons). By mid-2008 Roskosmos had already come to the conclusion that to launch the ACTS (later the PPTS) it would require a launcher more powerful than the Soyuz-2-3, and actually a family of launchers for different roles. The Zenit was excluded for political reasons, because it is Ukrainian. And the Proton was excluded for ecological reasons owing to its highly toxic propellants. In any case, it was soon to be withdrawn in favor of the Angara, a family of modular launchers that would be an interesting solution. However, the Angara would require some changes in order to "man-rate" it. And because it was being developed by Khrunichev, choosing it could leave the entire Russian space transportation infrastructures in the hands of a single company.

Consequently, in the summer of 2008 Roskosmos asked industry to offer options. Formal specifications were drawn up in February 2009 for a "Rus-M" launcher. Its study was awarded to Samara and Energiya in April, with a contract of 145 million rubles (€3.5 million). This launcher of reference would consist of three identical stages – a central core and two boosters – loaded with 180 tons of liquid oxygen-kerosene and powered by Energomash RD-180 engines for a total thrust of 9,000 kN at take-off. Above the core would be a stage loaded with 46.5 tons of cryogenic propellants and powered by four RD-0146 engines. The three first stage units would be produced by GPTs Makeyev and KB Mash in Miass, while Samara would act as prime contractor for the second stage. It would be able to put 23.8 tons in a low orbit at 51.7° when launched from the Vostochny cosmodrome (51.8°N, 128.2°E) to be built on the site of the Svobodniy cosmodrome, near Blagoveshensk, not far from the Amur River in eastern Siberia, scheduled for commissioning in 2015-2016.

Using the Rus-M stages as a base, a modular family could be built. With a core and four boosters, it could put 33-36 tons in low orbit, and by stretching the stages this could be increased to as much as 53-54 tons. A "light" version with a single first stage and the third stage of the Soyuz-2 as its second stage could place 6.5 tons into low orbit.

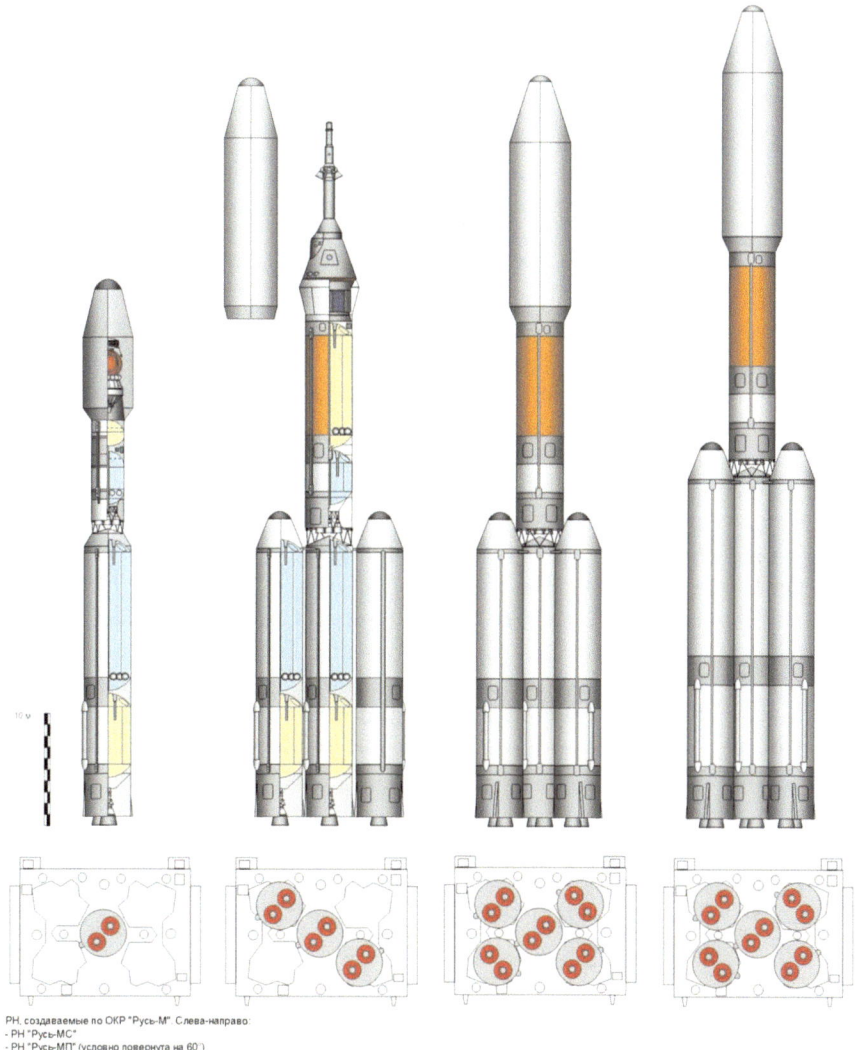

РН, создаваемые по ОКР "Русь-М". Слева-направо:
- РН "Русь-МС"
- РН "Русь-МП" (условно повернута на 60°)
- РН "Русь-МТ-35"
- РН "Русь-МТ-50"

The Rus-M family studied by Samara and Makeiev. From left to right: the lightweight Rus-MS version, the Rus-MP manned version, and the heavy Rus-MT, Rus-35 and Rus-MT-60 versions.

Samara even had concepts that could increase the capability to 100 or 150 tons in low orbit, which could revive lunar missions, since the design of the PPTS was to include such missions.

Preliminary studies on the Rus-M were completed in late 2010, and in October Roskosmos invited tenders for a technical phase budgeted at 1.63 billion rubles (€40 million) through 2012. However, on October 7, 2011, the entire Rus-M project was canceled because of its high projected cost. This left the less radical versions of the venerable Semyorka to soldier on.


## POST-SOYUZ IN EUROPE

Europe is also working on a successor for the Soyuz, although in this case with much less ambitious objectives. In November 2008 the ESA ministerial council met in The Hague, Netherlands, where the Ariane 5 program was adopted 21 years earlier. For the first time, Germany committed to investing more than France in the agency: €2.7 billion as against €2.3 billion. Also, the study of an Ariane 5ME (Mid-life Evolution) to supersede the Ariane 5ECA after 2015 was assigned for the first time to industry – namely to Astrium Space Transportation – rather than to CNES, as was the case for the previous versions. A budget of €357 million was allotted for these studies, which it was hoped would lead to a decision at the next council meeting in 2012 on possible development, estimated at €1.5 billion.

On May 17, 2007, after the election of Nicolas Sarkozy to the French presidency, François Fillon became prime minister, but did not give up his interest in the space sector. On January 23, 2009, after the council meeting in The Hague, Fillon wrote to Yannick d'Escatha, the president of CNES, Laurent Collet-Billon, chief executive of DGA (France's arms procurement agency), and Bernard Bigot, administrator general of the CEA (Atomic Energy Commission),[189] asking them to prepare a report to give the French government the information it required to prepare its position in regard to European space transport for the period 2020-2025. Having consulted with about 40 players in the sector, they submitted their report on May 18.

Jean-Yves Le Gall, CEO of Arianespace and Anatoly Perminov, director general of Roskosmos, signing an order for ten Soyuz launchers on September 20, 2009 in Sochi, in the presence of François Fillon, the French prime minister, and his Russian counterpart Vladimir Putin.

They advocated a quick study of a concept for a new expendable Ariane 6 launcher so that decisions could also be made relative to its comprehensive study and development at the ministerial council meeting in 2012.

The report presented this Ariane 6 as the single successor to both the Ariane 5 and Soyuz. For its definition, the report proposed two marketing scenarios. The first one, called "industrial", would require an ambitious policy to retain a 50% market share while satisfying governmental needs under optimal conditions. The second scenario, called "dynamic operational", would seek to ensure launch priority at the best cost in terms of governmental costs with a market presence limited to ensure the minimum rate needed to maintain reliability. The rapporteurs advocated this second scenario, in order to ensure "efficiency and transparency" in the management of public funds. The dual launch approach that was the foundation of Ariane's success for more than a quarter of a century would be abandoned, along with commercial ambitions, since the "industrial" option was not even explored in the final report. The objective would be to reduce the number of commercial satellites needed to maintain the rate, in order to develop resistance to the volatility of the market.

The report recommended a modular design for 3-6 tons in geostationary transfer orbit, bringing it closer to the concept and performance of the Ariane 4 (modularity, and 2-5-tons in geostationary transfer). The Ariane 6 would have a cryogenic upper stage powered by the Vinci engine. For the lower section there was a choice between the cryogenic or solid propulsion, which were already mastered in Europe, or even liquid oxygen-hydrocarbon (kerosene or methane), which could be developed. The projected development costs would be in the range of €3.5 to 8 billion.

Does this report really foreshadow the future of European space transportation for 2020-2025? True, it does provide for parallel operation of the Ariane 5 (ECA, later ME) and Soyuz through to that date, but beyond that the scenarios and assumptions held by the rapporteurs depict a European space transportation system reminiscent of the current American situation where priority for governmental missions has pushed the Atlas 5 and Delta 4 launchers out of the commercial field. The result has been a significant increase in the cost of their continued use, even after their consolidation within a single industrial structure – the exact opposite of the intended objective.

In addition, the withdrawal of the Europeans from commercial access to space would pose the question of the commercial sustainability of the European satellite industry, whose export success has been largely concurrent with that of Ariane. Even if there is no direct relationship of cause and effect, the decline of US giants in the satellite business Boeing and Lockheed Martin has corresponded to the withdrawal of their launchers from the market.

Finally, the report totally ignores the issue of launch capability for manned space flight, orbital infrastructure or exploration, taking the position that these activities can only be undertaken "in the context of global cooperation and independently of the need for access to space". Once again, there is a parallel with the US situation, where manned space flight is in crisis for lack of a launcher to supersede the space shuttle. Even without going as far as manned missions, the absence of the Europeans in the orbital infrastructure sector is surprising at a time when Russia, China, Japan,

The new dispensers developed by Astrium use weight-saving carbon technology, making it possible to carry six second generation Globalstar satellites.

The Arianespace launchers which operate from the Guiana Space Centre: Ariane 5GS (withdrawn in late 2009), Ariane 5ECA, Ariane 5ES, Soyuz and Vega.

Complex N°1 of Baikonur and its enormous flame trench.

and to a lesser extent India, are developing national launchers suited to these activities, precisely in order to maintain autonomous capability into 2020-2025. The overall investment and developments which led European industry to the highest international standard with the success of the ATV and the European presence in the International Space Station would be dead without launch capability, something not addressed in the report to François Fillon.

In November 2012, the ESA council will meet in Italy to discuss all these issues. Will they choose to reduce European space transportation to the minimum proposed in the French report? Or will they prefer to design a new future for the Ariane 5 and Soyuz, and prepare worthy successors by granting the European space sector the necessary resources to realize its ambitions?

For over half a century, the Semyorka family has been one of the gems of human access to space for satellites, space probes and manned spacecraft. And for 30 years, Ariane has provided access to commercial space to everyone and revolutionized the world of telecommunications and remote sensing.

In 2011, the Soyuz and Ariane began jointly operating from the same territory of French Guiana. This was the fruit of decades of technological and human efforts, the tenacity of a few visionaries, and the work of generations of engineers, technicians and workers whom history often made adversaries but nevertheless shared a passion for a common goal that was achieved in spite of everything.

This is perhaps the great lesson of the two stories of the Soyuz, and one that is highlighted with each launch of these technological gems from the sometimes hostile savannah of an obscure corner of the world, far from Siberia where Korolev nearly perished, quite far from those boundaries which once separated men. The impossible is only a question of scheduling and point of view. The most beautiful visions remain within the realm of the possible, as long as we have the strength to believe.

# Annexes

### FRANÇOIS CALAQUE

The future creator of Starsem was born in 1940, at Homécourt in Meurthe-et-Moselle, a small town in the mining and metallurgical area of Lorraine, on the border of Alsace-Lorraine, at that time annexed by the Reich. He grew up in the shadow of the blast furnaces that resumed operating after the liberation. Once he had passed his baccalaureate exam (secondary), and after completing his preparatory classes at the time that the Semyorka was born, he went to Châlons-sur-Marne (now Châlons-en-Champagne) to begin his engineering studies at the École Nationale Supérieure des Arts et Métiers.

Just after graduation, in 1964, he went to work for SEREB (Société pour l'Étude et la Réalisation d'Engins Balistiques) in charge of infrastructure studies for surface-to-surface strategic ballistics weapon systems (SSBS) and implementation of missile

A visit of the Les Mureaux site with minister Hubert Curien. Michel Delaye on the left and Henri Martre, at that time CEO of Aerospatiale, on the right.

François Calaque in front of the Ariane 5 launcher integration site at Les Mureaux.

A visit by the mayor of Paris to Les Mureaux in February 1994. From left to right standing in front of an integrated Ariane 5 tank: André Motet, Jacques-Luis Lions, president of CNES from 1984 to 1992, Luis Gallois, president and CEO of Aerospatiale, Michel Delaye, Jean-Marie Luton, director general of ESA, Jacques Chirac (mayor of Paris at that time, later French president), François Calaque, and Robert Claveyrolas.

launch sites for France's nuclear deterrent, for which deployment had been decided a year earlier. He was then sent to a desert region in Vaucluse on the Albion plateau to discreetly identify the best sites for the silos and to negotiate the purchase of land in secret. The task was all the more difficult because the planning maps of the zone had

Inauguration of the universal exposition in Seville, Spain, in April 1992, where a full-scale Ariane 4 model was erected: From left to right: Michel Delaye, François Calaque, Patrice de Lanversin, head of communications at Aerospatiale, and André Lebeau, then president of the council of Eumetsat and future president of CNES.

not been updated since the epoch of Napoleon III, and because some owners refused to speak anything but Provençal and tried to pass off their oak groves as truffle zones in order to raise the price of their parcel. But the mission was finally successful and the implementation work was able to begin in the spring of 1966.

In 1970, he joined the SNIAS (Société Nationale Industrial Aerospace, created by the merger of Sud Aviation, Nord Aviation and SEREB) where he was responsible for building the control network for the second SSBS system until 1977. Later, he became head of the project of the SX mobile strategic missile that the Army wished to develop as an extension of the SSBS. Following the elections of 1981 and the presidency of François Mitterrand, the project was curtailed. Meanwhile, the SNIAS became Aérospatiale, and in 1983 François Calaque was appointed deputy director of forecasting in the industrial department of the strategic and space systems division. In 1984, he became deputy technical and industrial director of the facilities at Les Mureaux in Yvelines, where the first and third stages of the Ariane launchers were produced. In 1986 he became director, and the establishment underwent significant development to accommodate the production of the main stage of the Ariane 5. Two years later, he was appointed director of launchers and space infrastructure. In 1992, he received the Medal of Merit, and this was awarded in February 1993 by Hubert Curien, the minister for research. He became director of the operational space center in 1994, and began discussions with his counterparts in Samara to market the Soyuz. In July 1996, he created the Starsem company and became its first president. He had to resign in April 1998 owing to illness, and died suddenly two months later.

## VICTOR EDUARDOVICH NIKOLAEV

Victor Nikolaev was born on May 3, 1960, a year before Gagarin rode a Semyorka into space. His entire family worked in aerospace. His father, Edouard Victorovich, worked at the Khrunichev (ZiKh) factory, which produced Myasishchev aircraft and later the Chelomei launchers. Not far from the plant, where many employees lived, a special French school was opened by a representative of the Embassy of France in 1967. Victor attended it, and came to master the French language perfectly. In 1977 he enrolled at faculty N°6 of the Moscow aviation institute, nominally devoted to "flying apparatuses" but which actually focused on space research and development. The head of the chair, and principal constructor of KB Salyut in Khrunichev, V. K. Karrask, directed Nikolaev's theses, which he defended in 1983. He was then sent to the Khrunichev plant. Karrask wanted him in the design bureau, but he went into production. He found himself working in the assembly hall of the Proton rocket and the Mir orbital station. Quickly, he became a foreman on Chelomei's Meteorit cruise missile (the employees called the airborne version A, "Annouchka").

In 1985, when Gorbachev arrived in power, main directorate N°13 of MOM was created and given responsibility for cooperation and marketing in the field of space.

Victor Nikolaev in front of a Soyuz launcher, both equipped with thermal protection suited to the rigors of the Kazakh climate.

Under the watchful eye of Yuri Gagarin, Victor Nikolaiev presides over the commission for the launch of the Galaxy 14 satellite in August 2005: 1, Alexandre Mezentsev; 2, Valeriy Kapitonov; 3, Yevgeny Kushnir; 4, Leonid Baranov; 5, Victor Nikolaev; 6, Doug Connelly (Orbital Sciences Corporation); 7, Yevgeny Khosnutdinov (interpreter); 8, Yevgeny Cherniy; 9, Vladimir Surzhko; 10, Yevgeny Sokolov; 11, Sergey Karassiov.

Victor Nikolaev making the victory sign for Starsem's latest success.

This was the beginning of Glavkosmos, under the direction of A. I. Dunaiev. This organization, which provided the link with the outside world, was intended to replace the Interkosmos council. Hence the manned space flight department was eventually transferred over. But for the 1987 flight with a Syrian cosmonaut, the project teams were still managed jointly. After that, there were flights with an Afghan, a Bulgarian, and a Frenchman (the 1988 Aragatz mission). For the latter, Victor Nikolaev's boss in Glavkosmos was Evgeny Bogomolov. Nikolaev also worked on a flight involving a Japanese journalist. But then the Soviet Union broke up in 1991. "It was", he said, "a disaster for the space industry, which became a mere department of the ministry of industry." It was a colossus with feet of clay. Consequently, Yuri Koptev prepared a folder for Boris Yeltsin, which led to the creation the Russian space agency, RKA, in February 1992. Until August, there were only two people officially at RKA: Koptev and Medvedchikov. In the spring of 1992, Victor Nikolaev helped them to forge ties with the West, including France (CNES) and Europe (ESA). When RKA offered to hire him, he readily accepted. Agreements were signed concerning the international orbital station (e.g. the docking section of the ATV in exchange for the DMS, data management system). But discussions also addressed the purely industrial issue of launchers. He explored opportunities for cooperation between Aerospatiale and the Samara TsSKB. This involved having a medium class launcher as a complement to the Ariane 5. The available launchers were the Proton, Soyuz, Tsyklon, Kosmos and Rokot. The United States, France, Germany and Italy were in the running to gain access to the launchers of the former Soviet Union.

In the fall of 1992, while Koptev was visiting the SEP museum in Vernon, he received the confirmation that the Ariane 4 would be withdrawn from service.

It was four years before the creation of Starsem in July 1996. It started as a purely industrial initiative, but became a government supported project. The contract with Globalstar was signed in September 1996. Until April 1998 the first CEO of Starsem was Victor Kuznetsov, formerly of NPO Lavochkin and Glavkosmos. Jean-Charles Vincent, who had just arrived from Arianespace, was the marketing director. Victor Nikolaev was in charge of Starsem's relationship with RKA. He had prepared all the documents for the creation of the company. In late October-early November 1996, he was on an assignment in France, and Koptev decided to let him stay there. The hiring process was completed on February 10, 1997. He set to work transforming the three options of the Globalstar contract into firm orders. In 1999 he took over from Victor Kuznetsov. Since then, Starsem has made over two dozen successful launches. The launches of the second generation satellites for the Globalstar constellation from Baikonur began in 2010. In 2011 the first launch from French Guiana carried the two test satellites for the European Galileo navigational system. The second generation MetOp weather satellite for Eumetsat is scheduled for launch in the autumn of 2012. And if need be, "we can accept clients who insist on taking off from Baikonur," said Victor Nikolaev. In addition, the Baikonur clean rooms are constantly maintained in perfect condition. The end of Starsem is not in sight.

## JEAN-YVES LE GALL

Although the family name "Le Gall" originated in Britanny, Jean-Yves was born in Marseille, on April 30, 1959, as the Semyorka was achieving success with the first lunar probes. He is the son of a merchant marine officer and a teacher. He did his secondary studies at Lycée Thiers before going to Paris to study at École Supérieure d'Optique, where he obtained his engineering degree in 1981, a pivotal year in which the space shuttle made its first orbital test flight and Ariane was declared operational.

Returning to Marseille, he entered the astronomy laboratory of CNRS (national center for scientific research) as a researcher in astronomy. He prepared his thesis on the Hipparcos astrometry missions and ISO (infrared space observatory) for ESA. In 1983 he got his doctorate in optics and signal processing. He recalls, "Soviet-French cooperation was important at that time, in particular with the VeGa program for a flyby of Halley's comet. We experienced the atmosphere of this cooperative program on a daily basis. This was a particularly instructive period for me."

He changed course in 1985 with an initial foray into the world of politics, joining the space branch of the general directorate of industry. When Jacques Chirac became prime minister of France in 1986, Le Gall worked for the ministry of industry. At a time when ESA was undertaking the development of Ariane 5, he was responsible for relations with the space industry. In 1988 France changed governments again, and Le Gall was appointed as the technical advisor to Paul Quilès, the minister of posts, telecommunications and space. He displayed so much enthusiasm for the space sector that he succeeded in having the portfolio transferred to his responsibility and retained it when he was appointed to the ministry of equipment housing, transport and space in 1991. At the height of Gorbachev's *glasnost*, he had the opportunity to accompany his minister to Baikonur to witness the launch of the Fobos 1 probe on a Proton rocket on July 4, 1988. On November 22 he returned to attend the launch of Jean-Loup Chrétien for the Aragatz mission to Mir, along with President François Mitterrand, which earned him a flight home in the Concorde from the runway of the Youbiléniy ("Jubilee") aerodrome where the Buran shuttle had landed for the first and last time one week earlier. He then negotiated an agreement for three additional manned missions at the unit price of $12 million. These missions were Antares with Michel Tognini in 1992, Altair with Jean-Pierre Haigneré in 1993, and Cassiopeia with Claudie André-Deshays (later Claudie Haigneré) in 1996.

In 1993, with the arrival of a new prime minister and the establishment of a new government, Le Gall left politics for industry. He joined Novespace, a subsidiary of CNES devoted to technology transfer, as managing director. He worked to increase international business, particularly with the United States, and met Michael Griffin, a future NASA administrator. He also participated in marketing payloads aboard the Samara TsSKB's Foton capsules. And in the area of parabolic flights, another key Novespace service, he oversaw the replacement in 1995 of the "Zero-G Caravel" by a much larger Airbus A300, which is still in service today.

In 1996 the government set out to reorganize the management of CNES, and replaced its general manager with three deputy managing directors. Jean-Yves Le

Jean-Yves Le Gall congratulating Anatoli Perminov, director general of Roskosmos, after signing the order for the first four Soyuz launchers for Guiana, at the Paris Air Show 2007.

Signing the order for Russian equipment for the Guiana site in April 2005: From left to right: Igor Barmine, general manager of KBOM, Jean-Yves Le Gall, Anatoli Perminov, director general of Roskosmos, and Aleksandr Kirilin, director general of the Samara Space Center.

Jean-Yves Le Gall in front of an Ariane 5 on July 5, 2008.

Gall was one of these three. He developed the strategic plan for the center, and took charge of communications and international relations. In this capacity, he represented France at ESA council meetings. He also negotiated other Franco-Russian manned missions: Pegasus with Léopold Eyharts in 1998, Perseus with Jean-Pierre Haigneré in 1999, and Andromède with Claudie Haigneré in 2001.

Jean-Yves Le Gall became CEO of Starsem in 1998, taking over from François Calaque who had to retire owing to illness. Despite the many obstacles set up by the American administration, which did not look kindly upon the emergence of a direct competitor to its Delta 2 launchers, Le Gall succeeded in gaining a foothold in the market for the launch of satellite constellations in low Earth orbit. In Europe, he also faced the hostility of some industrialists and politicians, but nonetheless managed to solidify the Soyuz launch business as an ESA reference, and prepared for its arrival in French Guiana. In 2001 he was appointed chief operating officer of Arianespace, then executive vice president in 2002, with Jean-Marie Luton becoming chairman of the board of directors. In 2007, Le Gall was made CEO of Arianespace, a position which he holds concurrently with that of CEO of Starsem.

Jean-Yves Le Gall presenting a model of Ariane 5 to Angela Merkel, the German chancellor, during the 2008 ILA Show in Berlin.

Jean-Yves Le Gall (on the right) and Roger-Maurice Bonnet, director of scientific programs of ESA (center), observing a Starsem Semyorka launch of the second pair of Cluster 2 satellites.

## Record of Semyorka versions

(does not include Block L failures)

| Version | N° of launches | Successes | Failures* | Success rate | Service entry | Name |
|---|---|---|---|---|---|---|
| R-7 | 8K71 | - | - | SS-6 | Sapwood | |
| 8K71 | 28 | 19 | 9 | 67,8 % | 1957 | R-7 |
| 8K72 | 26 | 18 | 8 | 69,2 % | 1958 | Luna |
| 8K73 | 0 | 0 | 0 | 0 | 1958 | project |
| 8K74 | 28 | 26 | 2 | 92,8 % | 1959 | R-7A |
| 8K78 | 40 | 29 | 11* | 72,5 % | 1960 | Molniya |
| 8K78M | 279 | 277 | 2** | 99,2 % | 1964 | Molniya M |
| 8A91 | 2 | 1 | 1 | 50 % | 1958 | Sputnik |
| 8A92 | 45 | 40 | 5 | 88,8 % | 1962 | Vostok-2 |
| 8A92M | 94 | 92 | 2 | 97,8 % | 1966 | Vostok-2M |
| 11A57 | 299 | 285 | 14 | 95,3 % | 1963 | Voskhod |
| 11A58 | 0 | 0 | 0 | 0 | 1958 | project |
| 11A59 | 2 | 2 | 0 | 100 % | 1963 | Poliot/IS |
| 11A510 | 2 | 2 | 0 | 100 % | 1965 | Kosmos/US |
| 11A511 | 32 | 30 | 2 | 93,7 % | 1966 | Soyuz |
| 11A511L | 3 | 3 | 0 | 100 % | 1971 | Soyuz L |
| 11A511M | 8 | 8 | 0 | 100 % | 1971 | Soyuz M |
| 11A511U | 744 | 724 | 20 | 97,3 % | 1973 | Soyuz U |
| 11A511U2 | 70 | 70 | 0 | 100 % | 1982 | Soyuz U2 |
| 11A511U | 6 | 6 | 0 | 100 % | 1999 | Soyuz U/Ikar |
| 11A511U | 4 | 4 | 0 | 100 % | 2000 | Soyuz U/Fregat |
| 11A511FG | 13 | 13 | 0 | 100 % | 2002 | Soyuz FG |
| 11A511FG | 9 | 9 | 0 | 100 % | 2002 | Soyuz FG/Fregat |
| 14A14.1a | 3 | 3 | 0 | 100 % | 2004 | Soyuz 2.1a |
| 14A14.1b | 2 | 2 | 0 | 100 % | 2006 | Soyuz 2.1b |
| Total | 1739 | 1663 | 76 | 95,6 % | | |

Taking into account the failures of block L: * 20 ** 12.

Record of Semyorka versions.

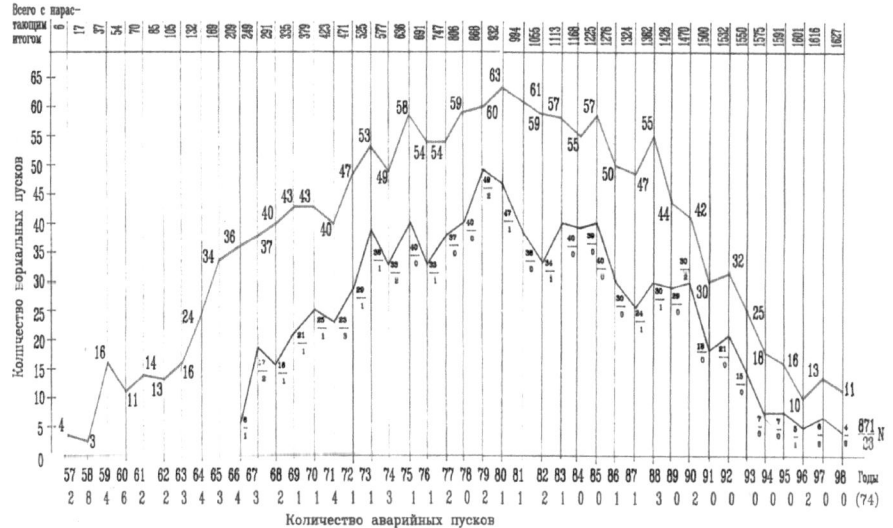

Semyorka launches.

During his twelve years in the European space transport industry, Jean-Yves Le Gall has faced and overcome some of the worst crises ever to have shaken the sector: the failure of the first flight of Ariane 5ECA, the collapse of the launch market, and aggressive competition. He nevertheless managed to win a wager that Europeans could retain their position as world leader, while succeeding in providing the most complementary range of launchers in the global market at the Guiana Space Centre: the Ariane 5, Soyuz and Vega.

# Chapter notes

## Chapter 1

1. For their role in the launch of Sputnik 1, G. A. Tiouline, Yuri A. Mozjorine, P. A. Agadjanov, G. S. Narimanov, A. V. Brykov, P. E. Eliashberg, I. K. Bajinov, and I. M. Yatsounsky of NII-4 received the Lenin Prize in December 1957. Similarly, A. I. Semenov, A. G. Mrykine, G. A. Tiouline, Yuri A. Mozjorine, V. I. Vozniouk, A. S. Kirillov, and V. A. Bokov received the Medal of Hero of Socialist Labor in 1961 for Gagarin's flight.
2. The group included A. M. Isaiev, V. F. Berglezov, I. I. Raikov, A. I. Tolstov, A. I. Tchechkov, A. M. Smirnov, A. V. Pallo, A. S. Raievsky, A. S. Kossiatov and L. I. Volkov. Later, they were joined by L. S. Dushkin, A. I. Dedov, A. A. Gukhman, G. F. Knorre, N. G. Chernyshyov and A. S. Bakaiev.
3. His team included V. A. Rudnitsky, A. N. Vasiliev, V. A. Timofeiev, Yuri E. Endeka and others who developed multiple versions of the Katyushas (BM-13-16, BM-8-48, etc.).
4. Commission members were P. N. Goremykin (Deputy Minister of Munitions), Y. L. Bibikov, (head of NII-1), I. G. Zubovitch (Deputy Minister of the power industry) and G. A. Ouger (section chief of the radar committee).
5. It included D. F. Ustinov (deputy minister of armaments), S. I. Vetochkine (head of the 7th department of the minister of armaments), I. A. Serov (KGB), S. N. Shishkin (Deputy Minister of the aviation industry), N. I. Vorontsov (Deputy Minister of communication means)V. P. Terentyev (deputy minister of the naval industry), N. I. Kotchnov (deputy minister of machinery and instrumentation), M. K. Soukov (GlavKislorod CEO), and generals P. F. Jigarev (air force), M. P. Vorobiev (military engineers) and V. I. Vinogradov (military logistics) of the ministry of defense.
6. G. M. Malenkov's deputies were D. F. Ustinov (minister of armaments) and I. G. Zubovich (first deputy of the ministry of the electrotechnical industry). The members of the Committee were N. D. Yakovlev (of the GAU), P. I. Kirpitchnikov (deputy director of Gosplan), A. I. Berg (deputy director of the council of radars and director of the TsNII-108), P. N. Goremykin (agricultural machinery minister), I. A. Serov (Minister of the Interior), and N. E. Nossovsky (head of the technical directorate of the ministry of armaments).
7. The first deputies were G. A. Tiouline (1965-1976), B. V. Balmont (1976-1980), O. D. Baklanov (1981-1983), V. N. Konovalov (1983-1987), V. K. Dogujiev (1987-1988), O. N. Shishkin (1988-1989) and R. R. Kiriuchin (1989-1991). Deputies included N. D.

Khokhlov, A. S. Matrenin (ICBM and launchers), G. M. Tabakov, V. N. Konovalov, (engines and SLBMs) V. K. Dogujiev, V. Ya. Litvinov 1965-1973, Yuri N. Koptev 1988-1991 (space), G. R. Oudarov, S. S. Vanine (ground facilities), L. I. Gusev (6 months), M. A. Brezhnev, B. V. Balmont, O. D. Baklanov, O. N. Shishkin, E. A. Jelonov, V. E. Sokolov (radio, on-board instrumentation and gyroscopes systems), E. V. Mazur, V. N. Sochine (building), V. V. Lobanov (administration), A. I. Dunaiev (Glavkosmos), A. E. Shestakov (technical directorate) and G. F. Grigorenko (security).

8.  It was headed by N. P. Poletaiev in 1947, A. Y. Sherbakov (former Deputy to Korolev) in 1949, P. N. Baikovsky in 1950, M. I. Duplichev in 1954, then V. P. Makeiev from 1955 to 1985.

9.  The designers were A. D. Sakharov, Yuri B. Borisovich Khariton, Y. B. Zeldovich, E. A. Negine, S. G. Kotchariantz.

10. The designers were N. L. Dukhov and V. A. Zuievsky.

11. The deputies were Marshall M. I. Nedeline, Deputy Minister of defense for armaments, D. F. Ustinov, Minister of armaments, S. P. Korolev, Technical Director, V. I. Vozniouk, Director of Kapustin Yar, etc. In December 1956, P. M. Zernov, S. P. Korolev, V. P. Glushko, N. A. Pilyugin, V. I. Kuznetsov, M. S. Riazansky, V. P. Barmine, V. P. Mishin, M. V. Keldysh, A. D. Sakharov, Ya. B. Zeldovitch, Yu B. Khariton and E. A. Negine then received the Medal of Socialist Labor Hero.

12. The signatories were V. M. Malyshev (minister of medium-sized machines), G. K. Zhukov (Marshal, Minister of defense from February 9), A. M. Vassilievsky (marshal, deputy of defense for the new technologies), P. V. Dementiev (Minister of the aviation industry), A. V. Domrachev (first deputy defense industry) and V. D. Kalmykov (Minister of radio engineering industry).

13. Headed by generals V. F. Zotov in 1949, B. V.Bytchevsky in 1957, M. G. Grigorenko 1959, K. M. Vertelov in 1971, N. V. Tchekov in 1979, V. S. Grigorkine in 1985, V. M. Zakimatov in 1987, Yuri V. Ovchinnikov in 1992, etc.

14. The committee, headed by V. M. Riabikov, included G. A. Titov (1st deputy), A. K. Repin, A. N. Shukin (head of the committee on science and technology), G. N. Pachkov, V. V. Illiouviev, P. I. Kalinuchkin and A. Kiassov. Titov and Shukin came from the 3rd glavka (TGU), while Repin was the former chief engineer and deputy director of the Air force who was imprisoned with A. I. Shalkhurin in 1946-1953, then rehabilitated after Stalin's death.

## Chapter 2

15. Commission composed of M.V. Keldysh (Scientific Director of NII-1 and head of the sector of applied mathematics of the Steklov Institute of Mathematics ), A. A. Dorodnitsyn (Deputy Head of the TsAGI), A. I. Makarevsky (deputy Chief of the TsAGI), B. N. Petrov (Institute of telemechanics and automatic systems), S. A. Lavochkin (Buria main designer), A. M. Lioulka (turbojet main designer), K. A. Rakhmatulin (head of NII-88 aerodynamics sector), B. S. Stechkin (deputy main designer of OKB-300), A. P. Vanichev (head of laboratory at NII-1), G. A. Tiouline (deputy director of NII-4) A. G. Mrykin (UZKA), N. N. Smirnitsky (UZKA), etc.

16. The thirteen Heroes of Socialist Labor were K. D. Buchuyev, S. O. Okhapkin and L. A. Voskresensky of OKB-1, V. A. Vitka and V. I. Kurbatov OKB-456, G. P. Glazkov, M.I. Borissenko and E. Ya. Boguslavsky NII-885, A. F. Bogomolov of OKB MEI, V. A. Rudnitsky of GSKB SpetzMach, two workers and a soldier. The Lenin Prize was awarded to S. P. Korolev, V. P. Mishin, B. E. Chertok, S. S. Kryukov, S. S. Lavrov, D. I. Kozlov, M.S. Khomiakov, M.K. Tikhonravov, G. Y. Maximov of OKB-1; V. P.

Glushko, G. N. List, V. L. Chabransky, S. P. Agafonov, Y. D. Soloviev, N. A. Jeltoukhine of OKB-456; M.S. Riazansky, N. I. Pilyugin, V. G. Sergeyev of NII-885; V. I. Kuznetsov, Z. M. Tsetsiour, D. K. Radkevich, N. V. Markichev, A. Y. Ichlinsky of NII-944; V. P. Barmine, V. A. Rudnitsky, Y. L. Troitsky of GSKB SpetzMach, G. A. Tiouline, Y. A. Mozjorine, P. A. Agadjanov, G. S. Narimanov, P. E. Eliashberg, I. K. Bajinov, O. V. Gurko, I. M. Yatsunsky, A. V. Brykov of NII-4; M. V. Keldysh, G. A. Skuridin, D. E. Okhotsimsky, T. M. Eneiev of the Academy of Sciences, A. P. Vanichev of NII-1, and others.

## Chapter 3

17. The Institute had four sectors:
    – Solid-fuel rockets (Y. A. Pobedonostsev, K. K. Glukharev, L. E. Schwartz).
    – Liquid-fuel rockets (M. K. Tikhonravov, A. I. Steniaev, A. G. Kostikov).
    – Winged rockets and aeronautical applications (V. I. Dudakov, P. P. Zuikov).
    – Mechanics and chemistry (I. S. Alexandrov, N. G. Chernyshyov).
18. Kostikov's responsibility in the purges of the RNII was not acknowledged by everyone. Despite official investigation commissions carried out by the ministry of defense in December 1966 and the defense sector to the central Committee in March 1989 and June 1991, some historians refuse to recognize Kostikov's responsibility. Nevertheless, during the Symposium of Space Propulsion organized by the Association of Aeronautics and Astronautics of France (AAAF) at Heraklion in May 2008, the historian of NPO EnergoMach indicated that there was no doubt on this issue.
19. According to Asif Siddiqi, historian of the Soviet cosmonauts working for NASA, the first Western organization to have identified Korolev as the "main designer" before his death was the Aerospace Information Division of the library of Congress in Washington DC. This reference was included in the report "Top Personalities in the Soviet Space Program" May 26, 1964.
20. Korolev was replaced by V. P. Mishin from 1966-1974, V. P. Glushko from 1974-1989, Y. P. Semenov from 1989-2005, N. N. Sevastianov from 2005-2008, then V. A. Lopota since 2008. Korolev's deputies included V. P. Mishin (1st Deputy from 1946-1966), L. A. Voskresensky, S. O. Okhapkine, K. D. Bushuyev, B. E. Chertok, M. V. Melnikov, S. S. Kryukov, A. P. Abramov, E. V. Shabarov, P. V. Tsybine, Y. I. Tregub, B. A. Dorofeiev, I. N. Sadovsky, R. A. Turkov, V. M. Kliutcharev, A. A. Borissenko,
21. The NKVD OKB had 65 people there, including designers A. D. Tcharomsky (diesel engines), B. S. Stechkin, A. M. Dobrotvorsky, M. A. Kolossov, A. S. Nazarov
22. There, Dobrotvorsky's OKB developed the MB-100 and MB-102 engines based on Klimov's M-105, a launch booster (JATO) based on Stechkin's developments, pulse jets, and from Glushko's model, the liquid propellant JATO RD-1 KHZ (300 kg of thrust).
23. On January 19, 1990 Radovsky's OKB regained its independence and became NPO EnergoMach. In March 1991 Radovsky was replaced by B. I. Katorgin who directed the enterprise until 2005. The director general was N. A. Pirogov from 2005-2009, and has been D. V. Pakhomov since 2009. The head of the OKB is V. K. Chvanov. The company has subsidiaries in Samara, Perm and St Petersburg.
24. The first deputies at the OKB were I. I. Abramov (1951-1957), M. A. Goloubev (1957-1960), A. D. Konopatov (1960-1965), G. I. Choursin (1965-1970), A. A. Golubev (1970-1985), S. P. Axenov (1985-1993).
25. It was led by I. I. Abramov (1957-1965 and from 1969-1976), B. A. Chevela (1965-1969), V. F. Soloviev (1976-1981), G. V. Kostin (1981-1993), A. I. Chassovskikh and A. V. Bondar.

26. With E. S. Gubenko, S. G. Pechkov, G. V. Petropavlovsky, P. A. Tunik and K. A. Kerimov, the latter was president of the state commission for flight from 1965 to 1991.
27. Since 2001 the director has been a brilliant young engineer: Y. M. Urlichich.
28. In March 1993 the Institute was renamed RNIIKP. In April 2006 it formed the holding company "Instrumentation and information systems for rockets and satellites" which included RNII KP, NIIKP, NII TP, NII FI, NII IT, NII PP, the NPO Orion and the OKB MEI.
29. His successors were K. A. Pobedonostsev, 1988-2004, Y. N. Bugaiev in 2004-2005, then A. S. Chebotarev in 2005.
30. Kuznetsov's successors were I. N. Sapozhnikov and A. P. Mezentsev. His deputies were N. V. Markitchev, Z. M. Tsetsior, N. N. Khlybov and D. K. Radkevich.

**Chapter 4**
31. Since 1962, the directors of plant N°1 have been A. T. Abramov from 1962-1966, A. Y. Linkov from 1966-1980, A. A. Chizhov from 1980-1996, A. N. Kirilline in 1997. Chief engineers have been M. K. Golubev from 1952-1957, G. A. Protsenko from 1957-1972, V. N. Mentiukov from 1972-1988, and N. P. Rodine from 1988.
32. This subsidiary, created in March 1964, was led by L. I. Kotenev from 1967-1972. Today, the director is G. Y. Sonis.
33. It was directed by B. G. Penzine from 1974-1987, then S. A. Petrenko in 1987.
34. The Director-General has successively been M. S. Jezlov from 1941-1950, M. L. Kononenko from 1950-1952, G. M. Popov from 1952-1953, P. P. Prudovsky from 1953-1956, P. D. Lavrentyev from 1956-1961, L. S. Chechenia from 1961-1982, B. V. Plotnikov from 1982-1986, I. L. Shitarev (1939) from 1986-2008, and N. F. Nikitin since 2008.
35. These engines were designed by chief constructors A. A. Bessonov, A. D. Shvetsov, A. A. Mikulin, V. Y. Klimov and N. D. Kuznetsov (M-11, M-17, M-26, M-34, AM-35, AM-37, AM-38, AM-39, AM-40, AM-42, VK-1, NK-4, NK-6, NK-8, NK-12, NK-22, NK-25, NK-144, NK-32-1, NK-86, NK-88, NK-89, NK-90, NK-92 and NK-93).
36. It was directed by A. A. Ovcharov from 1957-1961, M. R. Flissk from 1961-1966, N. A. Dondukov from 1966-1975, N. G. Trofimov from 1975-1994, G. A. Burmistrov from 1994-2001, V. N. Ovchinnikov in 2001-2005, and then E. P. Kocherov since 2005.
37. It has successively been directed by Y. D. Soloviev from April 1958 to December 1960, R.I. Zelenev from 1960-1975, A.F. Udalov 1975-1978, then A.A. Ganine in 1978.
38. It was then directed by A. S. Kirillov, V. I. Samsonov, V. V. Favorsky, and others.
39. It was directed by Y. P. Kienko from 1973-1983, then E. A. Rechetov from 1984-1994.
40. It was directed by I. P. Vetlov from 1974-1988, then V. V. Axenov from 1988-1992.
41. It was directed by Y. K. Khodarev from 1975-1978, then A. V. Sidorenko from 1978-1982 (vice president of AN).
42. Four US experiments (mutations of flies, vestibular apparatus of fish eggs, tissue culture of normal carrot cells and cells affected by a tumor), the French radiobiology experiment Biobloc 1 with plants and seeds, various species of fungi and bacteria, Chinese experiment with mouse cells (Biotherm 5), etc.
43. Kosmos 1645 on April 16, 1985; Kosmos 1744 on May 21, 1986; Kosmos 1841 on April 24, 1987; Foton 4 on April 14, 1988; Foton 5 on April 26, 1989; Foton 6 on April 11, 1990; Foton 7 on October 4, 1991; Foton 8 on October 8, 1992; Foton 9 on June 14, 1994; Foton 10 on February 16, 1995; Foton 11 on October 9, 1997; Foton 12 on September 9, 1999; Foton M1 on October 15, 2002; Foton M2 on May 31, 2005 and Foton M3 on September 14, 2007.

## Chapter 5

44.  It included K. N. Rudnev (Minister of the defense industry), M. V. Keldysh (Academy of sciences), S. P. Korolev (Technical Director), A. G. Mrykin (GURVO), F. A. Agaltsov (Air Force), V. A. Ambartsumyan (Director of the Biurakan Observatory), A. V. Belussov (head of the SKB-567), K. D. Bushuyev (Deputy of the OKB-1), I. T. Bulichev (Army Deputy for transmissions), A. G. Golovko (Navy), K. V. Gerchik (Director of Baikonur), A. Y. Ichlinsky (NII-944), G. N. Pashkov (Assistant of the VPK), L. A. Grishin (Deputy minister of the defense industry), A. I. Sokolov (Chief of NII-4), K. A. Kerimov (head of the space directorate of GURVO), A. I. Shokin (Deputy Minister of radio-electronics industry).

## Chapter 6

45.  His successors were General K. V. Gerchik (1958-1961), A. G. Zakharov (1961-1965), A. A. Kuruchin (1965-1973), V. I. Fadeiev (1973-1978), Y. N. Sergunin (1978-1983), Y. A. Zhukov (1983-1989), A. L. Kryzhko (1989-1992), A. A. Chumilin (1992-1997), L. T. Baranov (1997-2007), then O. V. Maidanovich (2007-2008).

46.  The military part was successively led by Y. A. Yashin (1975-1979), V. L. Ivanov (1979-1984), G. A. Kolesnikov (1984-1985), I. I. Oleinik (1985-1991), A. N. Perminov (1991-1993), Y. M. Zhuravlev (1993-1999), G. N. Kovalenko (1999-2003), A. A. Bachlakov (2003-2007), O. N. Ostapenko (2007-2008), then Oleg Maidanovich (2008), while the cosmodrome part was directed by V. A. Grin (1989-1990), A. F. Ovchinnikov (1990-1996), and then V. P. Pronikov (March to December 1997).

## Chapter 7

47.  In a presentation by Professor Leonid Sedov, of the Academy of Sciences, on August 2, 1955, to the 6th International Congress of Astronautics in Copenhagen. This resulted in Sedov, who had no connection with the program, being considered in the West as "the father of Sputnik" for several years.

48.  It should be pointed out that the Secretary of State's brother, Allen W. Dulles, was none other than the Director of the CIA and as such, since February 1956, had been in charge of American spy satellite program WS-117L which in 1959 gave rise to the Corona and Samos optical observation programs (Satellite and Missile Observation System) and Midas (Missile Detection Alarm System).

49.  Within the framework of a policy of "freedom of space" developed after 1955 by the National Security Council – and for which the details were not declassified until the 1990s - the Eisenhower administration had wished that the first artificial Earth satellite not be put into orbit by a US weapons system, and that it perform a harmless, peaceful mission, in order to create a legal precedent for any fly overs of national territories from orbit and thus defuse any possible future protest concerning fly overs of the USSR and other countries of the Communist block by spy satellites developed in secret agenda under the WS-117L program. With Sputnik the Soviets unknowingly offered this cover – which was to be provided by the Vanguard program, based on Viking and Aerobee civilian sounding rockets – to the United States on a silver platter.

50.  This idea however did develop in a different direction, since it was eventually included in a plan for lunar exploration submitted by Sergei Korolev and Mstislav Keldysh to the central committee of the CPSU on January 28, 1958. But there was never any question of a fireworks display, rather the explosion would have made possible the spectroscopic analysis of ejected material.

51.  In fact, it was a first attempt to send a probe to Venus, but it remained in Earth orbit with its upper stage.

52. It was the "permanent inter-departmental commission for the coordination and control of the scientific and theoretical work in the field of the organization and realization of interplanetary travel" implemented in the astronomical committee of the Academy of Sciences of the USSR by Decree No. 3-1517 of August 19, 1954 and placed under the direction of Leonid Sedov.

53. In fact, the Soviets recovered more of the payload gondolas from the balloons than the Americans did, and were able to glean an ample harvest of high quality film that they subsequently used on early models of their own spy satellites.

54. To identify and name the site, the CIA used the only accurate maps then available of Central Asia, namely those prepared 15 years earlier by Mil-Geo, the Wehrmacht's cartographic division.

55. Special National Intelligence Estimate Number 11-10-57, "The Soviet ICBM Program". Approved December 17, 1957.

56. The use of an ICBM as the first satellite launcher was not necessarily the obvious choice. Before using the Atlas and Titan ICBMS to put their satellites into orbit, the Americans used the Juno 1 (aka Jupiter C) derived from the Redstone SRBM, the Juno 2, derived from the Jupiter IRBM and variants of the Thor IRBM, all of smaller size. On the Soviet side, the MRBM R-12 and IRBM R-14 (aka SS-4 and SS-5) later served as a basis for two families of Kosmos launchers.

57. The US Atlas missile, designed by engineer Karel J. Bossart, a naturalized US citizen of Belgian origin, was based on a single stage, made very light by the use of a balloon tank (structural rigidity provided by pressurization) supplying three engines at liftoff. Upon reaching altitude, it dropped two engines, along with their structural elements and aerodynamic protective fairing, and continued on only one engine. This configuration is known as "one and a half stage."

58. Among the sites "identified" by the US Air Force were ammunition depots in the Urals, a Crimean war monument, and even a medieval tower.

59. The estimate is accurate because the real value at that time was 61% with 22 successful launches out of 36, not counting the aborted launches or failures of experimental warheads.

60. In fact, the R-7 missile weighed 274 tons and its 5 engines provided 3,900 kN of thrust at liftoff.

## Chapter 8

61. Glavnoe Upravleniyr po Sozdaniyu i Ispolzovaniyu Kosmicheskoy Tekhniki dlya Narodnogo Khozyaystva i Nauchniy Issledovaniye - Executive directorate for the creation and the use of space technologies for the economy and scientific research.

62. Ministervo Obshcheye Mashinostroeniya (federal ministry of mechanical constructions) created by Brezhnev in 1965 as an umbrella organization for the activities of the Soviet military industrial complex in the space and ballistic fields.

63. To preserve the illusion of an independent agency and the secrecy that continued to surround MOM's existence, Glavkosmos headquarters were located in Moscow at no. 9, Krasnoproletarskaya Street, while the Ministry was located on Muyskaya, one kilometer away.

64. After the flight of Syrian Cosmonaut Mohamed Farris in July 1987, Glavkosmos resumed international manned space flight operations which had been managed for the 10 previous years by the Interkosmos Council of the Academy of Sciences. Starting in 1988, Glavkosmos also oversaw all of the Soviet-French space cooperation programs that had existed since 1967, also under the aegis of Interkosmos.

65. In January 1991 Glavkosmos signed an agreement worth $350 million with ISRO for the transfer of Russian cryogenic technologies for GSLV, the Indian launch vehicle, which resulted in a two-year ban on Glavkosmos undertaking any commercial activity with the United States for not having followed the anti-proliferation guidelines of the MTCR (Missile Technology Control Regime). Neither India nor Russia was a member of MTCR at the time. Finally, Russia – namely Khrunichev – delivered only seven integrated 12KRB cryogenic stages to equip the first versions of the GSLV launcher, operated by ISRO since 2001.

66. Before the creation of Arianespace, in October 1972, President Nixon committed the United States to launching the satellites of their allies provided that they "respect the international agreements in force." This doctrine thus authorized the launch in 1975 and 1976 of the Franco-German Symphonie communication satellites, but prohibited their operational use for international telecommunications under the Intelsat convention, signed in 1964, thus giving this organization – largely dominated by the US company Comsat – a monopoly on international communications via satellite. By placing their satellites on Ariane, the operators of the 1980s bypassed this limitation, which then led to market liberalization and the expansion which followed.

67. "Entrusted" was the correct word because during the first two missions, the French were not present for the integration of the satellite on the launcher, or for the launch. Two technological SRET satellites, embedded as auxiliary payloads on the Molniya-M launches in April 1972 and June 1975, even lifted off from Plesetsk, although the existence of this cosmodrome was still officially denied by the Soviets. It was not until the launch of the astronomical satellite, Signe 3 on a Kosmos 3M in June 1977 that a French team was allowed to enter a Soviet launch site, in this case the Kapustin Yar range, near Volgograd.

68. At this time, Arthur M. Dula, a lawyer specialized in the space sector and intellectual property issues, was already known as an opponent of the 1967 Space Treaty, and a co-founder of several companies in the space sector, including Space Services Inc. (SSI) under the direction of former astronaut Donald "Deke" Slayton, who was the first person to obtain a federal license to carry out a private space launch, though he only sent up a sounding rocket - the Conestoga – from Matagorda Island, Texas in September 1982. Art Dula was also one of the founders of Spacehab Inc., which provided pressurized modules for the space shuttle. As executor of science fiction writer Robert A. Heinlein's will, he participated in the Heinlein Prize, which rewarded the best initiatives to develop the commercial use of space with private funds. In 2011, he became the head of Excalibur Almaz Ltd, which offers to market Almaz manned capsules developed in the 1970s by Vladimir Chelomei.

69. Taken over by Boeing in 2000.

70. The satellite, originally ordered in 1985 by Satellite Transponder Leasing Corp. (STLC), but which failed to ensure its funding, became a "white tail" (a designation derived from aviation terminology referring to an aircraft during assembly, or after completion, which has not yet been definitively attributed to an airline and therefore displays no livery or logo). In 1986, Intelsat took an option on the satellite and it was under this premise for an actual order that Hughes decided to reserve a launch for it on Proton. The international operator did not confirm the acquisition, and the satellite was launched on an Ariane 4 under the name of SBS-6, in October 1990, on behalf of Hughes Communications.

71. Tsentralnoe Spetsializirovannoe KB - Central specialized design bureau created in 1959 as subsidiary No. 3 of OKB-1 for development and testing on the Kuybyshev site (now Samara) and separated from NPO Energiya since 1975.

72. One of the top eight putsch participants was none other than the Minister Oleg Dmitrivich Baklanov, who had been a political architect of the program for the giant Energiya launcher for the Buran shuttle and the Polyus space combat station.

73. These included NPO Avtomatiki i Priborostroeniye, supplier of guidance equipment, engine manufacturer NPO Energomach, NPO Energiya (associate member), the supplier of materials and optics NPO Kompozit, the MZ Khrunichev and MZ Progress factories that manufactured Proton and Semyorka launchers and the design bureaus of KB Salyut and TsSKB, associated with them, constructor of space probes NPO Lavochkin, NPO Mashinostroeniya and NPO Prikladnoy Mekhaniki satellite builders, the TsNII Mashinostroeniya testing center, MAI (Moscow aviation institute), APO Polyot producer of Kosmos launchers, the NII-TP research center, metallurgist NPO Tekhnomach and equipment manufacturer NPO Tochnekh Proborov. There were also a few non-Russian members, such as the Estonian space agency and the Ukrainian and Uzbek equipment builders NPO Khardron and TashKB Mashinostroeniya.

74. Mashinostroeniya Zavod imeni Khrunicheva - Khrunichev production plant, sometimes referred to by the acronym ZiKH.

75. However in mid-1993 Lockheed began development of a range of small launchers, LLV (Lockheed Launch Vehicles) which flew between 1995 and 2001 under the names LLV, then LMLV (Lockheed Martin Launch Vehicles) and finally Athena.

76. Gosudartsveniy Kosmicheskiy Nauchno-Proïzvodstvenniy Tsentr imeni Khrunicheva - Khrunichev federal state center of space research and production.

77. Bought by Martin Marietta in 1993 and merged with Lockheed in 1995 to form Lockheed Martin Astro Space, now absorbed within Lockheed Martin Space Systems.

78. Element 66 is Dysprosium, from the Greek root "dysprôsitos" meaning "difficult to obtain." It is easy to understand why Motorola preferred to retain the name "Iridium."

79. This network has fewer satellites than Iridium because the selected orbits are higher and better optimized to cover zones, including the inter-tropical regions. The orbits selected by Iridium give optimal coverage at the poles and minimal coverage around the equator, where nonetheless its core target was located.

80. This was an error. The extremely rapid development of digital ground cell coverage in the second half of the 1990s, the inter-network roaming compatibility guaranteed by standards such as GSM, and ease of installation of cells on any kind of infrastructure where needed – mines, oil fields, etc. – made the use of the satellites for mobile phone connections obsolete even before the inauguration of Iridium services in November 1998 and Globalstar in October 1999. With barely 50,000 subscribers instead of the one million predicted, Iridium filed for bankruptcy in August 1999 and was sold for $25 million the following year, while over $5 billion had been sunk into the project, making it one of the largest industrial losses in the history of commercial space. Globalstar in turn, filed for bankruptcy in February 2002, and was sold in April 2004 for $43 million. In 2011, Iridium and Globalstar are still in business and are preparing to introduce their second generation of satellites.

81. Bought in 2002 by Northrop Grumman.

82. None of these projects was realized. Teledesic was reduced to 288 satellites in 1997, then to 30, before being officially abandoned in 2002.

83. Chang Zheng - Long March.

84. Initially "Satellite Probatoire pour l'Observation de la Terre" or "Earth observation test satellite," the program had been ridiculed by its opponents who believed that optical observation would soon be surpassed by radar observation – unaffected by cloud cover – and that public funds were poorly spent on this "Satellite Pour Occuper Toulouse,"

(satellite to keep Toulouse busy). Two months after its launch in February, Spot 1 was making front page headlines in all the papers by providing the first satellite images of the nuclear power plant at Chernobyl and its surrounding area. Designed for a 3-year mission, it continued to operate until November 2003.

85.   In the course of 144 missions, the Viking engine used on Ariane from 1979 to 2003, demonstrated flight reliability of 99.8%.

86.   A version with a cryogenic upper stage derived from Ariane's H10 was not selected. It would have had the capability to place 8,200 kg in geostationary transfer orbit, particularly for triple launches.

87.   In fact, Ariane 5's capacity of nearly 6 tons would have been ideal for double launches at the time of its introduction in 1996. Unfortunately, the failure of its maiden flight delayed its entry into the commercial market by 3 years and in the meantime, satellites continued to grow, so that improved performance soon became necessary.

88.   Not necessarily "small" launchers, but a smaller class of launchers. For example, scientific missions in low Earth orbit which would have required a medium sized launcher, such as Delta or an intermediary one such as Atlas, would now embark on small launchers such as Pegasus or Taurus. At the same time, the space probes that had been carried on intermediates, such as Atlas, or heavies, such as Titan would be small enough to be launched by a Delta.

89.   This program was abandoned.

90.   This program was also abandoned.

91.   DLA is also the acronym of the CNES launcher directorate.

92.   In the course of the development of Ariane 5, the EPS (storable propellant upper stage) underwent several major changes. Originally scheduled to carry 2-4 tons of liquid fuels (L2/L4), it was expanded to 5, and then to 7 tons (L5, L7) before reaching 9.7 tons (L9, 7). The version used on Ariane 5GS and ES can carry 10 tons of fuel, but its capacity is limited to 5.2 tons for ATV launches.

93.   The Italian concept stemmed from an effort undertaken in 1977 by the University of Rome to improve the American launcher Scout, operated in cooperation with NASA from the Italian San Marco platform, anchored off the coast of Kenya since 1967. The Eagle Scout, and then the Scout 2 from 1988 were studied in collaboration with BPD and initially involved adding two or four AAPs (Ariane solid strap-on boosters) from Ariane 3 to a Scout G1 launcher. Renamed the San Marco Scout in 1992 after the withdrawal of the US manufacturer Loral Vought Systems, the project moved towards a family of small modular launchers, based on a Zefiro motor loaded with 9 tons of propellant, derived from the AAP. The project was again renamed: Vega (Vettore Europeo di Generazione Avanzata, or advanced generation European vector) in mid-1993, with an eye toward its possible Europeanization.

**Chapter 9**

94.   In addition to cooperation in the field of launchers, considerable work was done in the field of space propulsion between SEP and the largest Russian engine builders, thanks to Marcel Pouliquen, then Chief Engineer of systems studies for the French industrial firm.

95.   The "Space Clipper" concept was taken up again in 1993 by Dassault Aviation – the manufacturer of Mirage and Rafale fighters as well as Falcon bizjets – for its "Talisman" project of a family of small launchers.

96.   Arianespace had even begun to consider purchasing an Airbus as a carrier aircraft.

97.   The Conestoga only made a single flight, on October 23, 1995, which resulted in a failure after 46 seconds, signaling the end of the program.

98. The indirect result was two launches, both aboard the Soyuz-U: the GFZ-1 geodetic microsatellite from Potsdam GeoForschungZentrum in April 1995 (taken aboard Mir by a Progress cargo and dropped from the station) and the DLR's Mirka capsule in October 1997.

99. The Kazakh capital was renamed Almaty in 1993 and the seat of Government transferred to the city of Astana (formerly Tselinograd) in 1998.

100. The term "Rus launcher," which occasionally appeared in the western press, resulted from confusion between the launcher and the program.

101. Finally adopted in 1996 in a revised version, the M51.

102. Called R2000 in the literature of the time.

103. When Spot 4 was launched by an Ariane 40 on March 24, 1998, its mass had increased to 2,759 kg.

104. They nearly succeeded in making that date, since flight 64 took off on June 17 and Ariane 4 made eight flights in 1994 - as expected - but with two failures (V63 and V70). The last two of its career.

105. All these satellites were launched under the name of Kosmos, a basic designation used since 1962 to serve as a cover to all Soviet military programs and a few programs of scientific research or applications. Invariably, the press releases relating to Kosmos satellite launches described them as proceeding with "the exploration of cosmic space." Upon returning their capsule of films to Earth, the spy satellites were reported to have "fulfilled their scientific mission objectives" and "brought back data which were being processed and studied."

106. Martin Marietta acquired General Dynamics' space business in December 1993. The merger was finalized the following May.

107. This group resulted from the merger of the French firm Matra Space and the British firm Marconi Space in 1990. It absorbed British Aerospace in 1993 before merging with DASA in 2000 to form Astrium Satellites.

108. The Ariane engine manufacturer was absorbed by Snecma in 1997.

109. Before taking charge of Aerospatiale's Espace & Défense branch in 1990, Michel Delaye worked in DMA (Délégation ministérielle pour l'armement or ministerial delegation for armaments in charge of procurement) then the DGA (Direction générale pour l'armement directorate general for armaments) which succeeded it in 1977. He was given responsibility for missiles, atomic and space affairs in 1970, then became director of the CAEPE (Centre d'achèvement et d'essais des propulseurs et engins or center for completion and testing of boosters and missile systems) at Saint-Médard-en-Jalles, near Bordeaux, in 1974, then director of the ballistic missile and space group and Director of the STEn (missile systems technical service) in 1979 and 1983, and finally Deputy Delegate General for armaments in 1986.

110. The API bylaws were not wasted, because entire sections of the agreement prepared with Arianespace were included in the bylaws of ILS (International Launch Services), the company created two years later by LKEI and Lockheed Martin for joint operation of the Proton and Atlas launchers.

111. Like all Soviet liquid-fuel missiles, the UR-100s were deployed in silos with their tanks filled and sealed. The corrosive nature of the fuels used, which included nitrogen peroxide, limited service life to 10 years. Therefore, the last missiles deployed in 1987 theoretically should have reached the end of their service life in 1997. In 1994, Khrunichev felt capable of extended this to 15 years, and therefore prolonging the availability of the Rokot until 2002. In fact, the missiles turned out to be in much better condition than expected. After beginning its commercial career in 2000, the Rokot should remain operational until at least 2014.

112. This contract signed with DBSi, an American company, for the launch of two triplets of small E-Sat messaging satellites, was finally canceled in 2000. In the meantime, Eurockot signed other launch contracts with Iridium, the German Aerospace Agency DLR, and Leo One, another stillborn constellation project.

113. At the same time, Arianespace was also approached by the US solid rocket motor manufacturer Alliant Techsystems, which produced SRMU (Solid Rocket Motor Upgrade) boosters for the Titan 4B launchers and proposed incorporating elements of Ariane 5 into its LCLS family of launchers (Low Cost Launch System) with which it hoped to compete for the US Air Force's EELV tender (Evolved Expendable Launch Vehicle) against Boeing, McDonnell Douglas and Lockheed Martin. The LCLS project did not get beyond the preliminary selection phase in December 1996.

114. Five to nine Vikings and one HM-7B, depending on the versions.

115. During the November 7 reorganization, his title was even explicitly changed to Delegate Minister for the Postal Services, Telecommunications and Space.

116. As the pace of launches was ramped up, due to the growth of the market and the delay in introducing Ariane 5, whose first flight had been postponed from October 1995 to June 1996, additional Ariane 4 launchers were ordered beyond the originally planned 71 units: a batch of five launchers in March 1995 (batch P9.5), and then ten more in January 1996 (batch P9.6). After the failure of Ariane 501 on June 4, 1996 and the postponement of the qualifying trials, two other batches were ordered: first 10 launchers in January 1997 (batch P9.8), and then 20 in December 1997 (batch P9.9). The possibility of a P9.10 batch was considered but rejected because the industry would not have been able to reduce the cost of production enough for Ariane 4 to remain competitive. Ariane 5 was declared operational after its third flight in December 1999. The 116th and last Ariane 4 was launched on February 15, 2003.

117. Although it is a European industrial firm, Alenia Spazio was not involved in the Ariane program. However, it participated in many US programs along with McDonnell Douglas, including the SpaceHab module for the space shuttle. Starting in 2001, Alenia even provided tanks for McDonnell Douglas Delta 2 launchers, which were absorbed by Boeing in August 1997.

118. The company European Space Industries (ESI), which should have resulted from this merger, never came to be. Negotiations were abandoned on the eve of the 1997 Paris Air Show, where the merger was to be announced.

119. Institut National de Recherche en Informatique et Automatique - French information technology research organization.

120. Successively Minister of the Postal Service, Telecommunications and Space, and then Minister for Infrastructures, Housing, Transport and Space, from May 1988 to April 1992, with the Socialist governments of prime ministers Michel Rocard and Edith Cresson.

121. Gosudartsvenny Nauchno-Proizvodstvenny Raketno-Kosmichesky Tsentr – state center for research and launcher production and space.

122. The acronym was rarely spelled out, but one of the first company brochures detailed it as: "Space Technology Alliance based on R-7 SEMyorka launch vehicles." As a sign of the times, today the name has been updated to "Société de Transport spatial ARiane-SEMiorka" or Ariane-Semyorka space transport company. The brand name was registered on March 29, 1996.

123. Actually, this second part of the Starsem's business was never developed.

**Chapter 10**

124. This LCA could not be marketed for less than $60 million, subject to the stipulation that production of the Ariane 4 had to be maintained. There was little chance that it would have been competitive with the McDonnell Douglas Delta 2 for $45 million or the Starsem Soyuz for $35 million.

125. Both related to a problem of delamination on a new model of composite fairing.

126. Future president of EADS Astrium.

127. According to a joke in vogue at the time in the industry: "A white [clean] room in Russia, was a room painted white."

128. The development of the World Wide Web, the main application of the Internet, surprised all the analysts. The number of sites jumped from 10,000 in early 1995, to 100,000 in January 1996 and over a million by April 1997.

129. KST's Eclipse project, typical of the proliferation of ideas that characterized the 1990s, included a reusable, automated launch vehicle weighing 320 tons at take-off, equipped with wings and powered by three RD-120 engines from NPO Energomash. The originality of the concept came from its take-off mode. The Eclipse, towed as a glider by a large aircraft such as a Boeing 747 or a Lockheed C-141 Starlifter, was to be released at an altitude of 14 km, and would then climb to 140 km using rocket propulsion, before ejecting its payload – itself propelled by a conventional apogee motor for injection into orbit - and glide back to earth. Before Teledesic, Motorola had chosen KST for Iridium, and had even planned to entrust it with the launch of 20 satellites for $89 million. The Eclipse project did not get beyond a few reduced-scale demonstrations using QF-106 target aircraft towed by a C-141.

130. François Auque of Aerospatiale, Jean-Marie Luton and Françoise Bouzitat of Arianespace.

131. The missions were Antarès (Michel Tognini in 1992), Altaïr (Jean-Pierre Haigneré in 1993) and Cassiopée (Claudie André-Deshays in 1996), all to Mir.

132. The Pégase mission (Léopold Eyharts in 1998) and the long-duration Perseus flight (Jean-Pierre Haigneré in 1999) on Mir, and the Andromède (Claudie Haigneré in 2001) mission to the International Space Station.

133. He was dismissed by Boris Yeltsin after the August 1998 crash when the ruble lost 60% of its value.

134. Only one option was exercised on the Boeing contract.

135. That agreement would also permit ILS to prepare the Proton launch of AsiaSat 3S, built by Hughes in California.

136. The failure occurred outside of any telemetry coverage.

137. Although initially it had been thought that the 6,180 kg probe and its radio isotopic generators had fallen near Easter Island, apparently that was where the block D-2 stage fell. The probe may in fact have disintegrated over the Atacama Desert, in northern Chile.

138. The designation was not used after Mars Express, and it was quickly forgotten that it had been "Mission F-1."

139. There were three others, placed in suborbital trajectories by Volna missiles from nuclear submarines in July 2001, July 2002 and October 2005, before termination of the program.

140. The Delta 3's two flights in August 1998 and May 1999, resulted in one failure and one sub-nominal performance, leading to the destruction of the Galaxy 10 satellite and reduction of the service life of the Orion 3 satellite. The third flight, in August 2000 carrying a satellite mock-up, was a complete success, but the launcher was abandoned shortly after in favor of the Delta 4, which did not fly until November 2002.

141. It was officially abandoned in 2002. Teledesic did no better, after multiple versions under successive prime contractors – Boeing (1997-1998), Motorola (1999-2000) and Alenia

(2002) – and being shrunk from 840 to 288, then 120 and finally 30 satellites, it was finally terminated in September 2002.

142. Dubbed "Hayabusa", the Muses-C probe was finally launched by the Japanese M-5 on May 9, 2003 and reached the asteroid 1998SF36, renamed Itokawa, on September 12, 2005. It eventually brought back the first asteroid dust particles to Earth on June 13, 2010.

143. The Fregat stage would have lifted the capsule – with its redesigned heat shield – into an elliptical orbit culminating at 27,000 km, before reducing the perigee to 60 km over the Pacific Ocean. The goal was an entry into the upper atmosphere at close to 10 km/s to decrease the apogee of the orbit to 1,100 km, for the purpose of simulating the capture in Martian orbit of a probe arriving at interplanetary speed after braking in the atmosphere of the planet.

144. Supported by the geochemist Claude Allègre - Minister of research and technology in the Socialist Government of Lionel Jospin from 1997 to 2000 - the project was estimated at over €500 million, just for the first PREMIER mission (preparation for the return of Martian samples and installation of experiments in a network), which was scheduled for launch in 2007. It did not survive the political alternation resulting from the presidential and legislative elections of May and June 2002. PREMIER was officially terminated in December 2002 as part of the redirection of CNES activities.

145. At the request of the European Commission, Aerospatiale ceded its own satellite division to Alcatel Space to avoid the creation of a European monopoly.

146. The acronym EADS was initially a provisional designation for the purpose of negotiating the rapprochement between the three companies. The acronym was kept, due to the inability to find a name likely to be acceptable to the different partners.

147. Several architectures along the lines of the DLA or ACLV concepts were proposed: EAP/P75/L10 or EAP/H25, or P75/P75/H18.

148. The Rokot was soon removed from the product line due to opposition from Khrunichev, a shareholder of ILS, the main competitor of Arianespace. It was replaced by Vega, after approval by ESA's Ariane program council in November 2000.

149. Mainly the Helios 2A spy satellite and the Syracuse 3 military communications satellite

150. It involved a super-synchronous transfer orbit culminating at 42,166 km to facilitate the twist maneuvers in orbit necessary to transfer it into an equatorial orbit from Baikonur. In fact, the farther these orbital inclination changes take place from the Earth, the lower the required velocity changes and propellant consumption.

## Chapter 11

151. Future Lockheed Martin.

152. Operator initially held by the Australian Government, privatized in 1994 under the name of Optus. It became Cable & Wireless Optus in 1998 then SingTel Optus in 2001, in the course of share buyouts by Cable & Wireless (Hong Kong) and SingTel (Singapore).

153. The island of Biak was also proposed to China and India as a launch site, but they both declined the offer.

154. Ariane 4 finally launched Palapa B5 in 1999 under the name of Telkom 1.

155. In late 2001, Khrunichev mentioned a launch site project for Angara, the successor to the Proton, in Cam Ranh, Viet Nam (12°N, 109.2°E), where Russia was preparing to vacate a former Soviet military airfield, operated since 1979 on the grounds of a former US military base abandoned in 1972. There was no follow-up and the site was converted into a civil airfield in 2004 and then into an international airport in 2007.

156. A misunderstanding was always possible since the Shavit was derived from a Jericho 2 ballistic missile, suspected of having nuclear capability.

157. Finally launched by an Ariane 40 on December 3, 1999.
158. Renamed Horus in the context of a Franco-German partnership set up during the Baden-Baden summit in December 1995, which stipulated a German contribution of 25% in the Helios 2 program conducted by France in exchange for a French contribution of 25% in the Horus program led by Germany. After having postponed signature of the intergovernmental agreement for more than two years, Germany renounced this cooperative action, leading to the official abandonment of the Horus program in August 1998.
159. Not selected in the context of France's 1996 military programming law.
160. A concept for a Soyuz cryogenic upper stage had been in the drawer since 1962 as a possible evolution of the Molniya (8K78L) launcher, but had never gotten off the drawing board. It was this version which became the basis for the Soyuz HO studies.
161. In Guiana, a "creek" is the size of a river, although in the case of the Malmanoury it is a coastal river which flows through forest and savannah for 25 km before dispersing in the coastal mangrove swamp. The term "river" is reserved for much larger water courses. For example, the Kourou River, bordering the city of the same name, is one of the shortest in Guiana, since it flows a distance of only 112 km. Its estuary is nevertheless as wide as the Potomac River where it passes Washington DC and its average discharge is about 10 times larger.
162. At the request of Germany.
163. Which it did not do, and the Vega program, in its original form, was abandoned in October. It was revived in November 2000 with different task sharing, under purely Italian leadership. The company Vegaspazio, one of whose objectives was the preservation of certain systemic French industry skills, was dissolved and replaced by ELV SpA, a subsidiary of ASI and Fiat Avio, which, among its other priorities, sought to develop the industrial skills in Italian industry that Vegaspazio was precisely intended to preserve. Claude Allègre became extremely unpopular due to the positions he took vis-a-vis teachers, and left the government of Lionel Jospin on March 27, 2000. Since 2006, he has received extensive media coverage by contesting the human contribution to climate warming and by denouncing the "precautionary principle".
164. Russian aerospace agency, which succeeded the RKA in June 1999.
165. Not to be confused with Christmas Island (Kiritimati) belonging to the Republic of Kiribati (former Gilbert Islands), over 10,800 km away, where Japan has a tracking station and at one time planned to build a landing site for its HOPE automatic space-plane project (H-2 Orbiting Plane) abandoned in 2003.
166. Boeing was never involved in the Aurora program or APSC projects but the interest shown by McDonnell Douglas – absorbed by Boeing in 1997 – for the Soyuz was known in 1996. In addition, Aurora's avionics, incorporating some elements of the Sea Launch Zenit – a consortium 40% owned by Boeing – led to confusion.
167. The agreement also made references to the Spacelift Australia project which foresaw deployment of Start launchers from the former Woomera space base in Southern Australia, where ELDO (European Launcher Development Organisation), one of the precursor organizations to ESA, launched ten test flights with its Europa launcher– including three unsuccessful orbital attempts – from 1964-1970. This project was not followed up either.
168. Christmas Island has long been a destination of choice for Indonesian boat people. In August 2001, the Australian Government used military force to block the Norwegian freighter MV Tampa which had rescued 438 refugees in international waters and wanted to disembark them on the island so they could receive care. The refugees were finally transferred to an Australian military vessel and under the pretext of "lack of space to welcome them to Australia" they were disembarked on the independent island of Nauru

in the Pacific, which incidentally is the smallest Republic in the world (21.3 sq.km). The case triggered a resounding diplomatic and political scandal.

169. The break-down was as follows: 145 million for the European industry, 110 million for the Russian industry and 20 million in taxes for Guiana.

170. As of 2005, this market was nevertheless the target of Land Launch's efforts, the company created by Sea Launch (Boeing, Norwegian shipowner Aker, RKK Energiya and Yuzhnoe/Yuzhmash) and Russian-Ukrainian partners grouped within the Space International Services consortium (Yuzhnoe/Yuzhmash, RKK Energiya, KBTM and TsENKI) to operate Zenit launchers from Baikonur.

171. Russian federal space agency, which succeeded Rosaviakosmos in March 2004.

172. Although built in Russia, the MBO was not part of the equipment purchased by Arianespace because it was an addition to the traditional Soyuz launch concept.

173. Soyouz Infrastructure included the Infrasoyouz consortium focused on first phase structural construction - formed by Vinci Construction Grands Projets and Nofrayane, the French Guiana subsidiary of Nord France - as well as German industrialist MT Aerospace (former MAN Technologie), the French electrical engineering firm Clemessy and ACIA the refrigeration and air conditioning consortium composed of French and Belgian branches of AximaFrench Crystal and Intemann of Austria.

174. Part of CSG, and in particular the ELA-3 zone, enjoys a special status granted by the US Department of State concerning the conditions of confidentiality and protection of intellectual property benefiting payloads developed in the United States. To maintain this status, it is important to properly supervise the Russian teams and limit their access to these areas.

## Chapter 12

175. In Greek mythology, Hylas was a young companion of Hercules during the expedition of the Argonauts. After Hylas 1, Avanti planned to develop a Hercules satellite, but it was rejected in favor of a Hylas 2 commissioned in December 2009.

176. The Giove A and B (Galileo In - Orbit Validation Element) demonstrators were launched by Starsem from Baikonur on December 28, 2005 and April 27, 2008.

177. MetOp-A was launched from Baikonur on October 19, 2006, aboard a Soyuz-2.1a equipped with a 4 m diameter ST fairing.

178. Joined on October 15, 2003, by the Chinese CZ-2F launcher which carried the experimental Shenzhou 5 spacecraft for the very first flight of "taikonaut" Yang Liwei.

179. In addition to the post-Columbia episode in 2003-2006, Soyuz was already the only operational manned launcher between the end of Apollo missions in July 1975 and the first flight of the Columbia Shuttle in April 1981, and between the destruction of Challenger in January 1986 and the return to flight of the shuttle in September 1988.

180. After Hermes, Ariane 5 was proposed to carry a CRV (Crew Return Vehicle) capsule studied by ESA in 1992-1995, then an ACRV (Assured Crew Return Vehicle) lifting-body vehicle based on the X-38 demonstrator developed jointly by NASA and ESA starting in 1996 and ended unilaterally by NASA in April 2002, after Europe had delivered all of the high technology elements which constituted its contribution. In November 2008, at The Hague, the ESA Ministerial Council approved a budget of €44 million for the study of an unmanned cargo capsule, the ARV (Advanced Reentry Vehicle), using the ATV service module. EADS Astrium extensively communicated on the possible evolution of the ARV to a manned capability, but the politicians only mentioned it in passing. In early 2011, under budgetary pressure, ESA decided not to pursue the activities on the ARV beyond system studies.

181. Although Dassault Aviation has had limited presence among the major players in the European space industry, the company has played an important role in safety systems for space launches. The company's expertise and skills in the hypersonic and supersonic field earned it a role in the Hermes and X-38 programs.

182. The only water landing of a manned Soyuz spacecraft took place on the flight of the Soyuz 23 mission, on October 16, 1976. The capsule came down in Lake Tengiz, a salt lake north of the Kazakhstan, during a snowstorm with the temperature at -20°C. The capsule was dragged under the surface by its parachute. Recovery took more than 9 hours but the crew was rescued. In February 1967, Kosmos 140, the third uninhabited Soyuz prototype, sank into the Aral Sea after perforation of its heat shield during re-entry, but was retrieved at a depth of 10 meters. Two unmanned capsules of the Soyuz L1 were also recovered in the Indian Ocean in September 1968 and October 1970 at the end of the Moon flyby missions of Zond 5 and Zond 8.

183. The NK-33 is an evolved version of the NK-15 engine developed to propel the first stage of the Korolev N-1 lunar rocket. To achieve the necessary thrust for liftoff of the 2,735-ton giant, thirty NK-15 launchers were needed. The management and steering difficulties of this propulsion block caused four failures of the N-1 from 1969 to 1972. The NK-33 was intended to power a derivative of the N-1, the N-1F, but the program was canceled in 1974 before that could take place. While all materials developed for the program had to be destroyed, approximately 150 NK-33 engines were conserved in a warehouse and "rediscovered" after the fall of the Soviet Union. In 1996-1997, 36 engines were sold to Aerojet for $1.1 million each. The American company upgraded them to western standards under the designation of AJ26. After having been considered for the propulsion of the first stage of the US Atlas 3 and Japanese GX launchers (eventually propelled by the RD-180 from NPO Energomash), as well as for the concept of the Kistler reusable launcher Aerospace K-1 (canceled in 2007), the AJ26 is expected to go into space at the end of 2011 on the Antares 2 launcher developed by Orbital Sciences Corp. (OSC) to ensure service to the International Space Station after the end of the US shuttle program. The first stage of the Antares 2, manufactured in Ukraine, will be powered by two AJ26's. In mid-2009, about forty NK-33's in flight condition remained available in Russia, and OPK Oboronprom, which took control of the manufacturer SNTK Kuznetsov in 2008, was preparing for the production start-up at Motorstroitel in Samara by 2014.

184. Two satellites launched in 1999, two others in 2003. Three new satellites, including one built in Europe by Thales Alenia Space, are planned for 2012 and 2013.

185. As the result of development started in 1997 in cooperation with Pratt & Whitney in the United States (which retains commercial rights outside of Russia) the RD-0146 is the first expander cycle Russian engine. The RD-0146U version will equip the cryogenic upper stages of Khrunichev's Angara launchers.

186. This very compact engine has the particularity of its combustion chamber being located in the center of its nozzle with a ring collar for the evacuation of gases. It was tested on the bench in 1999.

187. The General Director of RKK Energiya, Nikolai Sevastianov was fired on June 22, 2007, just as he was attending the Paris Air Show, due to differences with Roskosmos, including on the issue of the Kliper project.

188. At the time of this writing in mid-2011, this is now scheduled for launch in December 2012, or possibly in 2013.

189. Former chief of staff of Claudie Haigneré, French Minister Delegate for Research and New Technologies, 2002-2004.

# Semyorka launches

| Date | Vehicle | # | Payload | Codename |
|------|---------|---|---------|----------|
| **1957** | | | | |
| 05/15 B Failure | 8K71 (M1-5) | 1 | Dummy warhead | |
| 06/11 B Abort | 8K71 (M1-6) | | Dummy warhead | |
| 07/12 B Failure | 8K71 (M1-7) | 2 | Dummy warhead | |
| 08/21 B | 8K71 (M1-8) | 3 | Dummy warhead | |
| 09/07 B | 8K71 (M1-9) | 4 | Dummy warhead | |
| 10/04 B | 8K71 (M1-1PS) | 5 | PS-1 | Sputnik 1 |
| 11/03 B | 8K71 (M1-2PS) | 6 | PS-2 | Sputnik 2 |
| 12/31 B Abort | 8K71 (M1-11) | | Dummy warhead | |
| **1958** | | | | |
| 01/29 B Failure | 8K71 (M1-11) | 7 | Dummy warhead | |
| 03/12 B Abort | 8K71 (M1-6) | | Dummy warhead | |
| 03/29 B | 8K71 (M1-10) | 8 | Dummy warhead | |
| 04/04 B | 8K71 (M1-12) | 9 | Dummy warhead | |
| 04/27 B Failure | 8K91 (B1-1) | 10 | D1-1 | |
| 05/15 B | 8K91 (B1-2) | 11 | D1-2 | Sputnik 3 |
| 05/24 B Failure | 8K71 (B1-3) | 12 | Dummy warhead | |
| 07/10 B Failure | 8K71 (B1-4) | 13 | Dummy warhead | |
| 09/23 B Failure | 8K72 (B1-3) | 14 | Ye-1 #1 (Luna) | |
| 10/11 B Failure | 8K72 (B1-4) | 15 | Ye-1 #2 (Luna) | |
| 12/04 B Failure | 8K72 (B1-5) | 16 | Ye-1 #3 (Luna) | |
| 12/24 B Failure | 8K71 (B3-16) | 17 | Dummy warhead | |

| Date | Vehicle | No. | Payload | Notes |
|---|---|---|---|---|
| **1959** | | | | |
| 01/02 B | 8K72 (B1-6) | 18 | Ye-1 #4 (Luna) | Luna 1 |
| 02/17 B | 8K71 (041081) | 19 | *Dummy warhead #13* | |
| 02/21 B *Abort* | 8K71 (041082) | | *Dummy warhead #14* | |
| 03/25 B | 8K71 (13-18) | 20 | *Dummy warhead #15* | |
| 03/30 B *Failure* | 8K71 (13-20) | 21 | *Dummy warhead #16* | |
| 05/09 B | 8K71 (13-21) | 22 | *Dummy warhead #17* | |
| 05/30 B | 8K71 (13-22) | 23 | *Dummy warhead* | |
| 06/09 B *Failure* | 8K71 (13-23) | 24 | *Dummy warhead* | |
| 06/18 B *Failure* | 8K72 (I1-7) | 25 | Ye-1 #5 (Luna) | |
| 07/18 B | 8K71 (13-24) | 26 | *Dummy warhead* | |
| 07/30 B | 8K71 (041082) | 27 | *Dummy warhead #14* | |
| 08/13 B | 8K71 (13-25) | 28 | *Dummy warhead* | |
| 09/09 B *Abort* | 8K72 (I1-7A) | | Ye-1 #6 (Luna) | |
| 09/12 B | 8K72 (I1-7B) | 29 | Ye-1 #7 (Luna) | Luna 2 |
| 09/18 B | 8K71 (I1-1T) | 30 | *Dummy warhead* | |
| 10/04 B | 8K72 (I1-8) | 31 | Ye-2A #1 (Luna) | Luna 3 |
| 10/22 B | 8K71 (267432) | 32 | *Dummy warhead to the Equator* | |
| 10/25 B | 8K71 (267434) | 33 | *Dummy warhead to the Equator* | |
| 11/01 B | 8K71 (267431) | 34 | *Dummy warhead to the Equator* | |
| 11/20 B | 8K71 (I2-1T) | 35 | *Dummy warhead* | |
| 11/27 B | 8K71 (267433) | 36 | *Dummy warhead* | |
| 12/23 B | 8K74 (I1-1) | 37 | *Dummy warhead to 12.000 km* | |
| **1960** | | | | |
| 01/20 B | 8K74 (I1-2) | 38 | *Dummy warhead to 12.000 km* | |
| 01/24 B *Failure* | 8K74 (I1-3) | 39 | *Dummy warhead* | |
| 01/31 B | 8K74 (I1-4) | 40 | *Dummy warhead* | |
| 03/17 B | 8K74 (L1-5) | 41 | *Dummy warhead* | |
| 03/24 B | 8K74 | 42 | *Dummy warhead* | |
| 04/15 B *Failure* | 8K72 (I1-9) | 43 | Ye-3 #1 (Luna) | |
| 04/16 B *Failure* | 8K72 (L1-9A) | 44 | Ye-3 #2 (Luna) | |
| 05/15 B | 8K72 (L1-11) | 45 | Vostok-1KP | Sputnik 4 |

| Date | No. | Designation | Payload | Name |
|---|---|---|---|---|
| 06/04 B | 46 | 8K71 (L1-9) | Dummy warhead | |
| 07/05 B | 47 | 8K74 | Dummy warhead to the Equator | |
| 07/07 B | 48 | 8K74 | Dummy warhead to the Equator | |
| 07/28 B Failure | 49 | 8K72K (L1-10) | Vostok-1K #1 | |
| 08/19 B | 50 | 8K72K (L1-12) | Vostok-1K #2 | Sputnik 5 |
| 10/10 B Failure | 51 | 8K78 (L1-4) | 1M-1 (Mars) | |
| 10/14 B Failure | 52 | 8K78 (L1-5) | 1M-2 (Mars) | |
| 12/01 B | 53 | 8K72K (L1-13) | Vostok-1K #5 | Sputnik 6 |
| 12/22 B Failure | 54 | 8K72K (L1-13A) | Vostok-1K #6 | |
| **1961** | | | | |
| 01/14 B | 55 | 8K74 (L1-1T) | Dummy warhead | |
| 02/04 B Failure | 56 | 8K78 (L1-6) | 1VA #1 (Venera) | Sputnik 7 |
| 02/12 B | 57 | 8K78 (L1-7) | 1VA #2 (Venera) | Sputnik 8 & Venera 1 |
| 02/13 B | 58 | 8K74 (L1-3T) | Dummy warhead | |
| 02/27 B | 59 | 8K71 (L2-1) | Dummy warhead | |
| 03/09 B | 60 | 8K72K (Ye10314) | Vostok-3KA #1 | Sputnik 9 |
| 03/25 B | 61 | 8K72K (Ye10315) | Vostok-3KA #2 | Sputnik 10 |
| 04/12 B | 62 | 8K72K (Ye10316) | Vostok-3KA #3 | Vostok 1 |
| 04/14 B Failure | 63 | 8K74 (Ye15001-05) | Dummy warhead | |
| 06/15 B | 64 | 8K74 (Ye15001-06) | Dummy warhead | |
| 07/04 B | 65 | 8K74 (L2-2) | Dummy warhead | |
| 07/04 B | 66 | 8K74 (L2-4) | Dummy warhead | |
| 08/06 B | 67 | 8K72K (Ye10317) | Vostok-3KA #4 | Vostok 2 |
| 09/21 B | 68 | 8K74 (Ye15003-03) | Dummy warhead | |
| 11/29 B | 69 | 8K74 (L2-3) | Dummy warhead | |
| 12/11 B Failure | 70 | 8K72K (Ye10321) | 11F61 #1 (Zenit-2) | |
| **1962** | | | | |
| 04/26 B | 71 | 8K72K | 11F61 #2 (Zenit-2) | Kosmos-4 |
| 06/01 B Failure | 72 | 8K72K (Ye15000-01) | 11F61 #3 (Zenit-2) | |
| 07/02 B | 73 | 8K74 (T15001-03) | Dummy warhead | |
| 07/28 B | 74 | 8A92 (T15000-07) | 11F61 #4 (Zenit-2) | Kosmos 7 |
| 08/11 B | 75 | 8K72K (Ye10323) | Vostok-3KA #5 | Vostok 3 |

| | | | | |
|---|---|---|---|---|
| 08/12 B | 8K72K (Ye10322) | Vostok-3KA #6 | 76 | Vostok 4 |
| 08/25 B Failure | 8K78 (T103-12) | 2MV-1 #3 (Venera) | 77 | |
| 09/01 B Failure | 8K78 (T103-13) | 2MV-1 #2 (Venera) | 78 | |
| 09/12 B Failure | 8K78 (T103-14) | 2MV-2 #1 (Venera) | 79 | |
| 09/27 B | 8A92 (T15000-06) | 11F61 #7 (Zenit-2) | 80 | Kosmos 9 |
| 10/17 B | 8A92 (T15000-03) | 11F61 #5 (Zenit-2) | 81 | Kosmos 10 |
| 10/24 B Failure | 8K78 (T103-15) | 2MV-4 #1 (Mars) | 82 | |
| 11/01 B Failure | 8K78 (T103-16) | 2MV-4 #2 (Mars) | 83 | |
| 11/04 B Failure | 8K78 (T103-17) | 2MV-3 #1 (Mars) | 84 | |
| 22/12 B | 8A92 (T15000-10) | 11F61 #6 (Zenit-2) | 85 | Kosmos 12 |
| **1963** | | | | |
| 01/04 B Failure | 8K78 (T103-09) | Ye-6 #1 (Luna) | 86 | |
| 02/03 B Failure | 8K78 (G103-10) | Ye-6 #2 (Luna) | 87 | |
| 03/21 B | 8A92 (T15000-01) | 11F61 #9 (Zenit-2) | 88 | Kosmos 13 |
| 04/02 B | 8K78 (G103-11) | Ye-6 #3 (Luna) | 89 | Luna 4 |
| *04/22 B* | *8K74 (L1-4)* | *Dummy warhead* | *90* | |
| 04/22 B | 8A92 (T15000-08) | 11F61 #8 (Zenit-2) | 91 | Kosmos 15 |
| 04/28 B | 8A92 (Ye15000-02) | 11F61 #10 (Zenit-2) | 92 | Kosmos 16 |
| *05/18 B* | *8K74 (L1-6)* | *Dummy warhead* | *93* | |
| 05/24 B | 8A92 (Ye15000-12) | 11F61 #11 (Zenit-2) | 94 | Kosmos 18 |
| 06/14 B | 8K72K | Vostok-3KA #7 | 95 | Vostok 5 |
| 06/16 B | 8K72K | Vostok-3KA #8 | 96 | Vostok 6 |
| 07/10 B Failure | 8A92 (Ye15000-04) | 11F61 #12 (Zenit-2) | 97 | |
| *10/14 B* | *8K74 (L2-5)* | *Dummy warhead* | *98* | |
| 10/18 B | 8A92 (G15001-01) | 11F61 #13 (Zenit-2) | 99 | Kosmos 20 |
| 11/01 B | 11A59 | I-2B #102 | 100 | Polyot 1 |
| 11/11 B Failure | 8K78 (G103-18) | 3MV-1 #1 (Zond) | 101 | Kosmos 21 |
| 11/16 B | 11A57 (G15000-06) | 11F69 #1 (Zenit-4) | 102 | Kosmos 22 |
| 11/28 B Failure | 8A92 (G15001-02) | 11F61 #14 (Zenit-2) | 103 | |
| 12/19 B | 8A92 (G15001-03) | 11F61 #15 (Zenit-2) | 104 | Kosmos 24 |
| **1964** | | | | |
| 01/30 B | 8K72 | 2D #1 & 2 | 105 | Elektron 1 & 2 |

| Date | Vehicle | No. | Payload | Designation |
|---|---|---|---|---|
| 02/19 B Failure | 8K78 (T15000-19) | 106 | 3MV-1 #2 (Zond) | |
| 03/21 B Failure | 8K78 (T15000-20) | 107 | Ye-6 #4 (Luna) | |
| 03/27 B Failure | 8K78 (T15000-27) | 108 | 3MV-1 #3 (Zond) | Kosmos 27 |
| 04/02 B | 8K78 (T15000-28) | 109 | 3MV-1 #4 (Zond) | Zond 1 |
| 04/04 B | 8A92 (G15001-04) | 110 | 11F61 #16 (Zenit-2) | Kosmos 28 |
| 04/12 B | 11A59 | 111 | I-1B #112 | Polyot 2 |
| 04/20 B Failure | 8K78 (G15000-21) | 112 | E-6 #5 (Luna) | |
| 04/25 B | 8A92 (R15001-01) | 113 | 11F61 #19 (Zenit-2) | Kosmos 29 |
| 05/18 B | 11A57 (G15000-12) | 114 | 11F69 #2 (Zenit-4) | Kosmos 30 |
| *06/03 B* | *8K74* | *115* | *Dummy warhead* | |
| 06/04 B Failure | 8K78 (R103-34) | 116 | 11F67 #1 (Molniya-1) | |
| 06/10 B | 8A92 (R15001-02) | 117 | 11F61 #18 (Zenit-2) | Kosmos 32 |
| 06/23 B | 8A92 (G15001-05) | 118 | 11F61 #20 (Zenit-2) | Kosmos 33 |
| 07/01 B | 11A57 (T15000-04) | 119 | 11F69 #3 (Zenit-4) | Kosmos 34 |
| 07/11 B | 8K72 | 120 | 2D #3 & 4 | Elektron 3 & 4 |
| 07/15 B | 8A92 (R15001-03) | 121 | 11F61 #21 (Zenit-2) | Kosmos 35 |
| *07/27 B* | *8K74* | *122* | *Dummy warhead* | |
| 08/14 B | 8A92 (R15001-04) | 123 | 11F61 #22 (Zenit-2) | Kosmos 37 |
| 08/22 B | 8K78 (R103-36) | 124 | 11F67 #2 (Molniya-1) | Kosmos 41 |
| 08/28 B | 8A92M (T15000-25) | 125 | 11F614 #1 (Meteor) | Kosmos 44 |
| 09/09 B Abort | 11A57 | | 11F69 #4 (Zenit-4) | |
| 09/13 B | 11A57 (R15001-01) | 126 | 11F69 #4 (Zenit-4) | Kosmos 45 |
| 09/24 B | 8A92 (R15001-05) | 127 | 11F61 #23 (Zenit-2) | Kosmos 46 |
| 10/06 B | 11A57 (R15000-02) | 128 | Voskhod-3KV #1 | Kosmos 47 |
| 10/12 B | 11A57 (R15000-04) | 129 | Voskhod-3KV #2 | Voskhod 1 |
| 10/14 B | 8A92 (R15002-01) | 130 | 11F61 #24 (Zenit-2) | Kosmos 48 |
| 10/28 B | 8A92 (R15002-02) | 131 | 11F61 #25 (Zenit-2) | Kosmos 50 |
| 11/30 B | 8K78 (G15000-29) | 132 | 3MV-4 #2 (Zond) | Zond 2 |
| **1965** | | | | |
| 01/11 B | 8A92 (R15002-03) | 133 | 11F61 #26 (Zenit-2) | Kosmos 52 |
| 02/22 B | 11A57 (R15000-03) | 134 | Voskhod-3KD #2 | Kosmos 57 |
| 02/26 B | 8A92M (R15000-09) | 135 | 11F614 #2 (Meteor) | Kosmos 58 |
| 03/07 B | 11A57 (R15001-05) | 136 | 11F69 #5 (Zenit-4) | Kosmos 59 |

| Date | Vehicle | Spacecraft | No. | Name |
|---|---|---|---|---|
| 03/12 B Failure | 8K78 (R103-25) | Ye-6 #9 (Luna) | 137 | Kosmos 60 |
| 03/18 B | 11A57 (R15000-05) | Voskhod-3KD #4 | 138 | Voskhod 2 |
| 03/25 B | 8A92 (G15001-06) | 11F61 #17 (Zenit-2) | 139 | Kosmos 64 |
| 04/10 B Failure | 8K78 (R103-26) | Ye-6 #8 (Luna) | 140 | |
| 04/17 B | 11A57 (G15000-11) | 11F69 #6 (Zenit-4) | 141 | Kosmos 65 |
| 04/23 B | 8K78 (U103-35) | 11F67 #3 (Molniya-1) | 142 | Molniya-1 1 |
| 05/07 B | 8A92 (R15002-04) | 11F61 #27 (Zenit-2) | 143 | Kosmos 66 |
| 05/09 B | 8K78 (U103-30) | Ye-6 #10 (Luna) | 144 | Luna 5 |
| 05/25 B | 11A57 (R15001-04) | 11F69 #7 (Zenit-4) | 145 | Kosmos 67 |
| 06/08 B | 8K78 (U103-31) | Ye-6 #7 (Luna) | 146 | Luna 6 |
| 06/15 B | 8A92 (U15001-01) | 11F61 #29 (Zenit-2) | 147 | Kosmos 68 |
| 06/25 B | 11A57 (G15000-10) | 11F69 #8 (Zenit-4) | 148 | Kosmos 69 |
| 07/08 B Abort | 11A57 (U15001-01) | 11F69 #9 (Zenit-4) | | |
| 07/13 B Failure | 8A92 (R15002-05) | 11F61 #28 (Zenit-2) | 149 | |
| 07/18 B | 8K78 | 3MV-4 #3 (Zond) | 150 | Zond 3 |
| 08/03 B | 11A57 (U15001-01) | 11F69 #9 (Zenit-4) | 151 | Kosmos 77 |
| 08/14 B | 8A92 (U15001-02) | 11F61 #30 (Zenit-2) | 152 | Kosmos 78 |
| 08/25 B | 11A57 (Ye15001-06) | 11F69 #10 (Zenit-4) | 153 | Kosmos 79 |
| 09/04 B Abort | 8K78M (U103-27) | Ye-6 #11 (Luna) | | |
| 09/09 B | 11A57 (R15001-02) | 11F69 #11 (Zenit-4) | 154 | Kosmos 85 |
| 09/23 B | 11A57 (R15001-03) | 11F69 #12 (Zenit-4) | 155 | Kosmos 91 |
| 10/04 B | 8K78M (U103-27) | Ye-6 #11 (Luna) | 156 | Luna 7 |
| 10/14 B | 8K78 (U103-37) | 11F67 #4 (Molniya-1) | 157 | Molniya-1 2 |
| 10/16 B | 11A57 (U15001-04) | 11F69 #13 (Zenit-4) | 158 | Kosmos 92 |
| 10/28 B | 11A57 (U15001-03) | 11F69 #14 (Zenit-4) | 159 | Kosmos 94 |
| 11/12 B | 8K78 (U103-42) | 3MV-4 #4 (Venera) | 160 | Venera 2 |
| 11/16 B | 8K78 (U103-31) | 3MV-3 #1 (Venera) | 161 | Venera 3 |
| 11/23 B Failure | 8K78 (U103-30) | 3MV-4 #6 (Venera) | 162 | Kosmos 96 |
| 11/26 B Abort | 8K78 | 3MV-3 #2 (Venera) | | |
| 11/27 B | 8A92 (U15001-05) | 11F61 #31 (Zenit-2) | 163 | Kosmos 98 |
| 12/03 B | 8K78 (U103-28) | Ye-6 #12 (Luna) | 164 | Luna 8 |
| 12/10 B | 8A92 (U15001-04) | 11F61 #32 (Zenit-2) | 165 | Kosmos 99 |

| Date | Rocket | No. | Payload | Name |
|---|---|---|---|---|
| *12/14 P* | *8K74* | *166* | *Dummy warhead* | |
| 12/17 B | 8A92M (R15000-31) | 167 | 11F614 #3 (Meteor) | Kosmos 100 |
| *12/21 P* | *8K74* | *168* | *Dummy warhead* | |
| 28/12 B | 11A510 (G15000-01) | 169 | US #YeA 0110 | Kosmos 102 |
| **1966** | | | | |
| 01/07 B | 8A92 | 170 | 11F61 #36 (Zenit-2) | Kosmos 104 |
| 01/22 B | 8A92 | 171 | 11F61 #38 (Zenit-2) | Kosmos 105 |
| 01/31 B | 8K78M (U103-32) | 172 | Ye-6 #13 (Luna) | Luna 9 |
| 02/10 B | 8A92 | 173 | 11F61 #34 (Zenit-2) | Kosmos 107 |
| 02/19 B | 11A57 | 174 | 11F69 #15 (Zenit-4) | Kosmos 109 |
| 02/22 B | 11A57 (R15000-06) | 175 | Voskhod-3KV #5 | Kosmos 110 |
| 03/01 B Failure | 8K78M (N103-41) | 176 | Ye-6S #204 (Luna) | Kosmos 111 |
| 03/17 P | 8A92 | 177 | 11F61 #37 (Zenit-2) | Kosmos 112 |
| 03/21 B | 11A57 | 178 | 11F69 #16 (Zenit-4) | Kosmos 113 |
| 03/27 B Failure | 8K78 (N103-38) | 179 | 11F67 #5 (Molniya-1) | |
| 03/31 B | 8K78M (N103-42) | 180 | Ye-6S #206 (Luna) | Luna 10 |
| 04/06 P | 11A57 (U15001-02) | 181 | 11F69 #17 (Zenit-4) | Kosmos 114 |
| 04/20 B | 8A92? | 182 | 11F61 #35 (Zenit-2) | Kosmos 115 |
| 04/25 B | 8K78 (N103-39) | 183 | 11F67 #6 (Molniya-1) | Molniya-1 3 |
| 05/06 B | 8A92 (N15001-01) | 184 | 11F61 #39 (Zenit-2) | Kosmos 117 |
| 05/11 B | 8A92M | 185 | 11F614 #4 (Meteor) | Kosmos 118 |
| 05/17 B Failure | 11A57 | 186 | 11F69 #18 (Zenit-4) | |
| *05/27 B* | *8K74* | *187* | *Dummy warhead* | |
| 06/08 B | 11A57 | 188 | 11F61 #41 (Zenit-2) | Kosmos 120 |
| 06/17 P | 11A57 | 189 | 11F69 #19 (Zenit-4) | Kosmos 121 |
| *06/25 B* | *8K74* | *190* | *Dummy warhead* | |
| 06/25 B | 8A92M (N15000-21) | 191 | 11F614 #5 (Meteor) | Kosmos 122 |
| 07/14 B | 11A57 (N15001-14) | 192 | 11F61 #42 (Zenit-2) | Kosmos 124 |
| 07/20 B | 11A510 | 193 | US #YeA 0110 | Kosmos 125 |
| 07/28 B | 11A57 (N15001-01) | 194 | 11F69 #20 (Zenit-4) | Kosmos 126 |
| 08/08 B | 11A57 (N15001-13) | 195 | 11F69 #21 (Zenit-4) | Kosmos 127 |
| 08/24 B | 8K78M (N103-43) | 196 | Ye-6LF #101 (Luna) | Luna 11 |

| Date | Rocket | No. | Payload | Name |
|---|---|---|---|---|
| 08/27 B | 11A57 (N15001-03) | 197 | 11F69 #22 (Zenit-4) | Kosmos 128 |
| 09/16 B Failure | 8A92 | 198 | 11F61 #40 (Zenit-2) | |
| 10/14 P | 8A92 | 199 | 11F61 #33 (Zenit-2) | Kosmos 129 |
| 10/20 B | 8K78 (N103-40) | 200 | 11F67 #7 (Molniya-1) | Molniya-1 4 |
| 10/20 B | 11A57 (N15001-04) | 201 | 11F69 #23 (Zenit-4) | Kosmos 130 |
| 10/22 B | 8K78M (N103-44) | 202 | Ye-6LF #102 (Luna) | Luna 12 |
| 11/12 P | 11A57 | 203 | 11F69 #24 (Zenit-4) | Kosmos 131 |
| 11/19 B | 8A92 (N15001-08) | 204 | 11F61 #46 (Zenit-2) | Kosmos 132 |
| 11/28 B | 11A511 (U15000-02) | 205 | 11F615 #02A (Soyuz) | Kosmos 133 |
| 12/03 B | 11A57 (N15001-06) | 206 | 11F69 #25 (Zenit-4) | Kosmos 134 |
| 12/14 B Failure | 11A511 (U15000-01) | 207 | 11F615 #01A (Soyuz) | |
| 12/19 P | 8A92 | 208 | 11F61 #47 (Zenit-2) | Kosmos 136 |
| 12/21 B | 8K78M (N103-45) | 209 | Ye-6M #205 (Luna) | Luna 13 |

**1967**

| Date | Rocket | No. | Payload | Name |
|---|---|---|---|---|
| 01/19 P | 8A92 | 210 | 11F61 #43 (Zenit-2) | Kosmos 138 |
| 02/07 B | 11A511 (U15000-03) | 211 | 11F615 #03L (Soyuz) | Kosmos 140 |
| 02/08 P | 11A57 | 212 | 11F69 #26 (Zenit-4) | Kosmos 141 |
| 02/27 B | 8A92 (U15000-03) | 213 | 11F61 #45 (Zenit-2) | Kosmos 143 |
| 02/28 P | 8A92M | 214 | 11F614 #6 (Meteor) | Kosmos 144 |
| 03/13 P | 8A92 | 215 | 11F61 #44 (Zenit-2) | Kosmos 147 |
| 03/22 P | 11A57 | 216 | 11F69 #27 (Zenit-4) | Kosmos 150 |
| 04/04 P | 8A92 | 217 | 11F61 #48 (Zenit-2) | Kosmos 153 |
| 04/12 B | 11A57 (N15001-08) | 218 | 11F69 #28 (Zenit-4) | Kosmos 155 |
| 04/23 B | 11A511 (U15000-04) | 219 | 11F615 #04L (Soyuz) | Soyuz 1 |
| 04/27 P | 8A92M | 220 | 11F614 #7 (Meteor) | Kosmos 156 |
| 05/12 B | 8A92 | 221 | 11F61 #49 (Zenit-2) | Kosmos 157 |
| 05/16 B | 8K78M (Ya716-56) | 222 | Ye-6LS #111 (Luna) | Kosmos 159 |
| 05/22 P | 11A57 | 223 | 11F69 #29 (Zenit-4) | Kosmos 161 |
| 05/24 B | 8K78 | 224 | 11F67 #8 (Molniya-1) | Molniya-1 5 |
| 06/01 B | 11A57 (Ya15001-11) | 225 | 11F69 #30 (Zenit-4) | Kosmos 162 |
| 06/08 P | 11A57 (Ya15001-13) | 226 | 11F61 #50 (Zenit-2) | Kosmos 164 |
| 06/12 B | 8K78M (Ya716-70) | 227 | B-67 #310 (Venera) | Venera 4 |

| Date | Type | Payload | No. | Name |
|---|---|---|---|---|
| 06/17 B Failure | 8K78M (Ya716-71) | B-67 #311 (Venera) | 228 | Kosmos 167 |
| 06/20 P Failure | 11A57 | 11F69 #31 (Zenit-4) | 229 | Kosmos 168 |
| 07/04 B | 11A57 (Ya15001-05) | 11F61 #52 (Zenit-2) | 230 | |
| 07/21 B Failure | 11A57 (Ya15001-14) | 11F69 #32 (Zenit-4) | 231 | |
| 07/25 P | 8K74 | *Dummy warhead* | 232 | |
| 08/09 B | 11A57 (Ya15001-15) | 11F69 #33 (Zenit-4) | 233 | Kosmos 172 |
| 08/31 B | 8K78 (Ya716-81) | 11F67 #9 (Molniya-Yu) | 234 | Kosmos 174 |
| 09/01 P Failure | 11A57 | 11F61 #51 (Zenit-2) | 235 | |
| 09/11 P | 11A57 | 11F69 #34 (Zenit-4) | 236 | Kosmos 175 |
| 09/16 B | 11A57 (Ya15001-06) | 11F61 #53 (Zenit-2) | 237 | Kosmos 177 |
| 09/26 P | 11A57 | 11F61 #54 (Zenit-2) | 238 | Kosmos 180 |
| 10/03 B | 8K78 (Ya716-83) | 11F67 #10 (Molniya-1) | 239 | Molniya-1 6 |
| 10/11 P | 11A57 | 11F61 #55 (Zenit-2) | 240 | Kosmos 181 |
| 10/16 B | 11A57 | 11F69 #35 (Zenit-4) | 241 | Kosmos 182 |
| 10/22 B | 8K78 (Ya716-82) | 11F67 #11 (Molniya-1) | 242 | Molniya-1 7 |
| 10/25 P | 8A92M | 11F614 #8 (Meteor) | 243 | Kosmos 184 |
| 10/27 B | 11A511 (15000-06) | 11F615 #06A (Soyuz) | 244 | Kosmos 186 |
| 10/30 B | 11A511 (15000-05) | 11F615 #05L (Soyuz) | 245 | Kosmos 188 |
| 11/03 P | 11A57 | 11F69 #36 (Zenit-4) | 246 | Kosmos 190 |
| 11/25 P | 11A57 | 11F61 #58 (Zenit-2) | 247 | Kosmos 193 |
| 12/03 P | 11A57 | 11F69 #37 (Zenit-4) | 248 | Kosmos 194 |
| 12/16 P | 11A57 | 11F61 #57 (Zenit-2) | 249 | Kosmos 195 |

**1968**

| Date | Type | Payload | No. | Name |
|---|---|---|---|---|
| 01/16 P | 11A57 | 11F61 #59 (Zenit-2) | 250 | Kosmos 199 |
| 02/06 B | 11A57 | 11F69 #38 (Zenit-4) | 251 | Kosmos 201 |
| 02/07 B Failure | 8K78M (Ya716-57) | Ye-6LS #112 (Luna) | 252 | |
| 03/05 P | 11A57 | 11F61 #56 (Zenit-2) | 253 | Kosmos 205 |
| 03/14 P | 8A92M | 11F614 #9 (Meteor) | 254 | Kosmos 206 |
| 03/16 P | 11A57 | 11F69 #39 (Zenit-4) | 255 | Kosmos 207 |
| 03/21 B | 11A57 | 11F690 (Zenit-2M) | 256 | Kosmos 208 |
| 04/03 P | 11A57 | 11F61 #60 (Zenit-2) | 257 | Kosmos 210 |
| 04/07 B | 8K78M (Ya716-58) | Ye-6LS (Luna) | 258 | Luna 14 |

| Date | Rocket | Designation | No. | Name |
|---|---|---|---|---|
| 04/14 B | 11A511 (15000-08) | 11F615 #08L (Soyuz) | 259 | Kosmos 212 |
| 04/15 B | 11A511 (15000-07) | 11F615 #07L (Soyuz) | 260 | Kosmos 213 |
| 04/18 P | 11A57 | 11F69 #40 (Zenit-4) | 261 | Kosmos 214 |
| 04/20 B | 11A57 | 11F61 #62 (Zenit-2) | 262 | Kosmos 216 |
| 04/21 B | 8K78M (Ya716-84) | 11F67 #12 (Molniya-1) | 263 | Molniya-1 8 |
| 06/01 P | 11A57 | 11F61 #63 (Zenit-2) | 264 | Kosmos 223 |
| 06/04 B | 11A57 | 11F69 #41 (Zenit-4) | 265 | Kosmos 224 |
| 06/12 P | 8A92M | 11F614 #10 (Meteor) | 266 | Kosmos 226 |
| 06/18 B | 11A57 | 11F69 #42 (Zenit-4) | 267 | Kosmos 227 |
| 06/21 B | 11A57 | 11F690 (Zenit-2M) | 268 | Kosmos 228 |
| 06/26 P | 11A57 | 11F69 #43 (Zenit-4) | 269 | Kosmos 229 |
| 07/05 B | 8K78M (Ya716-85) | 11F67 #13 (Molniya-1) | 270 | Molniya-1 9 |
| 07/10 B | 11A57 | 11F61 #64 (Zenit-2) | 271 | Kosmos 231 |
| 07/16 P | 11A57 | 11F69 #44 (Zenit-4) | 272 | Kosmos 232 |
| 07/30 B | 11A57 | 11F69 #45 (Zenit-4) | 273 | Kosmos 234 |
| 08/09 B | 11A57 | 11F61 #61 (Zenit-2) | 274 | Kosmos 235 |
| 08/27 P | 11A57 | 11F69 #46 (Zenit-4) | 275 | Kosmos 237 |
| 08/28 B | 11A511 (15000-09) | 11F615 #09L (Soyuz) | 276 | Kosmos 238 |
| 09/05 B | 11A57 | 11F69 #47 (Zenit-4) | 277 | Kosmos 239 |
| 09/14 B | 11A57 | 11F61 #66 (Zenit-2) | 278 | Kosmos 240 |
| 09/16 P | 11A57 | 11F69 #48 (Zenit-4) | 279 | Kosmos 241 |
| 09/23 B | 11A57 | 11F690 (Zenit-2M) | 280 | Kosmos 243 |
| 10/05 B | 8K78M (Ya716-86) | 11F67 #14 (Molniya-1) | 281 | Molniya-1 10 |
| 10/07 P | 11A57 | 11F69 #49 (Zenit-4) | 282 | Kosmos 246 |
| 10/11 P | 11A57 | 11F61 #65 (Zenit-2) | 283 | Kosmos 247 |
| 10/25 B | 11A511 (15000-11) | 11F615 #11L (Soyuz) | 284 | Soyuz 2 |
| 10/26 B | 11A511 (15000-10) | 11F615 #10L (Soyuz) | 285 | Soyuz 3 |
| 10/31 B | 11A57 | 11F691 (Zenit-4M) | 286 | Kosmos 251 |
| 11/13 P | 11A57 | 11F61 #67 (Zenit-2) | 287 | Kosmos 253 |
| 11/21 P | 11A57 | 11F69 #50 (Zenit-4) | 288 | Kosmos 254 |
| 11/29 P | 11A57 | 11F61 #68 (Zenit-2) | 289 | Kosmos 255 |
| 12/10 B | 11A57 | 11F61 #69 (Zenit-2) | 290 | Kosmos 258 |

| No. | Date | Rocket | Payload | Name |
| --- | --- | --- | --- | --- |
| 291 | 12/16 B | 8K78M (Ya716-87) | 11F67 #15 (Molniya-Yu) | Kosmos 260 |
| **1969** | | | | |
| 292 | 01/05 B | 8K78M | V69 #300 (Venera) | Venera 5 |
| 293 | 01/10 B | 8K78M | V69 #331 (Venera) | Venera 6 |
| 294 | 01/12 P | 11A57 | 11F61 #70 (Zenit-2) | Kosmos 263 |
| 295 | 01/14 B | 11A511 (-5000-12) | 11F615A #12L (Soyuz) | Soyuz 4 |
| 296 | 01/15 B | 11A511 (-5000-13) | 11F615A #13L (Soyuz) | Soyuz 5 |
| 297 | 01/23 B | 11A57 | 11F691 (Zenit-4M) | Kosmos 264 |
| 298 | 02/01 P Failure | 8A92M | 11F614 #11 (Meteor-1) | |
| 299 | 02/25 P | 11A57 | 11F61 #71 (Zenit-2) | Kosmos 266 |
| 300 | 02/26 B | 11A57 | 11F69 #51 (Zenit-4) | Kosmos 267 |
| 301 | 03/06 P | 11A57 | 11F69 #52 (Zenit-4) | Kosmos 270 |
| 302 | 03/15 P | 11A57 | 11F69 #53 (Zenit-4) | Kosmos 271 |
| 303 | 03/22 P | 11A57 | 11F61 #72 (Zenit-2) | Kosmos 273 |
| 304 | 03/24 B | 11A57 | 11F69 #54 (Zenit-4) | Kosmos 274 |
| 305 | 03/26 P | 8A92M | 11F614 #12 (Meteor-1) | Meteor-1 1 |
| 306 | 04/04 P | 11A57 | 11F69 #55 (Zenit-4) | Kosmos 276 |
| 307 | 04/09 P | 11A57 | 11F61 #73 (Zenit-2) | Kosmos 278 |
| 308 | 04/11 B | 8K78M (Ya716-88) | 11F67 #16 (Molniya-1) | Molniya-1 11 |
| 309 | 04/15 P | 11A57 | 11F69 #56 (Zenit-4) | Kosmos 279 |
| 310 | 04/23 B | 11A57 | 11F691 (Zenit-4M) | Kosmos 280 |
| 311 | 05/13 P | 11A57 | 11F61 #74 (Zenit-2) | Kosmos 281 |
| 312 | 05/20 P | 11A57 | 11F69 #57 (Zenit-4) | Kosmos 282 |
| 313 | 05/29 B | 11A57 | 11F69 #58 (Zenit-4) | Kosmos 284 |
| 314 | 06/15 P | 11A57 | 11F69 #59 (Zenit-4) | Kosmos 286 |
| 315 | 06/24 P | 11A57 | 11F61 #75 (Zenit-2) | Kosmos 287 |
| 316 | 06/27 B | 11A57 | 11F69 #60 (Zenit-4) | Kosmos 288 |
| 317 | 07/10 P | 11A57 | 11F69 #61 (Zenit-4) | Kosmos 289 |
| 318 | 07/22 P | 11A57 | 11F61 #76 (Zenit-2) | Kosmos 290 |
| 319 | 07/22 B | 8K78M (Ya716-89) | 11F67 #17 (Molniya-1) | Molniya-1 12 |
| 320 | 08/16 B | 11A57 | 11F690 (Zenit-2M) | Kosmos 293 |
| 321 | 08/19 P | 11A57 | 11F69 #62 (Zenit-4) | Kosmos 294 |

| Date | Vehicle | No. | Payload | Name |
|---|---|---|---|---|
| 08/29 B | 11A57 | 322 | 11F69 #63 (Zenit-4) | Kosmos 296 |
| 09/02 P | 11A57 | 323 | 11F69 #64 (Zenit-4) | Kosmos 297 |
| 09/18 B | 11A57 | 324 | 11F69 #65 (Zenit-4) | Kosmos 299 |
| 09/24 P | 11A57 | 325 | 11F61 #77 (Zenit-2) | Kosmos 301 |
| 10/06 P | 8A92M | 326 | 11F614 #12 (Meteor-1) | Meteor-1 2 |
| 10/11 B | 11A511 (B15000-14) | 327 | 11F615A #14L (Soyuz) | Soyuz 6 |
| 10/12 B | 11A511 (Yu15000-19) | 328 | 11F615A #15L (Soyuz) | Soyuz 7 |
| 10/13 B | 11A511 (Yu15000-18) | 329 | 11F615A #16L (Soyuz) | Soyuz 8 |
| 10/17 P | 11A57 | 330 | 11F69 #66 (Zenit-4) | Kosmos 302 |
| 10/24 B | 11A57 | 331 | 11F690 (Zenit-2M) | Kosmos 306 |
| 11/12 P | 11A57 | 332 | 11F61 #78 (Zenit-2) | Kosmos 309 |
| 11/15 B | 11A57 | 333 | 11F69 #67 (Zenit-4) | Kosmos 310 |
| 12/03 P | 11A57 | 334 | 11F690 (Zenit-2M) | Kosmos 313 |
| 12/23 P | 11A57 | 335 | 11F692 (Zenit-4MK) | Kosmos 317 |
| **1970:** | | | | |
| 01/09 B | 11A57 | 336 | 11F690 (Zenit-2M) | Kosmos 318 |
| 01/21 P | 11A57 | 337 | 11F69 #68 (Zenit-4) | Kosmos 322 |
| 02/10 P | 11A57 | 338 | 11F69 #69 (Zenit-4) | Kosmos 323 |
| 02/19 P | 8K78M | 339 | 11F67 #18 (Molniya-1) | Molniya-1 13 |
| 03/04 P | 11A57 | 340 | 11F61 #79 (Zenit-2) | Kosmos 325 |
| 03/13 P | 11A57 | 341 | 11F61 #80 (Zenit-2) | Kosmos 326 |
| 03/17 P | 8A92M | 342 | 11F614 #13 (Meteor-1) | Meteor-1 3 |
| 03/27 P | 11A57 | 343 | 11F692 (Zenit-4MK) | Kosmos 328 |
| 04/03 P | 11A57 | 344 | 11F690 (Zenit-2M) | Kosmos 329 |
| 04/08 B | 11A57 | 345 | 11F69 #70 (Zenit-4) | Kosmos 331 |
| 04/15 P | 11A57 | 346 | 11F691 (Zenit-4M) | Kosmos 333 |
| 04/28 P | 8A92M | 347 | 11F614 #14 (Meteor-1) | Meteor-1 4 |
| 05/12 P | 11A57 | 348 | 11F61 #81 (Zenit-2) | Kosmos 344 |
| 05/20 B | 11A57 | 349 | 11F69 #71 (Zenit-4) | Kosmos 345 |
| 06/01 B | 11A511 (Yu15000-21S) | 350 | 11F615A #17L (Soyuz) | Soyuz 9 |
| 06/10 B | 11A57 | 351 | 11F69 #72 (Zenit-4) | Kosmos 346 |
| 06/17 P | 11A57 | 352 | 11F69 #73 (Zenit-4) | Kosmos 349 |

| Date | | Vehicle | Payload | No. | Name |
|---|---|---|---|---|---|
| 06/23 | P | 8A92M | 11F614 #15 (Meteor-1) | 353 | Meteor-1 5 |
| 06/26 | P | 8K78M | 11F67 #19 (Molniya-1) | 354 | Molniya-1 14 |
| 06/26 | B | 11A57 | 11F690 (Zenit-2M) | 355 | Kosmos 350 |
| 07/07 | B | 11A57 | 11F69 #74 (Zenit-4) | 356 | Kosmos 352 |
| 07/09 | P | 11A57 | 11F690 (Zenit-2M) | 357 | Kosmos 353 |
| 07/21 | P Failure | 11A57 | 11F69 #75 (Zenit-4) | 358 | |
| 08/07 | P | 11A57 | 11F69 #76 (Zenit-4) | 359 | Kosmos 355 |
| 08/17 | B | 8K78M | V70 #630 (Venera) | 360 | Venera 7 |
| 08/22 | B Failure | 8K78M | V70 #631 (Venera) | 361 | Kosmos 359 |
| 08/29 | P | 11A57 | 11F691 (Zenit-4M) | 362 | Kosmos 360 |
| 09/08 | P | 11A57 | 11F691 (Zenit-4M) | 363 | Kosmos 361 |
| 09/17 | B | 11A57 | 11F690 (Zenit-2M) | 364 | Kosmos 363 |
| 09/22 | P | 11A57 | 11F692 (Zenit-4MK) | 365 | Kosmos 364 |
| 09/29 | P | 8K78M | 11F67 #20 (Molniya-1) | 366 | Molniya-1 15 |
| 10/01 | B | 11A57 | 11F690 (Zenit-2M) | 367 | Kosmos 366 |
| 10/08 | B | 11A57 | 11F690 (Zenit-2M) | 368 | Kosmos 368 |
| 10/09 | B | 11A57 | 11F691 (Zenit-4M) | 369 | Kosmos 370 |
| 10/15 | P | 8A92M | 11F614 #16 (Meteor-1) | 370 | Meteor-1 6 |
| 10/30 | P | 11A57 | 11F691 (Zenit-4M) | 371 | Kosmos 376 |
| 11/11 | B | 11A57 | 11F690 (Zenit-2M) | 372 | Kosmos 377 |
| 11/24 | B | 11A511L | T2K #1 | 373 | Kosmos 379 |
| 11/27 | P | 8K78M | 11F67 #21 (Molniya-1) | 374 | Molniya-1 16 |
| 12/03 | P | 11A57 | 11F692 (Zenit-4MK) | 375 | Kosmos 383 |
| 12/10 | P | 11A57 | 11F690 (Zenit-2M) | 376 | Kosmos 384 |
| 12/15 | B | 11A57 | 11F691 (Zenit-4M) | 377 | Kosmos 386 |
| 12/18 | P | 8A92M | 11F619 #1 (Tselina-D) | 378 | Kosmos 389 |
| 12/25 | P | 8K78M | 11F67 #22 (Molniya-1) | 379 | Molniya-1 17 |
| **1971:** | | | | | |
| 01/12 | B | 11A57 | 11F691 (Zenit-4M) | 380 | Kosmos 390 |
| 01/20 | P | 8A92M | 11F614 #17 (Meteor-1) | 381 | Meteor-1 7 |
| 01/21 | B | 11A57 | 11F690 (Zenit-2M) | 382 | Kosmos 392 |
| 02/18 | P | 11A57 | 11F691 (Zenit-4M) | 383 | Kosmos 396 |

| No. | Date | Vehicle | Payload | Name |
|-----|------|---------|---------|------|
| 384 | 02/26 B | 11A511L | T2K #2 | Kosmos 398 |
| 385 | 03/03 B | 11A57 | 11F691 (Zenit-4M) | Kosmos 399 |
| 386 | 03/05 P Failure | 11A57 | 11F690 (Zenit-2M) | |
| 387 | 03/27 P | 11A57 | 11F691 (Zenit-4M) | Kosmos 401 |
| 388 | 04/02 P | 11A57 | 11F690 (Zenit-2M) | Kosmos 403 |
| 389 | 04/07 P | 8A92M | 11F619 #2 (Tselina-D) | Kosmos 405 |
| 390 | 04/14 P | 11A57 | 11F691 (Zenit-4M) | Kosmos 406 |
| 391 | 04/17 P | 8A92M | 11F614 #18 (Meteor-1) | Meteor-1 8 |
| 392 | 04/23 B | 11A511 (15000-25) | 11F615A8 #1L (Soyuz) | Soyuz 10 |
| 393 | 05/06 B | 11A57 | 11F690 (Zenit-2M) | Kosmos 410 |
| 394 | 05/18 B | 11A57 | 11F691 (Zenit-4M) | Kosmos 420 |
| 395 | 05/28 P | 11A57 | 11F691 (Zenit-4M) | Kosmos 424 |
| 396 | 06/06 B | 11A511 | 11F615A8 #21L (Soyuz) | Soyuz 11 |
| 397 | 06/11 P | 11A57 | 11F692 (Zenit-4MK) | Kosmos 427 |
| 398 | 06/24 B | 11A57 | 11F690 (Zenit-2M) | Kosmos 428 |
| 399 | 06/25 P Failure | 11A57 | 11F691 (Zenit-4M) | |
| 400 | 07/16 P | 8A92M | 11F614 #19 (Meteor-1) | Meteor-1 9 |
| 401 | 07/20 B | 11A57 | 11F691 (Zenit-4M) | Kosmos 429 |
| 402 | 07/23 P | 11A57 | 11F691 (Zenit-4M) | Kosmos 430 |
| 403 | 07/28 P | 8K78M | 11F67 #23 (Molniya-1) | Molniya-1 18 |
| 404 | 07/30 B | 11A57 | 11F690 (Zenit-2M) | Kosmos 431 |
| 405 | 08/05 B | 11A57 | 11F691 (Zenit-4M) | Kosmos 432 |
| 406 | 08/12 B | 11A511L | T2K #3 | Kosmos 434 |
| 407 | 08/19 B Failure | 11A57 | 11F691 (Zenit-4M) | |
| 408 | 09/14 P | 11A57 | 11F692 (Zenit-4MK) | Kosmos 438 |
| 409 | 09/21 P | 11A57 | 11F690 (Zenit-2M) | Kosmos 439 |
| 410 | 09/28 B | 11A57 | 11F691 (Zenit-4M) | Kosmos 441 |
| 411 | 09/29 P | 11A57 | 11F691 (Zenit-4M) | Kosmos 442 |
| 412 | 10/07 P | 11A57 | 11F690 (Zenit-2M) | Kosmos 443 |
| 413 | 10/14 B | 11A57 | 11F691 (Zenit-4M) | Kosmos 452 |
| 414 | 11/02 P | 11A57 | 11F691 (Zenit-4M) | Kosmos 454 |
| 415 | 11/19 P | 11A57 | 11F691 (Zenit-4M) | Kosmos 456 |

| | Date | Vehicle | Designation | Payload |
|---|---|---|---|---|
| 416 | 11/24 B | 8K78M | 11F628 #1 (Molniya-2) | Molniya-2 1 |
| 417 | 12/03 P Failure | 11A57 | 11F690 (Zenit-2M) | |
| 418 | 12/06 B | 11A57 | 11F691 (Zenit-4M) | Kosmos 463 |
| 419 | 12/10 P | 11A57 | 11F691 (Zenit-4M) | Kosmos 464 |
| 420 | 12/16 B | 11A57 | 11F691 (Zenit-4M) | Kosmos 466 |
| 421 | 12/19 P | 8K78M | 11F67 #24 (Molniya-1) | Molniya-1 19 |
| 422 | 12/27 P | 11A511M | 11F629 (Zenit-4MT/Orion) | Kosmos 470 |
| 423 | 12/29 P | 8A92 | 11F614 #20 (Meteor-1) | Meteor-1 10 |
| **1972:** | | | | |
| 424 | 01/12 B | 11A57 | 11F691 (Zenit-4M) | Kosmos 471 |
| 425 | 02/03 B | 11A57 | 11F690 (Zenit-2M) | Kosmos 473 |
| 426 | 02/16 B | 11A57 | 11F691 (Zenit-4M) | Kosmos 474 |
| 427 | 03/01 P | 8A92M | 11F619 #3 (Tselina-D) | Kosmos 476 |
| 428 | 03/04 P | 11A57 | 11F690 (Zenit-2M) | Kosmos 477 |
| 429 | 03/15 P | 11A57 | 11F691 (Zenit-4M) | Kosmos 478 |
| 430 | 03/27 B | 8K78M | V-72 #670 (Venera) | Venera 8 |
| 431 | 03/30 P | 8A92M | 11F614 #21 (Meteor-1) | Meteor-1 11 |
| 432 | 03/31 B Failure | 8K78M | V-72 #671 (Venera) | Kosmos 482 |
| 433 | 04/03 P | 11A57 | 11F691 (Zenit-4M) | Kosmos 483 |
| 434 | 04/04 P | 8K78M | 11F67 #25 (Molniya-1) / SRET-1 | Molniya-1 20 |
| 435 | 04/06 P | 11A57 | 11F690 (Zenit-2M) | Kosmos 484 |
| 436 | 04/07 B | 11A57 | 13KS Energiya | Interkosmos 6 |
| 437 | 04/14 B | 8K78M | SO-M #501 | Prognoz 1 |
| 438 | 04/14 P | 11A57 | 11F691 (Zenit-4M) | Kosmos 486 |
| 439 | 05/05 P | 11A57 | 11F692 (Zenit-4MK) | Kosmos 488 |
| 440 | 05/17 P | 11A57 | 11F690 (Zenit-2M) | Kosmos 490 |
| 441 | 05/19 P | 8K78M | 11F628 #2 (Molniya-2) | Molniya-2 2 |
| 442 | 05/25 P | 11A57 | 11F691 (Zenit-4M) | Kosmos 491 |
| 443 | 06/09 P | 11A57 | 11F691 (Zenit-4M) | Kosmos 492 |
| 444 | 06/21 P | 11A57 | 11F690 (Zenit-2M) | Kosmos 493 |
| 445 | 06/23 P | 11A57 | 11F692 (Zenit-4MK) | Kosmos 495 |

| | Date | Vehicle | Payload | Satellite |
|---|---|---|---|---|
| 446 | 06/26 B | 11A511 | 11F615A8 #33A (7K-T-3) | Kosmos 496 |
| 447 | 06/29 B | 8K78M | SO-M #502 | Prognoz 2 |
| 448 | 06/30 P | 8A92M | 11F614 #22 (Meteor-1) | Meteor-1 12 |
| 449 | 07/06 P | 11A57 | 11F691 (Zenit-4M) | Kosmos 499 |
| 450 | 07/13 P | 11A511M | 11F629 (Zenit-4MT/Orion) | Kosmos 502 |
| 451 | 07/19 P | 11A57 | 11F691 (Zenit-4M) | Kosmos 503 |
| 452 | 07/28 P | 11A57 | 11F690 (Zenit-2M) | Kosmos 512 |
| 453 | 08/02 P | 11A57 | 11F691 (Zenit-4M) | Kosmos 513 |
| 454 | 08/18 P | 11A57 | 11F692 (Zenit-4MK) | Kosmos 515 |
| 455 | 08/30 B | 11A57 | 11F690 (Zenit-2M) | Kosmos 517 |
| 456 | 09/02 P Failure | 11A57 | 11F692 (Zenit-4MK) | |
| 457 | 09/15 P | 11A57 | 11F690 (Zenit-2M) | Kosmos 518 |
| 458 | 09/16 B | 11A57 | 11F691 (Zenit-4M) | Kosmos 519 |
| 459 | 09/19 B | 8K78M | 5V95 (US-K/Oko #1) | Kosmos 520 |
| 460 | 09/30 P | 8K78M | 11F628 #3 (Molniya-2) | Molniya-2 3 |
| 461 | 10/04 P | 11A57 | 11F691 (Zenit-4M) | Kosmos 522 |
| 462 | 10/14 P | 8K78M | 11F67 #26 (Molniya-1) | Molniya-1 21 |
| 463 | 10/18 P | 11A57 | 11F690 (Zenit-2M) | Kosmos 525 |
| 464 | 10/26 P | 8A92M | 11F614 #23 (Meteor-1) | Meteor-1 13 |
| 465 | 10/31 P | 11A57 | 11F692 (Zenit-4MK) | Kosmos 527 |
| 466 | 11/25 B | 11A57 | 11F690 (Zenit-2M) | Kosmos 537 |
| 467 | 12/02 B | 8K78M | 11F67 #27 (Molniya-1) | Molniya-1 22 |
| 468 | 12/12 P | 8K78M | 11F628 #4 (Molniya-2) | Molniya-2 4 |
| 469 | 12/14 P | 11A57 | 11F691 (Zenit-4M) | Kosmos 538 |
| 470 | 12/27 P | 11A511M | 11F629 (Zenit-4MT/Orion) | Kosmos 541 |
| 471 | 12/28 P | 8A92M | 11F619 (Tselina-D #4) | Kosmos 542 |
| **1973:** | | | | |
| 472 | 01/11 B | 11A57 | 11F691 (Zenit-4M) | Kosmos 543 |
| 473 | 02/01 B | 11A57 | 11F690 (Zenit-2M) | Kosmos 547 |
| 474 | 02/03 B | 8K78M | 11F67 #28 (Molniya-1) | Molniya-1 23 |
| 475 | 02/08 P | 11A57 | 11F691 (Zenit-4M) | Kosmos 548 |
| 476 | 02/15 B | 8K78M | SO-M #503 | Prognoz 3 |

| Date | Vehicle | No. | Designation | Payload |
|---|---|---|---|---|
| 03/01 P | 11A57 | 477 | 11F692 (Zenit-4MK) | Kosmos 550 |
| 03/06 B | 11A57 | 478 | 11F691 (Zenit-4M) | Kosmos 551 |
| 03/20 P | 8A92M | 479 | 11F614 #24 (Meteor-1) | Meteor-1 14 |
| 03/22 P | 11A57 | 480 | 11F690 (Zenit-2M) | Kosmos 552 |
| 04/05 P | 8K78M | 481 | 11F628 #5 (Molniya-2) | Molniya-2 5 |
| 04/19 P | 11A57 | 482 | 11F692 (Zenit-4MK) | Kosmos 554 |
| 04/25 P | 11A57 | 483 | 11F690 (Zenit-2M) | Kosmos 555 |
| 05/05 P | 11A57 | 484 | 11F692 (Zenit-4MK) | Kosmos 556 |
| 05/18 P | 11A511U | 485 | 11F692 (Zenit-4MK) | Kosmos 559 |
| 05/23 P | 11A57 | 486 | 11F691 (Zenit-4M) | Kosmos 560 |
| 05/25 P | 11A57 | 487 | 11F690 (Zenit-2M) | Kosmos 561 |
| 05/29 P | 8A92M | 488 | 11F614 #25 (Meteor-1) | Meteor-1 15 |
| 06/06 P | 11A57 | 489 | 11F691 (Zenit-4M) | Kosmos 563 |
| 06/10 B | 11A57 | 490 | 11F691 (Zenit-4M) | Kosmos 572 |
| 06/15 B | 11A511 (S_5000-27) | 491 | 11F615A8 #36 (7K-T-4) | Kosmos 573 |
| 06/21 P | 11A57 | 492 | 11F690 (Zenit-2M) | Kosmos 575 |
| 06/27 P | 11A511M | 493 | 11F629 (Zenit-4MT/Orion) | Kosmos 576 |
| 07/04 P Failure | 11A57 | 494 | 11F691 (Zenit-4M) | |
| 07/11 P | 8K78M | 495 | 11F628 #6 (Molniya-2) | Molniya-2 6 |
| 07/25 P | 11A57 | 496 | 11F691 (Zenit-4M) | Kosmos 577 |
| 08/01 P | 11A57 | 497 | 11F690 (Zenit-2M) | Kosmos 578 |
| 08/21 P | 11A57 | 498 | 11F691 (Zenit-4M) | Kosmos 579 |
| 08/24 B | 11A57 | 499 | 11F691 (Zenit-4M) | Kosmos 581 |
| 08/30 P | 8K78M | 500 | 11F67 #29 (Molniya-1) | Molniya-1 24 |
| 08/30 B | 11A57 | 501 | 11F690 (Zenit-2M) | Kosmos 583 |
| 09/06 P | 11A57 | 502 | 11F691 (Zenit-4M) | Kosmos 584 |
| 09/21 P | 11A511U | 503 | 11F692 (Zenit-4MK) | Kosmos 587 |
| 09/27 B | 11A511 (15000-30) | 504 | 11F615A8 #37 (Soyuz) | Soyuz 12 |
| 10/03 P | 11A57 | 505 | 11F690 (Zenit-2M) | Kosmos 596 |
| 10/06 P | 11A57 | 506 | 11F692 (Zenit-4MK) | Kosmos 597 |
| 10/10 P | 11A57 | 507 | 11F691 (Zenit-4M) | Kosmos 598 |
| 10/15 B | 11A57 | 508 | 11F690 (Zenit-2M) | Kosmos 599 |

| Date | Rocket | Payload | No. | Name |
|---|---|---|---|---|
| 10/16 P | 11A57 | 11F691 (Zenit-4M) | 509 | Kosmos 600 |
| 10/19 P | 8K78M | 11F628 #7 (Molniya-2) | 510 | Molniya-2 7 |
| 10/20 P | 11A57 | 11F692 (Zenit-4MK) | 511 | Kosmos 602 |
| 10/27 P | 11A57 | 11F691 (Zenit-4M) | 512 | Kosmos 603 |
| 10/29 P | 8A92M | 11F619 (Tselina-D #5) | 513 | Kosmos 604 |
| 10/31 P | 11A511U | 12KS Bion-1 | 514 | Kosmos 605 |
| 11/02 P | 8K78M | 5V95 (US-K/Oko #2) | 515 | Kosmos 606 |
| 11/10 P | 11A57 | 11F692 (Zenit-4MK) | 516 | Kosmos 607 |
| 11/14 B | 8K78M | 11F67 #30 (Molniya-1) | 517 | Molniya-1 25 |
| 11/21 B | 11A57 | 11F691 (Zenit-4M) | 518 | Kosmos 609 |
| 11/28 P | 11A57 | 11F692 (Zenit-4MK) | 519 | Kosmos 612 |
| 11/30 B | 11A511 | 11F615A8 #34A (7K-T-6) | 520 | Kosmos 613 |
| 11/30 P | 8K78M | 11F67 #31 (Molniya-1M) | 521 | Molniya-1 26 |
| 12/17 P | 11A511M | 11F629 (Zenit-4MT/Orion) | 522 | Kosmos 616 |
| 12/18 B | 11A511 | 11F615A8 #33 (Soyuz) | 523 | Soyuz 13 |
| 12/21 P | 11A57 | 11F692 (Zenit-4MK) | 524 | Kosmos 625 |
| 12/25 P | 8K78M | 11F628 #8 (Molniya-2) | 525 | Molniya-2 8 |
| **1974:** | | | | |
| 01/24 P | 11A57 | 11F690 (Zenit-2M) | 526 | Kosmos 629 |
| 01/30 P | 11A57 | 11F692 (Zenit-4MK) | 527 | Kosmos 630 |
| 02/12 B | 11A57 | 11F692 (Zenit-4MK) | 528 | Kosmos 632 |
| 03/05 P | 8A92M | 11F614 #26 (Meteor-1) | 529 | Meteor-1 16 |
| 03/14 P | 11A57 | 11F690 (Zenit-2M) | 530 | Kosmos 635 |
| 03/20 B | 11A57 | 11F692 (Zenit-4MK) | 531 | Kosmos 636 |
| 04/03 B | 11A511U | 11F615A12 #71(7K-TM-1) | 532 | Kosmos 638 |
| 04/04 P | 11A57 | 11F692 (Zenit-4MK) | 533 | Kosmos 639 |
| 04/11 P | 11A57 | 11F690 (Zenit-2M) | 534 | Kosmos 640 |
| 04/12 B Failure | | 11F692 (Zenit-4MK) | 535 | |
| 04/20 P | 8K78M | 11F67 #32 (Molniya-1M) | 536 | Molniya-1 27 |
| 04/24 P | 8A92M | 11F614 #27 (Meteor-1) | 537 | Meteor-1 17 |
| 04/26 P | 8K78M | 11F628 #9 (Molniya-2) | 538 | Molniya-2 9 |
| 04/29 P | 11A57 | 11F692 (Zenit-4MK) | 539 | Kosmos 649 |

| No. | Date | Vehicle | Designation | Name |
|---|---|---|---|---|
| 540 | 05/15 B | 11A57? | 11F692 (Zenit-4MK) | Kosmos 652 |
| 541 | 05/15 P | 11A57? | 11F690 (Zenit-2M) | Kosmos 653 |
| 542 | 05/23 P Failure | 11A511U | 11F624 (Yantar-2K #1) | |
| 543 | 05/27 B | 11A511 (S15000-26) | 11F615A9 #61 (7K-T-8) | Kosmos 656 |
| 544 | 05/30 P | 11A57 | 11F692 (Zenit-4MK) | Kosmos 657 |
| 545 | 06/06 B | 11A57 | 11F690 (Zenit-2M) | Kosmos 658 |
| 546 | 06/13 P | 11A57 | 11F691 (Zenit-4M) | Kosmos 659 |
| 547 | 06/29 P | 11A511M | 11F629 (Zenit-4MT/Orion) | Kosmos 664 |
| 548 | 06/29 P | 8K78M | 5V95 (US-K/Oko #3) | Kosmos 665 |
| 549 | 07/03 B | 11A511 | 11F615A9 #62 (Soyuz) | Soyuz 14 |
| 550 | 07/09 P | 8A92M | 11F614ME #1 (Meteor-Priroda) | Meteor-1 18 |
| 551 | 07/12 P | 11A57 | 11F692 (Zenit-4MK) | Kosmos 666 |
| 552 | 07/23 P | 8K78M | 11F628 #10 (Molniya-2) | Molniya-2 10 |
| 553 | 07/25 B | 11A57 | 11F691 (Zenit-4M) | Kosmos 667 |
| 554 | 07/26 P | 11A57 | 11F690 (Zenit-2M) | Kosmos 669 |
| 555 | 08/06 B | 11A511U (K15000-11) | 11F732 #1L (7K-S-1) | Kosmos 670 |
| 556 | 08/07 P | 11A57 | 11F692 (Zenit-4MK) | Kosmos 671 |
| 557 | 08/12 B | 11A511U | 11F615A12 #72(7K-TM-2) | Kosmos 672 |
| 558 | 08/16 P | 8A92M | 11F619 (Tselina-D #6) | Kosmos 673 |
| 559 | 08/26 B | 11A511 | 11F615A9 #63 (Soyuz) | Soyuz 15 |
| 560 | 08/29 B | 11A57 | 11F692 (Zenit-4MK) | Kosmos 674 |
| 561 | 08/30 P Failure | 11A57 | 11F690 (Zenit-2M) | |
| 562 | 09/20 B | 11A57 | 11F690 (Zenit-2M) | Kosmos 685 |
| 563 | 10/18 P | 11A57 | 11F692 (Zenit-4MK) | Kosmos 688 |
| 564 | 10/22 P | 11A511U | 12KS Bion-2 | Kosmos 690 |
| 565 | 10/24 P | 8K78M | 11F67 #33 (Molniya-1M) | Molniya-1 28 |
| 566 | 10/25 B | 11A57 | 11F692 (Zenit-4MK) | Kosmos 691 |
| 567 | 10/28 P | 8A92M | 11F614 #28 (Meteor-1) | Meteor-1 19 |
| 568 | 11/01 P | 11A57 | 11F690 (Zenit-2M) | Kosmos 692 |
| 569 | 11/04 P | 11A511M | 11F629 (Zenit-4MT/Orion) | Kosmos 693 |
| 570 | 11/16 P | 11A57 | 11F692 (Zenit-4MK) | Kosmos 694 |
| 571 | 11/21 P | 8K78M | 11F637 #1 (Molniya-3) | Molniya-3 1 |

| Date | Rocket | No. | Designation | Name |
|---|---|---|---|---|
| 11/27 P | 11A57 | 572 | 11F690 (Zenit-2M) | Kosmos 696 |
| 12/02 B | 11A511U | 573 | 11F615A12 #73 (Soyuz) | Soyuz 16 |
| 12/13 P | 11A511U | 574 | 11F624 (Yantar-2K #2) | Kosmos 697 |
| 12/17 P | 8A92M | 575 | 11F614 #29 (Meteor-1) | Meteor-1 20 |
| 12/21 P | 8K78M | 576 | 11F628 #11 (Molniya-2) | Molniya-2 11 |
| 12/27 B | 11A57 | 577 | 11F692 (Zenit-4MK) | Kosmos 701 |
| **1975:** | | | | |
| 01/10 B | 11A511 (15000-31) | 578 | 11F615A8 #38 (Soyuz) | Soyuz 17 |
| 01/17 B | 11A57 | 579 | 11F690 (Zenit-2M) | Kosmos 702 |
| 01/23 P | 11A57 | 580 | 11F692 (Zenit-4MK) | Kosmos 704 |
| 01/30 P | 8K78M | 581 | 5V95 (US-K/Oko #4) | Kosmos 706 |
| 02/06 P | 8K78M | 582 | 11F628 #12 (Molniya-2) | Molniya-2 12 |
| 02/12 P | 11A57 | 583 | 11F692 (Zenit-4MK) | Kosmos 709 |
| 02/26 B | 11A57 | 584 | 11F692 (Zenit-4MK) | Kosmos 710 |
| 03/12 B | 11A57 | 585 | 11F692 (Zenit-4MK) | Kosmos 719 |
| 03/21 P | 11A511M | 586 | 11F629 (Zenit-4MT/Orion) | Kosmos 720 |
| 03/26 P | 11A57 | 587 | 11F690 (Zenit-2M) | Kosmos 721 |
| 03/27 B | 11A57 | 588 | 11F692 (Zenit-4MK) | Kosmos 722 |
| 04/01 P | 8A92M | 589 | 11F614 #30 (Meteor-1) | Meteor-1 21 |
| 04/05 B Failure | 11A511 (Kh15000-023) | 590 | 11F615A8 #39 (Soyuz) | Soyuz 18A |
| 04/14 P | 8K78M | 591 | 11F637 #2 (Molniya-3) | Molniya-3 2 |
| 04/16 B | 11A57 | 592 | 11F692 (Zenit-4MK) | Kosmos 727 |
| 04/18 P | 11A57 | 593 | 11F690 (Zenit-2M) | Kosmos 728 |
| 04/24 P | 11A57 | 594 | 11F692 (Zenit-4MK) | Kosmos 730 |
| 04/29 P | 8K78M | 595 | 11F67 #34 (Molniya-1M) | Molniya-1 29 |
| 05/21 B | 11A57 | 596 | 11F690 (Zenit-2M) | Kosmos 732 |
| 05/24 B | 11A511 | 597 | 11F615A8 #40 (Soyuz) | Soyuz 18 |
| 05/28 B | 11A57 | 598 | 11F692 (Zenit-4MK) | Kosmos 740 |
| 05/30 P | 11A57 | 599 | 11F690 (Zenit-2M/NX) | Kosmos 741 |
| 06/03 P | 11A57 | 600 | 11F692 (Zenit-4MK) | Kosmos 742 |
| 06/05 P | 8K78M | 601 | 11F67 #35 (Molniya-1M) | Molniya-1 30 |
| | | | SRET-2 | |

| No. | Date | Vehicle | Payload | Name |
|---|---|---|---|---|
| 602 | 06/12 P | 11A57 | 11F692 (Zenit-4MK) | Kosmos 743 |
| 603 | 06/20 P | 8A92M | 11F619 (Tselina-D #7) | Kosmos 744 |
| 604 | 06/25 P | 11A57 | 11F692 (Zenit-4MK) | Kosmos 746 |
| 605 | 06/27 P | 11A57 | 11F690 (Zenit-2M) | Kosmos 747 |
| 606 | 07/03 P | 11A57 | 11F692 (Zenit-4MK) | Kosmos 748 |
| 607 | 07/08 P | 8K78M | 11F628 #13 (Molniya-2) | Molniya-2 13 |
| 608 | 07/11 P | 8A92M | 11F632 #1 (Meteor-2) | Meteor-2 1 |
| 609 | 07/15 B | 11A511U | 11F615A12 #75 (Soyuz) | Soyuz 19 |
| 610 | 07/23 P | 11A57 | 11F690 (Zenit-2M) | Kosmos 751 |
| 611 | 07/31 P | 11A57 | 11F692 (Zenit-4MK) | Kosmos 753 |
| 612 | 08/13 B | 11A57 | 11F692 (Zenit-4MK) | Kosmos 754 |
| 613 | 08/22 P | 8A92M | 11F619 (Tselina-D #8) | Kosmos 756 |
| 614 | 08/27 P | 11A57 | 11F692 (Zenit-4MK) | Kosmos 757 |
| 615 | 09/02 P | 8K78M | 11F67 #36 (Molniya-1M) | Molniya-1 31 |
| 616 | 09/05 P | 11A511U | 11F624 (Yantar-2K #3) | Kosmos 758 |
| 617 | 09/09 P | 8K78M | 11F628 #14 (Molniya-2) | Molniya-2 14 |
| 618 | 09/12 P | 11A511M | 11F629 (Zenit-4MT/Orion) | Kosmos 759 |
| 619 | 09/16 B | 11A57 | 11F692 (Zenit-4MK) | Kosmos 760 |
| 620 | 09/18 P | 8A92M | 11F614 #31 (Meteor-1) | Meteor-1 22 |
| 621 | 09/23 P | 11A57 | 11F690 (Zenit-2M) | Kosmos 769 |
| 622 | 09/25 P | 11A511U | 11F635 (Zenit-4MKT/Fram) | Kosmos 771 |
| 623 | 09/29 B | 11A511U (F15000-21) | 11F732 #2L (7K-S-2) | Kosmos 772 |
| 624 | 10/01 B | 11A57 | 11F692 (Zenit-4MK) | Kosmos 774 |
| 625 | 10/17 P | 11A57 | 11F690 (Zenit-2M) | Kosmos 776 |
| 626 | 11/04 P | 11A57 | 11F692 (Zenit-4MK) | Kosmos 779 |
| 627 | 11/14 P | 8K78M | 11F637 #3 (Molniya-3) | Molniya-3 3 |
| 628 | 11/17 B | 11A511U | 11F615A9 #64 (Soyuz) | Soyuz 20 |
| 629 | 11/21 B | 11A57 | 11F690 (Zenit-2M) | Kosmos 780 |
| 630 | 11/25 B | 11A511U | 12KS Bion-3 | Kosmos 782 |
| 631 | 12/03 P | 11A57 | 11F690 (Zenit-2M) | Kosmos 784 |
| 632 | 12/16 B | 11A57 | 11F692 (Zenit-4MK) | Kosmos 786 |
| 633 | 12/17 P | 8K78M | 11F628 #15 (Molniya-2) | Molniya-2 15 |

| Date | Rocket | Designation | No. | Name |
|---|---|---|---|---|
| 12/22 B | 8K78M | SO-M #504 | 634 | Prognoz 4 |
| 12/25 P | 8A92M | 11F614 #32 (Meteor-1) | 635 | Meteor-1 23 |
| 12/27 P | 8K78M | 11F637 #4 (Molniya-3) | 636 | Molniya-3 4 |
| **1976:** | | | | |
| 01/07 P | 11A57 | 11F692 (Zenit-4MK) | 637 | Kosmos 788 |
| 01/22 B | 8K78M | 11F67 #37 (Molniya-1M) | 638 | Molniya-1 32 |
| 01/29 B | 11A57 | 11F690 (Zenit-2M) | 639 | Kosmos 799 |
| 02/11 P | 11A57 | 11F692 (Zenit-4MK) | 640 | Kosmos 802 |
| 02/20 P | 11A511U | 11F624 (Yantar-2K #4) | 641 | Kosmos 805 |
| 03/10 B | 11A57 | 11F692 (Zenit-4MK) | 642 | Kosmos 806 |
| 03/11 P | 8K78M | 11F67 #38 (Molniya-1M) | 643 | Molniya-1 33 |
| 03/16 P | 8A92M | 11F619 (Tselina-D #9) | 644 | Kosmos 808 |
| 03/18 B | 11A57 | 11F690 (Zenit-2M) | 645 | Kosmos 809 |
| 03/19 B | 8K78M | 11F67 #39 (Molniya-1M) | 646 | Molniya-1 34 |
| 03/26 P | 11A57 | 11F692 (Zenit-4MK) | 647 | Kosmos 810 |
| 03/31 P | 11A511M | 11F629 (Zenit-4MT/Orion) | 648 | Kosmos 811 |
| 04/07 P | 8A92M | 11F614 #33 (Meteor-1) | 649 | Meteor-1 24 |
| 04/09 P | 11A57 | 11F690 (Zenit-2M) | 650 | Kosmos 813 |
| 04/28 P | 11A57 | 11F692 (Zenit-4MK) | 651 | Kosmos 815 |
| 05/05 B | 11A57 | 11F692 (Zenit-4MK) | 652 | Kosmos 817 |
| 05/12 P | 8K78M | 11F637 #5 (Molniya-3) | 653 | Molniya-3 5 |
| 05/15 P | 8A92M | 11F651 #2-1 (Meteor-Priroda) | 654 | Meteor-1 25 |
| 05/20 B | 11A57 | 11F690 (Zenit-2M) | 655 | Kosmos 819 |
| 05/21 P | 11A511U | 11F635 (Zenit-4MKT/Fram) | 656 | Kosmos 820 |
| 05/26 P | 11A57 | 11F692 (Zenit-4MK) | 657 | Kosmos 821 |
| 06/08 B | 11A57 | 11F692 (Zenit-4MK) | 658 | Kosmos 824 |
| 06/16 P | 11A57 | 11F692 (Zenit-4MK) | 659 | Kosmos 833 |
| 06/24 P | 11A57 | 11F690 (Zenit-2M) | 660 | Kosmos 834 |
| 06/29 B | 11A57 | 11F692 (Zenit-4MK) | 661 | Kosmos 835 |
| 07/01 P Failure | 8K78M | 11F628 #16 (Molniya-2) | 662 | Kosmos 837 |
| 07/06 B | 11A511U | 11F615A8 #41 (Soyuz) | 663 | Soyuz 21 |
| 07/14 P | 11A57 | 11F690 (Zenit-2M) | 664 | Kosmos 840 |

| Date | Launcher | Payload | No. | Name |
|---|---|---|---|---|
| 07/22 P | 11A511U | 11F624 (Yantar-2K #5) | 665 | Kosmos 844 |
| 07/23 B | 8K78M | 11F67 #40 (Molniya-1M) | 666 | Molniya-1 35 |
| 08/04 P | 11A57 | 11F692 (Zenit-4MK) | 667 | Kosmos 847 |
| 08/12 P | 11A511U | 11F690 (Zenit-2M) | 668 | Kosmos 848 |
| 08/27 P | 8A92M | 11F619 (Tselina-D #10) | 669 | Kosmos 851 |
| 08/28 B | 11A511U | 11F692 (Zenit-4MK) | 670 | Kosmos 852 |
| 09/01 P Failure | 8K78M | 11F628 #17 (Molniya-2) | 671 | Kosmos 853 |
| 09/03 B | 11A511U | 11F692 (Zenit-4MK) | 672 | Kosmos 854 |
| 09/15 B | 11A511U | 11F615A12 #74 (Soyuz) | 673 | Soyuz 22 |
| 09/21 P | 11A511U | 11F629 (Zenit-4MT/Orion) | 674 | Kosmos 855 |
| 09/22 B | 11A511U | 11F690 (Zenit-2M) | 675 | Kosmos 856 |
| 09/24 P | 11A511U | 11F692 (Zenit-4MK) | 676 | Kosmos 857 |
| 10/04 P Failure | 11A511U | 11F635 (Zenit-4MKT/Fram) | 677 | |
| 10/10 B | 11A511U | 11F692 (Zenit-4MK) | 678 | Kosmos 859 |
| 10/14 B | 11A511U | 11F615A9 #65 (Soyuz) | 679 | Soyuz 23 |
| 10/16 P | 8A92M | 11F614 #34 (Meteor-1) | 680 | Meteor-1 26 |
| 10/22 P | 8K78M | 5V95 (US-K/Oko #5) | 681 | Kosmos 862 |
| 10/25 P | 11A511U | 11F692 (Zenit-4MK) | 682 | Kosmos 863 |
| 11/01 P | 11A511U | 11F690 (Zenit-2M) | 683 | Kosmos 865 |
| 11/11 B | 11A511U | 11F692 (Zenit-4MK) | 684 | Kosmos 866 |
| 11/23 P | 11A511U | 11F645 (Zenit-6/Argon) | 685 | Kosmos 867 |
| 11/25 B | 8K78M | SO-M #505 | 686 | Prognoz 5 |
| 11/29 B | 11A511U (Ye15000-071) | 11F732 #3L (7K-S-3) | 687 | Kosmos 869 |
| 12/02 P | 8K78M | 11F628 #18 (Molniya-2) | 688 | Molniya-2 16 |
| 12/09 P | 11A511U | 11F690 (Zenit-2M) | 689 | Kosmos 879 |
| 12/17 B | 11A511U | 11F692 (Zenit-4MK) | 690 | Kosmos 884 |
| 12/28 P | 8K78M | 11F637 #6 (Molniya-3) | 691 | Molniya-3 6 |
| **1977:** | | | | |
| 01/06 B | 11A511U | 11F692 (Zenit-4MK) | 692 | Kosmos 888 |
| 01/07 P | 8A92M | 11F632 #2 (Meteor-2) | 693 | Meteor-2 2 |
| 01/20 B | 11A511U | 11F690 (Zenit-2M) | 694 | Kosmos-889 |
| 02/07 B | 11A511U | 11F615A9 #66 (Soyuz) | 695 | Soyuz 24 |

| Date | Vehicle | Payload | No. | Name |
|---|---|---|---|---|
| 02/09 P | 11A511U | 11F692 (Zenit-4MK) | 696 | Kosmos 892 |
| 02/11 P | 8K78M | 11F628 #19 (Molniya-2) | 697 | Molniya-2 17 |
| 02/22 B Failure | 11A511U | 11F692 (Zenit-4MK) | 698 | |
| 02/27 P | 8A92M | 11F619 (Tselina-D #11) | 699 | Kosmos 895 |
| 03/03 P | 11A511U | 11F645 (Zenit-6/Argon) | 700 | Kosmos 896 |
| 03/10 P | 11A511U | 11F692 (Zenit-4MK) | 701 | Kosmos 897 |
| 03/17 P | 11A511U | 11F690 (Zenit-2M) | 702 | Kosmos 898 |
| 03/24 P | 8K78M | 11F67 #41 (Molniya-1M) | 703 | Molniya-1 36 |
| 04/05 P | 8A92M | 11F614 #35 (Meteor-1) | 704 | Meteor-1 27 |
| 04/07 P | 11A511U | 11F692 (Zenit-4MK) | 705 | Kosmos 902 |
| 04/11 P | 8K78M | 5V95 (US-K/Oko #6) | 706 | Kosmos 903 |
| 04/20 B | 11A511U | 11F690 (Zenit-2M) | 707 | Kosmos 904 |
| 04/26 P | 11A511U | 11F624 (Yantar-2K #6) | 708 | Kosmos 905 |
| 04/28 P | 8K78M | 11F637 #7 (Molniya-3) | 709 | Molniya-3 7 |
| 05/05 P | 11A511U | 11F692 (Zenit-4MK) | 710 | Kosmos 907 |
| 05/17 B | 11A511U | 11F692 (Zenit-4MK) | 711 | Kosmos 908 |
| 05/26 P | 11A511U | 11F635 (Zenit-4MKT/Fram) | 712 | Kosmos 912 |
| 05/31 B | 11A511U | 11F690 (Zenit-2M) | 713 | Kosmos 914 |
| 06/08 P | 11A511U | 11F692 (Zenit-4MK) | 714 | Kosmos 915 |
| 06/10 P | 11A511U | 11F629 (Zenit-4MT/Orion) | 715 | Kosmos 916 |
| 06/16 P | 8K78M | 5V95 (US-K/Oko #7) | 716 | Kosmos 917 |
| 06/22 B | 11A511U | 11F692 (Zenit-4MK) | 717 | Kosmos 920 |
| 06/24 B | 8K78M | 11F67 #42 (Molniya-1M) | 718 | Molniya-1 37 |
| 06/29 B | 8A92M | 11F651 #2-2 (Meteor-Priroda) | 719 | Meteor-1 28 |
| 06/30 P | 11A511U | 11F690 (Zenit-2M) | 720 | Kosmos 922 |
| 07/07 P | 8A92M | 11F619 (Tselina-D #12) | 721 | Kosmos 925 |
| 07/12 P | 11A511U | 11F692 (Zenit-4MK) | 722 | Kosmos 927 |
| 07/20 P | 8K78M | 5V95 (US-K/Oko #8) | 723 | Kosmos 931 |
| 07/20 B | 11A511U | 11F692 (Zenit-4MK) | 724 | Kosmos 932 |
| 07/27 P | 11A511U | 11F645 (Zenit-6/Argon) | 725 | Kosmos 934 |
| 07/29 P | 11A511U | 11F690 (Zenit-2M) | 726 | Kosmos 935 |
| 08/03 P | 11A511U | 12KS Bion-4 | 727 | Kosmos 936 |

| Date | Vehicle | Designation | No. | Payload |
|---|---|---|---|---|
| 08/10 B Failure | 11A511U | 11F692 (Zenit-4MK) | 728 | |
| 08/24 P | 11A511U | 11F692 (Zenit-4MK) | 729 | Kosmos 938 |
| 08/27 P | 11A511U | 11F690 (Zenit-2M) | 730 | Kosmos 947 |
| 08/30 P | 8K78M | 11F67 #43 (Molniya-1M) | 731 | Molniya-1 38 |
| 09/02 P | 11A511U | 11F635 (Zenit-4MKT/Fram) | 732 | Kosmos 948 |
| 09/06 P | 11A511U | 11F624 (Yantar-2K #7) | 733 | Kosmos 949 |
| 09/13 P | 11A511U | 11F690 (Zenit-2M) | 734 | Kosmos 950 |
| 09/16 P | 11A511U | 11F692M (Zenit-4MKM) | 735 | Kosmos 953 |
| 09/20 P | 8A92M | 11F619 (Tselina-D #13) | 736 | Kosmos 955 |
| 09/22 B | 8K78M | SO-M #506 | 737 | Prognoz 6 |
| 09/30 B | 11A511U | 11F692M (Zenit-4MKM) | 738 | Kosmos 957 |
| 10/09 B | 11A511U | 11F615A8 #42 (Soyuz) | 739 | Soyuz 25 |
| 10/11 P | 11A511U | 11F645 (Zenit-6/Argon) | 740 | Kosmos 958 |
| 10/28 P | 8K78M | 11F637 #8 (Molniya-3) | 741 | Molniya-3 8 |
| 12/04 P | 11A511U | 11F692M (Zenit-4MKM) | 742 | Kosmos 964 |
| 12/10 B | 11A511U | 11F615A8 #43 (Soyuz) | 743 | Soyuz 26 |
| 12/12 B | 11A511U | 11F690 (Zenit-2M) | 744 | Kosmos 966 |
| 12/14 P | 8A92M | 11F632 #3 (Meteor-2) | 745 | Meteor-2 3 |
| 12/20 P | 11A511U | 11F692M (Zenit-4MKM) | 746 | Kosmos 969 |
| 12/27 B | 11A511U | 11F690 (Zenit-2M) | 747 | Kosmos 973 |

**1978:**

| Date | Vehicle | Designation | No. | Payload |
|---|---|---|---|---|
| 01/06 P | 11A511U | 11F692 (Zenit-4MK) | 748 | Kosmos 974 |
| 01/10 B | 11A511U | 11F615A8 #44 (Soyuz) | 749 | Soyuz 27 |
| 01/10 P | 8A92M | 11F619 (Tselina-D #14) | 750 | Kosmos 975 |
| 01/13 P | 11A511U | 11F690 (Zenit-2M) | 751 | Kosmos 984 |
| 01/20 B | 11A511U (Ye15000-075) | 11F615A15 #102 (Progress) | 752 | Progress 1 |
| 01/24 P | 8K78M | 11F637 #9 (Molniya-3) | 753 | Molniya-3 9 |
| 01/24 B | 11A511U | 11F692M (Zenit-4MKM) | 754 | Kosmos 986 |
| 01/31 P | 11A511U | 11F692M (Zenit-4MKM) | 755 | Kosmos 987 |
| 02/08 P | 11A511U | 11F629 (Zenit-4MT/Orion) | 756 | Kosmos 988 |
| 02/14 B | 11A511U | 11F692M (Zenit-4MKM) | 757 | Kosmos 989 |
| 03/02 B | 11A511U | 11F615A8 #45 (Soyuz) | 758 | Soyuz 28 |

| No. | Date | Vehicle | Payload | Name |
|---|---|---|---|---|
| 759 | 03/02 P | 8K78M | 11F67 #44 (Molniya-1M) | Molniya-1 39 |
| 760 | 03/04 B | 11A511U | 11F690 (Zenit-2M) | Kosmos 992 |
| 761 | 03/10 P | 11A511U | 11F692M (Zenit-4MKM) | Kosmos 993 |
| 762 | 03/17 P | 11A511U | 11F690 (Zenit-2M) | Kosmos 995 |
| 763 | 03/30 B | 11A511U | 11F692M (Zenit-4MKM) | Kosmos 999 |
| 764 | 04/04 B | 11A511U (D15000-123) | 11F732 #4L (7K-S-4) | Kosmos 1001 |
| 765 | 04/06 B | 11A511U | 11F690 (Zenit-2M) | Kosmos 1002 |
| 766 | 04/20 P | 11A511U | 11F692M (Zenit-4MKM) | Kosmos 1003 |
| 767 | 05/05 P | 11A511U | 11F690 (Zenit-2M) | Kosmos 1004 |
| 768 | 05/12 P | 8A92M | 11F619 (Tselina-D #15) | Kosmos 1005 |
| 769 | 05/16 P | 11A511U | 11F692M (Zenit-4MKM) | Kosmos 1007 |
| 770 | 05/23 P | 11A511U | 11F635 (Zenit-4MKT/Fram) | Kosmos 1010 |
| 771 | 05/25 P | 11A511U | 11F690 (Zenit-2M) | Kosmos 1012 |
| 772 | 06/02 P | 8K78M | 11F67 #45 (Molniya-1M) | Molniya-1 40 |
| 773 | 06/10 B | 11A511U | 11F692M (Zenit-4MKM) | Kosmos 1021 |
| 774 | 06/12 P | 11A511U | 11F692M (Zenit-4MKM) | Kosmos 1022 |
| 775 | 06/15 B | 11A511U | 11F615A8 #46 (Soyuz) | Soyuz 29 |
| 776 | 06/27 B | 11A511U | 11F615A9 #67 (Soyuz) | Soyuz 30 |
| 777 | 06/28 P | 8K78M | 5V95 (US-K/Oko #9) | Kosmos 1024 |
| 778 | 07/02 B | 11A511U | 13KS Energiya #2 | Kosmos 1026 |
| 779 | 07/07 B | 11A511U (S15000-128) | 11F615A15 #101 (Progress) | Progress 2 |
| 780 | 07/14 P | 8K78M | 11F67 #46 (Molniya-1M) | Molniya-1 41 |
| 781 | 08/05 P | 11A511U | 11F624 (Yantar-2K #8) | Kosmos 1028 |
| 782 | 08/07 B | 11A511U | 11F615A15 #103 (Progress) | Progress 3 |
| 783 | 08/22 P | 8K78M | 11F67 #47 (Molniya-1M) | Molniya-1 42 |
| 784 | 08/26 B | 11A511U | 11F615A8 #47 (Soyuz) | Soyuz 31 |
| 785 | 08/29 P | 11A511U | 11F692M (Zenit-4MKM) | Kosmos 1029 |
| 786 | 09/06 P | 8K78M | 5V95 (US-K/Oko #10) | Kosmos 1030 |
| 787 | 09/09 P | 11A511U | 11F692M (Zenit-4MKM) | Kosmos 1031 |
| 788 | 09/19 P | 11A511U | 11F690 (Zenit-2M) | Kosmos 1032 |
| 789 | 10/03 P | 11A511U | 11F635 (Zenit-4MKT/Fram) | Kosmos 1033 |
| 790 | 10/03 B | 11A511U (Ye15000-152) | 11F615A15 #105 (Progress) | Progress 4 |

| Date | Vehicle | Number | Payload | Name |
|---|---|---|---|---|
| 10/06 P | 11A511U | 791 | 11F692M (Zenit-4MKM) | Kosmos 1042 |
| 10/10 P | 8A92M | 792 | 11F619 (Tselina-D #16) | Kosmos 1043 |
| 10/13 P | 8K78M | 793 | 11F637 #10 (Molniya-3) | Molniya-3 10 |
| 10/17 P | 11A511U | 794 | 11F690 (Zenit-2M) | Kosmos 1044 |
| 10/30 B | 8K78M | 795 | SO-M #507 | Prognoz 7 |
| 11/01 P | 11A511U | 796 | 11F629 (Zenit-4MT/Orion) | Kosmos 1046 |
| 11/15 P | 11A511U | 797 | 11F692M (Zenit-4MKM) | Kosmos 1047 |
| 11/21 P | 11A511U | 798 | 11F692M (Zenit-4MKM) | Kosmos 1049 |
| 11/28 P | 11A511U | 799 | 11F645 (Zenit-6/Argon) | Kosmos 1050 |
| 12/07 P | 11A511U | 800 | 11F692M (Zenit-4MKM) | Kosmos 1059 |
| 12/08 B | 11A511U | 801 | 11F690 (Zenit-2M) | Kosmos 1060 |
| 12/14 P | 11A511U | 802 | 11F690 (Zenit-2M) | Kosmos 1061 |
| 12/19 P | 8A92M | 803 | 11F619 (Tselina-D #17) | Kosmos 1063 |
| 12/23 P | 8A92M | 804 | 11F653 #1 Astrofyzika | Kosmos 1066 |
| 12/26 P | 11A511U | 805 | 11F692M (Zenit-4MKM) | Kosmos 1068 |
| 12/28 P | 11A511U | 806 | 11F629 (Zenit-4MT/Orion) | Kosmos 1069 |
| **1979:** | | | | |
| 01/11 P | 11A511U | 807 | 11F690 (Zenit-2M) | Kosmos 1070 |
| 01/13 P | 11A511U | 808 | 11F692M (Zenit-4MKM) | Kosmos 1071 |
| 01/18 P | 8K78M | 809 | 11F637 #11 (Molniya-3) | Molniya-3 11 |
| 01/25 B | 8A92M | 810 | 11F651 #2-3 (Meteor-Priroda) | Meteor-1 29 |
| 01/30 P | 11A511U | 811 | 11F692M (Zenit-4MKM) | Kosmos 1073 |
| 01/31 B | 11A511U (Ye15000-142) | 812 | 11F732 #5L (7K-ST-5) | Kosmos 1074 |
| 02/13 P | 8A92M | 813 | 11F619 (Tselina-D #18) | Kosmos 1077 |
| 02/16 P Failure | 11A511U | 814 | 11F690 (Zenit-2M) | |
| 02/22 P | 11A511U | 815 | 11F692M (Zenit-4MKM) | Kosmos 1078 |
| 02/25 B | 11A511U | 816 | 11F615A8 #48 (Soyuz) | Soyuz 32 |
| 02/27 P | 11A511U | 817 | 11F624 (Yantar-2K #9) | Kosmos 1079 |
| 03/01 P | 8A92M | 818 | 11F632 #4 (Meteor-2) | Meteor-2 4 |
| 03/12 B | 11A511U (Ye15000-162) | 819 | 11F615A15 #104 (Progress) | Progress 5 |
| 03/14 P | 11A511U | 820 | 11F692M (Zenit-4MKM) | Kosmos 1080 |
| 03/31 P | 11A511U | 821 | 11F690 (Zenit-2M) | Kosmos 1090 |

| Date | Vehicle | No. | Payload | Name |
|---|---|---|---|---|
| 04/10 B | 11A511U | 822 | 11F615A8 #49 (Soyuz) | Soyuz 33 |
| 04/12 P | 8K78M | 823 | 11F67 #48 (Molniya-1M) | Molniya-1 43 |
| 04/14 P | 8A92M | 824 | 11F619 (Tselina-D #19) | Kosmos 1093 |
| 04/20 P | 11A511U | 825 | 11F645 (Zenit-6/Argon) | Kosmos 1095 |
| 04/27 P | 11A511U | 826 | 11F693 (Yantar-4K1 #1) | Kosmos 1097 |
| 05/13 B | 11A511U (Zh15000-175) | 827 | 11F615A15 #106 (Progress) | Progress 6 |
| 05/15 P | 11A511U | 828 | 11F692M (Zenit-4MKM) | Kosmos 1098 |
| 05/17 P | 11A511U | 829 | 11F635 (Zenit-4MKT/Fram) | Kosmos 1099 |
| 05/25 P | 11A511U | 830 | 11F690 (Zenit-2M/NX) | Kosmos 1102 |
| 05/31 P | 11A511U | 831 | 11F645 (Zenit-6/Argon) | Kosmos 1103 |
| 06/05 P | 8K78M | 832 | 11F637 #12 (Molniya-3) | Molniya-3 12 |
| 06/06 B | 11A511U | 833 | 11F615A8 #50 (Soyuz) | Soyuz 34 |
| 06/08 P | 11A511U | 834 | 11F635 (Zenit-4MKT/Fram) | Kosmos 1105 |
| 06/12 P | 11A511U | 835 | 11F690 (Zenit-2M/NX) | Kosmos 1106 |
| 06/15 P | 11A511U | 836 | 11F645 (Zenit-6/Argon) | Kosmos 1107 |
| 06/22 P | 11A511U | 837 | 11F635 (Zenit-4MKT/Fram) | Kosmos 1108 |
| 06/27 P | 8K78M | 838 | 5V95 (US-K/Oko #11) | Kosmos 1109 |
| 06/28 B | 11A511U (Zh15000-192) | 839 | 11F615A15 #107 (Progress) | Progress 7 |
| 06/29 P | 11A511U | 840 | 11F645 (Zenit-6/Argon) | Kosmos 1111 |
| 07/10 B | 11A511U | 841 | 11F692M (Zenit-4MKM) | Kosmos 1113 |
| 07/13 P | 11A511U | 842 | 11F635 (Zenit-4MKT/Fram) | Kosmos 1115 |
| 07/20 P | 8A92M | 843 | 11F619 (Tselina-D #20) | Kosmos 1116 |
| 07/25 P | 11A511U | 844 | 11F692M (Zenit-4MKM) | Kosmos 1117 |
| 07/27 P | 11A511U | 845 | 11F690 (Zenit-2M/NX) | Kosmos 1118 |
| 07/31 P | 8K78M | 846 | 11F67 #49 (Molniya-1M) | Molniya-1 44 |
| 08/03 P | 11A511U | 847 | 11F629 (Zenit-4MT/Orion) | Kosmos 1119 |
| 08/11 B | 11A511U | 848 | 11F692M (Zenit-4MKM) | Kosmos 1120 |
| 08/14 P | 11A511U | 849 | 11F624 (Yantar-2K #9) | Kosmos 1121 |
| 08/17 P | 11A511U | 850 | 11F690 (Zenit-2M/NX) | Kosmos 1122 |
| 08/21 P | 11A511U | 851 | 11F635 (Zenit-4MKT/Fram) | Kosmos 1123 |
| 08/28 P | 8K78M | 852 | 5V95 (US-K/Oko #12) | Kosmos 1124 |
| 08/31 P | 11A511U | 853 | 11F645 (Zenit-6/Argon) | Kosmos 1126 |

| Date | Vehicle | Payload | No. | Name |
|---|---|---|---|---|
| 09/05 P | 11A511U | 17F41 Resurs-F1 #1 | 854 | Kosmos 1127 |
| 09/14 P | 11A511U | 11F692M (Zenit-4MKM) | 855 | Kosmos 1128 |
| 09/25 P | 11A511U | 12KS Bion-5 | 856 | Kosmos 1129 |
| 09/28 P | 11A511U | 11F645 (Zenit-6/Argon) | 857 | Kosmos 1138 |
| 10/05 P | 11A511U | 11F629 (Zenit-4MT/Orion) | 858 | Kosmos 1139 |
| 10/12 P Failure | 11A511U | 11F645 (Zenit-6/Argon) | 859 | |
| 10/20 P | 8K78M | 11F67 #50 (Molniya-1M) | 860 | Molniya-1 45 |
| 10/22 P | 11A511U | 11F645 (Zenit-6/Argon) | 861 | Kosmos 1142 |
| 10/26 P | 8A92M | 11F619 (Tselina-D #21) | 862 | Kosmos 1143 |
| 10/31 P | 8A92M | 11F632 #5 (Meteor-2) | 863 | Meteor-2 5 |
| 11/02 P | 11A511U | 11F624 (Yantar-2K #10) | 864 | Kosmos 1144 |
| 11/27 P | 8A92M | 11F619 (Tselina-D #22) | 865 | Kosmos 1145 |
| 12/12 P | 11A511U | 11F645 (Zenit-6/Argon) | 866 | Kosmos 1147 |
| 12/16 B | 11A511U (78018-912) | 11F732 #6L (Soyuz-T) | 867 | Soyuz-T 1 |
| 12/28 P | 11A511U | 11F692M (Zenit-4MKM) | 868 | Kosmos 1148 |
| **1980** | | | | |
| 01/09 P | 11A511U | 11F645 (Zenit-6/Argon) | 869 | Kosmos 1149 |
| 01/11 P | 8K78M | 11F67 #51 (Molniya-1M) | 870 | Molniya-1 46 |
| 01/24 P | 11A511U | 11F624 (Yantar-2K #11) | 871 | Kosmos 1152 |
| 01/30 P | 8A92M | 11F619 (Tselina-D #23) | 872 | Kosmos 1154 |
| 02/07 P | 11A511U | 11F645 (Zenit-6/Argon) | 873 | Kosmos 1155 |
| 02/12 P | 8K78M | 5V95 (US-K/Oko #13) | 874 | Kosmos 1164 |
| 02/21 P | 11A511U | 11F692M (Zenit-4MKM) | 875 | Kosmos 1165 |
| 03/04 P | 11A511U | 11F645 (Zenit-6/Argon) | 876 | Kosmos 1166 |
| 03/18 P Failure | 8A92M | 11F619 (Tselina-D #24) | 877 | |
| 03/27 B | 11A511U (Zh15000-200) | 11F615A15 #108 (Progress) | 878 | Progress 8 |
| 04/01 B | 11A511U | 11F692M (Zenit-4MKM) | 879 | Kosmos 1170 |
| 04/09 B | 11A511U | 11F615A8 #51 (Soyuz) | 880 | Soyuz 35 |
| 04/12 P | 8K78M | 5V95 (US-K/Oko #14) | 881 | Kosmos 1172 |
| 04/17 B | 11A511U | 11F692M (Zenit-4MKM) | 882 | Kosmos 1173 |
| 04/18 P Failure | 8K78M | 11F637 #13 (Molniya-3) | 883 | Kosmos 1175 |
| 04/27 B | 11A511U (Zh15000-210) | 11F615A15 #109 (Progress) | 884 | Progress 9 |

| Date | Vehicle | | No. | Payload | Name |
|---|---|---|---|---|---|
| 04/29 P | 11A511U | | 885 | 11F693 (Yantar-4K1 #2) | Kosmos 1177 |
| 05/07 P | 11A511U | | 886 | 11F645 (Zenit-6/Argon) | Kosmos 1178 |
| 05/15 P | 11A511U | | 887 | 11F629 (Zenit-4MT/Orion) | Kosmos 1180 |
| 05/23 P | 11A511U | | 888 | 11F635 (Zenit-4MKT/Fram) | Kosmos 1182 |
| 05/26 B | 11A511U | | 889 | 11F615A8 #52 (Soyuz) | Soyuz 36 |
| 05/28 P | 11A511U | | 890 | 11F645 (Zenit-6/Argon) | Kosmos 1183 |
| 06/04 P | 8A92M | | 891 | 11F619 (Tselina-D #25) | Kosmos 1184 |
| 06/05 B | 11A511U (Zh15000-196) | | 892 | 11F732 #7L (Soyuz-T) | Soyuz-T 2 |
| 06/06 P | 11A511U | | 893 | 17F41 Resurs-F1 #2 | Kosmos 1185 |
| 06/12 P | 11A511U | | 894 | 11F645 (Zenit-6/Argon) | Kosmos 1187 |
| 06/14 P | 8K78M | | 895 | 5V95 (US-K/Oko #15) | Kosmos 1188 |
| 06/18 B | 8A92M | | 896 | 11F651 #3-1 (Meteor-Priroda) | Meteor-1 30 |
| 06/21 P | 8K78M | | 897 | 11F67 #52 (Molniya-1M) | Molniya-1 47 |
| 06/26 P | 11A511U | | 898 | 11F645 (Zenit-6/Argon) | Kosmos 1189 |
| 06/29 B | 11A511U (P15000-232) | | 899 | 11F615A15 #110 (Progress) | Progress 10 |
| 07/02 P | 8K78M | | 900 | 5V95 (US-K/Oko #16) | Kosmos 1191 |
| 07/09 P | 11A511U | | 901 | 11F645 (Zenit-6/Argon) | Kosmos 1200 |
| 07/15 P | 11A511U | | 902 | 11F635 (Zenit-4MKT/Fram) | Kosmos 1201 |
| 07/18 P | 8K78M | | 903 | 11F637 #14 (Molniya-3) | Molniya-3 13 |
| 07/23 B | 11A511U | | 904 | 11F615A8 #53 (Soyuz) | Soyuz 37 |
| 07/24 P | 11A511U | | 905 | 11F645 (Zenit-6/Argon) | Kosmos 1202 |
| 07/31 P | 11A511U | | 906 | 17F41 Resurs-F1 #3 | Kosmos 1203 |
| 08/12 P | 11A511U | | 907 | 11F645 (Zenit-6/Argon) | Kosmos 1205 |
| 08/15 P | 8A92M | | 908 | 11F619 (Tselina-D #26) | Kosmos 1206 |
| 08/22 P | 11A511U | | 909 | 11F635 (Zenit-4MKT/Fram) | Kosmos 1207 |
| 08/26 P | 11A511U | | 910 | 11F624 (Yantar-2K #12) | Kosmos 1208 |
| 09/03 P | 11A511U | | 911 | 17F41 Resurs-F1 #4 | Kosmos 1209 |
| 09/09 P | 8A92M | | 912 | 11F632 #6 (Meteor-2) | Meteor-2 6 |
| 09/18 P | 11A511U | | 913 | 11F615A8 #54 (Soyuz) | Soyuz 38 |
| 09/19 P | 11A511U | | 914 | 11F645 (Zenit-6/Argon) | Kosmos 1210 |
| 09/23 P | 11A511U | | 915 | 11F629 (Zenit-4MT/Orion) | Kosmos 1211 |
| 09/26 P | 11A511U | | 916 | 11F635 (Zenit-4MKT/Fram) | Kosmos 1212 |

| No. | Date/Site | Vehicle | Payload | Designation |
|---|---|---|---|---|
| 917 | 09/28 B | 11A511U (P15000-219) | Progress 11 | 11F615A15 #111 (Progress) |
| 918 | 10/03 P | 11A511U | Kosmos 1213 | 11F645 (Zenit-6/Argon) |
| 919 | 10/10 P | 11A511U | Kosmos 1214 | 11F692M (Zenit-4MKM) |
| 920 | 10/16 P | 11A511U | Kosmos 1216 | 11F645 (Zenit-6/Argon) |
| 921 | 10/24 P | 8K78M | Kosmos 1217 | 5V95 (US-K/Oko #17) |
| 922 | 10/30 B | 11A511U | Kosmos 1218 | 11F693 (Yantar-4K1 #3) |
| 923 | 10/31 P | 11A511U | Kosmos 1219 | 11F645 (Zenit-6/Argon) |
| 924 | 11/12 P | 11A511U | Kosmos 1221 | 11F645 (Zenit-6/Argon) |
| 925 | 11/16 P | 8K78M | Molniya-1 48 | 11F67 #53 (Molniya-1M) |
| 926 | 11/21 P | 8A92M | Kosmos 1222 | 11F619 (Tselina-D #27) |
| 927 | 11/27 B | 11A511U | Soyuz-T 3 | 11F732 #8L (Soyuz-T) |
| 928 | 11/28 P | 8K78M | Kosmos 1223 | 5V95 (US-K/Oko #18) |
| 929 | 12/01 P | 11A511U | Kosmos 1224 | 11F645 (Zenit-6/Argon) |
| 930 | 12/16 P | 11A511U | Kosmos 1227 | 11F645 (Zenit-6/Argon) |
| 931 | 12/25 B | 8K78M | Prognoz 8 | SO-M #508 |
| 932 | 12/26 P | 11A511U | Kosmos 1236 | 11F624 (Yantar-2K #13) |

**1981**

| No. | Date/Site | Vehicle | Payload | Designation |
|---|---|---|---|---|
| 933 | 01/06 P | 11A511U | Kosmos 1237 | 11F645 (Zenit-6/Argon) |
| 934 | 01/09 P | 8K78M | Molniya-3 14 | 11F637 #15 (Molniya-3) |
| 935 | 01/16 P | 11A511U | Kosmos 1239 | 11F629 (Zenit-4MT/Orion) |
| 936 | 01/20 B | 11A511U | Kosmos 1240 | 11F624 (Yantar-2K #14) |
| 937 | 01/24 B | 11A511U (P15000-235) | Progress 12 | 11F615A15 #113 (Progress) |
| 938 | 01/27 P | 8A92M | Kosmos 1242 | 11F619 (Tselina-D #28) |
| 939 | 01/30 P | 8K78M | Molniya-1 49 | 11F67 #54 (Molniya-1M) |
| 940 | 02/13 P | 11A511U | Kosmos 1245 | 11F645 (Zenit-6/Argon) |
| 941 | 02/18 B | 11A511U | Kosmos 1246 | 11F660 (Yantar-1KFT #1) |
| 942 | 02/19 P | 8K78M | Kosmos 1247 | 5V95 (US-K/Oko #19) |
| 943 | 03/05 P | 11A511U | Kosmos 1248 | 11F624 (Yantar-2K #15) |
| 944 | 03/12 B | 11A511U | Soyuz-T 4 | 11F732 #10L (Soyuz-T) |
| 945 | 03/17 B | 11A511U | Kosmos 1259 | 11F645 (Zenit-6/Argon) |
| 946 | 03/22 B | 11A511U | Soyuz 39 | 11F615A8 #55 (Soyuz) |
| 947 | 03/24 P | 8K78M | Molniya-3 15 | 11F637 #16 (Molniya-3) |

| Date | Vehicle | No. | Payload | Satellite |
|---|---|---|---|---|
| 03/28 B Failure | 11A511U | 948 | 11F624 (Yantar-2K #16) | |
| 03/31 P | 8K78M | 949 | 5V95 (US-K/Oko #20) | Kosmos 1261 |
| 04/07 P | 11A511U | 950 | 11F645 (Zenit-6/Argon) | Kosmos 1262 |
| 04/15 B | 11A511U | 951 | 11F645 (Zenit-6/Argon) | Kosmos 1264 |
| 04/16 P | 11A511U | 952 | 11F645 (Zenit-6/Argon) | Kosmos 1265 |
| 04/28 B | 11A511U | 953 | | Kosmos 1268 |
| 05/14 B | 11A511U | 954 | 11F615A8 #56 (Soyuz) | Soyuz 40 |
| 05/14 P | 8A92M | 955 | 11F632 #7 (Meteor-2) | Meteor-2 7 |
| 05/18 B | 11A511U | 956 | 11F624 (Yantar-2K #17) | Kosmos 1270 |
| 05/19 P | 8A92M | 957 | 11F619 (Tselina-D #29) | Kosmos 1271 |
| 05/21 B | 11A511U | 958 | 11F645 (Zenit-6/Argon) | Kosmos 1272 |
| 05/22 P | 11A511U | 959 | 11F635 (Zenit-4MKT/Fram) | Kosmos 1273 |
| 06/03 P | 11A511U | 960 | 11F624 (Yantar-2K #18) | Kosmos 1274 |
| 06/09 P | 8K78M | 961 | 11F637 #17 (Molniya-3) | Molniya-3 16 |
| 06/16 P | 11A511U | 962 | 11F635 (Zenit-4MKT/Fram) | Kosmos 1276 |
| 06/17 B | 11A511U | 963 | 11F645 (Zenit-6/Argon) | Kosmos 1277 |
| 06/19 P | 8K78M | 964 | 5V95 (US-K/Oko #21) | Kosmos 1278 |
| 06/24 P | 8K78M | 965 | 11F67 #54 (Molniya-1M) | Molniya-1 50 |
| 07/01 B | 11A511U | 966 | 11F645 (Zenit-6/Argon) | Kosmos 1279 |
| 07/02 P | 11A511U | 967 | 17F41 Resurs-F1 #5 | Kosmos 1280 |
| 07/07 P | 11A511U | 968 | 11F645 (Zenit-6/Argon) | Kosmos 1281 |
| 07/10 B | 8A92M | 969 | 11F651 #2-4 (Meteor-Priroda) | Meteor-1 31 |
| 07/15 B | 11A511U | 970 | 11F624 (Yantar-2K #19) | Kosmos 1282 |
| 07/17 P | 11A511U | 971 | 11F645 (Zenit-6/Argon) | Kosmos 1283 |
| 07/29 P | 11A511U | 972 | 11F645 (Zenit-6/Argon) | Kosmos 1284 |
| 08/04 P | 8K78M | 973 | 5V95 (US-K/Oko #22) | Kosmos 1285 |
| 08/07 P | 8A92M | 974 | Interkosmos-22/Bulgaria-1300 | |
| 08/13 P | 11A511U | 975 | 11F624 (Yantar-2K #20) | Kosmos 1296 |
| 08/18 P | 11A511U | 976 | 11F645 (Zenit-6/Argon) | Kosmos 1297 |
| 08/21 B | 11A511U | 977 | 11F695 (Yantar-4K2 #1) | Kosmos 1298 |
| 08/27 P | 11A511U | 978 | 17F41 Resurs-F1 #6 | Kosmos 1301 |
| 09/04 B | 11A511U (78015-941) | 979 | 11F645 (Zenit-6/Argon) | Kosmos 1303 |

| Date | Vehicle | Payload | Designation |
|---|---|---|---|
| 09/11 P Failure | 8K78M | 11F637 #18 (Molniya-3) | Kosmos 1305 |
| 09/15 P | 11A511U | 11F645 (Zenit-6/Argon) | Kosmos 1307 |
| 09/18 P | 11A511U | 11F629 (Zenit-4MT/Orion) | Kosmos 1309 |
| 10/01 B | 11A511U | 11F645 (Zenit-6/Argon) | Kosmos 1313 |
| 10/09 P | 11A511U | 11F635 (Zenit-4MKT/Fram) | Kosmos 1314 |
| 10/14 P | 8A92M | 11F619 (Tselina-D #30) | Kosmos 1315 |
| 10/15 B | 11A511U | 11F645 (Zenit-6/Argon) | Kosmos 1316 |
| 10/17 P | 8K78M | 11F637 #19 (Molniya-3) | Molniya-3 17 |
| 11/01 P | 8K78M | 5V95 (US-K/Oko #23) | Kosmos 1317 |
| 11/03 P | 11A511U | 11F624 (Yantar-2K #21) | Kosmos 1318 |
| 11/13 B | 11A511U | 11F645 (Zenit-6/Argon) | Kosmos 1319 |
| 11/17 P | 8K78M | 11F67 #55 (Molniya-1M) | Molniya-1 51 |
| 12/04 B | 11A511U | 11F645 (Zenit-6/Argon) | Kosmos 1329 |
| 12/19 B | 11A511U | 11F624 (Yantar-2K #22) | Kosmos 1330 |
| 12/23 B | 8K78M | 11F67 #56 (Molniya-1M) | Molniya-1 52 |

**1982**

| Date | Vehicle | Payload | Designation |
|---|---|---|---|
| 01/12 P | 11A511U | 11F629 (Zenit-4MT/Orion) | Kosmos 1332 |
| 01/20 P | 11A511U | 11F645 (Zenit-6/Argon) | Kosmos 1334 |
| 01/30 B | 11A511U | 11F624 (Yantar-2K #23) | Kosmos 1336 |
| 02/16 P | 11A511U | 11F645 (Zenit-6/Argon) | Kosmos 1338 |
| 02/19 P | 8A92M | 11F619 (Tselina-D #31) | Kosmos 1340 |
| 02/26 P | 8K78M | 11F67 #57 (Molniya-1M) | Molniya-1 53 |
| 03/03 P | 8K78M | 5V95 (US-K/Oko #24) | Kosmos 1341 |
| 03/05 P | 11A511U | 11F645 (Zenit-6/Argon) | Kosmos 1342 |
| 03/17 P | 11A511U | 11F645 (Zenit-6/Argon) | Kosmos 1343 |
| 03/24 P | 8K78M | 11F637 #20 (Molniya-3) | Molniya-3 18 |
| 03/31 P | 8A92M | 11F619 (Tselina-D #32) | Kosmos 1346 |
| 04/02 B | 11A511U | 11F695 (Yantar-4K2 #2) | Kosmos 1347 |
| 04/07 P | 8K78M | 5V95 (US-K/Oko #25) | Kosmos 1348 |
| 04/15 P | 11A511U | 11F624 (Yantar-2K #24) | Kosmos 1350 |
| 04/21 B | 11A511U | 11F645 (Zenit-6/Argon) | Kosmos 1352 |
| 04/23 P | 11A511U | 11F635 (Zenit-4MKT/Fram) | Kosmos 1353 |

| Date | Vehicle | Note | No. | Payload | Name |
|---|---|---|---|---|---|
| 05/05 P | 8A92M | | 1011 | 11F619 (Tselina-D #33) | Kosmos 1356 |
| 05/13 B | 11A511U | | 1012 | 11F732 #11L (Soyuz-T) | Soyuz-T 5 |
| 05/15 P Failure | 11A511U | | 1013 | 11F645 (Zenit-6/Argon) | |
| 05/20 P | 8K78M | | 1014 | 5V95 (US-K/Oko #26) | Kosmos 1367 |
| 05/21 B | 11A511U | | 1015 | 11F645 (Zenit-6/Argon) | Kosmos 1368 |
| 05/23 B | 11A511U | (Ts15000-283) | 1016 | 11F615A15 #114 (Progress) | Progress 13 |
| 05/25 P | 11A511U | | 1017 | 17F41 Resurs-F1 #7 | Kosmos 1369 |
| 05/28 B | 11A511U | | 1018 | 11F660 (Yantar-1KFT #2) | Kosmos 1370 |
| 05/28 P | 8K78M | | 1019 | 11F67 #58 (Molniya-1M) | Molniya-1 54 |
| 06/02 B | 11A511U | | 1020 | 11F645 (Zenit-6/Argon) | Kosmos 1373 |
| 06/08 P | 11A511U | | 1021 | 17F41 Resurs-F1 #8 | Kosmos 1376 |
| 06/08 B | 11A511U | | 1022 | 11F693 (Yantar-4K1 #4) | Kosmos 1377 |
| 06/12 B Failure | 11A511U | | 1023 | 11F624 (Yantar-2K #25) | |
| 06/18 B | 11A511U | | 1024 | 11F645 (Zenit-6/Argon) | Kosmos 1381 |
| 06/24 B | 11A511U | | 1025 | 11F732 #9L (Soyuz-T) | Soyuz-T 6 |
| 06/25 P | 8K78M | | 1026 | 5V95 (US-K/Oko #27) | Kosmos 1382 |
| 06/30 P | 11A511U | | 1027 | 11F624 (Yantar-2K #26) | Kosmos 1384 |
| 07/06 P | 11A511U | | 1028 | 11F645 (Zenit-6/Argon) | Kosmos 1385 |
| 07/10 B | 11A511U | (Shch15000-318) | 1029 | 11F615A15 #117 (Progress) | Progress 14 |
| 07/13 P | 11A511U | | 1030 | 11F635 (Zenit-4MKT/Fram) | Kosmos 1387 |
| 07/21 B | 8K78M | | 1031 | 11F67 #59 (Molniya-1M) | Molniya-1 55 |
| 07/27 P | 11A511U | | 1032 | 11F645 (Zenit-6/Argon) | Kosmos 1396 |
| 08/03 P | 11A511U | | 1033 | 11F629 (Zenit-4MT/Orion) | Kosmos 1398 |
| 08/04 B | 11A511U | | 1034 | 11F693 (Yantar-4K1 #5) | Kosmos 1399 |
| 08/05 P | 8A92M | | 1035 | 11F619 (Tselina-D #34) | Kosmos 1400 |
| 08/19 B | 11A511U | | 1036 | 11F732 #12L (Soyuz-T) | Soyuz-T 7 |
| 08/20 P | 11A511U | | 1037 | 17F41 Resurs-F1 #9 | Kosmos 1401 |
| 08/27 P | 8K78M | | 1038 | 11F637 #21 (Molniya-3) | Molniya-3 19 |
| 09/01 B | 11A511U | | 1039 | 11F645 (Zenit-6/Argon) | Kosmos 1403 |
| 09/01 P | 11A511U | | 1040 | 11F645 (Zenit-6/Argon) | Kosmos 1404 |
| 09/08 P | 11A511U | | 1041 | 11F635 (Zenit-4MKT/Fram) | Kosmos 1406 |
| 09/15 P | 11A511U | | 1042 | 11F624 (Yantar-2K #27) | Kosmos 1407 |

| Date | Vehicle | Payload | No. | Name |
|---|---|---|---|---|
| 09/18 B | 11A511U (Ts15000-292) | 11F615A15 #112 (Progress) | 1043 | Progress 15 |
| 09/22 P | 8K78M | 5V95 (US-K/Oko #28) | 1044 | Kosmos 1409 |
| 09/30 P | 11A511U | 11F645 (Zenit-6/Argon) | 1045 | Kosmos 1411 |
| 10/14 B | 11A511U | 11F645 (Zenit-6/Argon) | 1046 | Kosmos 1416 |
| 10/31 B | 11A511U (Shch15000-335) | 11F615A15 #115 (Progress) | 1047 | Progress 16 |
| 11/02 B | 11A511U | 11F645 (Zenit-6/Argon) | 1048 | Kosmos 1419 |
| 11/18 B | 11A511U | 11F645 (Zenit-6/Argon) | 1049 | Kosmos 1421 |
| 12/03 P | 11A511U | 11F645 (Zenit-6/Argon) | 1050 | Kosmos 1422 |
| 12/08 B Failure | 8K78M | 11F67 #60 (Molniya-1M) | 1051 | Kosmos 1423 |
| 12/14 P | 8A92M | 11F632 #9 (Meteor-2) | 1052 | Meteor-2 9 |
| 12/16 B | 11A511U | 11F693 (Yantar-4K1 #6) | 1053 | Kosmos 1424 |
| 12/23 B | 11A511U (77024-777) | 11F645 (Zenit-6/Argon) | 1054 | Kosmos 1425 |
| 12/28 B | 11A511U2 | 11F694 (Yantar-4KS1 #1) | 1055 | Kosmos 1426 |

## 1983

| Date | Vehicle | Payload | No. | Name |
|---|---|---|---|---|
| 01/20 P | 8A92M | 11F619 (Tselina-D #35) | 1056 | Kosmos 1437 |
| 01/27 B | 11A511U | 11F645 (Zenit-6/Argon) | 1057 | Kosmos 1438 |
| 02/06 B | 11A511U | 11F624 (Yantar-2K #28) | 1058 | Kosmos 1439 |
| 02/10 P | 11A511U | 17F41 Resurs-F1 #10 | 1059 | Kosmos 1440 |
| 02/16 P | 8A92M | 11F619 (Tselina-D #36) | 1060 | Kosmos 1441 |
| 02/25 P | 11A511U | 11F693 (Yantar-4K1 #7) | 1061 | Kosmos 1442 |
| 03/02 P | 11A511U | 11F645 (Zenit-6/Argon) | 1062 | Kosmos 1444 |
| 03/11 P | 8K78M | 11F637 #22 (Molniya-3) | 1063 | Molniya-3 20 |
| 03/16 B | 11A511U (77024-778) | 11F645 (Zenit-6/Argon) | 1064 | Kosmos 1446 |
| 03/16 P | 8K78M | 11F67 #61 (Molniya-1M) | 1065 | Molniya-1 56 |
| 03/31 P | 11A511U | 11F645 (Zenit-6/Argon) | 1066 | Kosmos 1449 |
| 04/02 B | 8K78M | 11F658 #1 (Molniya-1T) | 1067 | Molniya-1 57 |
| 04/08 P | 11A511U | 11F645 (Zenit-6/Argon) | 1068 | Kosmos 1451 |
| 04/20 B | 11A511U (15000-372) | 11F732 #13L (Soyuz-T) | 1069 | Soyuz-T 8 |
| 04/22 P | 11A511U | 11F624 (Yantar-2K #29) | 1070 | Kosmos 1454 |
| 04/25 P | 8K78M | 5V95 (US-K/Oko #29) | 1071 | Kosmos 1456 |
| 04/26 B | 11A511U | 11F693 (Yantar-4K1 #8) | 1072 | Kosmos 1457 |
| 04/28 P | 11A511U | 11F635 (Zenit-4MKT/Fram) | 1073 | Kosmos 1458 |

| Date | Vehicle | Payload | Serial | Name |
|---|---|---|---|---|
| 05/06 B | 11A511U | 11F645 (Zenit-6/Argon) | 1074 | Kosmos 1460 |
| 05/17 P | 11A511U | 17F41 Resurs-F1 #11 | 1075 | Kosmos 1462 |
| 05/26 B | 11A511U | 11F693 (Yantar-4K1 #9) | 1076 | Kosmos 1466 |
| 05/31 P | 11A511U | 11F645 (Zenit-6/Argon) | 1077 | Kosmos 1467 |
| 06/07 P | 11A511U | 17F41 Resurs-F1 #12 | 1078 | Kosmos 1468 |
| 06/14 P | 11A511U | 11F645 (Zenit-6/Argon) | 1079 | Kosmos 1469 |
| 06/27 B | 11A511U (15000-379) | 11F732 #14L (Soyuz-T) | 1080 | Soyuz-T 9 |
| 06/28 P | 11A511U | 11F624 (Yantar-2K #30) | 1081 | Kosmos 1471 |
| 07/01 B | 8K78M | SO-M #509 | 1082 | Prognoz 9 |
| 07/05 P | 11A511U | 11F645 (Zenit-6/Argon) | 1083 | Kosmos 1472 |
| 07/08 P | 8K78M | 5V95 (US-K/Oko #30) | 1084 | Kosmos 1481 |
| 07/13 B | 11A511U | 11F645 (Zenit-6/Argon) | 1085 | Kosmos 1482 |
| 07/19 B | 8K78M | 11F67 #62 (Molniya-1M) | 1086 | Molniya-1 58 |
| 07/20 P | 11A511U | 17F41 Resurs-F1 #13 | 1087 | Kosmos 1483 |
| 07/24 B | 8A92M | 11F651 #3-2 (Resurs-OYe) | 1088 | Kosmos 1484 |
| 07/26 P | 11A511U | 11F645 (Zenit-6/Argon) | 1089 | Kosmos 1485 |
| 08/05 P | 11A511U | 17F41 Resurs-F1 #14 | 1090 | Kosmos 1487 |
| 08/09 P | 11A511U | 11F645 (Zenit-6/Argon) | 1091 | Kosmos 1488 |
| 08/10 B | 11A511U | 11F693 (Yantar-4K1 #10) | 1092 | Kosmos 1489 |
| 08/17 B | 11A511U (Ts15000-302) | 11F615A15 #119 (Progress) | 1093 | Progress 17 |
| 08/23 P | 11A511U | 11F645 (Zenit-6/Argon) | 1094 | Kosmos 1493 |
| 08/30 P | 8K78M | 11F637 #23 (Molniya-3) | 1095 | Molniya-3 21 |
| 09/03 P | 11A511U | 11F635 (Zenit-4MKT/Fram) | 1096 | Kosmos 1495 |
| 09/07 P | 11A511U | 11F693 (Yantar-4K1 #11) | 1097 | Kosmos 1496 |
| 09/09 P | 11A511U | 11F645 (Zenit-6/Argon) | 1098 | Kosmos 1497 |
| 09/14 P | 11A511U | 17F41 Resurs-F1 #15 | 1099 | Kosmos 1498 |
| 09/17 P | 11A511U | 11F645 (Zenit-6/Argon) | 1100 | Kosmos 1499 |
| 09/26 B Failure | 11A511U (Yu15000-363) | 11F732 #16L (Soyuz-T) | 1101 | Soyuz-T 10A |
| 10/14 B | 11A511U | 11F695 (Yantar-4K2 #3) | 1102 | Kosmos 1504 |
| 10/20 B | 11A511U (Ts15000-287) | 11F615A15 #118 (Progress) | 1103 | Progress 18 |
| 10/21 P | 11A511U | 11F645 (Zenit-6/Argon) | 1104 | Kosmos 1505 |
| 10/28 P | 8A92M | 11F632 #10 (Meteor-2) | 1105 | Meteor-2 10 |

| Date | Rocket | Payload | No. | Name |
|---|---|---|---|---|
| 11/17 P | 11A511U | 11F645 (Zenit-6/Argon) | 1106 | Kosmos 1509 |
| 11/23 P | 8K78M | 11F67 #63 (Molniya-1M) | 1107 | Molniya-1 59 |
| 11/30 P | 11A511U | 11F693 (Yantar-4K1 #12) | 1108 | Kosmos 1511 |
| 12/07 P | 11A511U | 11F645 (Zenit-6/Argon) | 1109 | Kosmos 1512 |
| 12/14 P | 11A511U | 12KS Bion-6 | 1110 | Kosmos 1514 |
| 12/21 P | 8K78M | 11F637 #24 (Molniya-3) | 1111 | Molniya-3 22 |
| 12/27 B | 11A511U | 11F660 (Yantar-1KFT #3) | 1112 | Kosmos 1516 |
| 12/28 P | 8K78M | 5V95 (US-K/Oko #31) | 1113 | Kosmos 1518 |
| **1984** | | | | |
| 01/11 P | 11A511U | 11F645 (Zenit-6/Argon) | 1114 | Kosmos 1530 |
| 01/13 P | 11A511U | 11F693 (Yantar-4K1 #13) | 1115 | Kosmos 1532 |
| 01/26 B | 11A511U (76032-791) | 11F645 (Zenit-6/Argon) | 1116 | Kosmos 1533 |
| 02/08 B | 11A511U | 11F732 #15L (Soyuz-T) | 1117 | Soyuz-T 10 |
| 02/16 P | 11A511U | 17F41 Resurs-F1 #16 | 1118 | Kosmos 1537 |
| 02/21 B | 11A511U | 11F615A15 #120 (Progress) | 1119 | Progress 19 |
| 02/28 P | 11A511U | 11F693 (Yantar-4K1 #14) | 1120 | Kosmos 1539 |
| 03/06 P | 8K78M | 5V95 (US-K/Oko #32) | 1121 | Kosmos 1541 |
| 03/07 B | 11A511U (76032-792) | 11F645 (Zenit-6/Argon) | 1122 | Kosmos 1542 |
| 03/10 P | 11A511U | 36KS Efir #1 | 1123 | Kosmos 1543 |
| 03/16 P | 8K78M | 11F67 #64 (Molniya-1M) | 1124 | Molniya-1 60 |
| 03/21 P | 11A511U | 11F645 (Zenit-6/Argon) | 1125 | Kosmos 1545 |
| 04/03 B | 11A511U | 11F732 #17L (Soyuz-T) | 1126 | Soyuz-T 11 |
| 04/04 P | 8K78M | 5V95 (US-K/Oko #33) | 1127 | Kosmos 1547 |
| 04/10 P | 11A511U | 11F693 (Yantar-4K1 #15) | 1128 | Kosmos 1548 |
| 04/15 B | 11A511U (Yu15000-005) | 11F615A15 #121 (Progress) | 1129 | Progress 20 |
| 04/19 P | 11A511U | 11F645 (Zenit-6/Argon) | 1130 | Kosmos 1549 |
| 05/07 B | 11A511U | 11F615A15 #116 (Progress) | 1131 | Progress 21 |
| 05/11 P | 11A511U | 11F645 (Zenit-6/Argon) | 1132 | Kosmos 1551 |
| 05/14 B | 11A511U2 | 11F694 (Yantar-4KS1 #2) | 1133 | Kosmos 1552 |
| 05/22 P | 11A511U | 11F635 (Zenit-4MKT/Fram) | 1134 | Kosmos 1557 |
| 05/25 P | 11A511U | 11F693 (Yantar-4K1 #16) | 1135 | Kosmos 1558 |
| 05/28 B | 11A511U | 11F615A15 #122 (Progress) | 1136 | Progress 22 |

| | | | | |
|---|---|---|---|---|
| 06/01 P | 11A511U | 1137 | 11F645 (Zenit-6/Argon) | Kosmos 1568 |
| 06/06 P | 8K78M | 1138 | 5V95 (US-K/Oko #34) | Kosmos 1569 |
| 06/11 B | 11A511U | 1139 | 17F116 (Zenit-8/Oblik) | Kosmos 1571 |
| 06/15 P | 11A511U | 1140 | 17F41 Resurs-F1 #17 | Kosmos 1572 |
| 06/19 P | 11A511U | 1141 | 11F645 (Zenit-6/Argon) | Kosmos 1573 |
| 06/22 P | 11A511U | 1142 | 17F41 Resurs-F1 #18 | Kosmos 1575 |
| 06/26 P | 11A511U | 1143 | 11F695 (Yantar-4K2 #4) | Kosmos 1576 |
| 06/29 P | 11A511U | 1144 | 17F116 (Zenit-8/Oblik) | Kosmos 1580 |
| 07/03 P | 8K78M | 1145 | 5V95 (US-K/Oko #35) | Kosmos 1581 |
| 07/17 B | 11A511U (Yu15000-006) | 1146 | 11F732 #18L (Soyuz-T) | Soyuz-T 12 |
| 07/19 P | 11A511U | 1147 | 17F41 Resurs-F1 #19 | Kosmos 1582 |
| 07/24 P | 11A511U | 1148 | 17F116 (Zenit-8/Oblik) | Kosmos 1583 |
| 07/27 P | 11A511U | 1149 | 17F116 (Zenit-8/Oblik) | Kosmos 1584 |
| 07/31 B | 11A511U | 1150 | 11F695 (Yantar-4K2 #5) | Kosmos 1585 |
| 08/02 P | 8K78M | 1151 | 5V95 (US-K/Oko #36) | Kosmos 1586 |
| 08/06 P | 11A511U | 1152 | 17F116 (Zenit-8/Oblik) | Kosmos 1587 |
| 08/10 P | 8K78M | 1153 | 11F67 #65 (Molniya-1M) | Molniya-1 61 |
| 08/14 B | 11A511U (15000-711) | 1154 | 11F615A15 #124 (Progress) | Progress 23 |
| 08/16 P | 11A511U | 1155 | 17F41 Resurs-F1 #20 | Kosmos 1590 |
| 08/24 P | 8K78M | 1156 | 11F67 #66 (Molniya-1M) | Molniya-1 62 |
| 08/30 P | 11A511U | 1157 | 17F41 Resurs-F1 #21 | Kosmos 1591 |
| 09/04 P | 11A511U | 1158 | 17F116 (Zenit-8/Oblik) | Kosmos 1592 |
| 09/07 P | 8K78M | 1159 | 5V95 (US-K/Oko #37) | Kosmos 1596 |
| 09/13 P | 11A511U | 1160 | 11F635 (Zenit-4MKT/Fram) | Kosmos 1597 |
| 09/25 P | 11A511U | 1161 | 11F695 (Yantar-4K2 #6) | Kosmos 1599 |
| 09/27 B | 11A511U | 1162 | 17F116 (Zenit-8/Oblik) | Kosmos 1600 |
| 10/04 P | 8K78M | 1163 | 5V95 (US-K/Oko #38) | Kosmos 1604 |
| 11/14 B | 11A511U | 1164 | 11F660 (Yantar-1KFT #4) | Kosmos 1608 |
| 11/14 B | 11A511U | 1165 | 17F116 (Zenit-8/Oblik) | Kosmos 1609 |
| 11/21 B | 11A511U | 1166 | 11F695 (Yantar-4K2 #7) | Kosmos 1611 |
| 11/29 P | 11A511U | 1167 | 17F116 (Zenit-8/Oblik) | Kosmos 1613 |
| 12/14 P | 8K78M | 1168 | 11F67 #67 (Molniya-1M) | Molniya-1 63 |

**1985**

| Date | Config | Payload | No. | Name |
|---|---|---|---|---|
| 01/09 B | 11A511U | 11F695 (Yantar-4K2 #8) | 1169 | Kosmos 1616 |
| 01/16 P | 8K78M | 11F637 #25 (Molniya-3) | 1170 | Molniya-3 23 |
| 01/16 B | 11A511U | 17F116 (Zenit-8/Oblik) | 1171 | Kosmos 1623 |
| 02/06 P | 11A511U | 17F116 (Zenit-8/Oblik) | 1172 | Kosmos 1628 |
| 02/27 B | 11A511U | 11F695 (Yantar-4K2 #9) | 1173 | Kosmos 1630 |
| 03/01 P | 11A511U | 17F116 (Zenit-8/Oblik) | 1174 | Kosmos 1632 |
| 03/25 B | 11A511U2 | 11F694 (Yantar-4KS1 #3) | 1175 | Kosmos 1643 |
| 04/03 B | 11A511U | 17F116 (Zenit-8/Oblik) | 1176 | Kosmos 1644 |
| 04/16 P | 11A511U | 34KS Foton #1 | 1177 | Kosmos 1645 |
| 04/19 P | 11A511U | 11F695 (Yantar-4K2 #10) | 1178 | Kosmos 1647 |
| 04/25 P | 11A511U | 17F116 (Zenit-8/Oblik) | 1179 | Kosmos 1648 |
| 04/26 B | 8K78M | SO-M #510 | 1180 | Prognoz 10 |
| 05/15 P | 11A511U | 17F116 (Zenit-8/Oblik) | 1181 | Kosmos 1649 |
| 05/22 P | 11A511U | 17F41 Resurs-F1 #22 | 1182 | Kosmos 1653 |
| 05/23 B | 11A511U | 11F695 (Yantar-4K2 #11) | 1183 | Kosmos 1654 |
| 05/29 P | 8K78M | 11F637 #26 (Molniya-3) | 1184 | Molniya-3 24 |
| 06/06 B | 11A511U2 (B15000-008) | 11F732 #19L (Soyuz) | 1185 | Soyuz-T 13 |
| 06/07 P | 11A511U | 17F41 Resurs-F1 #23 | 1186 | Kosmos 1657 |
| 06/11 P | 8K78M | 5V95 (US-K/Oko #39) | 1187 | Kosmos 1658 |
| 06/13 P | 11A511U | 17F116 (Zenit-8/Oblik) | 1188 | Kosmos 1659 |
| 06/18 P | 8K78M | 5V95 (US-K/Oko #40) | 1189 | Kosmos 1661 |
| 06/21 B | 11A511U (15000-417) | 11F615A15 #125 (Progress) | 1190 | Progress 24 |
| 06/21 P | 11A511U | 17F41 Resurs-F1 #24 | 1191 | Kosmos 1663 |
| 06/26 P | 11A511U | 17F116 (Zenit-8/Oblik) | 1192 | Kosmos 1664 |
| 07/03 P | 11A511U | 17F116 (Zenit-8/Oblik) | 1193 | Kosmos 1665 |
| 07/10 P | 11A511U | 12KS Bion-7 | 1194 | Kosmos 1667 |
| 07/15 B | 11A511U | 17F116 (Zenit-8/Oblik) | 1195 | Kosmos 1668 |
| 07/17 P | 8K78M | 11F637 #26 (Molniya-3) | 1196 | Molniya-3 25 |
| 07/19 B | 11A511U (15000-446) | 11F615A15 #126 (Progress) | 1197 | Kosmos 1669 |
| 08/02 P | 11A511U | 17F116 (Zenit-8/Oblik) | 1198 | Kosmos 1671 |
| 08/07 P | 11A511U | 17F41 Resurs-F1 #25 | 1199 | Kosmos 1672 |

| Date | Rocket | No. | Payload | Name |
|---|---|---|---|---|
| 08/08 B | 11A511U | 1200 | 11F660 (Yantar-1KFT #5) | Kosmos 1673 |
| 08/12 P | 8K78M | 1201 | 5V95 (US-K/Oko #41) | Kosmos 1675 |
| 08/16 P | 11A511U | 1202 | 11F695 (Yantar-4K2 #12) | Kosmos 1676 |
| 08/22 P | 8K78M | 1203 | 11F67 #68 (Molniya-1M) | Molniya-1 64 |
| 08/29 P | 11A511U | 1204 | 17F41 Resurs-F1 #26 | Kosmos 1678 |
| 08/29 B | 11A511U | 1205 | 11F695 (Yantar-4K2 #13) | Kosmos 1679 |
| 09/06 P | 11A511U | 1206 | 11F635 (Zenit-4MKT/Fram) | Kosmos 1681 |
| 09/17 B | 11A511U2 (Shch15000-007) | 1207 | 11F732 #20L (Soyuz-T) | Soyuz-T 14 |
| 09/19 P | 11A511U | 1208 | 17F116 (Zenit-8/Oblik) | Kosmos 1683 |
| 09/24 P | 8K78M | 1209 | 5V95 (US-K/Oko #42) | Kosmos 1684 |
| 09/26 P | 11A511U | 1210 | 17F116 (Zenit-8/Oblik) | Kosmos 1685 |
| 09/30 P | 8K78M | 1211 | 5V95 (US-K/Oko #43) | Kosmos 1687 |
| 10/03 P | 8A92M | 1212 | 11F697 #1L (Resurs-O1) | Kosmos 1689 |
| 10/03 P | 8K78M | 1213 | 11F637 #27 (Molniya-3) | Molniya-3 26 |
| 10/16 B | 11A511U | 1214 | 17F116 (Zenit-8/Oblik) | Kosmos 1696 |
| 10/22 P | 8K78M | 1215 | 5V95 (US-K/Oko #44) | Kosmos 1698 |
| 10/23 B | 8K78M | 1216 | 11F67 #69 (Molniya-1M) | Molniya-1 65 |
| 10/25 P | 11A511U | 1217 | 11F695 (Yantar-4K2 #14) | Kosmos 1699 |
| 10/28 P | 8K78M | 1218 | 11F67 #70 (Molniya-1M) | Molniya-1 66 |
| 11/09 P | 8K78M | 1219 | 5V95 (US-K/Oko #45) | Kosmos 1701 |
| 11/13 P | 11A511U | 1220 | 17F116 (Zenit-8/Oblik) | Kosmos 1702 |
| 12/03 P | 11A511U | 1221 | 17F116 (Zenit-8/Oblik) | Kosmos 1705 |
| 12/11 P | 11A511U | 1222 | 11F695 (Yantar-4K2 #15) | Kosmos 1706 |
| 12/13 P | 11A511U | 1223 | 17F41 Resurs-F1 #27 | Kosmos 1708 |
| 12/24 P | 8K78M | 1224 | 11F637 #28 (Molniya-3) | Molniya-3 27 |
| 12/27 P | 11A511U | 1225 | 36KS Efir #2 | Kosmos 1713 |

**1986**

| Date | Rocket | No. | Payload | Name |
|---|---|---|---|---|
| 01/08 P | 11A511U | 1226 | 17F116 (Zenit-8/Oblik) | Kosmos 1715 |
| 01/15 P | 11A511U | 1227 | 11F695 (Yantar-4K2 #16) | Kosmos 1724 |
| 01/28 B | 11A511U | 1228 | 17F116 (Zenit-8/Oblik) | Kosmos 1728 |
| 02/01 P | 8K78M | 1229 | 5V95 (US-K/Oko #46) | Kosmos 1729 |
| 02/04 P | 11A511U | 1230 | 17F116 (Zenit-8/Oblik) | Kosmos 1730 |

| Date | Launch vehicle | Payload | No. | Satellite |
|---|---|---|---|---|
| 02/07 B | 11A511U2 | 11F694 (Yantar-4KS1M #1) | 1231 | Kosmos 1731 |
| 02/26 P | 11A511U | 11F695 (Yantar-4K2 #17) | 1232 | Kosmos 1734 |
| 03/13 B | 11A511U2 (B15000-012) | 11F732 #21L (Soyuz-T) | 1233 | Soyuz-T 15 |
| 03/19 B | 11A511U2 (B15000-010) | 11F615A15 #134 (Progress) | 1234 | Progress 25 |
| 03/26 B Failure | 11A511U | 17F116 (Zenit-8/Oblik) | 1235 | |
| 04/09 B | 11A511U | 11F695 (Yantar-4K2 #18) | 1236 | Kosmos 1739 |
| 04/15 P | 8K78M | 17F116 (Zenit-8/Oblik) | 1237 | Kosmos 1740 |
| 04/18 P | 8K78M | 11F637 #29 (Molniya-3) | 1238 | Molniya-3 28 |
| 04/23 B | 11A511U2 (B15000-009) | 11F615A15 #136 (Progress) | 1239 | Progress 26 |
| 05/14 P | 11A511U | 17F116 (Zenit-8/Oblik) | 1240 | Kosmos 1742 |
| 05/21 B | 11A511U2 (B15000-013) | 11F732 #51 (Soyuz-TM) | 1241 | Soyuz-TM 1 |
| 05/21 P | 11A511U | 34KS Foton #2 | 1242 | Kosmos 1744 |
| 05/28 P | 11A511U | 17F41 Resurs-F1 #28 | 1243 | Kosmos 1746 |
| 05/29 B | 11A511U | 17F116 (Zenit-8/Oblik) | 1244 | Kosmos 1747 |
| 06/06 B | 11A511U | 11F695 (Yantar-4K2 #19) | 1245 | Kosmos 1756 |
| 06/11 P | 11A511U | 17F116 (Zenit-8/Oblik) | 1246 | Kosmos 1757 |
| 06/19 B | 11A511U | 17F116 (Zenit-8/Oblik) | 1247 | Kosmos 1760 |
| 06/19 P | 8K78M | 11F637 #30 (Molniya-3) | 1248 | Molniya-3 29 |
| 07/05 P | 8K78M | 5V95 (US-K/Oko #47) | 1249 | Kosmos 1761 |
| 07/10 P | 11A511U | 17F41 Resurs-F1 #29 | 1250 | Kosmos 1762 |
| 07/17 B | 11A511U | 11F695 (Yantar-4K2 #20) | 1251 | Kosmos 1764 |
| 07/24 P | 11A511U | 17F116 (Zenit-8/Oblik) | 1252 | Kosmos 1765 |
| 07/30 P | 8K78M | 11F67 #71 (Molniya-1M) | 1253 | Molniya-1 67 |
| 08/02 P | 11A511U | 14F40 Resurs-F1 #30 | 1254 | Kosmos 1768 |
| 08/06 B | 11A511U2 | 11F694 (Yantar-4KS1 #4) | 1255 | Kosmos 1770 |
| 08/21 P | 11A511U | 17F116 (Zenit-8/Oblik) | 1256 | Kosmos 1772 |
| 08/27 B | 11A511U | 11F695 (Yantar-4K2 #21) | 1257 | Kosmos 1773 |
| 08/28 P | 8K78M | 5V95 (US-K/Oko #48) | 1258 | Kosmos 1774 |
| 09/03 B | 11A511U | 17F116 (Zenit-8/Oblik) | 1259 | Kosmos 1775 |
| 09/05 P | 8K78M | 11F658 #2 (Molniya-1T) | 1260 | Molniya-1 68 |
| 09/17 B | 11A511U | 17F116 (Zenit-8/Oblik) | 1261 | Kosmos 1781 |
| 10/03 P | 8K78M | 5V95 (US-K/Oko #49) | 1262 | Kosmos 1783 |

| Date | Vehicle | No. | Payload | Designation |
|---|---|---|---|---|
| 10/06 B | 11A511U | 1263 | 11F660 (Yantar-1KFT #6) | Kosmos 1784 |
| 10/15 P | 8K78M | 1264 | 5V95 (US-K/Oko #50) | Kosmos 1785 |
| 10/20 P | 8K78M | 1265 | 11F637 #31 (Molniya-3) | Molniya-3 30 |
| 10/22 B | 11A511U | 1266 | 17F116 (Zenit-8/Oblik) | Kosmos 1787 |
| 10/31 P | 11A511U | 1267 | 14F40 Resurs-F1 #31 | Kosmos 1789 |
| 11/04 P | 11A511U | 1268 | 17F116 (Zenit-8/Oblik) | Kosmos 1790 |
| 11/13 B | 11A511U | 1269 | 11F695 (Yantar-4K2 #22) | Kosmos 1792 |
| 11/16 P | 8K78M | 1270 | 11F658 #3 (Molniya-1T) | Molniya-1 69 |
| 11/20 P | 8K78M | 1271 | 5V95 (US-K/Oko #51) | Kosmos 1793 |
| 12/04 B | 11A511U | 1272 | 17F116 (Zenit-8/Oblik) | Kosmos 1804 |
| 12/12 P | 8K78M | 1273 | 5V95 (US-K/Oko #52) | Kosmos 1806 |
| 12/16 P | 11A511U | 1274 | 11F695 (Yantar-4K2 #23) | Kosmos 1807 |
| 12/26 P | 8K78M | 1275 | 11F658 #4 (Molniya-1T) | Molniya-1 70 |
| 12/26 B | 11A511U2 | 1276 | 11F694 (Yantar-4KS1M #2) | Kosmos 1810 |

**1987**

| Date | Vehicle | No. | Payload | Designation |
|---|---|---|---|---|
| 01/09 B | 11A511U | 1277 | 11F695 (Yantar-4K2 #24) | Kosmos 1811 |
| 01/15 P | 11A511U | 1278 | 17F116 (Zenit-8/Oblik) | Kosmos 1813 |
| 01/16 B | 11A511U2 (B15000-011) | 1279 | 11F615A15 #135 (Progress) | Progress 27 |
| 01/22 P | 8K78M | 1280 | 11F637 #32 (Molniya-3) | Molniya-3 31 |
| 02/05 B | 11A511U2 (I15000-014) | 1281 | 11F732 #52 (Soyuz-TM) | Soyuz-TM 2 |
| 02/07 P | 11A511U | 1282 | 17F116 (Zenit-8/Oblik) | Kosmos 1819 |
| 02/19 P | 11A511U | 1283 | 17F116 (Zenit-8/Oblik) | Kosmos 1822 |
| 02/26 P | 11A511U | 1284 | 11F695 (Yantar-4K2 #25) | Kosmos 1824 |
| 03/03 B | 11A511U2 (I15000-016) | 1285 | 11F615A15 #137 (Progress) | Progress 28 |
| 03/11 P | 11A511U | 1286 | 17F116 (Zenit-8/Oblik) | Kosmos 1826 |
| 04/09 B | 11A511U | 1287 | 11F695 (Yantar-4K2 #26) | Kosmos 1835 |
| 04/16 B | 11A511U2 | 1288 | 11F694 (Yantar-4KS1M #3) | Kosmos 1836 |
| 04/21 B | 11A511U2 (I15000-015) | 1289 | 11F615A15 #127 (Progress) | Progress 29 |
| 04/22 P | 11A511U | 1290 | 17F116 (Zenit-8/Oblik) | Kosmos 1837 |
| 04/24 P | 11A511U | 1291 | 34KS Foton #3 | Kosmos 1841 |
| 05/05 B | 11A511U | 1292 | 17F116 (Zenit-8/Oblik) | Kosmos 1843 |
| 05/13 B | 11A511U | 1293 | 17F116 (Zenit-8/Oblik) | Kosmos 1845 |

| | | | | |
|---|---|---|---|---|
| 05/19 B | 11A511U2 (115000-018) | 11F615A15 #128 (Progress) | 1294 | Progress 30 |
| 05/21 P | 11A511U | 14F40 Resurs-F1 #32 | 1295 | Kosmos 1846 |
| 05/26 P | 11A511U | 11F695 (Yantar-4K2 #27) | 1296 | Kosmos 1847 |
| 05/28 P | 11A511U | 17F116 (Zenit-8/Oblik) | 1297 | Kosmos 1848 |
| 06/04 P | 8K78M | 5V95 (US-K/Oko #53) | 1298 | Kosmos 1849 |
| 06/12 P | 8K78M | 5V95 (US-K/Oko #54) | 1299 | Kosmos 1850 |
| 06/18 P Failure | 11A511U (77015-105) | 14F40 Resurs-F1 #33 | 1300 | |
| 07/04 P | 11A511U | 17F116 (Zenit-8/Oblik) | 1301 | Kosmos 1863 |
| 07/08 B | 11A511U | 11F660 (Yantar-1KFT #7) | 1302 | Kosmos 1865 |
| 07/09 P | 11A511U | 11F695 (Yantar-4K2 #28) | 1303 | Kosmos 1866 |
| 07/22 B | 11A511U2 (115000-019) | 11F732 #53 (Soyuz-TM) | 1304 | Soyuz-TM 3 |
| 08/03 B | 11A511U2 (115000-017) | 11F615A15 #138 (Progress) | 1305 | Progress 31 |
| 08/19 P | 11A511U | 17F116 (Zenit-8/Oblik) | 1306 | Kosmos 1872 |
| 09/03 P | 11A511U | 17F116 (Zenit-8/Oblik) | 1307 | Kosmos 1874 |
| 09/11 B | 11A511U2 | 11F694 (Yantar-4KS1M #4) | 1308 | Kosmos 1881 |
| 09/15 P | 11A511U | 14F40 Resurs-F1 #34 | 1309 | Kosmos 1882 |
| 09/17 P | 11A511U | 11F695 (Yantar-4K2 #29) | 1310 | Kosmos 1886 |
| 09/23 B | 11A511U2 (L15000-021) | 11F615A15 #139 (Progress) | 1311 | Progress 32 |
| 09/29 P | 11A511U | 12KS Bion-8 | 1312 | Kosmos 1887 |
| 10/09 B | 11A511U | 17F116 (Zenit-8/Oblik) | 1313 | Kosmos 1889 |
| 10/22 P | 11A511U | 11F695 (Yantar-4K2 #30) | 1314 | Kosmos 1893 |
| 11/11 B | 11A511U | 17F116 (Zenit-8/Oblik) | 1315 | Kosmos 1895 |
| 11/14 B | 11A511U | 11F660 (Yantar-1KFT #8) | 1316 | Kosmos 1896 |
| 11/20 B | 11A511U2 (L15000-022) | 11F615A15 #140 (Progress) | 1317 | Progress 33 |
| 12/07 B | 11A511U | 17F116 (Zenit-8/Oblik) | 1318 | Kosmos 1899 |
| 12/14 B | 11A511U | 11F695 (Yantar-4K2 #31) | 1319 | Kosmos 1901 |
| 12/21 B | 11A511U2 (L15000-020) | 11F732 #54 (Soyuz-TM) | 1320 | Soyuz-TM 4 |
| 12/21 P | 8K78M | 5V95 (US-K/Oko #55) | 1321 | Kosmos 1903 |
| 12/25 B | 11A511U | 17F116 (Zenit-8/Oblik) | 1322 | Kosmos 1905 |
| 12/26 P | 11A511U | 17F42 Resurs-F2 #1 | 1323 | Kosmos 1906 |
| 12/29 P | 11A511U | 17F116 (Zenit-8/Oblik) | 1324 | Kosmos 1907 |

**1988**

| Date | | Vehicle | Payload | Index | Name |
|---|---|---|---|---|---|
| 01/20 | B | 11A511U2 (L15000-025) | 11F615A15 #142 (Progress) | 1325 | Progress 34 |
| 01/26 | P | 11A511U | 17F116 (Zenit-8/Oblik) | 1326 | Kosmos 1915 |
| 02/03 | B | 11A511U | 11F695 (Yantar-4K2 #32) | 1327 | Kosmos 1916 |
| 02/18 | P | 11A511U | 14F40 Resurs-F1 #35 | 1328 | Kosmos 1920 |
| 02/19 | B | 11A511U | 17F116 (Zenit-8/Oblik) | 1329 | Kosmos 1921 |
| 02/26 | P | 8K78M | 5V95 (US-K/Oko #56) | 1330 | Kosmos 1922 |
| 03/10 | P | 11A511U | 17F116 (Zenit-8/Oblik) | 1331 | Kosmos 1923 |
| 03/11 | B | 8K78M | 11F658 #5 (Molniya-1T) | 1332 | Molniya-1 71 |
| 03/17 | B | 8A92M | IRS-1A | 1333 | 1st commercial flight |
| 03/17 | P | 8K78M | 11F658 #6 (Molniya-1T) | 1334 | Molniya-1 72 |
| 03/23 | B | 11A511U2 (L15000-026) | 11F615A15 #143 (Progress) | 1335 | Progress 35 |
| 03/24 | P | 11A511U | 11F695 (Yantar-4K2 #32) | 1336 | Kosmos 1935 |
| 03/30 | B | 11A511U2 | 11F694 (Yantar-4KS1 #5) | 1337 | Kosmos 1936 |
| 04/11 | P | 11A511U | 17F116 (Zenit-8/Oblik) | 1338 | Kosmos 1938 |
| 04/14 | P | 11A511U | 34KS Foton #4 | 1339 | Foton 4 |
| 04/20 | B | 8A92M | 11F697 #2L Resurs-O1 | 1340 | Kosmos 1939 |
| 04/27 | B | 11A511U | 17F116 (Zenit-8/Oblik) | 1341 | Kosmos 1941 |
| 05/12 | P | 11A511U | 11F695 (Yantar-4K2 #33) | 1342 | Kosmos 1942 |
| 05/13 | B | 11A511U2 (L15000-023) | 11F615A15 #144 (Progress) | 1343 | Progress 36 |
| 05/18 | B | 11A511U | 11F660 (Yantar-1KFT #9) | 1344 | Kosmos 1944 |
| 05/19 | B | 11A511U | 17F116 (Zenit-8/Oblik) | 1345 | Kosmos 1945 |
| 05/26 | P | 8K78M | 11F637 #33 (Molniya-3) | 1346 | Molniya-3 32 |
| 05/31 | P | 11A511U | 14F43 Resurs-F1 #36 | 1347 | Kosmos 1951 |
| 06/07 | B | 11A511U2 (I15000-027) | 11F732 #55 (Soyuz-TM) | 1348 | Soyuz-TM 5 |
| 06/11 | B | 11A511U | 17F116 (Zenit-8/Oblik) | 1349 | Kosmos 1952 |
| 06/22 | B | 11A511U | 11F695 (Yantar-4K2 #34) | 1350 | Kosmos 1955 |
| 06/23 | P | 11A511U | 17F116 (Zenit-8/Oblik) | 1351 | Kosmos 1956 |
| 07/07 | P | 11A511U | 14F43 Resurs-F1 #37 | 1352 | Kosmos 1957 |
| 07/09 | B Failure | 11A511U2 | 11F694 (Yantar-4KS1 #6) | 1353 | |
| 07/18 | P | 11A511U2 (L15000-024) | 11F615A15 #145 (Progress) | 1354 | Progress 37 |
| 07/27 | P Failure | 11A511U (78039-130) | 14F43 Resurs-F1 #38 | 1355 | |

| | Date | Rocket | Payload | No. | Name |
|---|---|---|---|---|---|
| | 08/08 B | 11A511U | 17F116 (Zenit-8/Oblik) | 1356 | Kosmos 1962 |
| | 08/12 P | 8K78M | 11F658 #7 (Molniya-1T) | 1357 | Molniya-1 73 |
| | 08/16 B | 11A511U | 11F695 (Yantar-4K2 #35) | 1358 | Kosmos 1963 |
| | 08/23 B | 11A511U (78039-381) | 17F116 (Zenit-8/Oblik) | 1359 | Kosmos 1964 |
| | 08/23 P | 11A511U | 17F42 Resurs-F2 #2 | 1360 | Kosmos 1965 |
| | 08/29 B | 11A511U2 (Ye15000-031) | 11F732 #56 (Soyuz-TM) | 1361 | Soyuz-TM 6 |
| | 08/30 P | 8K78M | 5V95 (US-K/Oko #57) | 1362 | Kosmos 1966 |
| | 09/06 P | 11A511U | 17F116 (Zenit-8/Oblik) | 1363 | Kosmos 1967 |
| | 09/09 P | 11A511U | 14F43 Resurs-F1 #39 | 1364 | Kosmos 1968 |
| | 09/09 B | 11A511U2 (76048-930) | 11F615A15 #146 (Progress) | 1365 | Progress 38 |
| | 09/15 P | 11A511U | 11F695 (Yantar-4K2 #36) | 1366 | Kosmos 1969 |
| | 09/22 P | 11A511U | 17F116 (Zenit-8/Oblik) | 1367 | Kosmos 1973 |
| | 09/29 P | 8K78M | 11F637 #34 (Molniya-3) | 1368 | Molniya-3 33 |
| | 10/03 P | 8K78M | 5V95 (US-K/Oko #58) | 1369 | Kosmos 1974 |
| | 10/13 P | 11A511U | 17F116 (Zenit-8/Oblik) | 1370 | Kosmos 1976 |
| | 10/25 P | 8K78M | 5V95 (US-K/Oko #59) | 1371 | Kosmos 1977 |
| | 10/27 P | 11A511U | 17F116 (Zenit-8/Oblik) | 1372 | Kosmos 1978 |
| | 11/11 B Failure | 11A511U2 | 11F694 (Yantar-4KS1 #7) | 1373 | |
| | 11/24 P | 11A511U | 17F116 (Zenit-8/Oblik) | 1374 | Kosmos 1981 |
| | 11/26 B | 11A511U2 (Ye15000-028) | 11F732 #57 (Soyuz-TM) | 1375 | Soyuz-TM 7 |
| | 11/30 B | 11A511U | 17F116 (Zenit-8/Oblik) | 1376 | Kosmos 1982 |
| | 12/08 P | 11A511U | 17F116 (Zenit-8/Oblik) | 1377 | Kosmos 1983 |
| | 12/16 P | 11A511U | 11F695 (Yantar-4K2 #37) | 1378 | Kosmos 1984 |
| | 12/22 P | 8K78M | 11F637 #35 (Molniya-3) | 1379 | Molniya-3 34 |
| | 12/25 B | 11A511U2 (Ye15000-029) | 11F615A15 #147 (Progress) | 1380 | Progress 39 |
| | 12/28 P | 8K78M | 11F658 #8 (Molniya-1T) | 1381 | Molniya-1 74 |
| | 12/29 B | 11A511U | 11F660 (Yantar-1KFT #10) | 1382 | Kosmos 1986 |
| **1989** | | | | | |
| | 01/12 P | 11A511U | 17F42 Resurs-F2 #3 | 1383 | Kosmos 1990 |
| | 01/18 B | 11A511U (77059-388) | 17F116 (Zenit-8/Oblik) | 1384 | Kosmos 1991 |
| | 01/28 B | 11A511U | 11F695 (Yantar-4K2 #38) | 1385 | Kosmos 1993 |
| | 02/10 B | 11A511U2 (Ye15000-032) | 11F615A15 #148 (Progress) | 1386 | Progress 40 |

| Serial | Date | Vehicle | Payload | Name |
|---|---|---|---|---|
| 1387 | 02/10 P | 11A511U | 17F116 (Zenit-8/Oblik) | Kosmos 2000 |
| 1388 | 02/14 P | 8K78M | 5V95 (US-K/Oko #60) | Kosmos 2001 |
| 1389 | 02/15 B | 8K78M | 11F658 #9 (Molniya-1T) | Molniya-1 75 |
| 1390 | 02/17 P | 11A511U | 17F116 (Zenit-8/Oblik) | Kosmos 2003 |
| 1391 | 03/02 P | 11A511U | 11F695 (Yantar-4K2 #39) | Kosmos 2005 |
| 1392 | 03/16 P | 11A511U | 17F116 (Zenit-8/Oblik) | Kosmos 2006 |
| 1393 | 03/16 B | 11A511U2 (T15000-034) | 11F615A15 #149 (Progress) | Progress 41 |
| 1394 | 03/23 B | 11A511U2 | 11F694 (Yantar-4KS1 #8) | Kosmos 2007 |
| 1395 | 04/06 P | 11A511U | 17F116 (Zenit-8/Oblik) | Kosmos 2017 |
| 1396 | 04/20 P | 11A511U | 11F695 (Yantar-4K2 #40) | Kosmos 2018 |
| 1397 | 04/26 P | 11A511U | 34KS Foton #5 | Foton 5 |
| 1398 | 05/05 P | 11A511U | 17F116 (Zenit-8/Oblik) | Kosmos 2019 |
| 1399 | 05/17 B | 11A511U | 11F695 (Yantar-4K2 #41) | Kosmos 2020 |
| 1400 | 05/24 B | 11A511U (77059-471) | 11F660 (Yantar-1KFT #11) | Kosmos 2021 |
| 1401 | 05/25 P | 11A511U | 14F43 Resurs-F1 #40 | Resurs-F1 1 |
|  |  |  | Pion-1 & 2 |  |
| 1402 | 06/01 P | 11A511U | 17F116 (Zenit-8/Oblik) | Kosmos 2025 |
| 1403 | 06/08 P | 8K78M | 11F637 #36 (Molniya-3) | Molniya-3 35 |
| 1404 | 06/16 B | 11A511U (77059-398) | 17F116 (Zenit-8/Oblik) | Kosmos 2028 |
| 1405 | 06/27 P | 11A511U | 14F43 Resurs-F1 #41 | Resurs-F1 2 |
| 1406 | 07/05 P | 11A511U | 17F116 (Zenit-8/Oblik) | Kosmos 2029 |
| 1407 | 07/12 P | 11A511U | 11F695 (Yantar-4K2 #42) | Kosmos 2030 |
| 1408 | 07/18 P | 11A511U | 14F43 Resurs-F1 #42 | Resurs-F1 3 |
|  |  |  | Pion-3 & 4 |  |
| 1409 | 07/18 B | 11A511U2 | 17F12 (Orletz-1/Don #1) | Kosmos 2031 |
| 1410 | 07/20 P | 11A511U | 17F116 (Zenit-8/Oblik) | Kosmos 2032 |
| 1411 | 08/02 P | 11A511U | 17F116 (Zenit-8/Oblik) | Kosmos 2035 |
| 1412 | 08/15 P | 11A511U | 17F42 Resurs-F2 #4 | Resurs-F2 1 |
| 1413 | 08/22 P | 11A511U | 17F116 (Zenit-8/Oblik) | Kosmos 2036 |
| 1414 | 08/23 B | 11A511U2 (T15000-037) | 11F615A55 #201 (Progress-M) | Progress-M 1 |
| 1415 | 09/05 B | 11A511U2 (T15000-036) | 11F732 #58 (Soyuz-TM) | Soyuz-TM 8 |
| 1416 | 09/06 P | 11A511U | 14F43 Resurs-F1 #43 | Resurs-F1 4 |

| Date | Vehicle | Serial | Payload | Name |
|---|---|---|---|---|
| 09/15 P | 11A511U | 1417 | 12KS Bion-9 | Kosmos 2044 |
| 09/22 B | 11A511U (77059-392) | 1418 | 17F116 (Zenit-8/Oblik) | Kosmos 2045 |
| 09/27 P | 8K78M | 1419 | 11F658 #10 (Molniya-1T) | Molniya-1 76 |
| 10/03 P | 11A511U | 1420 | 11F695 (Yantar-4K2 #43) | Kosmos 2047 |
| 10/17 P | 11A511U | 1421 | 17F116 (Zenit-8/Oblik) | Kosmos 2048 |
| 11/17 B | 11A511U2 (78039-725) | 1422 | 11F694 (Yantar-4KS1M #5) | Kosmos 2049 |
| 11/23 P | 8K78M | 1423 | 5V95 (US-K/Oko #61) | Kosmos 2050 |
| 11/28 P | 8K78M | 1424 | 11F637 #37 (Molniya-3) | Molniya-3 36 |
| 11/30 P | 11A511U | 1425 | 11F695 (Yantar-4K2 #44) | Kosmos 2052 |
| 12/20 B | 11A511U2 (T15000-039) | 1426 | 11F615A55 #202 (Progress-M) | Progress-M 2 |
| **1990** | | | | |
| 01/17 P | 11A511U | 1427 | 17F116 (Zenit-8/Oblik) | Kosmos 2055 |
| 01/23 P | 8K78M | 1428 | 11F637 #38 (Molniya-3) | Molniya-3 37 |
| 01/25 P | 11A511U | 1429 | 11F695 (Yantar-4K2 #45) | Kosmos 2057 |
| 02/11 B | 11A511U2 (T15000-038) | 1430 | 11F732 #60 (Soyuz-TM) | Soyuz-TM 9 |
| 02/28 B | 11A511U2 (T15000-040) | 1431 | 11F615A55 #203 (Progress-M) | Progress-M 3 |
| 03/22 P | 11A511U | 1432 | 17F116 (Zenit-8/Oblik) | Kosmos 2062 |
| 03/27 P | 8K78M | 1433 | 5V95 (US-K/Oko #62) | Kosmos 2063 |
| 04/03 P Failure | 11A511U | 1434 | 11F695 (Yantar-4K2 #46) | |
| 04/11 P | 11A511U | 1435 | 34KS Foton #6 | Foton 6 |
| 04/11 B | 11A511U2 | 1436 | 11F694 (Yantar-4KS1M #6) | Kosmos 2072 |
| 04/17 P | 11A511U | 1437 | 17F116 (Zenit-8/Oblik) | Kosmos 2073 |
| 04/26 P | 8K78M | 1438 | 11F658 #11 (Molniya-1T) | Molniya-1 77 |
| 04/28 P | 8K78M | 1439 | 5V95 (US-K/Oko #63) | Kosmos 2076 |
| 05/05 B | 11A511U2 (T15000-041) | 1440 | 11F615A15 #150 (Progress) | Progress 42 |
| 05/07 P | 11A511U | 1441 | 11F695 (Yantar-4K2 #47) | Kosmos 2077 |
| 05/15 B | 11A511U (77059-472) | 1442 | 11F660 (Yantar-1KFT #12) | Kosmos 2078 |
| 05/29 P | 11A511U | 1443 | 14F43 Resurs-F1 #44 | Resurs-F1 5 |
| 06/13 P | 8K78M | 1444 | 11F637 #39 (Molniya-3) | Molniya-3 38 |
| 06/19 P | 11A511U | 1445 | 17F116 (Zenit-8/Oblik) | Kosmos 2083 |
| 06/22 P | 8K78M | 1446 | 5V95 (US-K/Oko #64) | Kosmos 2084 |
| 07/03 P Failure | 11A511U | 1447 | 11F695 (Yantar-4K2 #48) | |

| Date | Rocket | Payload designation | No. | Name |
|---|---|---|---|---|
| 07/11 B | 11A511U (T15000-035) | 19KA30 Gamma #1L | 1448 | Gamma 1 |
| 07/17 P | 11A511U | 17F42 Resurs-F2 #5 | 1449 | Resurs-F2 2 |
| 07/20 P | 11A511U | 17F116 (Zenit-8/Oblik) | 1450 | Kosmos 2086 |
| 07/25 P | 8K78M | 5V95 (US-K/Oko #65) | 1451 | Kosmos 2087 |
| 08/01 B | 11A511U2 (T15000-043?) | 11F732 #61A (Soyuz-TM) | 1452 | Soyuz-TM 10 |
| 08/03 P | 11A511U | 11F695 (Yantar-4K2 #49) | 1453 | Kosmos 2089 |
| 08/11 P | 8K78M | 11F658 #12 (Molniya-1T) | 1454 | Molniya-1 78 |
| 08/15 B | 11A511U2 (T15000-042) | 11F615A55 #204 (Progress-M) | 1455 | Progress-M 4 |
| 08/16 P | 11A511U | 14F43 Resurs-F1 #45 | 1456 | Resurs-F1 6 |
| 08/28 P | 8K78M | 5V95 (US-K/Oko #66) | 1457 | Kosmos 2097 |
| 08/31 P | 11A511U | 17F116 (Zenit-8/Oblik) | 1458 | Kosmos 2099 |
| 09/07 P | 11A511U | 14F43 Resurs-F1 #46 | 1459 | Resurs-F1 7 |
| 09/21 P | 8K78M | 11F637 #40 (Molniya-3) | 1460 | Molniya-3 39 |
| 09/27 B | 11A511U2 (T15000-044) | 11F615A55 #206 (Progress-M) | 1461 | Progress-M 5 |
| 10/01 B | 11A511U2 | 17F12 (Orletz-1/Don #2) | 1462 | Kosmos 2101 |
| 10/16 P | 11A511U | 11F695 (Yantar-4K2 #50) | 1463 | Kosmos 2102 |
| 11/16 P | 11A511U | 17F116 (Zenit-8/Oblik) | 1464 | Kosmos 2104 |
| 11/20 P | 8K78M | 5V95 (US-K/Oko #67) | 1465 | Kosmos 2105 |
| 11/23 P | 8K78M | 11F658 #13 (Molniya-1T) | 1466 | Molniya-1 79 |
| 12/02 B | 11A511U2 (T15000-046?) | 11F732 #61 (Soyuz-TM) | 1467 | Soyuz-TM 11 |
| 12/04 B | 11A511U | 11F695 (Yantar-4K2 #51) | 1468 | Kosmos 2108 |
| 12/21 B | 11A511U2 (77017-739) | 11F694 (Yantar-4KS1 #9) | 1469 | Kosmos 2113 |
| 12/26 P | 11A511U | 17F116 (Zenit-8/Oblik) | 1470 | Kosmos 2120 |
| **1991** | | | | |
| 01/14 B | 11A511U2 (T15000-045) | 11F615A55 #205 (Progress-M) | 1471 | Progress-M 6 |
| 01/17 P | 11A511U | 17F116 (Zenit-8/Oblik) | 1472 | Kosmos 2121 |
| 02/07 P | 11A511U | 11F695 (Yantar-4K2 #52) | 1473 | Kosmos 2124 |
| 02/15 B | 11A511U (77053-483) | 11F660 (Yantar-1KFT #13) | 1474 | Kosmos 2134 |
| 02/15 P | 8K78M | 11F658 #14 (Molniya-1T) | 1475 | Molniya-1 80 |
| 03/06 P | 11A511U | 17F116 (Zenit-8/Oblik) | 1476 | Kosmos 2136 |
| 03/19 B | 11A511U2 (R15000-049) | 11F615A55 #208 (Progress-M) | 1477 | Progress-M 7 |
| 03/22 P | 8K78M | 11F637 #41 (Molniya-3) | 1478 | Molniya-3 40 |

| Date | Vehicle | No. | Payload | Name |
|---|---|---|---|---|
| 03/26 P | 11A511U | 1479 | 11F695 (Yantar-4K2 #53) | Kosmos 2138 |
| 05/18 B | 11A511U2 (R15000-051) | 1480 | 11F732 #62 (Soyuz-TM) | Soyuz-TM 12 |
| 05/21 P | 11A511U | 1481 | 17F42 Resurs-F2 #6 | Resurs-F2 3 |
| 05/24 P | 11A511U | 1482 | 11F695 (Yantar-4K2 #54) | Kosmos 2149 |
| 05/30 B | 11A511U2 (R15000-050) | 1483 | 11F615A55 #207 (Progress-M) | Progress-M 8 |
| 06/18 P | 8K78M | 1484 | 11F658 #15 (Molniya-1T) | Molniya-1 81 |
| 06/28 P | 11A511U | 1485 | 14F43 Resurs-F1 #47 | Resurs-F1 8 |
| 07/10 P | 11A511U | 1486 | 17F116 (Zenit-8/Oblik) | Kosmos 2152 |
| 07/10 B | 11A511U2 (77059-726) | 1487 | 11F694 (Yantar-4KS1 #14) | Kosmos 2153 |
| 07/23 P | 11A511U | 1488 | 14F43 Resurs-F1 #48 | Resurs-F1 9 |
| 08/01 P | 8K78M | 1489 | 11F658 #16 (Molniya-1T) | Molniya-1 82 |
| 08/20 B | 11A511U2 (G15000-047) | 1490 | 11F615A55 #210 (Progress-M) | Progress-M 9 |
| 08/21 P | 11A511U | 1491 | 17F42 Resurs-F2 #7 | Resurs-F2 4 |
| 08/29 B | 8A92M | 1492 | IRS-1B | |
| 09/17 P | 8K78M | 1493 | 11F637 #42 (Molniya-3) | Molniya-3 41 |
| 09/19 P | 11A511U | 1494 | 11F695 (Yantar-4K2 #55) | Kosmos 2156 |
| 10/02 B | 11A511U2 (R15000-054) | 1495 | 11F732 #63 (Soyuz-TM) | Soyuz-TM 13 |
| 10/04 P | 11A511U | 1496 | 34KS Foton #7 | Foton 7 |
| 10/09 B | 11A511U2 (P15000-048) | 1497 | 17F12 (Orletz-1/Don #3) | Kosmos 2163 |
| 10/17 B | 11A511U2 (R15000-055) | 1498 | 11F615A55 #211 (Progress-M) | Progress-M 10 |
| 11/20 P | 11A511U | 1499 | 11F695 (Yantar-4K2 #56) | Kosmos 2171 |
| 12/17 B | 11A511U (76032-494) | 1500 | 11F660 (Yantar-1KFT #14) | Kosmos 2174 |
| **1992** | | | | |
| 01/21 P | 11A511U (77053-985) | 1501 | 11F695 (Yantar-4K2 #57) | Kosmos 2175 |
| 01/24 P | 8K78M (78023-608) | 1502 | 5V95 (US-K/Oko #68) | Kosmos 2176 |
| 01/25 B | 11A511U2 (R15000-058) | 1503 | 11F615A55 #212 (Progress-M) | Progress-M 11 |
| 03/04 P | 8K78M (76047-567) | 1504 | 11F658 #17 (Molniya-1T) | Molniya-1 83 |
| 03/17 B | 11A511U2 (R15000-056) | 1505 | 11F732 #64 (Soyuz-TM) | Soyuz-TM 14 |
| 04/01 P | 11A511U2 (78023-988) | 1506 | 11F695 (Yantar-4K2 #58) | Kosmos 2182 |
| 04/08 B | 11A511U2 (77053-750) | 1507 | 11F694 (Yantar-4KS1 #15) | Kosmos 2183 |
| 04/19 B | 11A511U2 (R15000-059) | 1508 | 11F615A55 #213 (Progress-M) | Progress-M 12 |
| 04/29 P | 11A511U (78012-267) | 1509 | 17F42 Resurs-F2 #8 | Resurs-F2 5 |

| Date | No. | Vehicle | Payload | Name |
|---|---|---|---|---|
| 04/29 B | 1510 | 11A511U (77017-738) | 11F660 (Yantar-1KFT #15) | Kosmos 2185 |
| 05/28 P | 1511 | 11A511U (78023-986) | 11F695 (Yantar-4K2 #59) | Kosmos 2186 |
| 06/23 P | 1512 | 11A511U (78012-154) | 14F43 Resurs-F1 #49 | Resurs-F1 10 |
| 06/30 B | 1513 | 11A511U2 (U15000-062) | 11F615A55 #214 (Progress-M) | Progress-M 13 |
| 07/08 P | 1514 | 8K78M (77053-605) | 5V95 (US-K/Oko #69) | Kosmos 2196 |
| 07/24 P | 1515 | 11A511U (78012-989) | 11F695 (Yantar-4K2 #60) | Kosmos 2203 |
| 07/27 B | 1516 | 11A511U2 (R15000-053) | 11F732 #65 (Soyuz-TM) | Soyuz-TM 15 |
| 07/30 P | 1517 | 11A511U (76032-832) | 17F116 (Zenit-8/Oblik) | Kosmos 2207 |
| 08/06 P | 1518 | 8K78M (78039-558) | 11F658 #18 (Molniya-1T) | Molniya-1 84 |
| 08/15 B | 1519 | 11A511U2 (U15000-064) | 11F615A55 #209 (Progress-M) | Progress-M 14 |
| 08/19 P | 1520 | 11A511U (76024-157) | 14F43 Resurs-F1 #50 | Resurs-F1 11 |
|  |  |  | Pion-Germes-1 & 2 |  |
| 09/22 P | 1521 | 11A511U (78023-987) | 11F695 (Yantar-4K2 #61) | Kosmos 2210 |
| 10/08 P | 1522 | 11A511U | 34KS Foton #8 | Foton 8 |
| 10/14 P | 1523 | 8K78M (78012-641) | 11F637 #50L (Molniya-3) | Molniya-3 42 |
| 10/21 P | 1524 | 8K78M (77053-604) | 5V95 (US-K/Oko #70) | Kosmos 2217 |
| 10/27 B | 1525 | 11A511U2 (U15000-061) | 11F615A55 #215 (Progress-M) | Progress-M 15 |
| 11/15 P | 1526 | 11A511U | 14F43 Resurs-F1 #51 | Resurs 500 |
| 11/20 P | 1527 | 11A511U (78012-990) | 11F695 (Yantar-4K2 #62) | Kosmos 2220 |
| 11/25 P | 1528 | 8K78M (77059-565) | 5V95 (US-K/Oko #71) | Kosmos 2222 |
| 12/02 P | 1529 | 8K78M (78023-607) | 11F637 #56L (Molniya-3) | Molniya-3 43 |
| 12/09 B | 1530 | 11A511U2 (76032-495) | 11F694 (Yantar-4KS1 #16) | Kosmos 2223 |
| 12/22 B | 1531 | 11A511U2 (76032-314) | 17F12 (Orletz-1/Don #4) | Kosmos 2225 |
| 12/29 P | 1532 | 11A511U | 12KS Bion-10 | Kosmos 2229 |
| **1993** |  |  |  |  |
| 01/13 P | 1533 | 8K78M (78012-643) | 11F658 #19 (Molniya-1T) | Molniya-1 85 |
| 01/19 P | 1534 | 11A511U (77017-979) | 11F695 (Yantar-4K2 #63) | Kosmos 2231 |
| 01/24 B | 1535 | 11A511U2 (U15000-060) | 11F732 #101 (Soyuz-TM) | Soyuz-TM 16 |
| 01/26 P | 1536 | 8K78M (78023-609) | 5V95 (US-K/Oko #72) | Kosmos 2232 |
| 02/21 B | 1537 | 11A511U2 (U15000-068) | 11F615A55 #216 (Progress-M) | Progress-M 16 |
| 03/31 B | 1538 | 11A511U2 (N15000-069) | 11F615A55 #217 (Progress-M) | Progress-M 17 |
| 04/02 P | 1539 | 11A511U (78012-991) | 11F695 (Yantar-4K2 #64) | Kosmos 2240 |

| Date | Rocket | No. | Designation | Name |
|---|---|---|---|---|
| 04/06 P | 8K78M (78012-644) | 1540 | 5V95 (US-K/Oko #73) | Kosmos 2241 |
| 04/21 P | 8K78M (76024-659) | 1541 | 11F637 #57L (Molniya-3) | Molniya-3 44 |
| 04/27 B | 11A511U (76047-737) | 1542 | 11F660 (Yantar-1KFT #16) | Kosmos 2243 |
| 05/21 P | 11A511U (78012-830) | 1543 | 17F42 Resurs-F2 #9 | Resurs-F2 6 |
| 05/22 B | 11A511U2 (U15000-063) | 1544 | 11F615A55 #218 (Progress-M) | Progress-M 18 |
| 05/26 P | 8K78M (77044-660) | 1545 | 11F658 #20 (Molniya-1T) | Molniya-1 86 |
| 06/25 P | 11A511U (76024-156) | 1546 | 14F43 Resurs-F1 #52 | Resurs-F1 12 |
| 07/01 B | 11A511U2 (N15000-070) | 1547 | 11F732 #66 (Soyuz-TM) | Soyuz-TM 17 |
| 07/14 P | 11A511U (76024-994) | 1548 | 11F695 (Yantar-4K2 #65) | Kosmos 2259 |
| 07/22 P | 11A511U (76024-269) | 1549 | 14F43 Resurs-F1 #53 | Kosmos 2260 |
| 08/04 P | 8K78M (78024-659) | 1550 | 11F637 #58L (Molniya-3) | Molniya-3 45 |
| 08/10 P | 8K78M (78023-610) | 1551 | 5V95 (US-K/Oko #74) | Kosmos 2261 |
| 08/10 B | 11A511U2 (N15000-634) | 1552 | 11F615A55 #219 (Progress-M) | Progress-M 19 |
| 08/24 P | 11A511U (78012-155) | 1553 | 14F43 Resurs-F1 #54 | Resurs-F1 13 |
| 09/07 B | 11A511U2 (77044-356) | 1554 | 17F12 (Orletz-1/Don #4) | Kosmos 2262 |
| 10/11 B | 11A511U2 (77044-270) | 1555 | 11F615A55 #220 (Progress-M) | Progress-M 20 |
| 11/05 B | 11A511U2 (76024-496) | 1556 | 11F694 (Yantar-4KS1 #16) | Kosmos 2267 |
| 12/22 P | 8K78M | 1557 | 11F658 #21 (Molniya-1T) | Molniya-1 87 |
| **1994** | | | | |
| 01/08 B | 11A511U2 (N15000-071) | 1558 | 11F732 #67 (Soyuz-TM) | Soyuz-TM 18 |
| 01/28 B | 11A511U2 (N15000-635) | 1559 | 11F615A55 #221 (Progress-M) | Progress-M 21 |
| 03/17 P | 11A511U (76032-993) | 1560 | 11F695 (Yantar-4K2 #66) | Kosmos 2274 |
| 03/22 B | 11A511U2 (76032-992) | 1561 | 11F615A55 #222 (Progress-M) | Progress-M 22 |
| 04/28 B | 11A511U2 (78023-751) | 1562 | 11F694 (Yantar-4KS1 #17) | Kosmos 2280 |
| 05/22 B | 11A511U2 (76024-355) | 1563 | 11F615A55 #223 (Progress-M) | Progress-M 23 |
| 06/07 P | 11A511U (78012-831) | 1564 | 17F116 (Zenit-8/Oblik) | Kosmos 2281 |
| 06/14 P | 11A511U | 1565 | 34KS Foton #9 | Foton 9 |
| 07/01 B | 11A511U2 (N15000-072) | 1566 | 11F732 #68 (Soyuz-TM) | Soyuz-TM 19 |
| 07/20 P | 11A511U (77044-995) | 1567 | 11F695 (Yantar-4K2 #67) | Kosmos 2283 |
| 07/29 B | 11A511U (77031-497) | 1568 | 11F660 (Yantar-1KFT #17) | Kosmos 2284 |
| 08/05 P | 8K78M | 1569 | 5V95 (US-K/Oko #75) | Kosmos 2286 |
| 08/23 P | 8K78M (77044-661) | 1570 | 11F637 #60L (Molniya-3) | Molniya-3 46 |

| Date | | Vehicle | No. | Payload | Name |
|---|---|---|---|---|---|
| 08/25 | B | 11A511U2 (N15000-636) | 1571 | 11F615A55 #224 (Progress-M) | Progress-M 24 |
| 10/03 | B | 11A511U2 (Ya15000-074) | 1572 | 11F732 #69 (Soyuz-TM) | Soyuz-TM 20 |
| 11/11 | B | 11A511U2 (Ya15000-638) | 1573 | 11F615A55 #225 (Progress-M) | Progress-M 25 |
| 12/14 | P | 8K78M (77031-672) | 1574 | 11F658 #22 (Molniya-1T) | Molniya-1 88 |
| 12/29 | B | 11A511U2 (76032-762) | 1575 | 11F694 (Yantar-4KS1 #18) | Kosmos 2305 |
| **1995** | | | | | |
| 02/15 | B | 11A511U2 (Ya15000-641) | 1576 | 11F615A55 #226 (Progress-M) | Progress-M 26 |
| 02/16 | P | 11A511U | 1577 | 34KS Foton #10 | Foton 10 |
| 03/14 | B | 11A511U2 (V15000-075) | 1578 | 11F732 #70 (Soyuz-TM) | Soyuz-TM 21 |
| 03/22 | P | 11A511U | 1579 | 11F695 (Yantar-4K2 #68) | Kosmos 2311 |
| 04/09 | B | 11A511U2 | 1580 | 11F615A55 #227 (Progress-M) | Progress-M 27 |
| 05/24 | P | 8K78M | 1581 | 5V95 (US-K/Oko #76) | Kosmos 2312 |
| 06/29 | P | 11A511U | 1582 | 11F695 (Yantar-4K2 #69) | Kosmos 2314 |
| 07/20 | B | 11A511U2 | 1583 | 11F615A55 #228 (Progress-M) | Progress-M 28 |
| 08/02 | P | 8K78M (N15000-294) | 1584 | SO-M #511 (Prognoz-M2) | Interbol 1 |
| 08/09 | P | 8K78M (77031-674) | 1585 | 11F637 #59L (Molniya-3) | Molniya-3 47 |
| 09/03 | B | 11A511U2 (V15000-076) | 1586 | 11F732 #71 (Soyuz-TM) | Soyuz-TM 22 |
| 09/26 | P | 11A511U | 1587 | 17F42 Resurs-F2 #10 | Resurs-F2 7 |
| 09/29 | B | 11A511U2 | 1588 | 11F694 (Yantar-4KS1 #19) | Kosmos 2320 |
| 10/08 | B | 11A511U2 (V15000-645) | 1589 | 11F615A55 #229 (Progress-M) | Progress M-29 |
| 12/18 | B | 11A511U2 (15000-647) | 1590 | 11F615A55 #230 (Progress-M) | Progress M-30 |
| 12/28 | B | 8K78M (B15000-040) | 1591 | IRS-1C + Skipper | |
| **1996** | | | | | |
| 02/21 | B | 11A511U2 (15000-651) | 1592 | 11F732 #72 (Soyuz-TM) | Soyuz-TM 23 |
| 03/14 | P | 11A511U | 1593 | 11F695 (Yantar-4K2 #70) | Kosmos 2331 |
| 05/05 | B | 11A511U2 | 1594 | 11F615A55 #231 (Progress-M) | Progress-M 31 |
| 05/14 | B Failure | 11A511U (78051-368) | 1595 | 11F660 (Yantar-1KFT #18) | |
| 06/20 | P Failure | 11A511U | 1596 | 11F695 (Yantar-4K2 #71) | |
| 07/31 | B | 11A511U2 | 1597 | 11F615A55 #232 (Progress-M) | Progress-M 32 |
| 08/14 | P | 8K78M | 1598 | 11F658 #23 (Molniya-1T) | Molniya-1 89 |
| 08/17 | B | 11A511U2 | 1599 | 11F732 #73 (Soyuz-TM) | Soyuz-TM 24 |
| 08/29 | P | 8K78M | 1600 | SO-M #512 (Prognoz-M2) | Interbol 2 |

Magion-5

| Date | Vehicle | Payload | No. | Name |
|---|---|---|---|---|
| 10/24 P | 8K78M (71612-697) | 11F637 #62L (Molniya-3) | 1601 | Molniya-3 48 |
| 11/19 B | 11A511U2 | 11F615A55 #233 (Progress-M) | 1602 | Progress-M 33 |
| 12/24 P | 11A511U (15000-050) | 12KS Bion-11 | 1603 | Bion 11 |
| **1997** | | | | |
| 02/10 B | 11A511U | 11F732 #74 (Soyuz-TM) | 1604 | Soyuz-TM 25 |
| 04/06 B | 11A511U | 11F615A55 #234 (Progress-M) | 1605 | Progress-M 34 |
| 04/09 P | 8K78M (76032-647) | 5V95 (US-K/Oko #77) | 1606 | Kosmos 2340 |
| 05/14 P | 8K78M | 5V95 (US-K/Oko #78) | 1607 | Kosmos 2342 |
| 05/15 B | 11A511U | 17F12 (Orletz-1/Don #5) | 1608 | Kosmos 2343 |
| 07/05 B | 11A511U | 11F615A55 #235 (Progress-M) | 1609 | Progress-M 35 |
| 08/05 B | 11A511U | 11F732 #75 (Soyuz-TM) | 1610 | Soyuz-TM 26 |
| 09/24 P | 8K78M | 11F658 #24 (Molniya-1T) | 1611 | Molniya-1 90 |
| 10/05 B | 11A511U | 11F615A55 #237 (Progress-M) | 1612 | Progress-M 36 |
| | | Inspector | | |
| 10/09 P | 11A511U | 34KS Foton #11 | 1613 | Foton 11 |
| 11/18 P | 11A511U | 14F43M Resurs-F1M #1 | 1614 | Resurs-F1M 1 |
| 12/15 P | 11A511U | 11F695 (Yantar-4K2 #71) | 1615 | Kosmos 2348 |
| 12/20 B | 11A511U | 11F615A55 #236 (Progress-M) | 1616 | Progress-M 37 |
| **1998** | | | | |
| 01/29 B | 11A511U | 11F732 #76 (Soyuz-TM) | 1617 | Soyuz-TM 27 |
| 02/17 B | 11A511U | 11F660 (Yantar-1KFT #19) | 1618 | Kosmos 2349 |
| 03/14 B | 11A511U (S15000-650) | 11F615A55 #240 (Progress-M) | 1619 | Progress-M 38 |
| 05/07 P | 8K78M (78051-676) | 5V95 (US-K/Oko #79) | 1620 | Kosmos 2351 |
| 05/14 B | 11A511U | 11F615A55 #238 (Progress-M) | 1621 | Progress-M 39 |
| 06/24 P | 11A511U | 11F695 (Yantar-4K2 #72) | 1622 | Kosmos 2358 |
| 06/25 B | 11A511U | 11F694 (Yantar-4KS1 #20) | 1623 | Kosmos 2359 |
| 07/01 P | 8K78M | 11F637 #50 (Molniya-3) | 1624 | Molniya-3 49 |
| 08/13 B | 11A511U | 11F732 #77 (Soyuz-TM) | 1625 | Soyuz-TM 28 |
| 09/28 P | 8K78M | 11F658 #25 (Molniya-1T) | 1626 | Molniya-1 91 |
| 10/25 B | 11A511U (15000-660) | 11F615A55 #239 (Progress-M) | 1627 | Progress-M 40 |

**1999**

| Date | Vehicle | No. | Payload | Name |
|---|---|---|---|---|
| 02/09 B | 11A511U/Ikar (S15000-058) | 1628 | Globalstar × 4 | Starsem (ST-1) |
| 02/20 B | 11A511U (M15000-662) | 1629 | 11F732 #78 (Soyuz-TM) | Soyuz-TM 29 |
| 03/15 B | 11A511U/Ikar (S15000-059) | 1630 | Globalstar × 4 | Starsem (ST-2) |
| 04/02 B | 11A511U | 1631 | 11F615A55 #241 (Progress-M) | Progress-M 41 |
| 04/15 B | 11A511U/Ikar (S15000-060) | 1632 | Globalstar × 4 | Starsem (ST-3) |
| 07/08 P | 8K78M | 1633 | 11F637 #51 (Molniya-3) | Molniya-3 50 |
| 07/16 B | 11A511U (15000-667) | 1634 | 11F615A55 #242 (Progress-M) | Progress-M 42 |
| 08/18 P | 11A511U | 1635 | 11F695 (Yantar-4K2/Kobalt) | Kosmos 2365 |
| 09/07 P | 11A511U | 1636 | 34KS Foton #12 | Foton 12 |
| 09/22 B | 11A511U/Ikar (S15000-061) | 1637 | Globalstar × 4 | Starsem (ST-4) |
| 09/28 P | 11A511U | 1638 | 14F43M Resurs-F1M #2 | Resurs-F1M 2 |
| 10/18 B | 11A511U/Ikar (M15000-062) | 1639 | Globalstar × 4 | Starsem (ST-5) |
| 11/22 B | 11A511U/Ikar (M15000-063) | 1640 | Globalstar × 4 | Starsem (ST-6) |
| 12/27 P | 8K78M | 1641 | US-KS/Oko-1 | Kosmos 2368 |

**2000**

| Date | Vehicle | No. | Payload | Name |
|---|---|---|---|---|
| 02/01 B | 11A511U (A15000-669) | 1642 | 11F615A55 #250 (Progress-M) | Progress-M1 1 |
| 02/08 B | 11A511U/Fregat (A15000-079) | 1643 | IRDT-1 | Starsem (ST-7) |
| 03/20 B | 11A511U/Fregat | 1644 | Dumsat | Starsem (ST-8) |
| 04/04 B | 11A511U | 1645 | 11F732 #204 (Soyuz-TM) | Soyuz-TM 30 |
| 04/24 B | 11A511U | 1646 | 11F615A55 #252 (Progress-M) | Progress-M1 2 |
| 05/03 B | 11A511U (A15000-649) | 1647 | Yantar-4KS2/Neman | Kosmos 2370 |
| 07/09 B | 11A511U/Fregat (A15000-069) | 1648 | Cluster-2 (Samba & Salsa) | Starsem (ST-9) |
| 08/06 B | 11A511U (K15000-668) | 1649 | 11F615A55 #251 (Progress-M) | Progress-M1 3 |
| 08/09 B | 11A511U/Fregat (A15000-070) | 1650 | Cluster-2 (Tango & Rumba) | Starsem (ST-10) |
| 09/29 B | 11A511U | 1651 | 11F660 (Yantar-1KFT/Kometa) | Kosmos 2373 |
| 10/16 B | 11A511U (K15000-085) | 1652 | 11F615A55 #243 (Progress-M) | Progress-M 43 |
| 10/31 B | 11A511U (A15000-666) | 1653 | 11F732 #205 (Soyuz-TM) | Soyuz-TM 31 |
| 11/16 B | 11A511U (K15000-671) | 1654 | 11F615A55 #253 (Progress-M) | Progress-M1 4 |

**2001**

| Date | Vehicle | No. | Payload | Name |
|---|---|---|---|---|
| 01/24 B | 11A511U (K15000-673) | 1655 | 11F615A55 #254 (Progress-M) | Progress-M1 5 |
| 02/26 B | 11A511U (15000-670) | 1656 | 11F615A55 #244 (Progress-M) | Progress-M 44 |

| Date/Pad | Launch vehicle | No. | Payload | Name |
|---|---|---|---|---|
| 04/28 B | 11A511U (15000-674) | 1657 | 11F732 #206 (Soyuz-TM) | Soyuz-TM 32 |
| 05/20 B | Soyuz-FG (K15000-001) | 1658 | 11F615A55 #255 (Progress-M) | Progress-M1 6 |
| 05/29 P | 11A511U | 1659 | 11F695 (Yantar-4K2/Kobalt) | Kosmos 2377 |
| 07/20 P | 8K78M | 1660 | Molniya-3K | Molniya-3 51 |
| 08/21 B | 11A511U (M15000-065) | 1661 | 11F615A55 #245 (Progress-M) | Progress-M 45 |
| 09/14 B | 11A511U (F15000-677) | 1662 | Progress-M-SO-1 & Pirs | |
| 10/21 B | 11A511U (15000-672) | 1663 | 11F732 #207 (Soyuz-TM) | Soyuz-TM 33 |
| 10/25 P | 8K78M (77013-687) | 1664 | Molniya-3K | Molniya-3 52 |
| 11/26 B | Soyuz-FG (F15000-002) | 1665 | 11F615A55 #256 (Progress-M) | Progress-M1 7 |
| **2002** | | | | |
| 02/25 P | 11A511U | 1666 | 11F695 (Yantar-4K2/Kobalt) | Kosmos 2387 |
| 03/21 B | 11A511U (F15000-678) | 1667 | 11F615A55 #257 (Progress-M) | Progress-M1 8 |
| 04/01 P | 8K78M | 1668 | US-KS/Oko-1 | Kosmos 2388 |
| 04/25 B | 11A511U (F15000-675) | 1669 | 11F732 #208 (Soyuz-TM) | Soyuz-TM 34 |
| 06/26 B | 11A511U (Ye15000-679) | 1670 | 11F615A55 #246 (Progress-M) | Progress-M 46 |
| 09/25 B | Soyuz-FG (Ye15000-003) | 1671 | 11F615A55 #258 (Progress-M) | Progress-M1 9 |
| 10/15 P Failure | 11A511U (15000-066) | 1672 | 34KS Foton-M1 | |
| 10/30 B | Soyuz-FG (Ye15000-004) | 1673 | 11F732 #211 (Soyuz-TM) | Soyuz-TMA 1 |
| 12/24 P | 8K78M (77045-685) | 1674 | US-KS/Oko-1 | Kosmos 2393 |
| **2003** | | | | |
| 02/02 B | 11A511U (Ye15000-680) | 1675 | 11F615A55 #247 (Progress-M) | Progress-M 47 |
| 04/02 P | 8K78M | 1676 | 11F658 (Molniya-1T) | Molniya-1 92 |
| 04/26 B | Soyuz-FG (Ye15000-006) | 1677 | 11F732 #212 (Soyuz-TM) | Soyuz-TMA 2 |
| 06/02 B | Soyuz-FG/Fregat (Ye15000-005) | 1678 | Mars Express | Starsem (ST-11) |
| 06/08 B | 11A511U (D15000-681) | 1679 | 11F615A55 #259 (Progress-M) | Progress-M1 10 |
| 06/19 P | 8K78M | 1680 | Molniya-3K | Molniya-3 53 |
| 08/12 B | 11A511U | 1681 | 17F12 (Orletz-1/Don) | Kosmos 2399 |
| 08/29 B | 11A511U (D15000-682) | 1682 | 11F615A55 #248 (Progress-M) | Progress-M 48 |
| 10/18 B | Soyuz-FG (D15000-007) | 1683 | 11F732 #213 (Soyuz-TM) | Soyuz-TMA 3 |
| 12/27 B | Soyuz-FG/Fregat (D15000-008) | 1684 | Amos-2 | Starsem (ST-12) |
| **2004** | | | | |
| 01/29 B | 11A511U (D15000-683) | 1685 | 11F615A55 #260 (Progress-M) | Progress-M1 11 |

| | Date | Vehicle | Designation | Name |
|---|---|---|---|---|
| 1686 | 02/18 P | 8K78M | 11F658 (Molniya-1T) | Molniya-1 93 |
| 1687 | 04/19 B | Soyuz-FG (Zh15000-009) | 11F732 #214 (Soyuz-TM) | Soyuz-TMA 4 |
| 1688 | 05/25 B | 11A511U (D15000-684) | 11F615A55 #249 (Progress-M) | Progress-M 49 |
| 1689 | 08/11 B | 11A511U (D15000-685) | 11F615A55 #350 (Progress-M) | Progress-M 50 |
| 1690 | 09/24 P | 11A511U (78036-844) | 11F695M (Yantar-4K2/Kobalt-M) | Kosmos 2410 |
| 1691 | 10/14 B | Soyuz-FG (Zh15000-012) | 11F732 #215 (Soyuz-TM) | Soyuz-TMA 5 |
| 1692 | *11/08 P* | *Soyuz-2.1a* | *17F116 (Zenit-8/Oblik)* | |
| 1693 | 12/23 B | 11A511U (Zh15000-092) | 11F615A55 #351 (Progress-M) | Progress-M 51 |
| **2005** | | | | |
| 1694 | 02/28 B | 11A511U (Zh15000-093) | 11F615A55 #352 (Progress-M) | Progress-M 52 |
| 1695 | 04/15 B | Soyuz-FG (Zh15000-014) | 11F732 #216 (Soyuz-TM) | Soyuz-TMA 6 |
| 1696 | 05/31 B | 11A511U (Zh15000-091) | 34KS Foton-M2 | |
| 1697 | 06/16 B | 11A511U (Zh15000-094) | 11F615A55 #353 (Progress-M) | Progress-M 53 |
| 1698 | 06/21 P Failure | 8K78M (77046-693) | Molniya-3K | Molniya-3 54 |
| 1699 | 08/13 B | Soyuz-FG/Fregat (Zh15000-011) | Galaxy-14 | Starsem (ST-13) |
| 1700 | 09/02 B | 11A511U (78036-576) | 11F660 (Yantar-1KFT/Kometa) | Kosmos 2415 |
| 1701 | 09/08 B | 11A511U (Zh15000-095) | 11F615A55 #354 (Progress-M) | Progress-M 54 |
| 1702 | 10/01 B | Soyuz-FG (P15000-017) | 11F732 #217 (Soyuz-TM) | Soyuz-TMA 7 |
| 1703 | 11/09 B | Soyuz-FG/Fregat (Zh15000-010) | Venus Express | Starsem (ST-14) |
| 1704 | 12/21 B | 11A511U (F15000-080) | 11F615A55 #355 (Progress-M) | Progress-M 55 |
| 1705 | 12/28 B | Soyuz-FG/Fregat (P15000-015) | Giove-A | Starsem (ST-15) |
| **2006** | | | | |
| 1706 | 03/30 B | Soyuz-FG (P15000-018) | 11F732 #218 (Soyuz-TM) | Soyuz-TMA 8 |
| 1707 | 04/24 B | 11A511U (P15000-100) | 11F615A55 #356 (Progress-M) | Progress-M 56 |
| 1708 | 05/03 P | 11A511U (78031-800) | 11F695M (Yantar-4K2/Kobalt-M) | Kosmos 2420 |
| 1709 | 06/15 B | 11A511U (P15000-096) | Resurs-DK #1L | Resurs-DK |
| 1710 | 06/24 B | 11A511U (Ts15000-101) | 11F615A55 #357 (Progress-M) | Progress-M 57 |
| 1711 | 07/21 P | 8K78M | US-KS/Oko-1 | Kosmos 2422 |
| 1712 | 09/14 B | 11A511U (78041-490) | 17F12 (Orletz-1/Don) | Kosmos 2423 |
| 1713 | 09/18 B | Soyuz-FG (Ts15000-023) | 11F732 #219 (Soyuz-TM) | Soyuz-TMA 9 |
| 1714 | 10/19 B | Soyuz-2.1a/Fregat (Zh15000-003) | Metop-A | Starsem (ST-16) |
| 1715 | 10/23 B | 11A511U (Ts15000-102) | 11F615A55 #358 (Progress-M) | Progress-M 58 |

| Date | No. | Launch vehicle | Payload | Name |
|---|---|---|---|---|
| 12/24 B | 1716 | Soyuz-2.1a/Fregat (76033-135) | 14F112 Meridian #11L | Meridian 1 |
| 12/27 B | 1717 | Soyuz-2.1b/Fregat (P15000-001) | Corot | Starsem (ST-17) |
| **2007** | | | | |
| 01/18 B | 1718 | 11A511U (Ts15000-107) | 11F615A55 #359 (Progress-M) | Progress-M 59 |
| 04/07 B | 1719 | Soyuz-FG (Ts15000-019) | 11F732 #220 (Soyuz-TM) | Soyuz-TMA 10 |
| 05/12 B | 1720 | 11A511U (Ts15000-104) | 11F615A55 #360 (Progress-M) | Progress-M 60 |
| 05/29 B | 1721 | Soyuz-FG/Fregat (Ts15000-021) | Globalstar × 4 | Starsem (ST-18) |
| 06/07 P | 1722 | 11A511U (76033-137) | 11F695M (Yantar-4K2/Kobalt-M) | Kosmos 2427 |
| 08/02 B | 1723 | 11A511U (Sh15000-108) | 11F615A55 #361 (Progress-M) | Progress-M 61 |
| 09/14 B | 1724 | 11A511U (P15000-098) | 34KS Foton-M3 | |
| 10/10 B | 1725 | Soyuz-FG (Ts15000-020) | 11F732 #221 (Soyuz-TM) | Soyuz-TMA 11 |
| 10/21 B | 1726 | Soyuz-FG/Fregat (Ts15000-022) | Globalstar × 4 | Starsem (ST-19) |
| 10/23 P | 1727 | 8K78M | US-KS/Oko-1 | Kosmos 2430 |
| 12/14 B | 1728 | Soyuz-FG/Fregat (Sh15000-025) | Radarsat-2 | Starsem (ST-20) |
| 12/23 B | 1729 | 11A511U (Ts15000-105) | 11F615A55 #362 (Progress-M) | Progress-M 62 |
| **2008** | | | | |
| 02/05 B | 1730 | 11A511U (Ts15000-106) | 11F615A55 #363 (Progress-M) | Progress-M 63 |
| 04/08 B | 1731 | Soyuz-FG (Sh15000-024) | 11F732 #222 (Soyuz-TM) | Soyuz-TMA 12 |
| 04/26 B | 1732 | Soyuz-FG/Fregat (P15000-016) | Giove-B | Starsem (ST-21) |
| 05/14 B | 1733 | 11A511U (Sh15000-110) | 11F615A55 #364 (Progress-M) | Progress-M 64 |
| 07/26 P | 1734 | Soyuz-2.1b (77057-145) | 14F137 #1 (Persona) | Kosmos 2441 |
| 09/10 B | 1735 | 11A511U (Sh15000-111) | 11F615A55 #365 (Progress-M) | Progress-M 65 |
| 10/12 B | 1736 | Soyuz-FG (Sh15000-026) | 11F732 #223 (Soyuz-TM) | Soyuz-TMA 13 |
| 11/14 P | 1737 | 11A511U (77057-146) | 11F695M (Yantar-4K2/Kobalt-M) | Kosmos 2445 |
| 11/26 B | 1738 | 11A511U (Sh15000-114) | 11F615A55 #401 (Progress-M) | Progress-M 01M |
| 12/02 P | 1739 | 8K78M (77046-693) | US-KS/Oko-1 | Kosmos 2446 |
| **2009** | | | | |
| 02/10 B | 1740 | 11A511U (Yu15000-115) | 11F615A55 #366 (Progress-M) | Progress-M 66 |
| 03/26 B | 1741 | Soyuz-FG (Yu15000-027) | 11F732 #224 (Soyuz-TM) | Soyuz-TMA 14 |
| 04/29 P | 1742 | 11A511U (78075-152) | 11F695M (Yantar-4K2/Kobalt-M) | Kosmos 2450 |
| 05/07 B | 1743 | 11A511U (Ts15000-113) | 11F615A55 #402 (Progress-M) | Progress-M 02M |
| 05/22 P | 1744 | Soyuz-2.1a/Fregat | 14F112 Meridian #12L | Meridian 2 |

| Date | No. | Launch vehicle | Payload designation | Payload |
|---|---|---|---|---|
| 05/27 B | 1745 | Soyuz-FG (Yu15000-030) | 11F732 #225 (Soyuz-TM) | Soyuz-TMA 15 |
| 07/24 B | 1746 | 11A511U (Yu15000-112) | 11F615A55 #367 (Progress-M) | Progress-M 67 |
| 09/17 B | 1747 | Soyuz-2.1b/Fregat (P15000-002) | Meteor-M #1 Sterkh-2, Tatiana-2, Sumbandila, Ugatusat, Iris, Blits | Meteor-M #1 |
| 09/30 B | 1748 | Soyuz-FG (B15000-029) | 11F732 #226 (Soyuz-TM) | Soyuz-TMA 16 |
| 10/15 B | 1749 | 11A511U (Yu15000-120) | 11F615A55 #403 (Progress-M) | Progress-M 03M |
| 11/10 B | 1750 | 11A511U (B15000-121) | Progress-M-SO2 & Poisk | |
| 11/20 P | 1751 | 11A511U (76043-811) | 14F138 Liana Lotos-S #801 | Kosmos 2455 |
| 11/21 B | 1752 | Soyuz-FG (B15000-031) | 11F732 #227 (Soyuz-TM) | Soyuz-TMA 17 |
| **2010** | | | | |
| 02/03 B | 1753 | 11A511U (Yu15000-117) | 11F615A55 #404 (Progress-M) | Progress-M 04M |
| 04/02 B | 1754 | Soyuz-FG (Yu15000-028) | 11F732 #228 (Soyuz-TM) | Soyuz-TMA 18 |
| 04/16 B | 1755 | 11A511U (76029-169) | 11F695 (Yantar-4K2/Kobalt-M) | Kosmos 2462 |
| 04/28 B | 1756 | 11A511U (B15000-118) | 11F615A55 #405 (Progress-M) | Progress-M 05M |
| 06/16 B | 1757 | Soyuz-FG (B15000-032) | 11F732 #229 (Soyuz-TM) | Soyuz-TMA 19 |
| 06/30 B | 1758 | 11A511U (B15000-119) | 11F615A55 #406 (Progress-M) | Progress-M 06M |
| 09/10 B | 1759 | 11A511U (B15000-122) | 11F615A55 #407 (Progress-M) | Progress-M 07M |
| 09/30 P | 1760 | 8K78M (78041-699) | US-KS/Oko-1 | Kosmos 2469 |
| 10/08 B | 1761 | Soyuz-FG (B15000-035) | 11F732 #701 (Soyuz-TM) | Soyuz-TMA 01M |
| 10/18 B | 1762 | Soyuz-2.1a/Fregat (B15000-007) | Globalstar 2G × 6 | Starsem (ST-22) |
| 10/27 B | 1763 | Soyuz-FG (I15000-123) | 11F615A55 #408 (Progress-M) | Progress-M 08M |
| 11/02 P | 1764 | Soyuz-2.1a/Fregat | 14F112 Meridian #13L | Meridian -3 |
| 12/15 B | 1765 | Soyuz-FG (B15000-034) | 11F732 #230 (Soyuz-TM) | Soyuz-TMA 20 |
| **2011** | | | | |
| 01/28 B | 1766 | 11A511U (I15000-126) | 11F615A55 #409 (Progress-M) | Progress-M 09M |
| 02/26 P | 1767 | Soyuz-2.1b/Fregat (77024-208) | 14F143 Glonass-K1 #701 | Kosmos 2471 |
| 04/05 B | 1768 | Soyuz-FG (I15000-036) | 11F732 #231 (Soyuz-TM) | Soyuz-TMA 21 |
| 04/27 B | 1769 | 11A511U (Yu15000-116) | 11F615A55 #410 (Progress-M) | Progress-M 10M |
| 05/04 P | 1770 | Soyuz-2.1a/Fregat (76012-230) | 14F112 Meridian #14L | Meridian 4 |
| 06/08 B | 1771 | Soyuz-FG (I15000-037) | 11F732 #702 (Soyuz-TM) | Soyuz-TMA 02M |
| 06/21 B | 1772 | 11A511U (I15000-128) | 11F615A55 #411 (Progress-M) | Progress-M 11M |

| No. | Date | Vehicle | Payload | Designation |
|---|---|---|---|---|
| 1773 | 06/27 B | 11A511U (76012-222) | 11F695 (Yantar-4K2/Kobalt-M) | Kosmos 2472 |
| 1774 | 07/13 B | Soyuz-2.1a/Fregat (Yu15000-007) | Globalstar-2G × 6 | Starsem (ST-23) |
| 1775 | 08/24 B Failure | 11A511U (L15000-132) | 11F615A55 #412 (Progress-M) | Progress-M 12M |
| 1776 | 10/02 P | Soyuz-2.1b/Fregat | 14F113 Glonass-M #742 | Kosmos 2474 |
| 1777 | 10/21 G | Soyuz-2.1b/Fregat (B15000-001) | Galileo IOV-1 & 2 | Arianespace (VS-01) |
| 1778 | 10/30 B | 11A511U (115000-129) | 11F615A55 #413 (Progress-M) | Progress-M 13M |
| 1779 | 11/14 B | Soyuz-FG (I15000-038) | 11F732 #232 (Soyuz-TM) | Soyuz-TMA 22 |
| 1780 | 11/28 P | Soyuz-2.1b/Fregat | 14F113 Glonass-M #746 | Kosmos 2478 |
| 1781 | 12/16 G | Soyuz-2.1a/Fregat (Sh15000-004) | Pléiades-1, Elisa × 4, SSOT | Arianespace (VS-02) |
| 1782 | 12/21 B | Soyuz-FG (L15000-039) | 11F732 #703 (Soyuz-TM) | Soyuz-TMA 03M |
| 1783 | 12/23 P Failure | Soyuz-2.1b/Fregat | 14F112 Meridian #15L | Meridian 5 |
| 1784 | 12/28 B | Soyuz-2.1a/Fregat (B15000-009) | Globalstar-2G × 6 | Starsem (ST-24) |

**2012**

| No. | Date | Vehicle | Payload | Designation |
|---|---|---|---|---|
| 1785 | 01/25 B | 11A511U (115000-127) | 11F615A55 #414 (Progress-M) | Progress-M 14M |
| 1786 | 04/20 B | 11A511J (L15000-135) | 11F615A55 #415 (Progress-M) | Progress-M 15M |
| 1787 | 05/15 B | Soyuz-FG (L15000-041) | 11F732 #705 (Soyuz-TM) | Soyuz-TMA 04M |
| 1788? | 05/17 P | 11A511U | 11F695 (Yantar-4K2/Kobalt-M) | Kosmos 24?? |
| 1789? | 07/15 B | Soyuz-FG | 11F732 #706 (Soyuz-TM) | Soyuz-TMA 05M |
| 1790? | 07/?? B | Soyuz-2.1a/Fregat | Metop-B | Starsem (ST-25) |
| 1791? | 07/31 B | 11A511U | 11F615A55 #416 (Progress-M) | Progress-M 16M |
| 17 | 08/?? B | Soyuz-2.1b | Resurs-P #1 | Resurs-P 1 |
| 17 | 08-09/?? B | Soyuz-2.1a/Fregat | Globalstar-2G × 6 | Starsem (ST-26) |
| 17 | 3rd Q P | Soyuz-2.1b/Fregat | 14F143 Glonass-K2 | Kosmos 24?? |
| 17 | 09/10 B | Soyuz-2.1b | Bion-M #1 + AIST | |
| 17 | 09/?? P | Soyuz-2.1v | Calibration spheres + AIST | |
| 17 | 09/28 G | Soyuz-2.1b/Fregat | Galileo IOV-3 & 4 | Arianespace (VS-03) |
| 17 | 10/15 B | Soyuz-FG | 11F732 #707 (Soyuz-TM) | Soyuz-TMA 06M |
| 17 | 11/01 B | 11A511U | 11F615A55 #417 (Progress-M) | Progress-M 17M |
| 18 | 11-12/?? B | Soyuz-2.1b/Fregat | Meteor-M #2 & | |
| | | | Baumanetz-2 & MKA-FKI #2 (Monika) | |
| 18 | 12/05 B | Soyuz-FG | 11F732 #708 (Soyuz-TM) | Soyuz-TMA 07M |
| 18 | 12/?? G | Soyuz-2.1a/Fregat | Pléiades-2 | Arianespace (VS-04?) |

| Date/Site | Rocket | | Payload | Operator |
|---|---|---|---|---|
| 12/26 B | 11A511U | 18 | 11F615A55 #418 (Progress-M) | Progress-M 18M |
| ?? B | Soyuz-FG/Fregat | 18 | Kanopus-V #1 | |
| | | | BKA, MKA-FKI #1 (Zond-PP), ADS-1B, TET-1 | |
| ?? P | Soyuz-2.1b/Fregat | 18 | 14F112 Meridian | Meridian |
| ?? P | Soyuz-2.1b | 18 | 14F137 (Persona) | Kosmos |
| ?? B | Soyuz | 18 | Bars | Kosmos |
| **2013** | | | | |
| 01-03/?? G | Soyuz-2.1b/Fregat | 18 | O3B × 4 | Arianespace (VS-?) |
| 01-03/?? G | Soyuz-2.1b/Fregat | 18 | O3B × 4 | Arianespace (VS-?) |
| 04/02 B | Soyuz-FG | 18 | 11F732 #709 (Soyuz-TM) | Soyuz-TMA 08M |
| 04/?? G | Soyuz-2.1b/Fregat | 18 | Galileo FOC-1 & 2 | Arianespace (VS-?) |
| 04/26 B | 11A511U | 18 | 11F615A55 #419 (Progress-M) | Progress-M 19M |
| 05/?? G | Soyuz-2.1b/Fregat | 18 | Sentinel-1A | Arianespace (VS-?) |
| 05/29 B | Soyuz-FG | 18 | 11F732 #710 (Soyuz-TM) | Soyuz-TMA 09M |
| 07/30 B | 11A511U | 18 | 11F615A55 #420 (Progress-M) | Progress-M 20M |
| 08/?? G | Soyuz-2.1b/Fregat | 18 | Gaia | Arianespace (VS-?) |
| 08/?? G | Soyuz-2.1b/Fregat | 18 | Galileo FOC-3 & 4 | Arianespace (VS-?) |
| 09/30 B | Soyuz-FG | 18 | 11F732 #711 (Soyuz-TM) | Soyuz-TMA 10M |
| 10/23 B | 11A511U | 18 | 11F615A55 #421 (Progress-M) | Progress-M 21M |
| 10-12/?? B | 11A511U | 18 | 34KS Foton-M4 | |
| 11/29 B | Soyuz-FG | 18 | 11F732 #712 (Soyuz-TM) | Soyuz-TMA-11M |
| 12/?? G | Soyuz-2.1b/Fregat | 18 | Galileo FOC-5 & 6 | Arianespace (VS-?) |
| ?? B | Soyuz-2.1b | 18 | Resurs-P #2 | Resurs-P 2 |
| ?? P | Soyuz-2.1b/Fregat | 18 | 14F113 Glonass-M | Kosmos |
| ?? P | Soyuz-2.1b/Fregat | 18 | 14F112 Meridian | Meridian |
| ?? P | Soyuz-2.1v | 18 | Mikhailo Lomonosov | Meridian |

B = Baykonur
P = Plesetsk
G = French Guiana
In italics: suborbital flights (excluding failed orbital flights)

# Bibliography

History of Russian-Soviet Cosmonautics

Afanassiev I., Neizvestnye Korabli [The Unknown spacecraft], Moscow, Znanie, 1991.

Afanassiev I. and Lavrenov A., Bolchoi Kosmitcheskiy Klub [The Grand Cosmic Club], Moscow, RTS Soft, 2006.

Bajinov I. K., "O Rabotakh M. K. Tikhonravova [The Works of M. K. Tikhonravov]," Novosti i Kosmonavtiki, N°8/2003, Moscow, August 2003.

Baturine Y. M., Mirovaya Pilotiruemaya Kosmonavtika [Global Piloted Cosmonautics], Moscow, RTSoft, 2005.

Chertok Boris, Rakety i Ludi [Rockets and Men], Moscow, Machinostroenie, t. I, 1995, t. II (Fili, Podlipki, Tiuratam), 1996, t. III (Goryachie Dni Kholodnoi Voiny [The hot days of the cold war]), 1997, t. IV (Lunaya Gonka [La course à la Lune]), 1999.

Clark Philip, The Soviet Manned Space Program, London, Salamander Books, 1988.

Davydov I. V., Triumf i Tragediya Sovetskoi Kosmonavtiki [Triumph and tragedy of Soviet Cosmonautics], Moscow, Globus, 2000.

Evteev I., Operejaya Vremya [Beyond Time], Moscow, Bioinformservice, 2002.

Favorski V. and Mecheryakov I., Voenno-Kosmicheskie Sily [Military Space Forces], St Petersburg-Moscow, Tipografie/Nauka and Vestnik Vozduchnovo Flota, t. I, 1997, t. II, 1998, t. III, 2001.

Glushko Valentin P., Entsiklopediya Kosmonavtika [Encyclopedia of Cosmonautics], Moscow, Sovetskaya Entsiklopediya, 1985.

Ivanov S. (directed by), Space Weapons, Moscow, The XXI Century Encyclopedia, coll. « "Russia's Arms and Technologies," 2002.

Lardier Christian, L'Astronautique soviétique, [Soviet Astronautics] Paris, Armand Colin, 1992.

Lardier Christian, "L'histoire de Sputnik 1," Bulletin d'information de l'IFHE, N°2, Paris, Institut français d'histoire de l'espace, October 2007.

Mecheriakov I. V., V Mire Kosmonavtiki [In the World of Cosmonautics], Nijni-Novgorod, Russki Kupets, 1996.

Michine Valentin P., Ot Sozdaniya Ballisticheskikh Raket k Raketno-Kosmicheskomu, Moscow, Inform-Znanie, 1998.

Michine Valentin P., Pochemy My Ne Sletali na Lunu [Why We Didn't Go to the Moon], Moscow, Znanie, coll. "Astronomie et Cosmonautique," 1990.

Minaiev A. V., Sovietskaya Voennaya Moch – ot Stalina do Gorbatcheva [Soviet Military Power from Stalin to Gorbachev], Moscow, Military Parad, 1999.

Mozjorine Y. A., Kosmonavtika SSSR, Moscow, Machinostroenie, 1986.

Patruchev V. S., "Vospominania voennogo ispiotatelia (1953-1957)," Press kit N°2 (22) for the launch of the Progress M52 satellite, Moscow, Tsenki, 2005.

Rauchenbakh B. V. (directed by), Issledovaniya po Istorii i Teorii Razvitiya Aviatsionnoi i Raketno-Kosmitcheskoi Nauki i Tekhniki N°8-10, Moscow, Nauka, 2001.

Sagdeev Roald, The Making of a Soviet Scientist, New York, John Wiley & Sons, 1994.

Sergeev I. D. (directed by), Voenny Entsiklopeditcheski Slovar RVSN [Encyclopedic dictionary of the RVSN], Moscow, BRE, 1999.

Siddiqi Asif A., Challenge to Apollo: The Soviet Union and the Space Race, 1945-1974, Washington, NASA History Division, 2000.

Spassky N. (directed by), Raketno-kosmitcheskaya tekhnika [Space Rocket Technology], Moscow, Military Parad, coll. "Russia's Arms Catalog," 1996-1997.

Tarassenko M., Voennye Aspekty Sovetskoi Kosmonavtiki [The Military Aspects of Soviet Cosmonautics], Moscow, Nikol, 1992.

Zaretsky Yuri, "Kak rojdaioutsia chedevry" Press kit N°13 (64) for the launch of the Radarsat 2 satellite, Moscow, Tsenki, 2007.

Encyclopédie Universalis de l'espace, Paris, Universalis, 1986.

## THE WESTERN VISION

Carlier Claude, and Gilli Marcel, Les Trente Premières Années du CNES [First Thirty Years of CNES], Paris, La Documentation française/CNES, 1994.

Devereux Tony, "The Russian Space Rockets," Flight International, August 17, 1961.

Dupas Alain, "Les essais de fusées soviétiques dans le Pacifique," [Soviet Rocket Tests in the Pacific] Air & Cosmos, N°171, Paris, October 29, 1966.

Johnson Nicholas L., The Soviet Year in Space 1990, Colorado Springs, Teledyne Brown Engineering, 1990.

Maurel Patrick, L'Escalade du cosmos [Climbing to the Cosmos]. Paris, Bordas Découverte, 1972.

Maurel Patrick, "L'effort spatial soviétique," [The Soviet Space Effort]Aviation Magazine International, N°453, Paris, 15 October 1966.

Ordway Frederick I., and Von Braun, Wernher, "Histoire mondiale de l'astronautique," (French edition of History of Rocketry and Space Travel), Paris, Larousse/Paris-Match, 1968 (1966).

Parfitt, J. A., "Soviet Space Boosters," Flight International, April 13 & 20, 1967.

Peebles Curtis, Guardians, Strategic Reconnaissance Satellites, Shepperton (Surrey), Ian Allan Ltd, 1987.

Pocock Chris, The U2 Skyplane Toward the Unknown, Atglen (Pennsylvania), Schiffer Military History, 2000.

Sheldon Charles S., "The Soviet space program, A growing enterprise," TRW Space Log 1968, Redondo-Beach (California), 1969.

"Soviet Capabilities and Probable Programs in the Guided Missile Field," National Intelligence Estimate, N°11-6-54, Washington DC, Office of National Estimates, October 5, 1954.

"Soviet Guided Missile Capabilities and Probable Programs," National Intelligence Estimate, N°11-12-55, Washington DC, Office of National Estimates, December 20, 1955.

"Soviet Capabilities and Probable Programs in the Guided Missile Field," National

Intelligence Estimate, N°11-5-57, Washington DC, Office of National Estimates, 12 March 1957.

"A New Realm of Flight," Flight International, 18 October 1957.

"Soviet Capabilities in Guided Missiles and Space Vehicles," National Intelligence Estimate, N°11-5-58, Washington DC, Office of National Estimates, 19 August 1958.

"The First Man-Made Planet," Flight International, 9 January 1959.

"Soviet Capabilities in Guided Missiles and Space Vehicles," National Intelligence Estimate, N°11-5-59, Washington DC, Office of National Estimates, 3 November 1959.

"Soviet Technical Capabilities in Guided Missiles and Space Vehicles," National Intelligence Estimate, N°11-5-61, Washington DC, Office of National Estimates, 25 April 1961.

'Strength and Deployment of Soviet Long Range Ballistic Missile Forces,' National Intelligence Estimate, 1 I-8/I-6 1, Washington DC, Office of National Estimates, 21 September 1961.

'Terminal Range Facilities of the Tyuratam Missile Test Range USSR,' NPIC/R-124/62, Langley (Virginie), CIA, 1962.

'The Soviet Space Program,' National Intelligence Estimate, N°11-1-62, Washington DC, Office of National Estimates, 5 December 1962.

'Development of the Kapustin Yar/Vladimirovka and Tyuratam Missile Test Center, USSR, 1957-1963,' CIA/NPIC Photographic intelligence report, Langley (Virginie), CIA, 1964.

'The Soviet Space Program,' National Intelligence Estimate, N°11-1-65, Washington DC, Office of National Estimates, 27 January 1965.

'Les Russes présentent leur Vostok,' [The Russians Present their Vostok] Interavia, Vol. 20, N°7, Genève, July 1965.

'Le Vostok est au Bourget,' [The Vostok at the Paris Air Show] Aviation Magazine International, N°468, Paris, 1er June 1967.

**Korolev**

Astachenkov P. T., Orbity Glavnogo Konstruktora [The Orbits of the Chief Constructor], Moscow, DOSAAF, 1973.

Golovanov Ya. K., Korolev: Mify i Fakty [Korolev: facts and fancy], Moscow, Nauka, 1994.

Harford Jim, Korolev, New York, John Wiley & Sons, Inc., 1997.

Ichlinsky A. You., Akademik S. P. Korolev - Utchenyi, Ingenier, Tchelovek [The Academician S. P. Korolev - scientist, engineer, politician, and person], Moscow, Nauka, 1986.

Keldysh M. V., Tvorcheskoe Nasledie S. P. Koroleva [The Creative Heritage of S. P. Korolev], Moscow, Nauka, 1980.

Koroleva Natalia, Otets [Father], Moscow, Nauka, 2001-2002.

Kuznetsky M. I., Korolev i troujeniki Baïkonour [Korolev and the Baikonur workers], Krasnoznamenk, Vladi, 2006.

Vetrov G. S., S. P. Korolev i Ego Delo [S. P. Korolev and his actions], Moscow, Nauka, 1998.

**History of Russian-Soviet launchers**

Afanassiev I., 'R-12, Sandalovoe Derevo [R-12, Sandal's Tree],' Hobby Magazine, Moscow, EksPrint, 1997.

Demyanko Y. G., Yadernye Raketnye Dvigateli [Nuclear Rocket Engines], Moscow, OOO Norma-Inform, 2001.

Drogovoz I., Raketnye Voiska SSSR [The USSR and its Army of Rockets], Moscow/Minsk, AST/Kharvest, 2005.

Dyadine G. V., Pamyatnye Starty [The Launch of a Memory], TsIPK, 2001.

Evstafiev M. D., Dolgii Puti K Buria [The Long Road to Buria], Moscow, Vuzovskaya Kniga, 1999.

Filine V. M., Put k Energii, Moscow, Logos, 2001.

Gubarev V., Yujnuï Start [The Yujnoe Launches], Nekos, 1998.

Karpenko A. V., Utkine A. F. and Popov A. D., Otvetchestvennye Strategitcheskye raketnye Kompleksy Spravotchnik [national strategic Rocket complexes, a reference book], St Petersburg, Nevskii Bastion, 1999.

Umansky S. P., Rakety Nositeli, Kosmodromy [Rockets, Launchers, Cosmodromes], Moscow, Restart+, 2001.

Panatov G. S., Samaliot TANTK G. M. Beriev 1945-1968 [The Airplanes of the TANTK G. M. Beriev 1945-1968], Moscow, Restart +, 2001.

Perminov A. N., Kosmitcheskie Voïska [The Army in Space], Moscow, 2003.

Pervov M., Mejkontinetalnye Ballisticheskie Rakety SSSR i Rossi [Intercontinental Ballistic Rockets of the Soviet Union and Russia], Moscow, Krasny Proletary, 1998.

Pervov M., Raketnye Kompleksy RVSN, Moscow, OAO Tipografie Novosti, 1999.

Pervov M., Zenitnoe Raketnoe Orujie Protivovozduchnoi Oborony Strany [Zenith Rockets for anti-Aircraft Defense of the Country], Moscow, Aviarus-XXI, 2001.

Porochkov V. V. et Semenov N. L., 'Khronika pervykh pouskov rakety R-7,' Press kit N°3 (23) for the launch of the Express AM-2 satellites, Moscow, Tsenki, 2005.

Yakovlev V. N., Raketniy Ch'it Otetchestva [Protecting the Nation with Rockets], Moscow, TsIPK RVSN, 1999.

Spaceports Worldwide, Baikonur, Plesetsk, Kourou, Samara, Samara Space Center, 2005.

**Baikonur**

Baranov L. T. (dir.), Kosmodrom Baïkonur, 50 Kosmitcheskikh Liot [The Baikonur Cosmodrome, 50 Years in Space], Moscow, Forces Spatiales, 2005.

Boltenko A. S., Lapidus B. G. and Prouss O. P. (directed by), Rokot Kosmodrom, Vospominania veteranov Baïkonour [The Growling of Baikonur, Memories of Veterans of Baikonur], Kiev, OOO Izdatelskii Dom Michanka, 2006.

Chatalov Dimitri, [History of the Founding of the Soviet Cosmodrome AT Baikonur], Oslo, 46th International Astroautics Congress, October 1995.

Chestopalov N. F., Sovietskie Voennye Stroiteli [The Soviet Military Builders], Moscow, Voennoe Izdatelstvo, 1988.

Gertchik K. V. (dir.), Nezabyvaemyï Baïkonour [Unforgettable Baikonur], Moscow, 1998.

Gertchik K. V., Proryv v kosmos, Otcherki ob ispytateliakh, spetsialistakh i stroïteliakh kosmodroma Baïkonour [Penetrating the Cosmos, Essay on the experimenters, specialists and builders of the Baikonur cosmodrome], Moscow, TOO Veles, 1994.

Gertchik K. V., Baïkonur, Pamyat Serdtsa [Baikonur, Memory of the Heart], Moscow, Terra, 2001.

Gertchik K. V., Vzgliad Skvoz Gody [Looking Across the Years], Moscow, IPO Profizdat, 2001.

Ivanov V. L., Povsednevnaya Jizn [Daily Life]. Moscow, Molodaya Gvardia, 2006.

Khrenov V. A. (dir.), Moï Baïkonour [My Baikonur], Moscow, Geroï Otetchesv, 2007.

Kourouchine A. A. (dir.), Glazami Otchevidtsev, Vospominania veteranov Baïkonour [Eyewitnesses: memories of Baikonur veterans], t. II, Moscow, The council of cosmodrome veterans, 1994.

Kourouchine A. A., C Baïkonour k Lune, Marsu, Venere: vospominaniia veteranov Baïkonur [From Baikonur toward the Moon, Mars and Venus], Moscow, Graal, 2001.

Kuznetsky M. I. and Strajeva I. V., Baikonur, Tchudo XX Veka [Baikonur, the miracle of the 20th century], Sovremennyi pissatel, 1995.

Kuznetsky M. I., Baikonur. Korolev. Yangel, Voronej, IPF Voronej, 1997.

Kuznetsky M. I., Gagarin na Kosmodrome Baïkonur, Krasnoznamenk, Vladi, 2001.

Kuznetsky M. I., Tvortzy raketno-kosmitcheskoï tekhniki na Kosmodrome Baïkonur [The creators of space rocket technology at the Baikonur cosmodrome], Krasnoznamenk, Vladi, 2004.

Marinine I. and Tverskoï O., 'Kosmodrom Baïkonur – sevodnia i zavtra,' Novosti i Kosmonavtiki, N°2/2005, Moscow, February 2005.

Markov Y. and Marinine I., 'Pervyï Kosmoport planety Zemlia,' Novosti i Kosmonavtiki, N°5/2005, Moscow, July 2005.

Menchikov V. A. (dir.), Glazami Otchevidtsev, Vospominania veteranov Baïkonur [Eye-witnesses: memories of Baikonur veterans], t. III, Moscow, Kosmo, 1997.

Menchikov, V. L., Baïkonur, Moya Bol i Liubov [Baikonur, my pain and my love], Moscow, 1994.

Menchikov, V. L., Baïkonur-Moskva-Iubilenyï [Baikonur-Moscow-Jubilee], Moscow, Re-start, 2003.

Moltchanov A. F. and Puchkariov A. A., Kosmitcheskaya Gavan [Havana and Space], Machinostroenie, 1982.

Netchessa Ya. V., 'Vitoribus Gloria,' Press kit N°6 (26) for the launch of the Foton M satellite, Moscow, Tsenki, 2005.

Omelko V., Kosmodrom Baïkonur, Moscow, OmV-Lutch, 2000.

Pitalev German, Vypolnit k sroku, Ob ispolnenii dolojit [Accomplished on schedule, Report on Accomplished Mission], Moscow, Demiourg-ART, 2005.

Porochkov V. V., 'Baïkonur – Unikalnyï kosmitcheskiï port,' Rossiski Kosmos, N°January 1-3, February, March 2006.

Porochkov V. V., Raketno-kosmitcheskiï podvig Baïkonoura [The Baikonur Space Rocket Exploit], Moscow, Patriot, 2007.

Tverdovsky V. N., Kosmodrom, Machinostroenie, 1976.

Villain Jacques (dir.), Baïkonour, la porte des étoiles [Baikonur, the Gateway to the Stars], Paris, Armand Colin, 1994.

Voslsky A. P. (Abramov A. P.) (directed by), Kosmodrom, Moscow, Voennoe Izdatelstva, 1977.

Zakharov A. G., Kak eto bylo, Vospominania hatchalnik kosmodroma Baïkonour [It Was Like This, Memoir of the Director of the Baikonur Cosmodrome], 1996.

Zavalichine A. P. (dir.), Glazami Otchevidtsev, Vospominania veteranov Baïkonour [Eyewitnesses: the memories of Baikonur veterans], t. I, Moscow, DOSAAF, 1991.

Baïkonour, Otcherki, Poemy, Stikhi, Khronika [Baikonur, Essay, Poems, Verses, Journals], Alma-Ata, Jazouchy, t. I, 1984, t. II, 1986, t. III, 1987.

Kosmodrom Baïkonour, V Natchal Pouti [The Baikonur Cosmodrome. The Start of the Path], Moscow, Council of cosmodrome veterans, 1992.

Baikonur 45 Years, Baikonur, town of Baikonur, 2000.

Glavni stroitel Baïkonour [The Chief Constructor of Baikonur], Moscow, Moskovskie Outchbniki CD press, 2004.

Baïkonour-50, Istoria kosmodroma v vospominaniakh veteranov [Baikonur-50, History of the Cosmodrome as the Veterans Remember It], Moscow, Novosti, 2005.

Baikonur, the advent of a new century, Moscow, Military Parad, 2005.

'Plochadki Baikonoura [The Zones of Baikonur]», Kosmodrom, N°9 and N°12, Baikonur, 2002.

**Opening of the market and cooperation with Europe**

Bjilianskaya Lioudmila, 'Russian launch vehicles on the world market: A case-study of International joint ventures,' Space Policy, Vol. 13, N°4, Maidens (Ayrshire), November 1997.

Calaque François, Evolution des lanceurs dans la décennie prochaine [Launcher Developments in the Coming Decade], Propositions, Les Mureaux, Aerospatiale Espace & Défense, 1994.

Chenard Stéphane, Space Industries and Markets in Russia and Other Countries of the Former Soviet Union, Paris, Euroconsult, t. I (Space Policy & Industry in CIS Countries, From Space Power to Space Market), 1993, t. IV (Space Transportation in CIS Countries, Prospects to 2000), 1993.

Isakowitz Steven J., Hopkins Joseph P., and Hopkins Joshua B., International Reference Guide to Space Launch Systems, Third Édition, Reston (Virginie), AIAA, 1999.

Wilson Andrew, ESA Achievements, More Than Thirty Years of Pioneering Space Activity, Noordwijk, ESA Publications Division/ESTEC, 2005.

A Proposal for a European Strategy in the Launch Sector, Paris, ESA Directorate of Launchers, 2000.

Development and Competitiveness of European Space Transportation Industries, Report of the Industry's High Level Group to the European Commission, Paris, 1996.

'L'industrie spatiale soviétique,' [The Soviet Space Industry]La Lettre d'Arianespace, N°49, Evry, Arianespace, November 1990.

'Arianespace et les constellations,' [Arianespace and the Constellations] La Lettre d'Arianespace, N°114, Evry, Arianespace, October 1996.

TsSKB/Ariane: Joint-venture underway to market Soyuz-2K,' The Orbital Launcher Report, Vol. 2, N°7, Paris, April 1996.

'Starsem: Russian-French joint-venture to be set up in July,' The Orbital Launcher Report, Vol. 2, N°8, Paris, May 1996.

'Starsem: French-Russian joint-venture to market 3 vehicles,' The Orbital Launcher Report, Vol. 2, N°9, Paris, June/July 1996.

'Arianespace: and CSG offered to other launch systems,' The Orbital Launcher Report, Vol. 3, N°4, Paris, January 1997.

'Starsem: Soyuz proposed to launch revived Cluster mission,' The Orbital Launcher Report, Vol. 3, N°5, Paris, February 1997.

'Starsem: Soyuz improvements under study,'
The Orbital Launcher Report, Vol. 3, N°6, Paris, March/April 1997.

'Starsem s'installe sur le marché des lanceurs,' [Starsem Penetrates the Launcher Market] Aéronautique Business, N°25, 2 December 1999.

'Pari réussi pour le lanceur Soyouz-Fregat,' [Winning Wager for the Soyuz-Fregat Launcher]Aéronautique Business, N°30, 17 February 2000.

**Other launch-site projects**

Bellin Sylvie, Chenard Stéphane, and Villain Rachel, Asia-Pacific Space Programs & Industry, Prospects to 2005, Paris, Euroconsult, 1995.

'Australia Spaceport Breakthrough: APSC-Starsem Deal for Soyuz,' Space Fax Daily Global Édition, N°1027, Kallua-Kona (Hawaii), 20 June 1997.

'Kourou devient l'enjeu de la coopération franco-russe,' [Kourou is the Issue of the Franco-Russian Cooperative Project]Aéronautique Business, N°57, 3 May 2001.

'Soyuz: poker menteur à Kourou,' [Soyuz: Bluff-Poker at Kourou]Aéronautique Business, N°72, 24 January 2002.

**Corporate publications**

Gorokhov Alexei, GKNPTs imeni Khrunichev a, 90 let [Khrunichev ,at 90], Moscow Polygon press, 2006, t. I (Filevskie Krylya), t. II (Filevskie Orbity).

Jiltsov S., GKNPTs imeni Khrunichev a, 80 let [Khrunichev at 80], Moscow, Voenniy Parad, 1996.

Kobelev V. N., TSKBTM 50 ans, Nazemloe Obordourovanie [Ground Equipment], Moscow, 1997.

Koniukhov S. N. (directed by), Prizvany Vremenem [50 years of KB Yujnoe], ART-Press 2004, t. I (From opponent to international cooperation), t. II (Rockets and satellites).

Korneyev N. M. and Neustroiev V. N., V. P. Barmine: osnovnye etapy jizni i deiatelnosti [V. P. Barmine: the main stages of his life and work], Moscow, KBOM, 1999.

Pallo-Korustine V., Dniepropetrovskiy Raketno-Kosmitcheskiy Tsentr [The Dniepropetrovsk space-rocket center], Dniépropetrovsk, 1994.

Semenov Yuri P., RKK Energiya imeni S. P. Koroleva, Korolev, RKK Energiya, 1996.

Semenov Yuri P., RKK Energia Imeni S.P. Koroleva, Na Rubezhe Dvukh Vekov 1996-2001, Korolev, RKK Energiya, 2001.

Titov A. O., KB Motor: 45 liot i vsia Jizn [KB Motor: 45 years and a lifetime], OOO Oko-Inform, 2006.

Khrunichev State Research and Production Center, Moscow, GKNPTs Khrunichev , 1998.

Khrunichev State Research and Production Space Center, Moscow, GKNPTs Khrunichev , 2001.

Na zemle i v kosmos [On the Earth and in the Cosmos] Moscow, KBOM, 2001.

NPO Machinostroenia, Moscow, NPO Machinostroenia, 2001

NPO Machinostroenia, 60 let samootverjennovo trouda vo imia mira 1944-2004 [NPO Machinostroenie, 60 years of devotion to work, 1944-2004], Moscow, Orujie i Tekhnologii, 2004.

RKK Energiya imeni S. P. Koroleva, Korolev, RKK Energiya, 1994.

**Websites**

Capdevilla, Didier, 'Le lanceur Soyuz en Guyane,' [The Soyuz Launcher in French Guiana]Capcom Espace (www.capcomespace.net), Sisteron, 2004-2010.

Day Dwayne, A., 'Tinker, Taylor, Satellite, Spy,' The Space Review (www.thespacereview.com), Bethesda (Maryland), 2007.

# Index of names

**USSR, Russia and Ukraine**
V. V. Aborenkov (1901-1954), 76
A. P. Abramov (1919-1998), 391
A. S. Abramov (1911-1999), 45
I. I. Abramov (1915-1982), 19, 391
A. T. Abramov (?), 392
G. N. Abramovich (1911-1996), 4, 10
V. I. Adasko (1933-1993), 159
S. A. Afanaseiev (1918-2001), 19, 198
P. A. Agadjanov (1923-2001), 389, 391
S. P. Agafonov (1918-1993), 391
F. A. Agaltsov (1900-1980), 145, 393
R. N. Akhmetov (1948), 111
N. A. Akkerman (?), 142
V. V. Alaverdov (1939), 20
A. P. Alexandrov (1903-1994), 80
I. S. Alexandrov (?), 391
P. S. Alexandrov (1907-1964), 76
S. M. Alexeiev (1909-1993), 84, 145
N. N. Alexeiev (1914-1980), 198
V. B. Alexeiev (1933), 128
V. K. Algunov (?), 142
B. S. Aliochin (1955), 332
Ya. I. Alksnis (1897-1938), 75
G. E. Alpaidze (1916-2006), 203
S. R. Ambartsumian (1911-1971), 10
V. A. Ambartsumian (1908-1996), 393
G. P. Anchakov (1937), 110, 260
V. A. Andreiev (1942), 19
Y. G. Antonov (?), 110
O. F. Antufiev (?), 20
A. I. Apeksimov (?), 108
R. F. Appazov (1920-2008), 15
Y. P. Artiukhin (1930-1998), 128
K. K. Artseulov (1891-1980), 72

S. P. Axenov (1937), 391
V. V. Axenov (1935), 392

A. A. Bachlakov (1957), 393
G. F. Baidukov (1907-1994), 18
P. N. Baikovsky (1913-1995), 390
I. K. Bajinov (1928), 389, 391
A. S. Bakaiev (1895-1977), 389
A. G. Bakanov (1932), 88
M. I. Bakhtiukov (?), 48
O. D. Baklanov (1932), 19, 26, 389, 390, 396
Bakurine (?), 12
G. M. Balandin (1881-1963), 71
B. V. Balmont (1927), 19, 389, 390
L. T. Baranov (1949), 393
V. P. Barmine (1909-1993), 11, 13, 17, 22, 42,
    91, 99, 101, 102, 145, 147, 149, 199, 390,
    391
I. V. Barmine (1943), 103, 335
A.K. Baulin (?), 98
V. A. Bechenov (1906-1993), 119
P. I. Belayev (1925-1970), 161
A. V. Belussov (1922-1996), 91, 393
B. N. Belussov (1930-1998), 128
I. S. Belussov (1928-2005), 18
N. G. Belov (?), 75, 76
N. I. Belov (1912-1982), 89
O. M. Belozerkovsky (1925), 110
L. E. Benkovich (?), 114
G. T. Beregovoy (1921-1995), 167
A. I. Berg (1893-1979), 389
V. F. Berglezov (?), 389
L. P. Beria-Gegetchoria (1899-1953), 10, 18,
    22, 31, 41, 76, 91
S. L. Beria (1924-2000), 17

A. A. Bessonov (1892-1983), 392
Ya. L. Bibikov (1902-1976), 389
S. S. Biriuzov (1904-1964), 196
M. R. Bisnovat (1905-1977), 4, 21
A. A. Blagonravov (1894-1975), 31, 79, 147
I. D. Bogatchev (?), 152
S. P. Bogdanovsky (1922), 85
A. F. Bogomolov (1913-2009), 92, 93, 142, 390
E. N. Bogomolov ( ?), 382
E. Ya. Boguslavsky (1917-1969), 12, 89, 90, 92, 390
V. A. Bokov (1921), 142, 389
V. F. Bolkhovitinov (1899-1970), 3, 4, 77
A. V. Bondar (1962), 391
M. M. Bondariuk (1908-1969), 4, 31, 137
N. I. Boravenkov (1920-1985), 19
M. I. Borissenko (1917-1984), 90, 390
A. A. Borissenko (1930-2004), 391
N. N. Borissov (?), 203
I. T. Borissov (?), 114
K. D. Buchuyev (1914-1978), 85, 107, 108, 147, 390, 391, 393
V. S. Budnik (1913-2007), 12, 15, 25
Y. N. Bugaiev (?), 392
N. A. Bulganin (1895-1975), 18, 38
I. T. Bulitchev (1897-1999), 42, 393
M. N. Burdaiev (1932), 128
G. A. Burmistrov (1941), 392
L. I. Brezhnev (1906-1982), 91, 112, 160, 187, 196
M. A. Brezhnev (1918-1973), 390
A. M. Brezinsky (?), 18
A. V. Brykov (1921-2007), 389, 391
Y. S. Bykov (1906-1970), 152
V. F. Bykovsky (1934), 145, 166
B. V. Bytchevsky (1902-1972), 390

A. D. Charomsky (1899-1982), 391
A. I. Chassovskikh (1958), 391
A. S. Chebotarev (1955), 392
A. I. Chechkov (?), 389
N. V. Chekov (1931-1996), 390
V. N. Chelomei (1914-1984), 4, 80, 82, 85, 88, 96, 102, 117, 146, 160, 163, 164, 193, 194, 195, 196, 229, 380
B. I. Cheranovsky (1896-1960), 72
K. U. Chernenko (1911-1985), 237
F. L. Cherniavsky (?), 199

V. S. Chernomyrdin (1938), 244, 245, 270, 293
N. G. Chernychev (1906-1953), 3, 75, 389, 391
B. E. Chertok (1912-2011), 10, 15, 16, 21, 86, 89, 95, 98, 107, 108, 188, 390, 391
L. S. Chetchenia (1913-1996), 392
A. V. Chetchin (1937), 110
B. A. Chevela (?), 391
A. A. Chijov (1934), 392
G. I. Chursin (1921), 391

V. A. Davydov (1953), 20
A. I. Dedov (?), 389
A. N. Dedov (1907-?), 75
G. I. Degtiarenko (?), 17, 90
P. A. Degtiarev (1903-1977), 27, 29
P. V. Dementiev (1907-1977), 390
D. G. Diatlov (1907-?), 13, 41
I. F. Dibrov (?), 197
A. M. Dobrotvorsky (?), 391
G. T. Dobrovolsky (1928-1971), 170
V. A. Dobrynin (1895-1978), 86
V. Kh. Dogujiev (1935), 19, 389, 390
L. I. Dolinov (1930), 203
A. V. Domratchev (1906-1961), 390
N. A. Dondukov (1928-1983), 392
A. A. Dorodnitsyn (1910-1994), 390
B. A. Dorofeiev (1927–1999), 391
G. D. Dorokhin (1905-1962), 20
L. S. Dushkin (1910-1990), 4, 75, 77, 83, 111, 389
V. I. Dudakov (1902-1975), 391
N. L. Dukhov (1904-1964), 390
A. I. Dunaiev (1934), 19, 20, 237, 382, 390
M. I. Duplichev (1912-1993), 390

N. I. Efremov (1906-1961), 76
V. A. Egorov (1930-2001), 65
B. B. Egorov (1937-1994), 160
R. P. Eïdeman (1895-1937), 75
P. E. Eliasberg (1914-?), 389, 391
A. P. Elisseiev (1913), 25
A. S. Elisseiev (1934), 166, 169, 170
E. I. Eller (?), 91
Y. E. Endeka (1902-1949), 389
T. M. Eneiev (1926-1993), 65, 391
A. A. Epishev (1908-1985), 196
A. I. Eremeiev (?), 21

V. G. Ermolaiev (1908-1944), 78
M. M. Esmin (?),86
S. V. Essenkov (1929), 203

V. I. Fadeiev (1923-1990), 393
V. V. Favorsky (1924), 392
P. I. Fedorov (1898-1945), 3
V. P. Fedorov (1915-1943), 77
K. P. Feoktistov (1926-2009), 160
A. V. Filipchenko (1928), 169
P. V. Flerov (1907-1976), 72
M. R. Flissk (1904-1966), 392
Florensky (?), 12
G. E. Fomin (?), 110, 282
Y. A. Fomin (?), 19
I. F. Frolov (1908-1983), 4
V. A. Frolov (?), 19
V. I. Frumson (?), 65

Y. A. Gagarin (1934-1968), 108, 144, 187, 216
L. M. Gaidukov (1911-1999), 11, 13, 15, 106
E. N. Galin (?), 92
M. L. Gallaï (1914-1998), 149
A. A. Ganin (1936), 392
K. V. Gerchik (1918-2001), 393
A. M. Ginzburg (1911-2000), 25
G. S. Giritsky (1893-1966), 84
G . P. Glazkov (1911-1993), 96, 390
Y. N. Glazkov (1939-2008), 128
V. P. Gluchko (1908-1989), 3, 4, 7, 11, 12, 13,
    15, 17, 21, 25, 26, 34, 42, 45, 66, 72, 75, 76,
    77, 79, 80, 82, 83, 84, 85, 88, 100, 102, 106,
    116, 137, 140, 142, 145, 147, 149, 188, 199,
    235, 390, 391
K. K. Glukharev (1896-1970), 22, 391
M. A. Golubev (1914-1990), 391
A. A. Golubev (1924), 391
M. K. Golubev (?), 392
P. M. Golovinov (?), 21
A. G. Golovko (1906-1962), 393
A. M. Goltsman (?), 22
P. D. Goltsov (?), 199
L. R. Gonor (1906-1969), 20, 86
M. S. Gorbachev (1931), 196, 237, 243
V. V. Gorbatko (1934), 169
S. G. Gorshkov (1910), 196
P. N. Goremykin (1902-1976), 19, 389
Goriunov (?), 12
P. A. Gorchakov (?), 19

V. A. Goviadinov (?), 14, 90
V. G. Grabin (1899-1980), 13, 80, 105
V. D. Grafov (?), 48
A. A. Gretchko (1903-1976), 196
V. F. Gribanov (?), 19
L. A. Grichin (1920-1960), 19, 99, 393
V. A. Grichmanovsky (1927), 92
P. A. Gridtchin (?), 108
M. G. Grigorenko (1906-1990), 390
G. F. Grigorenko (1918-2007), 390
M. G. Grigoriev (1918-1981), 197
V. S. Grigorkin (?), 390
P. D. Grigorovich (1883-1938), 71
V. A. Grin (1946), 203, 393
V. S. Grizodubova (1910-1993), 76
M. M. Gromov (1899-1985), 76
P. D. Gruchin (1906-1993), 87
A. A. Gubarev (1931), 128, 170
E. S. Gubenko (1911-1959), 392
M. T. Guchi (?), 199
N. B. Guerassimov (1927-2009), 19
N. F. Guerassiuta (1919-1987), 15
A. A. Gukhman (?), 389
V. I. Gulaiev (1937-1990), 128
M. I. Gurevich (1892-1976), 112
O. V. Gurko (?), 391
I. M. Gurovich (1915-1991), 40
G. Ya. Guskov (1918-2002), 45
L. I. Gussiev (1922), 92, 390
I. I. Gvai (1900-1960), 76

A. Y. Ichlinsky (1913-2003), 99, 142, 391, 393
N. Ya. Ilin (1901-1937), 75
S. V. Iliuchin (1894-1977), 72
V. V. Illiuviev (?), 390
V. N. Iordansky (1900-1974), 21
A. G. Iossifian (1905-1993), 22, 158
A. M. Isaiev (1908-1971), 4, 20, 34, 78, 88,
    145, 163, 164, 389
G. I. Ivanov (1904-?), 76
I. G. Ivanov (?), 18
V. L. Ivanov (1936), 393
N. E. Ivanov (1927-2006), 92
V. N. Ivanov (?), 19
O. G. Ivanovsky (1922), 108
A. G. Ivtchenko (1903-1968), 88

L. M. Kaganovich (1893-1991), 76
M. P. Kalianova (1908-1992), 75

P. I. Kalinuchkin (?), 390
V. D. Kalmykov (1908-1974), 98, 196, 390
N. P. Kamanin (1908-1982), 145
S. S. Kamynin (1927-1986), 33
V. A. Kapitonov (1939), 110
P. L. Kapitza (1894-1984), 22
A. G. Karas (1918-1979), 6, 8
A. G. Karpenko (?), 66
V. K. Karrask (1928-2004), 380
M. M. Kasianov (1957), 326, 332
B. I. Katorgin (1934), 391
M. V. Keldysh (1911-1978), 5, 31, 33, 65, 66,
     82, 102, 142, 145, 147, 149, 196, 390, 391,
     393
K. A. Kerimov (1917-2003), 6, 8, 19, 120,
     152, 158, 166, 392, 393
V. B. Kharin (?), 92
Y. B. Khariton (1904-1996), 390
B. I. Khlebnikov (1922), 48
N. N. Khlybov (1918), 98, 392
V. N. Khodakov (1933-2012), 20
Y. K. Khodarev (1922-2008), 392
N. D. Khokhlov (1918-1999), 390
M. S. Khomiakov (1921), 390
S. A. Khristianovich (1908-2000), 31
N. S. Khruchtchev (1894-1971), 24, 79, 112,
     146, 160, 187, 196, 213, 214
M. V. Khrunichev (1901-1961), 41, 65, 66
E. V. Khrunov (1933-2000), 166, 169
V. A. Khrustalev (?), 158
B. A. Kiassov (?), 390
Y. P. Kienko (1934-2005), 392
S. P. Kirienko (1962), 293
A. N. Kirillin (1950), 111, 392
A. S. Kirillov (1924-1987), 389, 392
R. R. Kiriuchin (?), 99, 389
P. I. Kirpitchnikov (1903-1980), 389
A. I. Kiselev (1938), 245, 269
I. G. Kisselev (?), 20
G. V. Kissunko (1918-1998), 146
L. D. Kizim (1941), 128
I. I. Klebanov (1951), 329
I. T. Kleimenov (1899-1938), 72, 75, 83, 100
P. K. Kliaritsky (?), 14
V. Ya. Klimov (1892-1962), 86, 391, 392
V. M. Kliutcharev (1919-1991), 85, 391
G. F. Knorre (?), 389
G. A. Kolesnikov (1936), 393
G. M. Kolesnikov (1936), 128

M. Ya. Kolessov (?), 203
Ya. P. Koliako (?), 85
K. P. Kolobenkov (?), 19
P. I. Kolodin (1930), 170
M. A. Kolossov (?), 391
V. A. Kolytchev (?), 19
V. M. Komarov (1927-1967), 160, 166
B. A. Komissarov (1918-1999), 15, 19
M. L. Kononenko (?), 392
A. D. Konopatov (1922-2004), 88, 391
B. M. Konoplev (1912-1960), 17, 27, 91
V. N. Konovalov (1925), 19, 389, 390
Y. N. Koptev (1940), 19, 20, 269, 275, 281,
     293, 327, 329, 330, 382, 390
L. K. Korneyev (1895-1972), 75, 76
N. S. Koroleva (1935), 71
S. P. Korolev (1907-1966), 7, 13, 15, 17, 20,
     22, 24, 25, 30, 31, 41, 42, 43, 65, 66, 67, 71,
     72, 74, 75, 76, 77, 78, 79, 80, 81, 82, 84, 85,
     87, 89, 93, 98, 99, 100, 101, 102, 105, 108,
     111, 112, 116, 117, 128, 137, 142, 145, 146,
     147, 149, 152, 153, 159, 160, 164, 187, 188,
     193, 196, 215, 220, 229, 390, 391, 393
P. Ya. Korolev (1877-1929), 71
S. A. Kosberg (1903-1965), 86, 87, 88, 116,
     140, 147, 153
V. S. Kossenko (?), 22
A. S. Kossiatov (1910-1994), 389
A. N. Kosygin (1904-1980), 160, 196
A. G. Kostikov (1899-1950), 4, 75, 76, 77, 83,
     100, 391
P. I. Kostin (1904-1970), 12, 20
G. V. Kostin (1934), 391
P. T. Kostin (?), 119, 121
S. G. Kotchariantz (1909-1993), 390
E. P. Kotcherov (?), 392
P. P. Kotcherov (1907), 19
N. I. Kotchnov (?), 389
V. A. Kotelnikov (1908-2005), 93
L. I. Kotenev (?), 392
N. I. Kotenkova-Ermolaieva (1920-1999), 78
G. N. Kovalenko (?), 393
I. M. Kozhemiako (1921-1994), 199
D. I. Kozlov (1919-2009), 78, 105, 106, 107,
     108, 110, 111, 112, 120, 121, 199, 281, 328,
     390
V. D. Kozlov (?), 282
P. V. Kozlov (1909-?), 90
V. I. Krainov (?), 110

A. Ya. Kramarenko (1942-2002), 128
A. Z. Krasnov (1907-1994), 31
S. S. Kriukov (1918-2005), 108, 390, 391
V. D. Kriutchkov (?), 19
N. I. Krupnov (1905-?), 11, 13, 21
A. L. Kryjko (1938), 393
N. I. Krylov (1903-1972), 196
V. N. Kubassov (1935), 169, 170
B. I. Kudrevich (1884-1960), 97
P. N. Kulechov (1908-2001), 18
V. I. Kurbatov (1916-2004), 390
E. M. Kurilo (?), 15
A. A. Kuruchin (1922-2006), 393
L. D. Kutchma (1938), 294
N. N. Kuznetsov (1903-1983), 6, 7, 11, 18
N. D. Kuznetsov (1911-1995), 82, 88, 117, 137, 163, 392
V. I. Kuznetsov (1913-1991), 12, 17, 22, 42, 91, 92, 96, 98, 99, 102, 110, 145, 390, 391
V. V. Kuznetsov (?), 281, 382

G. E. Langemak (1898-1938), 74, 75, 83, 100
A. I. Lapchin (?), 20
V. V. Lapchin (?), 99
V. L. Lapygin (1925-2002), 96
S. A. Lavochkin (1900-1960), 21, 24, 31, 34, 41, 82, 84, 86, 101, 112, 390
M. A. Lavrentiev (1900-1980), 66
P. D. Lavrentiev (1905-1979), 392
I. V. Lavrov (1920-1995), 66
S. S. Lavrov (1923-2004), 15, 390
V. G. Lazarev (1928-1990), 170
S. M. Lechenko (1904-1974), 41
A. A. Leonov (1934), 161, 170
N. S. Lidorenko (1916-2009), 70
A. Ya. Linkov (?), 392
S. N. Liuchin (1902-1978), 72
A. M. Liulka (1908-1984), 4, 163, 390
B. V. Lipatov (?), 18
M. I. Lissun (1935-2012), 128
G. N. List (1901-1994), 3, 391
V. Ya. Litvinov (1910-1983), 108, 112, 114, 390
V. V. Lobanov (?), 390
L. G. Loïtsiansky (1900-1991), 96
V. A. Lopota (1950), 391
G. E. Lozino-Lozinsky (1909-2001), 175
V. N. Lujin (1906-1955), 76
A. I. Lurie (1901-1980), 96

O. V. Madanovich (1964), 393
A. I. Makarevsky (1904-1979), 390
O. G. Makarov (1933-2004), 170
L. E. Makarov (?), 19
V. P. Makeiev (1924-1985), 24, 80, 105, 108, 110, 390
G. M. Malenkov (1902-1988), 18, 389
R. Ya. Malinovsky (1898-1967), 196
V. A. Malychev (1902-1957), 18, 31, 65, 390
N. V. Markitchev (1918-1972), 391, 392
Y. D. Masliukov (1937), 18
V. F. Matiachin (?), 35
A. S. Matrenin (1924-2002), 19, 389
A. V. Matveiev (?), 19
A. A. Maximov (1923-1990), 42, 156, 199
G. Y. Maximov (1926-2001), 65, 144, 390
N. D. Maximov (1909- ?), 90
E. V. Mazur (1910-1982), 390
M. G. Medkov (1907-?), 78
A. I. Medvedchikov (1946), 279, 293, 327, 382
E. L. Mejeritsky (1941), 96
P. I. Melechin (1919-1995), 20
B. N. Melioransky (1938), 111
A. T. Melnikov (?), 19
M. V. Melnikov (1919-1996), 36, 44, 87, 88, 140, 391
V. N. Mentiukov (?), 392
I. A. Merkulov (1913-1991), 111
G. M. Merzliakov (?), 198
A. P. Mezentsev (1946), 392
V. M. Miassichtchev (1902-1978), 31, 33, 41, 380
V. P. Michin (1917-2001), 12, 98, 108, 110, 128, 142, 188, 229, 390, 391
V. I. Mikerin (?), 20
K. I. Mikhailov (?), 99
G. K. Mikheiev (?), 197, 203
A. A. Mikulin (1895-1985), 86, 392
A. I. Mikoyan (1905-1970), 4, 41, 86, 146
L. M. Miloslavsky (?), 65
I. A. Mirzakhanov (1887-1960), 11
N. F. Moisseiev (1949), 20
A. M. Mokin (?), 19
V.V. Molodtsov (?), 142
V. M. Molotov-Skriabine (1890-1986), 76
B. N. Morozov (?), 203, 204
M. N. Moskalenko (1888-1980), 71
K. S. Moskalenko (1902-1985), 145

Y. A. Mozjorin (1920-1998), 6, 8, 389, 391
A. G. Mrykin (1905-1972), 6, 8, 42, 99, 389, 390, 393

A. D. Nadiradze (1914-1987), 21, 203
G. S. Narimanov (1922-1983), 142, 389, 391
A. S. Nazarov (1899-1987), 391
M. I. Nedeline (1902-1960), 38, 42, 66, 99, 187, 390
E. A. Negin (1921-1998), 390
V. V. Nekrassov (1931-1995), 119
A. I. Nesterenko (1908-1995), 10, 42, 187
N. F. Nikitin (?), 392
E. L. Nikolaï (1880-1950), 97
V. A. Nikolaev (?), 65
A. G. Nikolaev (1929-2004), 145, 169
V. E. Nikolaïev (1960), 293, 380, 382
A. I. Nossov (1913-1960), 15
N. E. Nossovsky (1905-1978), 11, 389
A. A. Novikov (1900-1976), 10
A. E. Nudelman (1912-1996), 13

V. V. Ofitserov (?), 22
S. O. Okhapkin (1910-1980), 108, 390, 391
D. E. Okhotsimsky (1921-2005), 33, 65, 391
I. I. Oleïnik (1937), 393
O. N. Ostapenko (?), 393
N. N. Ostriakov (1904-1946), 97
V. D. Ostrumov (?), 19
A. A. Ovtcharov (1917), 392
A. F. Ovtchinnikov (1950-1996), 393
Y. V. Ovtchinnikov (?), 390
V. N. Ovtchinnikov (1936), 392

G. N. Pachkov (1909-1993), 18, 41, 42, 149, 196, 390, 393
D. V. Pakhomov (1963), 391
A. V. Pallo (1912-2001), 76, 77, 389
V. M. Panferov (?), 21
P. I. Parchin (1899-1970), 11, 19, 100
S. P. Parniakov (1913-1987), 43
V. I. Patsaiev (1933-1971), 170
S. G. Pechkov (?), 392
B. G. Penzin (?), 392
Ya. I. Perelman (1882-1942), 82
A. N. Perminov (1945), 20, 348, 393
L. D. Perychkova (?), 86
V. M. Petlyakov (1891-1942), 77, 84
S. A. Petrenko (?), 392

B. S. Petropavlovsky (1898-1933), 76, 85
G. V. Petropavlovsky (?), 392
A. Ya. Petruchenko (1942-1992), 128
N. I. Petrov (1898-1950), 10
B. N. Petrov (1913-1980), 45, 142, 390
V. P. Petrov (1902-1984), 194
B. V. Petrovsky (1908-2004), 82
N. A. Pilyugin (1908-1982), 12, 16, 25, 42, 89, 90, 92, 94, 86, 98, 102, 110, 142, 145, 147, 164, 188, 390, 391
N. A. Pirogov (1951), 391
T. D. Pitzkhelauri (1941), 162
B. V. Plotnikov (?), 392
Y. A. Pobedonostsev (1907-1973), 3, 6, 7, 13, 72, 76, 79, 107, 108, 391
K. A. Pobedonostsev (1932-2008), 392
F. N. Poïda (1906-1979), 76
D. Y. Poletaïev (?), 241
N. P. Poletaiev (1905-1990), 390
A. K. Polevik (?), 14
A. I. Poliarny (1902-1991), 27, 75
N. N. Polikarpov (1892-1944), 4, 111
V. I. Polikovsky (1904-1965), 4
V. L. Ponomareva (1933), 162
A. S. Popov (1859-1906), 94
G. M. Popov (?), 392
P. R. Popovich (1930-2009), 128, 145
N. S. Porvatkin (1932-2009), 128
P. N. Potekhin (?), 20
V. V. Putin (1952), 332
A. V. Pravoslavnov (?), 103
P. Z. Prestensky (?), 197
A. V. Pridantsev (?), 48
N.D. Prokopov (?), 142
V. P. Pronikov (?), 393
G. A. Protsenko (?), 392
I. S. Prudnikov (1919-2005), 85
P. P. Prudovsky (?), 392

E. N. Rabinovich (?), 19
S. E. Rachkov (?), 13, 20
V. S. Rachuk (1936), 88
D. K. Radkevich (?), 391, 392
V. P. Radovsky (1920-2001), 85, 391
A. S. Raievsky (1911-2000), 389
I. I. Raïkov (1918-1999), 389
Kh. A. Rakhmatulin (1905-1988), 390
B. V. Rauchenbakh (1915-2001), 65
A. A. Raspletine (1908-1967), 164

M. F. Rechetnev (1924-1996), 82, 105
E. A. Rechetov (?), 392
A. K. Repin (1903-1976), 390
V. M. Riabikov (1907-1974), 11, 18, 41, 42, 66, 390
Y. V. Riabuchkin (1917-1996), 119
E. F. Riazanov (1923-1975), 65
M. S. Riazansky (1909-1987), 12, 17, 19, 42, 45, 67, 79, 88, 90, 91, 92, 95, 98, 102, 145, 147, 188, 390, 391
V. Ya. Riazantsev (?), 204
N. P. Rodin (?), 392
I. A. Rosselevich (1918-1991), 158
S. G. Rozental (?), 4
K. N. Rudnev (1911-1980), 65, 78, 145, 393
V. A. Rudnitsky (1910-2004), 12, 48, 389, 390, 391
N. N. Rukavishnikov (1932-2002), 170
I. P. Rumiantsev (?), 19

I. N. Sadovsky (1919-1993), 85, 391
V. M. Saïgak (1938-1996), 110
A. V. Sakhanitsky (1897), 21
A. D. Sakharov (1921-1989), 390
M. E. Salmanov (1899- ?), 90
V. I. Samsonov (1924), 392
I. N. Sapojnikov (1929), 392
B. M. Saprykin (1907- ?), 21
G. V. Sarafanov (1942-2005), 128
Y. S. Sarymov (?), 99
G. S. Saveliev (1910- ?), 90
L. E. Schwartz (1905-1945), 75, 76, 391
L. I. Sedov (1907-1999), 65, 66, 393, 394
A. S. Selivanov (?), 92
A. I. Semenov (1908-1973), 6, 18, 149, 389
Y. P. Semenov (1935), 85, 391
I. D. Serbin (1910-1981), 196
Y. N. Sergunin (1927-1993), 393
A. N. Sergueiev (?), 18
V. G. Sergueiev (1914-2009), 391
J. D. Sergeichik (1939), 162
I. A. Serov (1905-1990), 3, 389
V. I. Sevastianov (1935), 169, 170
N. N. Sevastianov (1961), 391, 404
D. D. Sevruk (1908-1994), 21, 24, 78, 84
E. V. Shabarov (1922-2003), 391
V. L. Shabransky (1917-1992), 15, 391
A. I. Shakhurin (1904-1975), 6, 10, 390
Ya. M. Shapiro (1902-1994), 13

V. D. Shargorodsky (?), 92
V. A. Shatalov (1927), 167, 169, 170
A. G. Shekhman (?), 3, 48
A. Ya. Sherbakov (1901-1978), 77, 111, 390
P. P. Sherbakov (?), 203
A. E. Shestakov (1935), 390
E. S. Shetinkov (1907-1976), 74, 78
V. I. Sheulov (1922), 158
S. N. Shishkin (1902-1981), 42, 389
O. N. Shishkin (1934), 19, 389, 390
I. L. Shitarev (1939), 392
D. A. Shitov (1902-?), 77
V. K. Shitov (?), 15
N. F. Shlygov (1922-1999), 156
N. S. Shniakin (1901-1996), 25, 84
A. I. Shokin (1909-1988), 149, 393
G. S. Shonin (1935-1997), 169
S. F. Shtanko (1922-1981), 202
G. M. Shubnikov (1903-1965), 40
A. N. Shukin (1900-1990), 390
V. A. Shuliakovsky (?), 19
M. F. Shum (?), 110
S. A. Shumakov (?), 20
A. A. Shumilin (1936), 393
L. F. Shumny (?), 108
V. K. Shvanov (1936), 391
A. D. Shvetsov (1892-1953), 86, 392
A. V. Sidorenko (1917-1982), 392
M. M. Sidorov (?), 48
S. F. Sigaiev (1918-1993), 19
E. V. Sinelchikov (1910-1990), 13, 20, 25
M. V. Sinelchikov (?), 20
G. A. Skuridin (1927-1991), 66, 391
B. M. Slonimer (1902-1972), 75, 76
N. N. Smirnitsky (1918-1993), 15, 18, 390
A. M. Smirnov (?), 389
L. V. Smirnov (1916-2001), 18, 19, 91, 196, 198
V. N. Sochin (?), 390
A. I. Sokolov (1910-1976), 6, 7, 18, 149, 393
V. E. Sokolov (?), 390
E. I. Sokolov (1940-2009), 381
A. M. Soldatenkov (1927), 110
Y. D. Soloviev (1917), 391, 392
V. F. Soloviev (?), 391
V. N. Soloviev (1924-2012), 194
I. B. Solovieva (1937), 162
G. Ya. Sonis (?), 392
R. E. Sorkin (1910-1983), 3

A. S. Spiridonov (1903-1976), 19
J. V. Staline-Gjugachvili (1879-1953), 76, 107, 390
A. I. Steniaev (?), 391
E. N. Stepanov (1937), 128
N. S. Stepantchenko (?), 197
B. S. Stetchkin (1891-1969), 390, 391
G. M. Strekalov (1940-2004), 188
M. A. Subbotchev (?), 19
N. A. Sudakov (?), 13
V. A. Sudetz (1904-1981), 196
P. O. Sukhoï (1895-1975), 84, 86
F. G. Sukhomlinov (?), 16
M. K. Sukov (?), 22, 389
N.A. Sulimovsky 1904-1961), 19
K. N. Surjin (1903-1976), 10
Ya. E. Suslov (?), 199
P. A. Sysoiev (1911-1991), 19
V. A. Sytchev (?), 18

G. M. Tabakov (1912-1993), 19,
N. I. Tarassov (?), 198
V. V. Terechkova (1937), 145
V. P. Terentiev (?), 389
N. I. Tikhomirov (1870-1930), 76
M. K. Tikhonravov (1900-1974), 3, 31, 33, 65, 66, 72, 77, 79, 390, 391
N. V. Timochuk (?), 21
V. A. Timofeiev (1907-1963), 14, 389
G. A. Tiuline (1914-1990), 6, 7, 8, 15, 147, 389, 390, 391
S. V. Tiuvelin (1960), 111
A. A. Titkin (1948), 20
G. S. Titov (1935-2000), 161
G. A. Titov (1909-1980), 390
V. G. Titov (1947), 188
V. F. Tolubko (1914-1989), 199
A. I. Tolstov (?), 389
A.V. Toptcheiev (1907-1962), 66
I. I. Toropov (1907-1977), 86
Ya. I. Tregub (1918-2007), 15, 391
A. T. Tretiakov (1899-1978), 112
V. N. Tretiakov (1906-1993), 22, 98
K. I. Tritko (?), 12, 20
N. G. Trofimov (1926), 392
Y. L. Troitsky (1914-1985), 48, 391
Y. N. Trufanov (1925-2008), 85
K. I. Trunov (1896-1975), 76, 79
Y. V. Trunov (1938), 96

F. A. Tsander (1887-1933), 72, 75
Z. M. Tsetsiur (1914-1990), 98, 391, 392
K. E. Tsiolkovsky (1857-1935), 72, 75, 78, 79, 80, 82, 85, 208
P. V. Tsybine (1905-1992), 21, 391
M. N. Tukhatchevsky (1893-1937),75
S. K. Tumansky (1901-1973), 164
R. A. Turkov (1901-1976), 391
P. A. Tunik (1914-1976), 392
A. N. Tupolev (1888-1972), 29, 41, 72, 77, 112
A. F. Tveretsky (1904-1992), 15

A. F. Udalov (?), 392
G. R. Udarov (1904-1990), 390
G. A. Uger (1905-1972), 389
N. L. Umansky (1908-1967), 3, 13, 20, 84
Y. M. Urlititch (1962), 392
A. V. Ussenkov (1935), 19
D. F. Ustinov (1908-1984), 10, 11, 18, 20, 24, 105, 112, 196, 389, 390
V. F. Utkin (1923-2000), 110, 203

S. S. Vanin (1927), 20, 390
B. L. Vannikov (1897-1962), 10, 11, 18, 65
R. B. Vannikov (?), 11
A. P. Vanitchev (1916-1993), 390, 391
A. K. Vanitsky (1925), 19
A. A. Vassiliev (1921-1973), 7, 18
A. N. Vassiliev (?), 48, 389
V. P. Vassiliev (?), 92
A. M. Vassilievsky (1895-1977), 390
V. D. Vatchnadze (1929), 19, 85
I. M. Vedeneiev (1912-1992), 88
E. A. Verbin (?), 19
K. A. Verchinin (1900-1973), 196
K. M. Vertelov (1923-1997), 390
I. P. Vetlov (?), 392
S. I. Vetotchkin (1905-1991), 19, 23, 41, 389
V. M. Vinogradov (1906- ?), 21
V. I. Vinogradov (1935- ?), 389
K. M. Vintsentini (1907-1991), 71, 78
V. A. Vitka (1900-1989), 84, 390
V. D. Vladimirov (?), 10
S. M. Vladimirsky (1908-1989), 41
I. I. Volkotrubenko (1898-1986), 11
L. I. Volkov (?), 389
V. N. Volkov (1935-1971), 169, 170
B. V. Volynov (1934), 169

M. P. Vorobiev (1896-1957), 389
V. A. Vorobiev (?), 203
K. E. Vorochilov (1881-1969), 76
G. I. Voronin (1906-1987), 145, 164
A. F. Voronov (1930-1993), 170
N. I. Vorontsov (?), 389
L. A. Voskressensky (1903-1965), 12, 15, 108, 188, 390, 391
A. N. Voznessensky (?), 21
V. I. Vozniuk (1907-1976), 10, 18, 37, 389, 390

Y. A. Yachin (1930), 203, 393
A. S. Yakovlev (1906-1989), 84, 86
N. D. Yakovlev (1898-1972), 11, 17, 18, 20, 389
M. K. Yangel (1911-1971), 31, 78, 80, 85, 91, 99, 101, 102, 105, 110, 116, 117, 147, 189, 193, 194, 195, 196, 220
I. M. Yatsunsky (1916-1983), 65, 389, 391
V. I. Yazdovsky (1913-1999), 145
B. N. Yeltsin (1931-2007), 243, 245, 249, 272, 286, 288, 382, 400
V. M. Yibshitz (?), 203
S. I. Yunochev (?), 20
E. S. Yurassov (1921-2008), 18

A. G. Zakharov (1921), 145, 393
A. A. Zakharov (?), 21
V. M. Zakimatov (?-2009), 390
Ya. B. Zeldovich (1914-1987), 390
R. I. Zelenev (?), 392
P. M. Zernov (1905-1964), 30, 390
E. A. Zhelonov (?), 390
N. A. Zheltukhin (1915), 391
M. S. Zhezlov (1898-1960), 392
P. F. Zhigarev (1900-1963), 389
M. D. Zholudev (?), 204
B. P. Zhukov (1912-2000), 80
G. K. Zhukov (1896-1974), 390
Y. A. Zhukov (1933), 393
Y. M. Zhuravlev (1941), 393
V. F. Zotov (?), 390
A. P. Zubov (1916-2001), 19
I. G. Zubovich (1901-1956), 19, 389
B. G. Zudin (?), 203, 204
V. D. Zudov (1942), 128
V. S. Zuiev (1908-1984), 77
V. A. Zuevsky (1918-1972), 390
P. P. Zuikov (1900-1988),391

S. A. Zverev (1912-1978), 149, 196

**GERMANY**
Werner Albring (1914-2007), 16
Erich Apel (?), 16
Manfred von Ardenne (1907-1997), 10
Brunholf Baade (1904-1969), 17
Manfred Bischoff (1942), 278
Kurt Blasig (?), 16
Josef Blass (?), 16
Wernher von Braun (1912-1977), 229
Robert Döpel (?), 10
Gustav Hertz (1887-1975), 10
Hans Hoch (?), 16
Helmut Gröttrup (1916-1981), 16
Heinz Jaffke (?), 16
Alois Jasper (?), 16
Helmut Kohl (1930), 278
Franz Lange (?), 16
Kurt Magnus (1912-?), 16
Emil Mende (?), 16
W. Quessel (?), 16
Heinz Pose (1905-1975), 10
Oswald Putze (?), 17
Nikolaus Riehl (1901-1990), 10
Hans Rössing (?), 17
Rozen-Plenter (?), 32
Wilhelm Schütz (?), 16
Max Steenbeck (1904-1981), 10
Peter-Adolf Theissen (?), 10
Karl Umpfenback (?), 16
Max Volmer (?), 10
Waldemar Wolff (?), 16
Heinz Zeise (?), 16

**FRANCE**
Claude Allègre (1937), 323, 324, 327, 401, 402
Frédéric d'Allest (1940), 269
François d'Aubert (1943), 334
François Auque (1956), 286, 328, 400
Édouard Balladur (1929), 276
Joël Barre (1955), 328
Alain Bensoussan (1940), 279, 280
Bernard Bigot (1950), 372
Charles Bigot (1932),268, 269, 270, 278, 286, 293
Jean-Michel Boucheron (1948), 328
Françoise Bouzitat (1949-2009), 400
Thierry Breton (1955), 334

Christian Cabal (1943-2008), 327
François Calaque (1940-1998), 111, 254, 255, 257, 265, 277, 281, 282, 289, 291, 326, 377, 379, 385
Armand Carlier (1949), 285
Jacques Chirac (1932), 196, 276, 278, 328, 383
Jean-Loup Chrétien (1938), 196, 383
Laurent Collet-Billon (1950), 372
Philippe Couillard (1945), 291
Claire Coulbeaux (?), 293
Maurice Couve de Murville (1907-1999), 229
Édith Cresson (1934), 399
Hubert Curien (1924-2005), 379
Yannick d'Escatha (1948), 335, 337, 372
Philippe de Fontaine Vive Curtaz (1959), 334
Michel Delaye (1934), 257, 266, 268, 272, 273, 277, 286, 291
Roland Deschamps (?), 269
Jean-Jacques Dordain (1946), 327, 333, 334, 335, 337
Michel Doubovick (1967), 293
Guy Dubau (?), 281, 282
Alain Dupas (?), 220
Léopold Eyharts (1957), 385, 400
Patrick Eymar (1947), 257, 259
François Fillon (1954), 276, 277, 278, 279, 280, 372
Gary Filmon (1942), 319
François Flohic (1920), 229
Louis Gallois (1944), 273, 278, 293
Charles de Gaulle (1890-1970), 187, 196, 229, 257
Claude Goumy (1940), 268
Claudie Haigneré, née André-Deshays (1957), 328, 330, 383, 385, 400, 404
Jean-Pierre Haigneré (1948), 383, 385, 400
Joël Hamelin (?), 324
Catherine Ivanov (?), 128
Lionel Jospin (1937), 328, 401, 402
Alain Juppé (1945), 276
Gilles Le Chatelier (1964), 327
Jean-Yves Le Gall (1959), 279, 292, 293, 311, 320, 328, 334, 335, 348, 383, 385, 388
André Lebeau (1932), 274, 278
Jean-Daniel Lévi (1941), 266, 274, 278
Gérard Longuet (1946), 276
Bernard Luciani (1962), 293
Jean-Marie Luton (1942), 286, 293, 311, 321, 385, 400

Jacques Marmain (?), 232
Pierre Marx (?), 251
Yves Michot (1941), 293
Patrick Millet (?), 282
François Mitterrand (1916-1996), 196, 276, 379, 383
Jérôme Paolini (1961), 276, 277, 278
René Pellat (1936-2003), 266
Georges Pompidou (1911-1974), 123, 187, 196, 229
Marcel Pouliquen (1945-2010), 397
Paul Quilès (1942), 292, 383
Jean-Pierre Raffarin (1948), 330, 332, 334
Henri Revol (1936), 327
Paul-Aimé Richard (?), 72
Michel Rocard (1930), 399
José Rossi (1944), 276
Jacques Rossignol (1940), 328
Nicolas Sarkozy (1955), 372
Roger-Gérard Schwartzenberg (1943), 327, 328
Jean Sollier (1933), 268
Denis Thirion (?), 128
Jacques Tiziou (1939), 230, 232
Michel Tognini (1949), 383, 400
Dominique de Villepin (1953), 348
Jean-Charles Vincent (?), 281, 382
Mathieu Weiss (1970), 293

**COUNTRIES IN THE COMMUNIST SPHERE**
F. Castro (1926), 196
N. Ceaucescu (1918-1989), 196
G. Gussak (1913-1991), 196
Y. Kádár (1912-1989), 196
W. Stoph (1914-1999), 196
L. Svoboda (1895-1979), 196
Y. Tsedenbal (1916-1991), 196

**UNITED STATES**
Paul Allen (1953), 249
Rawson Bennett (1905-1967), 211
Robert E. Berry (1928), 294
Karel J. ⸢Charlie⸥ Bossart (1904-1975), 394
George H. W. Bush (1924), 249
William Jefferson ⸢Bill⸥ Clinton (1946), 252
Vance D. Coffman (1944), 276
Arthur M. ⸢Art⸥ Dula (1947), 243, 395
Allen W. Dulles (1893-1969), 393

John F. Dulles (1888-1959), 211, 212
Dwight D. Eisenhower (1890-1969), 211
Bill Gates (1955), 249, 290
Daniel S.ˑDan¤ Goldin (1940), 252
Albert A.ˑAl¤ Gore (1948), 245, 270
Michael Griffin (1949), 383
James Hagerty (1909-1981), 211
Robert A. Heinlein (1907-1988), 395
William A. Holaday (?), 211
Elmer Hutchisson (1902-1983), 211
Joseph Kaplan (1902-1991), 208
Philip Kauffman (1936), 235
Craig McCaw (1949), 249, 290
Elon Musk (1971), 349
Richard M. Nixon (1913-1994), 395

Francis Gary Powers (1929-1977), 224
Charles S. Sheldon (1917-1981), 233
S. Fred Singer (1924), 216
Asif Siddiqi (?), 391
Donald K.ˑDeke¤ Slayton (1924-1993), 395
Richard H. Truly (1937), 252
Fred L. Whipple (1906-2004), 216

**REST OF THE WORLD**
Mohamed Farris (1951), 394
David Kwon (?), 326
Rakesh Sharma (1949), 238
Hiroyuki Yamaga (1962), 235
Yang Liwei (1965), 403

# Photo credits

© Air & Cosmos, 197, 221

© Arianespace, 372, 385, 386a, 308a, 366a, 366b

© ASPC, 179

© Astrium, 253, 254, 263, 264, 268, 273, 298a, 374

© Astrium/ESA, 357a

© Aviatsia & Kosmonavtika, 59b

© Boeing, 247, 318

© CADB, 141b, 154, 176b

© CIA, 225a, 225b, 227, 228

© Dassault/Aviaplans, 250b

© Demange/Gamma/Starsem, 295

© Dessin Christian Lardier, 32, 63b, 64a

© Dessin Jacques Tiziou, 230

© EADS, 250a

© ESA, 287b, 291, 301a, 301b, 302a, 386b

© ESA, Photo Cluster, 303

© ESA, Photo D. Ducros, 306a, 369

© ESA, Photo J. Huart, 324

© ESA, Photo P. Sebirot, 277, 305b, 333

© ESA, Photo S. Corvaja, 45, 46b, 47a, 209, 210, 226, 261b, 270, 280, 281a, 281b, 282a, 282b, 283, 284, 300a, 300b, 305a, 306b, 306c, 307, 308b, 308c, 309a, 309b, 310b, 312a, 312b, 313, 314, 323, 331, 336a, 336b, 343, 341a, 340a, 340b, 342, 343b, 344, 345a, 345b, 346, 351a, 353, 354a, 354b, 355, 358, 359, 360a, 360b, 361, 362, 363, 364a, 364b, 365, 375b

© ESA/ASI – Star City, 357b

© ESA-CNES-Arianespace, Photo D. Ducros, 338a, 338b, 339, 375a

© ESA-CNES-Arianespace/Service Optique CSG, 279, 341b, 351b

© ESA-Service Optique CSG, 251

© Eurockot Launch Services, 271

© Space Forces, 56

© France Soir, DR, 229

© Imperial War Museum, 7, 8

© KBOM, 50a, 51a, 51b, 52, 54a

© Khrunichev, 245

© Lockheed Martin, 248

© Baikonur Museum, 38, 39a, 39b, 49

© VNIIEF Museum, 62b

© NASA, 213, 223, 296, 352

© National Park Service, 215

© NPO Energomach, 28, 33a, 33b, 34

© NPO Lavochkin, 32b

© Alexandre Chliadinsky Photo, 139, 140, 147, 151, 155, 158, 161, 164, 165, 167, 168.

© Christian Lardier Photo, 37, 47b, 54b, 85, 107b, 109a, 109b, 110, 113a, 113b, 113c, 114a, 114b, 115a, 115b, 116, 118a, 118b, 119, 122, 127a, 178, 181, 182, 183, 191a, 191b, 191c, 192, 242b, 337

© Jacob Terweij Photo, 131b

© Pascal Photo, 231a

© Stefan Barensky Photo, 247a, 384a

© RKK Energya, 23, 24, 25, 26, 27, 29, 30, 46a, 64b, 67a, 67b, 68a, 68b, 69a, 69b, 120a, 120b, 123, 138, 141a, 142, 143, 145, 146, 149, 150, 162, 180

© Roscosmos, 176a, 177, 190b, 192b

© Starsem, 260, 266, 287a, 294, 297, 299b, 302b, 322

© TASS, 168

# About the Authors

**Christian Lardier** was born in 1952. He has been a member of the Cosmos Club de France (C2F) since 1965, Space Editor of *Espace & Civilisation* from 1978 to 1981, *Aviation Magazine* from 1983 to 1992, and *Air & Cosmos* from 1994 to 2012. He is an academician and member of the history committee of the International Academy of Astronautics (IAA), a senior member of the Association Aéronautique et Astronautique de France (AAAF), a co-founder and president since 2007 of the Institut Français d'Histoire de l'Espace (IFHE), and a co-founder of the Association Planet Mars (APM). He authored the book *L'Astronautique Soviétique* in 1992 and co-authored (with Stefan Barensky) *The Soyuz Launch Vehicle – The two lives of an engineering triumph* (in French in 2010 and in English in 2012).

**Stefan Barensky** was born in 1965 and is a professional science and technology writer. He has reported on space technologies and industries since 1991 as an editor for multiple French and international science, trade and politics publications such as *Science & Vie*, *Interavia*, *Air & Cosmos*, *Aero Defense News* and *European Voice*. A former space transportation analyst at Euroconsult and Launchspace, and editor-in-chief of the monthly newsletter *Orbital Launcher Report*, he also witnessed the gradual Europeanization of Soyuz from the inside as an editorial consultant to Aerospatiale (now Astrium), Arianespace, CNES and ESA.